About Island Press

Island Press is the only nonprofit organization in the United States whose principal purpose is the publication of books on environmental issues and natural resource management. We provide solutions-oriented information to professionals, public officials, business and community leaders, and concerned citizens who are shaping responses to environmental problems.

In 2003, Island Press celebrates its nineteenth anniversary as the leading provider of timely and practical books that take a multidisciplinary approach to critical environmental concerns. Our growing list of titles reflects our commitment to bringing the best of an expanding body of literature to the environmental community throughout North America and the world.

Support for Island Press is provided by The Nathan Cummings Foundation, Geraldine R. Dodge Foundation, Doris Duke Charitable Foundation, Educational Foundation of America, The Charles Engelhard Foundation, The Ford Foundation, The George Gund Foundation, The Vira I. Heinz Endowment, The William and Flora Hewlett Foundation, Henry Luce Foundation, The John D. and Catherine T. MacArthur Foundation, The Andrew W. Mellon Foundation, The Moriah Fund, The Curtis and Edith Munson Foundation, National Fish and Wildlife Foundation, The New-Land Foundation, Oak Foundation, The Overbrook Foundation, The David and Lucile Packard Foundation, The Pew Charitable Trusts, The Rockefeller Foundation, The Winslow Foundation, and other generous donors.

The opinions expressed in this book are those of the author(s) and do not necessarily reflect the views of these foundations.

THE KRUGER EXPERIENCE

The Kruger Experience

ECOLOGY AND MANAGEMENT OF
SAVANNA HETEROGENEITY

Edited by
Johan T. du Toit
Kevin H. Rogers
Harry C. Biggs

Foreword by Anthony R. E. Sinclair and Brian H. Walker

ISLAND PRESS
Washington • Covelo • London

Copyright © 2003 Island Press
All rights reserved under International and Pan-American Copyright Conventions. No part of this book may be reproduced in any form or by any means without permission in writing from the publisher: Island Press, 1718 Connecticut Avenue, N.W., Suite 300, Washington, DC 20009.

ISLAND PRESS is a trademark of The Center for Resource Economics.

Library of Congress Cataloging-in-Publication Data
The Kruger experience : ecology and management of savanna heterogeneity / edited by Johan du Toit, Harry Biggs, Kevin Rogers.
 p. cm.
 ISBN 1-55963-981-4 (cloth : alk. paper) — ISBN 1-55963-982-2 (pbk. : alk. paper)
 1. Savanna ecology—South Africa—Kruger National Park. 2. Ecological heterogeneity—South Africa—Kruger National Park. 3. Ecosystem management—South Africa—Kruger National Park. I. Du Toit, Johan. II. Biggs, Harry. III. Rogers, Kevin H.
 QH195.S6K78 2003
 577.4'8'0968271—dc21
 2003011114

British Cataloguing-in-Publication Data available

Printed on recycled, acid-free paper ♻
Manufactured in the United States of America
09 08 07 06 05 04 03 10 9 8 7 6 5 4 3 2 1

CONTENTS

PREFACE ... ix

FOREWORD ... xiii
Anthony R. E. Sinclair and Brian H. Walker

PART I. The Historical and Conceptual Framework ... 1

1. The Kruger National Park: A Century of Management and Research ... 3
   David Mabunda, Danie J. Pienaar, and Johan Verhoef

2. Biotic and Abiotic Variability as Key Determinants of Savanna Heterogeneity at Multiple Spatiotemporal Scales ... 22
   Steward T. A. Pickett, Mary L. Cadenasso, and Tracy L. Benning

3. Adopting a Heterogeneity Paradigm: Implications for Management of Protected Savannas ... 41
   Kevin H. Rogers

4. An Adaptive System to Link Science, Monitoring, and Management in Practice ... 59
   Harry C. Biggs and Kevin H. Rogers

PART II. A Template for Savanna Heterogeneity ... 81

5. The Abiotic Template and Its Associated Vegetation Pattern ... 83
   Freek J. Venter, Robert J. Scholes, and Holger C. Eckhardt

6. Biogeochemistry: The Cycling of Elements ... 130
   Mary C. Scholes, Robert J. Scholes, Luanne B. Otter, and Andrew J. Woghiren

7. Fire as a Driver of Ecosystem Variability ... 149
   Brian W. van Wilgen, Winston S. W. Trollope, Harry C. Biggs, André L. F. Potgieter, and Bruce H. Brockett

8. Surface Water Availability: Implications for Heterogeneity and Ecosystem Processes ... 171
   Angela Gaylard, Norman Owen-Smith, and Jessica Redfern

9. River Heterogeneity: Ecosystem Structure, Function, and Management ... 189
   Kevin H. Rogers and Jay O'Keeffe

PART III. Interactions between Biotic Components — 219

   10. Interactions between Species and Ecosystem Characteristics — 221
       Robert J. Naiman, Leo Braack, Rina Grant, Alan C. Kemp, Johan T. du Toit, and Freek J. Venter

   11. Vegetation Dynamics in the Kruger Ecosystem — 242
       Robert J. Scholes, William J. Bond, and Holger C. Eckhardt

   12. Insects and Savanna Heterogeneity — 263
       Leo Braack and Per Kryger

   13. Birds: Responders and Contributors to Savanna Heterogeneity — 276
       Alan C. Kemp, W. Richard J. Dean, Ian J. Whyte, Suzanne J. Milton, and Patrick C. Benson

   14. Large Herbivores and Savanna Heterogeneity — 292
       Johan T. du Toit

   15. Rainfall Influences on Ungulate Population Dynamics — 310
       Norman Owen-Smith and Joseph Ogutu

   16. Kruger's Elephant Population: Its Size and Consequences for Ecosystem Heterogeneity — 332
       Ian J. Whyte, Rudi van Aarde, and Stuart L. Pimm

   17. Wildlife Diseases and Veterinary Controls: A Savanna Ecosystem Perspective — 349
       Roy G. Bengis, Rina Grant, and Valerius de Vos

   18. Large Carnivores and Savanna Heterogeneity — 370
       Michael G. L. Mills and Paul J. Funston

PART IV. Humans and Savannas — 389

   19. Anthropogenic Influences at the Ecosystem Level — 391
       Stefanie Freitag-Ronaldson and Llewellyn C. Foxcroft

   20. Beyond the Fence: People and the Lowveld Landscape — 422
       Sharon Pollard, Charlie Shackleton, and Jane Carruthers

   21. Heterogeneity and Management of the Lowveld Rivers — 447
       Jay O'Keeffe and Kevin H. Rogers

   22. Integration of Science: Successes, Challenges, and the Future — 469
       Harry C. Biggs

   23. Reflections on the Kruger Experience and Reaching Forward — 488
       Michael G. L. Mills, Jane Lubchenco, William Robertson IV, Harry C. Biggs, and David Mabunda

CONTRIBUTORS — 503

INDEX — 507

PREFACE

Late one afternoon in July 1902, James Stevenson-Hamilton halted on the rim of the Drakensberg escarpment to take in the breathtaking view that stretched eastward below him, not realizing that he was looking out over what was to become one of the world's premier national parks. He then ordered his wagon team on and descended into the lowveld to take up his post as chief ranger of Sabi Game Reserve. In doing so he set in motion a process that, among many other things, produced this book a century later. Sabi Game Reserve became Kruger National Park, and in April 2002 some 115 ecologists gathered from all parts of the world at a symposium at Berg en Dal in the south of Kruger to discuss the material to be included in this, the first scientific book on 100 years of management and research in 20,000 km^2 of African savanna.

Some may find it surprising that this book was not produced sooner, but the past few decades of South Africa's history have been turbulent. Research in Kruger proceeded in comparative isolation through the apartheid era until the mid-1990s, although South African National Parks staff dedicatedly maintained an unbroken tradition of management, monitoring, and research. Collaborations with university academics and postgraduate students, particularly from the University of Pretoria and the University of the Witwatersrand, were fundamentally important for bringing the results of research in Kruger into the peer-reviewed literature. Then there was the example set by A. R. E. Sinclair and his colleagues, who in 1979 and 1995 produced two highly successful books on the Serengeti ecosystem of East Africa. Clearly something similar had to be done for Kruger, and after a few years of discussions around campfires, in various airport lounges, and in recesses at international conferences, it has happened.

This book uses the crosscutting theme of ecological heterogeneity to draw a thread of continuity through each chapter as a novel and experimental attempt to structure each author's interpretation of ecological patterns in space and time. Our aim is to bring out aspects of the Kruger experience that provide new insights into the ecology and management of African savannas in particular and large protected areas in general. We have divided the chapters into four parts.

Part I provides the historical context to research and management in Kruger, introduces the theme of heterogeneity, and explains the current philosophy in Kruger for linking science with management. The historical context is framed specifically in terms of management and research, but those interested in the broader aspects of Kruger's history will find references to other excellent historical works. A conceptual framework is woven with the premise that heterogeneity is the ultimate source of biodiversity and is the basis for ecosystem resilience. It follows therefore that heterogeneity must be the focus of ecosystem management, and management programs must be designed with a scientific understanding of the factors that drive heterogeneity.

Part II describes the template of natural components and processes, as influenced by management, that determine the present state of the Kruger ecosystem. Very simply, Kruger is divided down the middle of its long north–south axis into granitic soils in the west and basaltic clays in the east; annual rainfall decreases from about 600 mm in the south to about 400 mm in the north. Interactions between soil type, rainfall, vegetation, fire, and herbivory are textbook topics in savanna ecology, but the Kruger experience has more to offer: years of documented management of animal populations through introductions and culling, together with the management of fire and water, and therefore herbivory.

Part III is concerned with how species interact within the ecosystem to generate further heterogeneity across space and time. The taxa covered are those that have been best studied, and omissions should be viewed as pointers to rewarding avenues for future research. These chapters were written by field biologists, each with a sweat-soaked passion for studying their particular subject in their own way. Integrating all this information is a challenge, but readers will appreciate that among the marvelous features of ecology are the characters of its practitioners and the diversity of their approaches to interpreting patterns and processes in nature.

Part IV deals with humans as key components of savanna ecosystems, as powerful shareholders of Kruger's future, and as managers and scientists with the responsibility of conserving a global treasure. Kruger is embedded in a human-dominated matrix, with an elongated shape that exacerbates edge effects, and its major rivers all arise in or near South Africa's industrially and agriculturally developed hinterland. It is therefore both vulnerable and valuable to its surrounding human environment. This part of the book deals with the need for scientists and managers to understand and appreciate the human factor in conserving a large protected area, to increase its value to a wider range of beneficiaries, and to thereby reduce its vulnerability to human challenges. It is imperative that scientists working in Kruger integrate their research efforts to maximize the efficiency with which a locally accessible and relevant knowledge base can be developed to serve stakeholder needs. It is without doubt that a fundamental need in the South African lowveld is the protection of Kruger's ecological resilience as a safeguard for the sustainability and adaptive capacity of the wider social-ecological system.

Researchers and managers in Kruger have come and gone over the years, and nobody who has had the privilege of living and working in Kruger could avoid becoming emotionally bonded to its distinctive atmosphere and spectacular landscapes. The rich store of information that the authors of this book have mined was built up over a century by many dedicated people. It is inevitable that some will feel aggrieved if their contributions have not been mentioned or held in high enough esteem, and others may argue that important aspects of Kruger's natural history have not been covered adequately. However, the aim of this book is not to provide a definitive compendium on Kruger. It should be viewed as a selected extract from the Kruger experience, highlighting what the editors and authors consider significant contributions to improving our understanding and management of savanna ecosystems.

For the sake of brevity we refer throughout the book to South African National Parks as SANParks and to the Kruger National Park as Kruger. Specific information on Kruger's datasets can be found at http://www.parks-sa.co.za (follow the link to Conservation and then to Scientific Services).

Acknowledgments

Much of the information used in this book is available by virtue of the foresight and persistence of Dr. S. C. J. Joubert, who established the annual ecological surveys while he was in charge of research in Kruger. The production of this book would not have been possible without the solid support of SANParks, and the present management of Kruger in particular, and the Andrew W. Mellon Foundation generously provided funding for the initial planning and production costs. Staff of the Mammal Research Institute at the University of Pretoria and the Center for Water in the Environment at the University of the Witwatersrand provided additional assistance. Each chapter was refereed by at least two experts in the field, and we are deeply indebted to the following people for improving the quality of the book: Alan Anderson, Angela Arthington, Tracy Benning, John Blair, Gay Bradshaw, Charles Breen, Jane Carruthers, Oliver Chadwick, Sarah Cleaveland, Mike Coughenour, David Cumming, Julian Derry, Holly Dublin, Patrick Duncan, Christo Fabricius, Hervé Fritz, Jean-Michel Gaillard, Wayne Getz, Peter Groffman, Anthony Hall-Martin, Ken Hodgkinson, Andrew Illius, Clive Jones, Peter Jones, Jon Keeley, Richard Kock,

Hans Kruuk, Nigel Leader-Williams, Mervyn Mansell, Richard Margoluis, Sue Milton, James Murombedzi, Bob Naiman, Tim O'Connor, Steward Pickett, Stephen Pyne, Robin Reid, Belinda Reyers, Brian Richter, Peter Ryan, Clarke Scholtz, Tony Sinclair, Joe Slocombe, Rob Slotow, Martin Thoms, Brian Walker, and David Ward. We also thank Lize Fisher, Don Ntsala, and Sandra McFadyen for producing some of the figures. The copyediting skills of Meg and Alan Kemp proved invaluable in preparing the final manuscript for submission.

FOREWORD

Anthony R. E. Sinclair and Brian H. Walker

Key criteria for setting conservation priorities, with particular relevance to developing countries, were formulated in 1998 at a meeting in Malawi of parties to the International Convention on Biological Diversity. The nine "Malawi Principles" recognized that conservation was essentially a matter of societal choice. They embody two fundamental propositions. First, all stakeholders should be involved in the process of developing conservation management plans. Underlying this is the reality that most of the world's biodiversity occurs in tropical regions and is owned and administered by developing countries. The development and advancement of the peoples of such countries must be taken into account, or conservation problems will be ignored. In particular, if those who benefit from conservation are not the same people who bear its costs, then there will be no commitment from local peoples to embrace the conservation ethic, protect natural areas, and conserve biodiversity. The fundamental objective of conservationists therefore is to demonstrate that societies benefit in the long run from understanding how ecosystems work.

Second, the Malawi Principles recognize that the ecosystem is the unit of management rather than the species. Traditionally, conservation has focused on single species, particularly those that are endangered. Yet all these species need habitat and other resources; often the loss of such resources is the conservation problem, and so key resources must be conserved within the context of the whole ecosystem.

The publication of this book is timely because it complements the intent of the Malawi Principles in a country that has to confront the trade-off between development and environment. Kruger is one of a few outstanding protected areas, such as Yellowstone and Serengeti, with a long history of conservation. In this book we learn from Kruger's history not only how the philosophy and ethics of conservation have evolved but also how conservation may serve human society.

The early efforts in conservation management by Stevenson-Hamilton in the then Sabi Game Reserve addressed simple, short-term issues, ones that necessitated some active intervention by the warden. From these small beginnings there evolved over the next 70 years the philosophy that Kruger's managers had to control events both on the boundaries and within the park. In effect, Kruger was hermetically sealed off from its surroundings. For better or worse the system had to be self-contained. Fences were constructed, migrations stopped, the seasonal flux of surface water ameliorated with boreholes and dams, seasonal grass fires tamed and managed, and populations of animals manipulated to match predetermined levels. This period is now aptly named the command and control era.

We participated in a symposium in Kruger in 1982 that foreshadowed changes in attitude. Since that time, science, management, and philosophy have moved on, and, to their credit, so have the ideas of Kruger's scientists and managers. This book reflects the modern ethic. Mistakes may have been made in the past, but with hindsight we can learn from them; they become the unintended experiments that have taught us much. For example, we have understood from other areas in Africa, such as Serengeti, the role that migration of large ungulates plays in providing resilience and stability to an ecosystem. Kruger has provided the test for this idea through the manipulations mentioned earlier.

Ecosystem management is all about resilience. This book reflects the concept of resilience through two important themes. First, any ecosystem is continually changing through both abiotic influences of climate and the biotic interactions of species and human populations. History shows that attempts to prevent change merely build to an eventual catastrophic perturbation. Managers and scientists now emphasize that we need to incorporate ecosystem change in our management plans for protected areas.

The second theme, a major one for this book, emphasizes that spatial patchiness at different scales also enables the system to absorb sudden disturbances. Changes in time and space therefore provide the heterogeneity that leads to both high biodiversity and greater resilience of the system.

The study of heterogeneity through adaptive management allows us to understand the links between different parts of the system. We need to know about the main processes that drive the system and the feedbacks that can stabilize or destabilize it. What is adaptive management? It is not merely adapting management to changing conditions; it is applying different management strategies in different parts of the system so that from the outcomes we can decide which one provides best practice. The Kruger team has grappled with how to apply emerging theory in practical management of an entrusted region in an uncertain world. Their development and use of the notion of thresholds of potential concern is an excellent example of this and is worthy of consideration for other conservation regions.

The most profound philosophical change that permeates this book is the realization that protected areas are embedded in a matrix of human-dominated ecosystems. Resilience theory tells us that to understand and manage a system at some particular scale it is necessary to understand and take into account the dynamics of the system at scales above and below the one of primary interest. If we want to maintain a protected area, then we must understand how the human system works and how it influences the park. Thus, as we see from several chapters in the book, all rivers flowing into Kruger are radically influenced by human activity upstream; rivers are the lifeblood of the whole ecosystem, and so factors such as nutrient loading and rates of flow and sedimentation are crucial.

In turn, protected areas such as Kruger provide essential baselines against which to measure the impacts of human society on ecosystems. They provide early warnings of instability. We can evaluate changes in human agroforestry systems only by reference to a natural system. Indeed, protected areas provide the insurance policy for society.

No long-term management strategy is effective without the involvement of all stakeholders, particularly those who live in the immediately adjacent areas. Education and cooperation are needed for the long-term success of a park. Without this engagement, conservation will be ignored. The book recognizes this necessity throughout its chapters, especially those on the socioeconomic issues in regions surrounding Kruger.

We hope that programs to monitor current management will be maintained in the future so that we can assess whether they are meeting their objectives. We should remember that had we known what was present on many of the continents in the 1700s or 1800s, we would have a better idea of what to conserve and how to conserve it. In a hundred years, future generations may wish that we had been wiser. The extraordinary long-term information on the larger Kruger ecosystem that the editors and authors present in this book makes it a prime example of how future conservation in a human context should proceed. There is much more to learn, but this book is a good start.

PART I
The Historical and Conceptual Framework

Kruger's sheer size and its history of management, monitoring, and research over the past century frame in spatial and temporal terms the contribution it offers to savanna ecology. One book cannot convey it all, so a conceptual theme has been chosen to emphasize the relevance of research in Kruger to an understanding of how biodiversity arises from and is maintained by ecological heterogeneity at multiple spatial and temporal scales. No claims are made that Kruger is more heterogeneous than other parks, that savannas are more heterogeneous than other ecosystems, or that Kruger is a center for research on ecological heterogeneity. There will always be other candidates, depending on the purpose of inquiry and the relevant spatial and temporal scales. Part I of this book sets the focus by outlining the human historical background to what Kruger is today and presenting ecological heterogeneity in a conceptual framework to aid in interpreting the structural and functional complexity of the ecosystem.

It is easy to accept that ecosystems are heterogeneous in that they clearly change from one place and time to the next. In Kruger this is obvious to anyone driving from the wooded hills in the south to the open plains in the north, walking down a catena from open savanna on the upland crest to dense riverine bushveld in the valley bottom, or living at a ranger station and experiencing the wet and dry seasons as they come and go. What is less obvious is how to deal with this heterogeneity when conducting research or implementing management, but this first part of the book introduces the concepts and presents the Kruger experience as a case study of evolving understanding and conservation practice. The chapters in Part I sketch the past and introduce a modern perspective of ecosystem heterogeneity as a context for the book and as a platform for integrated science and management in the future.

The early years in Kruger were dedicated to establishing a human order: shooting predators, erecting fences, introducing watering points, and so on, in what seemed to be a wilderness in need of some control. A pioneering approach was understandable then, yet land managers around the world are still strongly influenced today by a desire to control ecosystems, homogenize landscapes, and reduce the effects of environmental variability. In the chapters that follow it is argued that ecosystem management must undergo a paradigm shift to accommodate heterogeneity as the basis for biodiversity. For the natural order to prevail, the goals of ecosystem management must be redesigned to enable natural flux rather than to enforce stasis, which means that the mindsets of people and institutions must change. Kruger's managers and scientists have risen to the challenge by forging a novel approach to protected area management in which science, monitoring, and management are linked in a strategic adaptive system. Their work in progress is presented here in the spirit of shared learning and illustrates a commitment to ensuring the ecological integrity of Kruger through the vagaries of at least another century.

Chapter 1

The Kruger National Park: A Century of Management and Research

David Mabunda, Danie J. Pienaar,
and Johan Verhoef

In this chapter we provide a brief historical overview of people and events that made Kruger the world-renowned park it is today. It has been said that those who do not honor their past do not deserve their future, but an in-depth analysis of some 40,000 years of history is not possible in one chapter. However, we did feel it necessary to include some early history because it shows how long humans have interacted with this ecosystem. The different eras were chosen to show when human impacts on the system, political power, and management or research philosophy changed. These changes were seldom abrupt and usually had a developing period or overlapped and sometimes coincided with increased technology or the influence of certain people (Figure 1.1).

The Hunter-Gatherer Period

Archaeologists also use the phrase "Stone Age" for this period because of the stone tools that were used during this period. Deacon and Deacon (1999) dated the divisions of the Stone Age in relation to the present as follows: Earlier Stone Age, 2.5 million–250,000 years before present (BP); Middle Stone Age, 250,000–22,000 years BP; Late Stone Age, 22,000–2,000 years BP; and Iron Age, 2,000 years BP to the colonial period.

The Earlier and Middle Stone Age people and the San (or Bushman) of the Later Stone Age period lived in this area for many thousands of years and are thought to have had little impact on the natural processes and populations. The San, the last remaining group of the Stone Age (Deacon and Deacon 1999), were hunters and gatherers and possibly scavenged from the prey of carnivores. They led a nomadic life in small groups, wandering through the area following

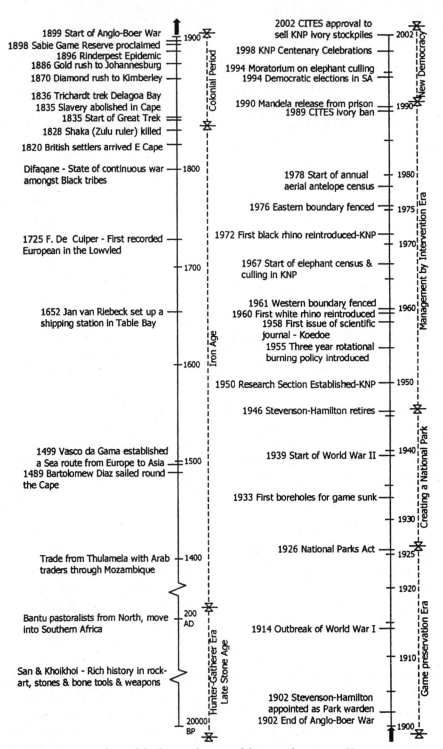

FIGURE 1.1. A timeline of the known history of the area that is now Kruger National Park.

migrating game herds (Plug 1982). They used the bow and arrow and microlithic tools and left a rich heritage of their rock paintings of animals and humans in numerous shelters in rocky outcrops in Kruger as well as deposits of ash, bone, small stone tools, and ostrich eggshell beads. They would have witnessed the arrival of a different cultural group who herded cattle, sheep, and goats, planted crops, and worked metal about 2,000 years ago.

Humans affect the environment in two ways: through physical presence in high numbers and in an intangible social manner through decision-making, induced conflict, religion, and so forth. The hunter-gatherer peoples surely possessed these characteristics, but population densities were so low that it is generally accepted that early humans did not shape the environment in a permanent way; rather, the environment at that time shaped them. Low-density occupation and low-intensity resource use of the Stone Age hunter-gatherers probably would have constituted a low-impact period in Kruger's history.

Farmers, Metalworkers, and Traders: The Iron Age (AD 200–1836)

Archaeological research has demonstrated that Iron Age communities had settled in southern Africa by at least AD 200 (Hall 1987), and by about AD 400 the first Bantu-speaking people started settling in the present-day Kruger area along the Letaba River. They possessed metalworking skills, traded, and had a residential lifestyle based on pastoralism. In the next 1,000 years additional groups settled along the Luvuvhu, Letaba, Olifants, Sabie, and Crocodile rivers. Population numbers are thought to have peaked around 15,000 during this period, resulting in localized homogenization of the ecosystem. They constructed villages, collected wood for fire and building material, cleared bush for grazing areas, prepared lands for agriculture, and stayed in an area until resources were depleted (Plug 1982). They hunted in formidable groups, often using fire and game pits to capture bigger animals. Hunting was still a major survival strategy because irregular and erratic rainfall and indigenous diseases limited herding and cropping (Plug 1989). Climatic fluctuations probably led to fluctuating densities of human settlements, with associated periods of higher and lower impact on the environment. Although it was probably a popular hunting locale, the Kruger area is considered to have been marginal or transitional in terms of cultural-historical occupation and farming, with a noticeable influence of human and livestock diseases such as nagana and malaria.

By the twelfth and thirteenth centuries there was active trade in ivory, skins, slaves, and gold between Mapungubwe along the Limpopo River and Arab traders who used the Sofala port in Mozambique (Huffman 1996). From Thulamela, a fifteenth-century site in the northern Kruger, these activities were continued until approximately 1650 (Kusel 1992). However, trade continued from other centers thereafter, and of significance are the references to ivory trade: Ferreira (2002), for instance, reports that ivory export via Inhambane amounted to 26,000 kg in 1768.

When Francois de Cuiper, the first recorded European to set foot in the lowveld, undertook his expedition from Delagoa Bay in 1725 to an area just north of the Crocodile River in the present-day Kruger, he found many black settlements. A hundred years later the situation looked very different, probably as a result of warfare and disease associated with climatic change. The period between 1800 and 1835 was a time of upheaval and changes in black political power south of the Limpopo River. This was a state of continuous war known as the Difaqane or Mfecane. This was also the time when Shaka, ruler of the Zulu nation, conquered many other black tribes and dispersed others toward Swaziland, the South African lowveld, Mozambique, and Zimbabwe.

The Colonial Period: Pioneers and Hunters (1836–1902)

In 1652 Europeans colonized the cape and introduced both a strong market economy and firearms, starting the overexploitation of wildlife (Carruthers 1995). Religion also played a role as Christianity excluded beliefs in the intrinsic power and value of nature, as believed by hunter-gatherers, and commanded its followers to tame and civilize nature in the service of humankind.

Early in the nineteenth century white people started exploring the area north of the Vaal River, and Louis Trichardt was the first white Voortrekker to trek through the present-day Kruger to Delagoa Bay (Maputo) in Mozambique in 1836. During this journey they lost all their cattle to nagana, carried by the dreaded tsetse fly, and most of the party succumbed to malaria. They recorded only a few small black settlements with hardly any cattle in the lowveld.

More white Voortrekkers trekked out of the Cape Colony and settled in the Transvaal to escape British rule, and political power was wrested from the resident African groups. Rural white Afrikaners and black Africans used wildlife as a resource and depended on produce from the environment for their existence. This was in stark contrast to the increasing number of British sportsmen who killed game for pleasure and trophies and documented their adventures (Cumming 1850; Harris 1838; Selous 1881). British tradition determined that sportsmen were gentlemen, and these upper classes scorned those who hunted commercially or for their own consumption. The rural Afrikaners found it difficult to believe that people would kill animals solely for amusement and waste the byproducts (Anderson 1888).

The period 1836–1902, including the Anglo-Boer War, was characterized by uncontrolled hunting for meat, skins, and ivory. This decimated the game populations in the lowveld (the low-lying area in which Kruger is situated), and campaigns began for the conservation of wild animals. As far back as 1858 laws to regulate hunting were proclaimed by the South African Republic. They were not successful in stopping or even slowing down the slaughter. The rinderpest epizootic that erupted in 1896 decimated both wildlife and domestic stock, and the government suspended all hunting restrictions to aid impoverished rural communities (Carruthers 1995).

After years of campaigning by various people for the creation of a game reserve between the Sabie and Crocodile rivers (Carruthers 1995), and with the looming Anglo-Boer War, President Paul Kruger eventually signed the proclamation creating the Sabi Game Reserve in 1898. The war was fought from 1899 to 1902 over political rights for foreigners and the gold riches in the Transvaal (Pakenham 1991). The British scorched-earth war policy of burning farms and homesteads and establishing concentration camps, in which many more Afrikaner and African women and children died than men on the battlefields, created much animosity against them (Pretorius 2001).

Game Preservation Era (1902–1925)

After the Anglo-Boer War, formal protection of game in the lowveld started in 1902 with the appointment of James Stevenson-Hamilton as warden of the Sabi Game Reserve (Figure 1.2). Stevenson-Hamilton was a Scottish professional soldier who had risen to the rank of major during the war. The instructions that Stevenson-Hamilton received with his appointment were vague and amounted to stopping hunting activities in the area and turning it into a game sanctuary. The British colonial administrators had a long history of European game preservation that centered around the creation of game sanctuaries to be used as exclusive hunting grounds by sportsmen and gentlemen (Carruthers 1995).

From 1902 to 1926 the emphasis was on the protection and rebuilding of these game populations. Stevenson-Hamilton was a good choice to lay the foundations of the new game reserve in that he was intelligent, a good leader, articulate, observant, and an efficient administrator (Carruthers 2001). In 1903 the area between the Sabie River and the Olifants River was added to the Sabi Game Reserve, and the Shingwitsi Game Reserve (an area between the Letaba and Luvuvhu rivers) was proclaimed.

At proclamation, these reserves housed low game numbers as a result of excessive hunting and the ravages of the 1896 rinderpest epidemic. Elephant and white rhino were locally extinct. Stevenson-Hamilton worked persistently to achieve his goals and appointed white game rangers assisted by black game scouts to patrol the area, arrest poachers, and enforce the law. He was opposed by farmers, hunters, and land companies (Carruthers 2001). This was a difficult task because the area was huge and there were no roads or infrastructure.

FIGURE 1.2. Three wardens of the Kruger National Park, key personalities in its history of research and management. James Stevenson-Hamilton (bottom) was the first park warden (1902–1946) and played a pioneering role in establishing the park's legal status and infrastructure. Dr. Uys de Villiers ("Tol") Pienaar (top left) began working in Kruger in 1955, and over the next 32 years he rose through the ranks to park warden while consistently building a tradition of pragmatic management based on research. David Mabunda (top right) is the present park warden, having been appointed in 1998, and is responsible for steering Kruger through a transition phase to align the park's management system with the new principles of governance in South Africa.

For instance, Major A. A. Fraser and 10 game scouts had to control the whole of the Shingwitsi Reserve, an area of about 800,000 ha.

Management actions included predator control and veld (range) burning to enhance the distribution of game, and Stevenson-Hamilton started keeping rainfall records. He also moved out the many isolated black families who lived in and were not employed by the reserve, earning him the unflattering nickname *Skukuza* ("he who sweeps clean") (Carruthers 1995). This policy of creating parks and moving indigenous people out of the area was followed in many other parts of the world (Burnham 2000), causing animosity from neighboring rural communities.

Stevenson-Hamilton became increasingly concerned about the lack of adequate protection provided by the provincial ordinances for the game reserves as pressures by commercial farmers and mining houses grew to deproclaim parts for commercial interests (Carruthers 1995). With help from some influential people he started lobbying to have the reserves proclaimed national parks. After much lobbying behind the scenes, this eventually happened after the Nationalist party came into power in 1924 and passed the National Parks Act (1926), when the Sabi and Singwitsi reserves were amalgamated and named the Kruger National Park (Carruthers 1995; Pienaar 1990).

Creating a National Park (1926–1946)

The new legislation provided for a Board of Trustees to be appointed, and the era of exclusive power of the warden was over. This also meant that the public obtained access, and the first three tourist cars entered in 1927. The state undertook to pay for management and maintenance of the new national park, but development had to be financed from tourist income. This necessitated the construction of roads and tourist accommodation facilities.

After initially using the South African Railways to manage tourism, in 1931 the board appointed outside contractors to provide catering and trade to tourists because of the lack of internal funds. These concessions continued until 1955, when the board again took them over after continuous complaints by the public concerning poor service. Initial accommodations were rustic, and Stevenson-Hamilton was determined to provide visitors with a wilderness experience. He fiercely resisted any upgrading of accommodation, being concerned that it would overcivilize the park (Carruthers 1995).

The stabilization of water resources to distribute game more evenly and counter the perceived desiccation of the lowveld was started in 1933 when the first six boreholes were sunk, signaling the start of more permanent form of management intervention. In 1938 after a foot-and-mouth epidemic among domestic stock in the region, the state veterinarians ordered the destruction of all cloven-hoofed domestic stock that were kept for milk and food in and around the park. This action was incomprehensible to the local people and unforgivable, also leaving Stevenson-Hamilton with a lasting mistrust of scientists (Joubert 1986).

Although Stevenson-Hamilton did not collect systematic scientific data or compile species checklists, he was a keen observer and wrote many scientific and popular publications, journals, reports, and books about Kruger and its animals (e.g., *South African Eden* [1937] and *Wild Life in South Africa* [1947]).

Stevenson-Hamilton retired in 1946 at age 79, after 44 years of building Kruger years into an internationally known and respected conservation area. The early history of Kruger and Stevenson-Hamilton's life up to this point has been comprehensively documented by Carruthers (1995 and 2001) and Pienaar (1990).

The Era of Management by Intervention (1946–1990)

It was during this period that discrete management and research functions emerged, and that specialist service divisions dealing, for instance, with technical and tourism services, developed.

Management

Colonel J. A. B. Sandenberg took over from Stevenson-Hamilton as warden in 1946. He outlawed controlled burning of grass and reintroduced carnivore control in parts of the park. Kruger was in a dry cycle, and in late 1950 the Letaba River stopped flowing for the first time in history (Chapter 21, this volume). Managers were concerned about game leaving the unfenced park in search of water and about localized overgrazing. More water-points were added evenly through the park (Chapter 8, this volume) to offset these problems and to attract game for tourists (Joubert 1986).

In 1955 tourist numbers exceeded 100,000 for the first time, marking the end of a quiet and romantic era. Kruger was becoming an institution run increasingly on business principles. Warden L. Steyn retired in 1961, the last self-trained warden-conservationist. He was replaced by Dolf Brynard, head of the Research Section, and park management was seen as having a firmer scientific base.

To cope with the proposed extensive development of road networks and tourist facilities, a Technical Services department was established in 1958. It completed fencing of the park boundaries for veterinary and disease control purposes, demanded by the National Department of Agriculture: the southern boundary along the Crocodile River in 1959, the western boundary in 1961, the eastern boundary in 1976, and the short northern boundary in 1980. The fence curbed the spread of diseases to domestic stock in the adjoining areas, kept dangerous animals from marauding outside, and facilitated boundary patrolling for poaching control.

The fence turned the park into an ecological island for large mammals. It prevented certain populations from moving seasonally and thereby escaping natural pressures such as water scarcity and droughts, fire effects on grazing, disease epidemics, and local predation. However, the fact that their access to water outside the park had been cut off led managers to attempt to stabilize the water situation in Kruger by drilling boreholes and building dams. This water stabilization policy had negative ecological consequences that became apparent decades later (Chapter 8, this volume), such as encouraging zebra and lion buildups in areas designated for roan antelope conservation. The boundary fence also abetted populations of large herbivores such as elephant, buffalo, and hippo. These no longer left the park and were not shot or snared in the surrounding buffer areas. In turn, concern for the impact of these large herbivores on the environment led the park managers to control their numbers through culling operations.

The first complete aerial census on elephant and buffalo was carried out in 1967, with 6,586 elephants and 15,758 buffalo being counted. Managers were surprised at this rapid growth because Stevenson-Hamilton (1905) had thought there were no elephants in 1903 and that only 10 had crossed into the park from Mozambique by 1905. Acceptable upper and lower population limits were set for these species and an annual culling program commenced. A decision was also made to use the meat and byproducts from culling, and a certified abattoir was erected in Kruger. Culling techniques were honed to conform to high ethical and animal welfare standards. The management motto of Kruger became "management by intervention" (Pienaar 1983).

During this time more people were being crowded into the rural areas west of Kruger through various government resettlement schemes, including forced removals. Industry, commercial forestry, and agriculture were developing in the upper catchments adjoining Kruger. Environmental impacts became bigger, especially a decrease in flow in perennial rivers, with strong agricultural irrigation boards dividing available water. Park managers had no legal right to claim water for the environment, and attempted to manage the situation by building dams in the park.

Reintroduction of species that were extinct in Kruger was a priority and was very successful in the case of rhinos. White rhino were first reintroduced from Natal in 1961, and today the Kruger population numbers about 5,000, the largest in the world. Black rhino reintroduction started in 1972, and although it is not yet completed, the population already numbers about 400, one of the largest in Africa (Pienaar 1994).

In the 1960s and 1970s rhino and elephant populations were being decimated in most reserves in postcolonial Africa (Caughley et al. 1990). Conservation managers in Kruger successfully resisted this onslaught, thanks largely to brave black field rangers on the ground.

The voting public were allies of Kruger, and through a public outcry the government was stopped from allowing coal mining in the park in the late 1970s. Infrastructure was established, roads built, and tourism facilities constructed so that by the end of this period the development footprint in Kruger was much the same as today. Research work continued, as did active refinement of policies, but Kruger's management was increasingly criticized for being insular.

Research

The retirement of Stevenson-Hamilton opened the way for changes. With the strong support of new member Dr. Rudolph Bigalke, director of the National Zoological Gardens, the board decided in 1950 to establish a research section in Kruger (Carruthers 1995, 2001). Dr. T. G. Nel was appointed in 1950 as a senior research officer, followed the next year by Manie van der Schijf as assistant biologist. The government made specific annual grants for this purpose, and by 1962 a research imperative was explicitly mentioned in the National Parks Act.

Similar changes were witnessed in East Africa with the opening of the Serengeti Research Institute in 1966 (Adams and McShane 1992). In South Africa, however, local scientists coordinated research, whereas elsewhere in Africa the work was done by expatriates (Rogers 2002).

Early research priorities related to the impact of fences on animal migrations, range burning, population studies, and their interdependence with ecological factors. This promoted understanding and interpretation of natural processes governing the functioning of Kruger's ecosystems (Joubert 1986). The main research projects were management oriented, and monitoring programs were implemented to measure the effect of management strategies. For the first time specific management policies were drafted, initially related to fire management, water provisioning, and predator control (Joubert 1986). The research section started collecting baseline information on the vegetation, geology, and soils and systematically cataloging species.

In 1954 veld-burning experiments began, as did the development of an extensive network of firebreak roads, which eventually divided the park into more than 400 burn blocks. A rotational burning regime was introduced, but by 1992 a strong wilderness lobby managed to change this rotational fire strategy to a natural one in which managers hoped lightning would drive ignitions (Biggs and Potgieter 1999). The use of fire for management purposes has been debated for many decades, and the park's burning policy is still being updated.

Scientists (initially mostly staff) in Kruger conducted groundbreaking research on and established safe procedures for the chemical immobilization of wild animals. This allowed safe translocation of game to other parks, effective techniques for administering drugs for disease control, and radiocollaring of animals for research projects. Population studies were undertaken on most larger mammal species in the Kruger ecosystem. Examples are lion (Smuts 1982), elephant (Hall-Martin 1984), zebra (Smuts 1976), wildebeest (Whyte and Joubert 1988), impala (Fairall 1972), roan antelope (Joubert 1975), and kudu (Owen-Smith 1984). Integrated research on topics such as predator-prey studies (Pienaar 1969), browser interactions (du Toit 1988), and mammal distribution (Pienaar 1963) was also done, climatic cycles were described (Gertenbach 1980), chemical game capture techniques developed (Pienaar 1968b), fire behavior investigated (Chapter 7, this volume), impact and control of disease epizootics described (Pienaar 1968a), and vegetation landscapes delineated (Gertenbach 1983). Aerial game census techniques were developed, and all large mammals were surveyed annually from 1978 onward (Joubert 1984).

A key person in this era was Dr. Uys de Villiers ("Tol") Pienaar, who started as a junior ranger in 1955 and rose through the ranks of biologist, director in charge of research and wildlife management for all national parks, and park warden of Kruger to chief director of the South African National Parks (SANParks). He had a wide scientific interest and was largely responsible for guiding research directions in this era. He established the reference museum in Skukuza and started cataloging plant, fish, reptile, mammal, amphibian, insect, and bird diversity in Kruger. Dr. Pienaar also played a leading role in research on chemical capture and immobilization of wild animals. He wrote numerous scientific papers and reference books and a book about the history of Kruger. He was the driving force behind pragmatic management by intervention, and he campaigned tirelessly to improve the worsening water situation in Kruger's perennial rivers. He retired in 1991 after 35 years of service, of which 32 years were spent in Kruger.

In 1958 the National Parks Board launched its scientific journal, *Koedoe*, where research conducted in or important to national parks was published. In earlier years many articles were in Afrikaans. A total of 771 articles have been published to date in *Koedoe*. Park managers and researchers did a fine job of documenting policy changes and the reasoning behind them in many internal documents held in the Skukuza archives (Joubert 1986), although monitoring targets were not always explicit.

South African Politics

In 1948 the National Party won the general election and stayed in power until 1994. The Party enforced grand apartheid (separation) policies and created independent "states" (homelands) for black people. The South African National Parks, as a parastatal institution, followed these policies. Black people were used mostly as laborers and not promoted into higher positions. Tourist accommodation facilities were also segregated, and blacks were encouraged to visit Manyeleti Game Reserve, an inferior nature reserve in the Gazankulu homeland designated for black people.

The National Party government used national parks to build an identity and sense of unity among Afrikaans-speaking white South Africans and to foster Afrikaner nationalism (Carruthers 1995). This was done as part of an active uplift program after the ravages of the Great Depression and the two great wars had created many poor white Afrikaners. Racial discrimination against blacks, segregation policies forcing them into unsustainable ethnic homelands, second-rate education, and very limited access to national parks all meant that no sense of ownership for national parks was built among black South Africans. This physical and psychological separation from the natural environment is a challenge that present park managers have to address as a matter of urgency. For black South Africans this was a sad chapter in the history of conservation.

The New Democracy: Black Empowerment (1990–2002)

The 1990s were a decade of rapid change for South Africa and its national parks. Major sociopolitical transformation, and a strong paradigm shift in ecosystem science and management, contributed to windows of opportunity that not only promoted the role of blacks in society, but also allowed innovations in management and research, extension of conservation estate, and far-reaching policy renewals.

Political Changes

Between 1985 and 1989 the country's isolation had intensified and pressure had grown for a democratically elected government. On February 2, 1990, President Frederik W. de Klerk released Nelson Mandela from prison and legalized all banned political parties. The first free elections were held in 1994, and the African National Congress (ANC) became the elected government. Initially the ANC government did not take a clear stand on the role of national parks, and some politicians made statements to newspapers that Kruger should be handed to local black cattle farmers. Nelson Mandela mapped the future relationship between national parks and the government in 1998 at the Kruger Centenary Celebration. He said that the conservation responsibility rested with "new leaders" and that he would like to see them build viable partnerships with neighboring communities.

In 1991 Dr. Nganani Enos Mabuza became the first black board member, and in 1995 a new demographically representative board was appointed. In 1997 Mavuso Msimang was appointed as the first black chief executive officer of SANParks and, in 1998, David Mabunda became the first black director of Kruger. These developments played a key role in changing government perception, with national parks being viewed as important national assets that attract international ecotourists.

Transformation is under way. Gender equality, affirmative action, and equal opportunities have become management objectives with clearly set targets. Whereas initially there were no blacks or women in Kruger management positions, in 2002 there were 36 white men, 20 black men, 11 black women, and 4 white women (South African National Parks 2002).

BOX 1.1
Kruger at a Glance

The Kruger National Park is situated in the lowveld (Chapter 20, this volume) of northeastern South Africa, bordering Mozambique in the east and touching on Zimbabwe in the north (Figure 1.3). It is an elongated park of about 2 million ha, roughly 350 km from north to south, with an average width of 60 km. The Crocodile River in the south, the Luvuvhu and Limpopo rivers in the north, and the Lebombo hills in the east form natural boundaries. The park is bordered on the west mainly by high-density communal areas and by private and provincial game reserves.

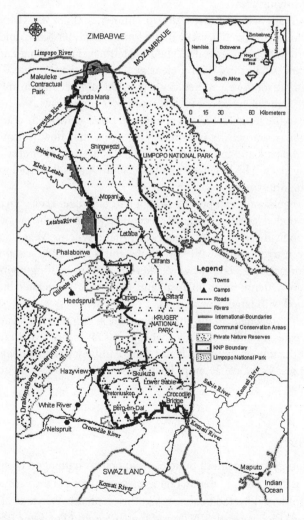

FIGURE 1.3. Kruger as it is today, embedded in a matrix of agricultural land and conserved wildlife areas (including the adjacent Limpopo National Park, which has been recently proclaimed in Mozambique).

There is an east-west altitude gradient, with basalt plains about 200 m above sea level in the east rising to 700 m in the granitic hills in the southwestern areas. Kruger is geologically split down its long axis, with the undulating western parts underlain by granite and the more level eastern plains underlain by basalt. The rhyolitic Lebombo hills in the far east, the granitic mountainous terrain in the southwest between Malelane and Pretoriuskop, and the sandstone hills northeast of Punda Maria show a diversity in geological parent material that is translated into distinctive associated biota.

Kruger straddles two climatic transitional zones: the tropical and subtropical north and the temperate south. Summer temperatures regularly exceed 35°C, and winter temperatures are moderate. Frost rarely occurs and is limited to low-lying areas.

Kruger falls in the southern African summer rainfall zone. Rain falls mostly from October to March, usually in the form of thunderstorms, and the period from April to September is dry. The long-term average annual rainfall for the whole park is 530 mm, with a clear gradient that varies from 730 mm per annum in the southwest at Pretoriuskop to 400 mm per annum at Pafuri in the northeast. Rainfall cycles of 15–20 years are recognizable, with 7–10 years being wetter than the long-term average, followed by a 7- to 10-year drier period.

Kruger is drained by five perennial rivers that flow from west to east through the park and into Mozambique and a large number of seasonal rivers of varying sizes. The larger seasonal rivers have pools that hold water during all but the driest years.

The vegetation in all but the wettest part of Kruger is classified as subarid to arid wooded savanna; botanists recognize at least eight major vegetation zones, subdivided into 35 landscapes or 11 land systems. Vegetation structure varies from open plains with low shrubs and a sparse tree canopy to closed gallery forest along certain rivers. There is also undulating open low woodland savanna, dense shrub on deep sand, and sparsely vegetated broken shrubland dotted with large baobab trees.

This heterogeneity in the abiotic template at different spatial and temporal scales creates a habitat diversity that supports an impressive array of species. Thus far the following species have been identified: 147 mammals, 505 birds, 119 reptiles, 49 fishes, 34 amphibians, 1,980 plants, and many thousands of invertebrates.

Management

Environmental management policies were being challenged by system fluctuations such as droughts and floods, changing perceptions of the ecosystem, and outside stakeholders. This led to changes in the elephant (Whyte et al. 1999), water provisioning (Pienaar et al. 1997), and fire (Biggs and Potgieter 1999) policies to make them more appropriate for a complex adaptive system. The scorching droughts of the early 1990s, with the lowest annual rainfall yet recorded for the park, and the February 2000 floods, during which the Sabie River burst its banks and flooded a third of Skukuza, attested to human inability to control nature. A huge wildfire raged through the park on September 4, 2001 and killed 24 people as well as elephants, white rhinos, and other species. These events illustrated that management actions can lessen or exacerbate the impact of these natural forces but that rigid policies are seldom appropriate. Incomplete information, outside influences, and difficulties in predicting environmental change, coupled to actions whose consequences are unpredictable and may only manifest many years later, all contributed to modifications of the rigid policies of the past. Kruger management has adopted an adaptive management process that promotes learning by doing, based on best available knowledge, as the most appropriate tool to manage the park in an ever-changing environment (Chapter 3, this volume).

There has always been an unusually close research-management link in Kruger, for two possible reasons. In the 1950s environmental management was an emerging science, and newly appointed scientists and managers learned together and from each other; the first degree in wildlife management in South Africa was awarded in 1965. Second, since 1961 at least one of the two most influential posts in ecosystem management, the park warden and the head of conservation management, has been drawn from the ranks of Kruger researchers. With the 1998 appointment as park warden of David Mabunda, whose background is in business and education, the SANParks board implemented its decision that Kruger should be run on sound business principles.

The management of large national parks such as Kruger is complex. Impacts and stakeholders all operate at different scales, from local to international. Decisions made inside the park have sociopolitical and economic impacts outside the park, and the long boundaries mean that there are many neighboring communities influenced by and influencing biodiversity and other management actions inside the park.

In an attempt to address this challenge, a Social Ecology Section was established in Kruger in 1995 that sought to involve neighboring communities more effectively. Important communication channels have been established, but there is still a long road ahead to foster a sense of pride and ownership among the park's neighbors (Chapter 20, this volume). The future survival of national parks such as Kruger will be guaranteed only by unconditional support of the black majority of South Africa's population. The challenge to managers is to make national parks relevant to the daily lives of all South Africans.

An exciting new development is the establishment of the Great Limpopo Transfrontier Park, which will join the Kruger with the Limpopo National Park in Mozambique and the Gonarhezou National Park in Zimbabwe. This will create a conservation area that covers 36,000 km^2, with prospects of even more land being added to it in the future. A joint management board has been established to manage this megapark, and management and tourism plans have been drafted. In 2002 about 1,000 head of different animal species (including 50 elephants) were translocated from Kruger to the Limpopo National Park as the start of a 4-year restocking program.

Research

Research in Kruger has benefited from these changes. More overseas scientists are pursuing research opportunities in the park, and it has become possible to secure international research funds. Exciting collaborative research programs, many reflected in the contents of this book, have been started with overseas and local academics.

In response to the worsening river situation in Kruger, the far-reaching Kruger Rivers Research Program was launched in 1990. Park management realized that the only way to mitigate the impact from beyond the borders of Kruger was to interact with external agencies and structures. This was done successfully in the Rivers Program to the extent that national water legislation enacted in 1998 was environmentally sound and supported the park's river management aims.

The Kruger management plan was updated in 1997, with clear research and management objectives. The new mission statement for the Kruger National Park reads, "To maintain **biodiversity** in all its natural facets and fluxes and to provide **human benefits** in keeping with the mission of SANParks in a manner which detracts as little as possible from the **wilderness qualities** of the Kruger National Park" (Braack 1999). Biodiversity is seen in its broadest definition, which also includes structure and processes (Noss 1990). The new management and research philosophy follows the postnormal trend (Blignaut 2002) that views systems as complex, unstable, open, and often with random reactions, incorporating these principles in management strategies and in the research focus.

An important research shift was toward more integrated programs that elucidate ecosystem function, compared with the more traditional species-based understanding. The threat of alien organisms such as bovine tuberculosis, a honeybee parasite (the *Varroa* mite), and many invasive alien plants necessitates research on the impact of these aliens on populations and processes and on effectiveness of control. Increasing demands on park resources and escalating impacts of development around Kruger necessitate new approaches in research. To meet all these challenges, the research section recently has been organized into three branches: Species Research, which concentrates more on traditional species projects; Systems Research, which aims at understanding broader ecosystem functioning; and Environmental Impact Research, which deals with human influences, often originating outside Kruger's boundaries. Because efficient collection, storage, analysis, and presentation of data are crucial, a strong quantitative ecology section with geographic information system (GIS) capability has been established.

Conclusion

Over a century, much has been achieved in Kruger. As we now view the situation, we see that mistakes have been made. Undoubtedly more will be made in equally good faith in the future. The commitment and dedication that our founding fathers displayed, the hardships they had to endure, and the sacrifices they made to build this world-renowned park can be applauded by citizens of South Africa and, indeed, the world. We honor the efforts and sacrifices of all the people who created this magnificent park, not least the historically deprived black communities. There are many challenges to face, above all building a broader constituency for conservation among all South Africans, not just the affluent. This goal must be balanced with immediate needs and aspirations of neighboring communities.

A new business model is being implemented in Kruger, and noncore activities (such as shops and restaurants) are being handed over to concessionaires to allow park management to concentrate on its core business, which it sees as biodiversity conservation, public benefits, and constituency building. To do this effectively, management must be based on a solid scientific foundation and on structures that allow one to adapt and respond quickly to an ever-changing system.

References

Adams, J., and T. McShane. 1992. *The myth of wild Africa*. New York: W.W. Norton.
Anderson, A. A. 1888. *Twenty-five years in a wagon*. London: Chapman & Hall, London. Reproduced by Cape Town: Struik, 1974.
Biggs, H. C., and A. L. F. Potgieter. 1999. Overview of the fire management policy of the Kruger National Park. *Koedoe* 42:101–110.
Blignaut, J. N. 2002. The search for a new economic system continues. *South African Journal of Economic and Management Sciences* 5:271–276.
Braack, Leo. 1999. *A revision of parts of the management plan for the Kruger National Park*, Vol. 7. *An objectives hierarchy for the management of the KNP*. Internal report. Skukuza: South African National Parks.
Burnham, P. 2000. *Indian country, God's country: native Americans and the national parks*. Washington, DC: Island Press.
Carruthers, J. 1995. *The Kruger National Park: A social and political history*. Pietermaritzburg: University of Natal Press.
Carruthers, J. 2001. *Wildlife and warfare: The life of James Stevenson-Hamilton*. Pietermaritzburg: University of Natal Press.
Caughley, G., H. Dublin, and I. Parker. 1990. Projected decline of the African elephant. Biological Conservation 54:157–164.
Cumming, R. G. 1850. *Five Years of a Hunter's Life in the Far Interior of Southern Africa*. London: John Murray.
Deacon, H. J., and J. Deacon. 1999. *Human beginnings in South Africa: uncovering the secrets of the Stone Age*. Cape Town: David Philip Publishers.
du Toit, J. 1988. *Patterns of resource use within the browsing ruminant guild in the central*

Kruger National Park. Unpublished Ph.D. thesis, University of the Witwatersrand, Johannesburg.

Fairall, N. 1972. Behavioral aspects of the reproductive physiology of the impala, *Aepyceros melampus* (Licht.). *Zoologica Africana* 7:167–174.

Ferreira, O. J. O. 2002. Montana in Zoutpansberg: 'n Portuguese handelsending van Inhambane se besoek aan Schoemansdal, 1855–1856. Pretoria: Protea Boekhuis.

Gertenbach, W. P. D. 1980. Rainfall patterns in the Kruger National Park. *Koedoe* 23:35–43.

Gertenbach, W. P. D. 1983. Landscapes of the Kruger National Park. *Koedoe* 26:9–122.

Hall, M. 1987. *The changing past: farmers, kings and traders in southern Africa 200–1860*. Cape Town: David Philip Publishers.

Hall-Martin, A. J. 1984. Conservation and management of elephants in the Kruger National Park, South Africa. Pages 1–20 in D. H. M. Jackson and P. Cummings (eds.), *The status and conservation of Africa's elephants and rhinos*. Gland, Switzerland: IUCN.

Harris, W. C. 1838. *The wild sports of southern Africa*. London: John Murray.

Huffman, T. N. 1996. *Snakes and crocodiles: power and symbolism in ancient Zimbabwe*. Johannesburg: Witwatersrand University Press.

Joubert, S. C. J. 1975. *The population ecology of the roan antelope,* Hippotragus equinus equinus *(Desmarest, 1804), in the Kruger National Park*. Unpublished D.Sc. thesis, University of Pretoria, South Africa.

Joubert, S. C. J. 1984. A monitoring program for an extensive national park. Pages 201–212 in N. Owen-Smith (ed.), *Management of large mammals in African conservation areas*. Pretoria: Haum Education Publishers.

Joubert, S. C. J. 1986. *Masterplan for the management of the Kruger National Park*, Vols. 1–6. Unpublished document, Skukuza archives, Kruger National Park, South Africa.

Kusel, M. M. 1992. A preliminary report on settlement layout and gold melting at Thula Mela, a Late Iron Age site in the Kruger National Park. *Koedoe* 35:55–64.

Noss, R. F. 1990. Indicators for monitoring biodiversity: a hierarchical approach. *Conservation Biology* 4:355–364.

Owen-Smith, N. 1984. Demography of greater kudu populations in the Kruger National Park in relation to rainfall. *Acta Zoologica Fennica* 172:197–199.

Pakenham, T. 1991. *The Boer War*. Cardinal, UK: Sphere Books Ltd.

Pienaar, D. J. 1994. Kruger's diversity enriched. *Custos* May:22–25.

Pienaar D. J., H. Biggs, A. Deacon, W. Gertenbach, S. Joubert, F. Nel, L. van Rooyen, and F. Venter. 1997. A revised water-distribution policy for biodiversity maintenance in the KNP. In Leo Braack (ed.), *A revision of parts of the management plan for the Kruger National Park*, Vol. 8. Pages 157–183. Skukuza: South African National Parks.

Pienaar, U. de V. 1963. The large mammals of the Kruger National Park: their distribution and present-day status. *Koedoe* 6:1–37.

Pienaar, U. de V. 1968a. Epidemiology of anthrax in wild animals and the control of anthrax epizootics in the Kruger National Park, South Africa. *Federal Proceedings* 26:1496–1502.

Pienaar, U. de V. 1968b. Recent advances in the field immobilization and restraint of wild ungulates in South African national parks. *Acta Zoologica et Pathologica* 46:17–38.

Pienaar, U. de V. 1969. Predator-prey relationships amongst the larger mammals of the Kruger National Park. *Koedoe* 12:108–176.

Pienaar, U. de V. 1983. Management by intervention: the pragmatic/economic option. Pages 23–36 in R. N. Owen-Smith (ed.), *Management of large African mammals in conservation areas.* Pretoria: Sigma Press.

Pienaar, U. de V. 1990. *Neem uit die verlede.* Pretoria: Sigma Press.

Plug, I. 1982. Man and animals in the prehistory of the Kruger National Park. *Transvaal Museum Bulletin* 18:9–10.

Plug, I. 1989. Aspects of life in the Kruger National Park during the Early Iron Age. *South African Archaeological Society, Goodwin Series* 6:62–68.

Pretorius, F. 2001. *Scorched earth.* Cape Town: Human and Rousseau.

Rogers, P. J. 2002. *Global governance/governmentality, wildlife conservation and protected area management: a comparative study of eastern and southern Africa.* Paper presented at the 43rd Annual International Studies Association Convention, New Orleans, LA.

Selous, F. C. 1881. *A hunter's wanderings in Africa.* London: Macmillan.

Smuts, G. L. 1976. Population characteristics of Burchell's zebra (*Equus burchelli antiquorum*, H. Smith, 1841) in the Kruger National Park. *South African Journal of Wildlife Research* 6:99–112.

South African National Parks. 2000. *Visions of change: social ecology and South African National Parks.* Pretoria: South African National Parks.

South African National Parks. 2002. *Annual report: 2002.* Pretoria: South African National Parks.

Stevenson-Hamilton, J. 1905. *Report on the Government Game Reserves for the year ended 30th June 1905.* Unpublished internal memorandum, South African National Parks, Skukuza.

Whyte, I. J., H. C. Biggs, A. Gaylard, and Leo Braack. 1999. A new policy for the management of the Kruger National Parks elephant population. *Koedoe* 42:111–132.

Whyte, I. J., and S. C. J. Joubert. 1988. Blue wildebeest population trends in the Kruger National Park and the effects of fencing. *South African Journal of Wildlife Research* 18:78–87.

Chapter 2
Biotic and Abiotic Variability as Key Determinants of Savanna Heterogeneity at Multiple Spatiotemporal Scales

STEWARD T. A. PICKETT,
MARY L. CADENASSO, AND TRACY L. BENNING

Heterogeneity is the degree of difference among a set of things. However, such a core definition is only a starting place; it is not adequate for a comprehensive understanding or management of heterogeneity in an ecological system as complex as Kruger, which is one of the few unconstrained landscapes remaining where spatial heterogeneity and ecological response can operate freely over large spaces and long times. The variety of geologic substrates (Chapter 5, this volume), the extreme variability in wet-dry cycles (Chapter 15, this volume), the rich megafauna (Chapter 11, this volume), and the role of fire (Chapter 7, this volume) combine to make Kruger a globally significant laboratory for studying the role of heterogeneity in ecosystems (Figure 2.1). Heterogeneity is the ultimate source of biodiversity (Pickett 1998), and therefore heterogeneity must be the ultimate focus of ecological management and restoration (Chapter 3, this volume). The variety of causes of heterogeneity in Kruger and its fundamental role in ecosystem resilience all demand a rigorous examination of how the general concept of heterogeneity applies in Kruger.

Ecological heterogeneity has three key features: the types of resources and environmental constraints, including substrates, organisms, materials, energy, and information (Kolasa and Rollo 1991; Pickett and Rogers 1997); spatially explicit configuration of these resources and constraints (Wiens 2000); and a focal organism, assemblage, or process for which heterogeneity is relevant. Identifying the focus ensures that the study of heterogeneity moves beyond describing structure or pattern to include the functional role of heterogeneity. In other words, as ecologists study the great pair of scientific inquiry, structure and function (or pattern and process), heterogeneity emerges as a key link between the two.

FIGURE 2.1. A low-level oblique aerial photograph of a section of northern Kruger on granitic lithology taken near the end of the wet season. Heterogeneity is apparent in the stature, density, and shade (indicative of greenness in the original color image) of vegetation, the presence of drainage and fracture lines, and areas of bare and vegetated soils.

The core definition of heterogeneity is relevant to all causes and effects (cf. Pickett and Cadenasso 2002). However, to use this concept in the real world and to determine when, at what scales, and for what phenomena heterogeneity is meaningful, ecologists have to specify additional things about heterogeneity and the context in which it is applied. Heterogeneity, like gravity, is everywhere, but how one deals with it depends on what ecological phenomena one is interested in and for what time and space scales questions are posed. In the case of gravity, depending on whether one is trying to articulate the ideal behavior of bodies in a frictionless environment or to build a bridge or an airplane, the solution to the ever-present phenomenon of gravity is different. So it is with heterogeneity. To evaluate heterogeneity, we follow the same approach used to model any ecological phenomenon (Jax et al. 1998; Pickett and Cadenasso 2002) and specify the parts of the system, how they are functionally connected, their hierarchical structure, their boundaries, their scales of resolution, and the processes that exist within them. This recipe is a key to understanding the different workings of heterogeneity in different places, for different organisms, and for different processes.

Kolasa and Rollo (1991) identified principles of origin, scale, and pattern that provide the context for our analysis of heterogeneity:

- Heterogeneity can be deterministic, random, or chaotic in origin, and many kinds of processes and agents can produce heterogeneity. Agents of heterogeneity include physical ones, such as geology, fire, or flood, and biotic ones, such as organisms and their mediation of resources. Different agents behave differently, and physical and biological sources of heterogeneity can interact.
- Agents create heterogeneity on specific scales. However, the effects of heterogeneity can appear either on the scale of action or on different scales. Therefore, sources and patterns of heterogeneity can be hierarchically arranged.
- Heterogeneity may be continuous or discontinuous and be expressed as gradients, patchworks, or graded patchworks. Patterns of heterogeneity may repeat at some interval, or patterns may be unique and not repeat.

The study of heterogeneity in landscapes is advanced by a framework. Frameworks are a tool to place disparate approaches, disciplines, perspectives, and study sites in a common conceptual system (Pickett et al. 1994). In general, frameworks identify the possible causal factors, scales, and interactions that structure a system or process. In particular, the framework for heterogeneity in Kruger will account for a comprehensive array of causes of and responses to heterogeneity; link the abstract, scale-independent definition of heterogeneity to specific systems and help analyze the functional significance of heterogeneity; and provide an overarching conceptual model that can be used to understand how heterogeneity is created, maintained, or transformed by physical and organismal agents.

In this chapter, we present a general framework for heterogeneity and demonstrate how it can be applied to the specific landscapes and assemblages of Kruger. The integrative framework accounts for both physical processes and organisms that contribute to heterogeneity of the savanna landscape. Physical and biological causes of heterogeneity will be compiled in a conceptual framework to evaluate the similarities and differences between the two. The framework guides the operationalization of the extremely general concept of heterogeneity so that it can be applied meaningfully.

Key Components of Heterogeneity

To operationalize the principles and features of heterogeneity, we identify the key aspects as agent, substrate, controller, and responder. Heterogeneity consists of patterns generated by the interaction of agents, substrates, and responders (Figure 2.2),

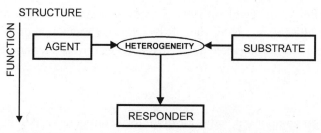

FIGURE 2.2. Heterogeneity is determined by an agent and substrate that are primarily responsible for the spatial pattern and determine whether the spatial pattern elicits a response in some process. The arrow from heterogeneity, as a spatial template, points toward the responder, indicating a causal relationship in that direction. The relationships or transformations represented by the arrows can be affected by other factors or processes in the system that act as controllers. The horizontal axis represents the structural dimension of heterogeneity, and the vertical axis represents the functional dimension. The significance of heterogeneity is the result of the functional relationship of a responder to heterogeneity at specified scales. Not all heterogeneity is significant at all scales.

and the interactions are controlled by a variety of factors. The general concept of heterogeneity becomes ecologically meaningful when it is given operational life by specifying what agents modify what substrates, what controls the modification, and what organisms or processes respond to the resulting spatial template. In this section, we define these four features and give some examples from Kruger to illustrate the definitions. Later, we will show how they can be connected in conceptual or other kinds of models to understand how heterogeneity comes to be, how it changes, and what its effects may be in specific contexts.

Agent

Agents create, maintain, or transform structural or functional features of a system. Agents include organismal activities and physical and chemical transformations. For example, termites can be an agent of heterogeneity. Their nest building creates areas that differ structurally and functionally from adjacent areas. There is a mound, a ring of eroded material, and a bare area cleared of litter by the foraging termites. Secondary effects may include trampling and herbivory by other animals that use termite mounds and nutrient dynamics altered by the activity of the termites and other animals (Griffioen and O'Connor 1990). A further example of a biotic agent of heterogeneity is isolated savanna trees. Such trees are nodes of enhanced nutrient concentration that may result from increased atmospheric deposition, congregation of animals and their associated defecation, cycles of litter concentration and decomposition, and altered abiotic conditions caused by both light and moisture dynamics (Belsky 1984; Belsky and Canham 1994). Floodwaters are an example of a physical agent that alters system heterogeneity by removing vegetation or redistributing sediment within the macrochannel banks of Kruger rivers (Heritage et al. 2001).

Substrate

The second component is the substrate for heterogeneity. A substrate is the entity the agent acts upon. Substrates are bounded in a spatial arena specified by the researcher. Substrates can be biological or physical. For example, a tree is a substrate for elephant disturbance, and alluvial reaches of Kruger rivers are substrates for flood action. Substrates have three-dimensional structure (Pickett and White 1985b; Breen et al. 1988) and vary in their susceptibility to transformation (Shachak and Brand 1991). For example, alluvial sections of the Kruger rivers are more readily modified by severe floods than are bedrock-controlled sections of river channels (Rountree et al. 2000). Notably, the scale of substrates varies tremendously in Kruger, for example, from the geological template (Chapter 5, this volume) to fissures in riparian rock outcrops. And each of these scales affects different processes and organisms.

Controller

Controllers affect the action of an agent on a substrate or the resultant transition between states of a substrate. In particular, controllers determine the spatial dispersion of the agent and the intensity of its action and the sensitivity and dispersion of the substrate. For example, the dispersion of grazers or browsers as an agent of heterogeneity differentially affects the impact of grazing or browsing across a landscape (Chapter 8, this volume). Such dispersion may be controlled by the degree of drought, for example. Further controls include the density and social structure of animals.

On the substrate side of the equation, vegetation condition and nutrient status of different species may control the impact of an animal agent of heterogeneity by controlling the sensitivity of the substrate (Chapter 11, this volume). For example, young trees free of disease or prior damage may resist elephant damage, whereas those that have been weakened by fire or drought may be easily damaged (Laws 1970; Scholes 1985).

Responder

The final component needed to understand heterogeneity is some entity or process that responds to the spatial differentiation. Although structural heterogeneity may exist in an arena, if there is no component of a system that responds to that heterogeneity, then the spatial heterogeneity is functionally neutral. It is especially important to specify a responder and to relate it to the scale of the spatial heterogeneity because what is an agent in one place or scale may be a substrate, controller, or responder to heterogeneity at other places or scales (Allen and Hoekstra 1992). For example, vegetation in the macrochannel can play all four roles: it is a substrate when removed by floodwaters, an agent of heterogeneity when it collects sediment, a controller when modifying flow patterns and hence flood damage, and a responder when establishing on new sediment.

Modeling Heterogeneity as a Process

Although modeling heterogeneity often starts by recognizing a pattern, it is completed by assessing how other ecological structures or processes affect or respond to the observed heterogeneity. Structural heterogeneity is the pattern of spatial variation at a given scale, whereas functional heterogeneity is the spatial pattern to which ecological entities may respond (Figure 2.2; Kolasa and Rollo 1991). Understanding the functional significance of heterogeneity is a pressing task of contemporary ecology (Hansson et al. 1995). Together, the structural and functional axes of heterogeneity combine to form a unified perspective (Pickett et al. 2000). Three features of heterogeneity must be understood when constructing models: the modes of change in heterogeneity, the scaling relationships of the action creating heterogeneity and response, and the contrasting behavior of biological or physical agents of heterogeneity.

To build a model of heterogeneity as a spatially and temporally dynamic process, the parts of a system must be specified, their relationships identified, their internal hierarchical structure determined, and their scales delimited. Once such a system is conceived and bounded in time and space, a model can be articulated. The four components—agent, substrate, controller, and responder—must be combined into such a model to understand the creation and impact of heterogeneity and to allow for the specificity of heterogeneity for any one system, scale, or situation. The basic interactions in any complex system model are the particular links or flows that govern specific changes in the system components (Figure 2.3).

Modes of Change in Heterogeneity

The conversion of one physical state of a substrate to another state of that substrate is the fundamental dynamic of heterogeneity. *State* here simply means a condition of some part of the system. Although thresholds often exist in the changes in ecological systems, gradual transitions are as likely. For purposes of showing how models of heterogeneity can be constructed, we focus on the extremes of gradients between states, or changes with clear thresholds. In gen-

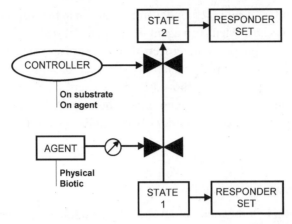

FIGURE 2.3. A general representation of the generation of heterogeneity and biotic response to heterogeneity. This abstract schema can be applied to any situation. The interaction between an agent and a substrate is controlled by one or more factors. The dial in the figure indicates that the action and control take place on a specified scale or range of scales. A bowtie represents a control gate on the transition from one state to another in the substrate. The agent drives, and the controller modifies, the transition from state 1 to state 2 of the substrate. The controller can modify the behavior of the agent or the susceptibility of the substrate. Agents can be physical or biological and can act in the short term or be persistent or slowly changing factors. Sets of responders, which may be processes or organisms, are sensitive to state 1 or state 2 of the substrate. Sets of responders may differ compositionally, comprising different species or processes, or may differ quantitatively, in magnitude or rate. The horizontal flow chain at the bottom of the figure represents the basic alphabet of heterogeneity, from which all more complex relationships are built.

eral, what constitutes a state change in the sense we use here is suggested by the research or management question and embodied in definitions supporting the model. For example, in Kruger scientists focus on thresholds of potential concern arrived at through a consensus dialog to guide management actions (Chapters 3 and 22, this volume). There are three modes of net change in heterogeneity in an arena over time: agents can create new states, maintain existing states, or transform existing states. Transformation can yield new spatial elements of heterogeneity or obliterate existing heterogeneity, resulting in more uniform environments. The mechanisms underlying these net effects must be identified in models of heterogeneity (Pickett et al. 1987). We show how models can be used to expose the mechanisms driving transformations in heterogeneity.

In the case of termite mounds, the two states are defined as intact soil and soil disturbed by nest building. The agent of change is the nesting of a termite colony. This basic unit of interaction may seem unduly simple. However, it is just such simplicity that allows larger, more comprehensive models to be con-

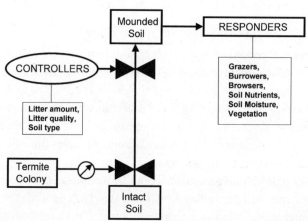

FIGURE 2.4. A flow chain model illustrating the application of the abstract schema of Figure 2.3 to the case of a termite colony creating heterogeneity. A termite colony acts on the intact soil state of the substrate to generate the state of mounded soil. The action of a colony may be modified by soil type, and the transition from intact to mounded soil may be modified by the amount and quality of litter available to the termites. Once a mound is created, grazer, burrower, and browser animals, soil nutrients and water, and vegetation may respond differently to the mound than to the intact soil.

structed and the origins of complex heterogeneity to be separated into their component mechanisms. In other words, the flow from one state to another mediated by an agent is the basic alphabet in which the understanding of heterogeneity is written (Figure 2.3).

The model of the effects of termites on heterogeneity is improved by adding vegetation as a controller. The volume and suitability of woody or herbaceous litter and whether litter remains standing or falls depend on the composition and architecture of the vegetation. Other aspects of vegetation may act as responders to termite-generated heterogeneity (Figure 2.4). Responders may be plant species that are rare or absent on intact soil (Griffioen and O'Connor 1990) and that capitalize on the soil disturbed by the creation of the mound.

The modes of change in heterogeneity can be represented using flow chain models (Shachak and Jones 1995). The flow chain approach has taken the core definition of heterogeneity and, through specification, shown it to represent a comprehensive, mechanistic, ecological system (Figure 2.3). Flow models show that the local transformations in physical state are the source of heterogeneity. However, such models are not spatially explicit. Knowledge of the explicit spatial configuration of various landscape elements is a key to understanding heterogeneity (Pickett and Cadenasso 1995; Wiens 2000).

Scaling of Action and Response

The conceptual model of agent, substrate, controls, and responders must be specified at a stated scale or range of scales. Agents of heterogeneity typically act over some range of scales. At coarser scales, their effects may disappear or be averaged out (O'Neill et al. 1986). For example, termite mounds may be invisible in a region observed at a grain size of square kilometers despite the known importance of termites in savanna landscapes. At scales finer than the mound size, mounds are part of the background.

Even though termite mounds may not be obvious at coarser scales of observation, they may still play a functional role in a large landscape. For example, trails of larger animals that visit termite mounds may be functionally significant for the movement or concentration of resources. Therefore, an element of heterogeneity may appear as a structure on one scale but as a functional feature on another scale.

Contrasting Behavior of Organismal and Physical Agents

Although the physical environment at a variety of scales is crucial to the generation of heterogeneity, organisms are especially important because they can react to and amplify physical components of heterogeneity. Organisms as agents of heterogeneity have different energetic, material, and informational controls than abiotic factors. For example, the mass of water in a flood contains potential energy that may be dissipated. In contrast, organisms have genetic controls, allometric relationships, behavioral repertoires, learning, and decision making that may yield a very different kind of heterogeneity from physical agents. Even among organisms, different groups may have different effects. For example, herbivores (Chapter 14, this volume) and carnivores (Chapter 18, this volume) may affect heterogeneity quite differently. These features of organisms as agents of heterogeneity may affect the scale, intensity, localization, repetition, and amplification of heterogeneity. The megaherbivores of Kruger may be especially significant agents (Chapter 11, this volume) because of their multiscalar effects (Chapter 8, this volume).

A major difference between physical and organismal agents of heterogeneity resides in how their effects are scaled. So far we have focused on the generation of heterogeneity within small spatial extents or by the creation of fine-scale features of the environment, such as termite mounds or damage to individual trees. However, many physical agents operate over immense spatial scales or represent long times. At the coarsest scale are tectonic uplift or subsidence, geology, climate, and geomorphology, which usually have very slow rhythms of change (Dollar et al. in prep.). At finer scales, nested within the coarse scale drivers, slope geomorphology determines the form, length, and steepness of land units and thereby governs gradients that drive local to medium-scale distribution of water, nutrients, toxics, and organisms (Chapter 5, this volume). The controls that exist on the different scales outlined here show the savanna landscape to be a complex, nested hierarchy of controls on heterogeneity, including terrestrial and aquatic components (Cadenasso et al., in prep.).

The final step in improving models of the effect of heterogeneity in ecosystems is to deal explicitly with space. Spatial elements of an area that can be altered to produce heterogeneity must be specified and related to one another in an explicit three-dimensional array. Without such specification, heterogeneity cannot be functionally understood. Adding the spatial dimensions and configuration to flow models also permits the spatial controllers of changes in state to be exposed. This approach is developed next, using the key processes of ecological engineering, ecological boundaries, and patch dynamics.

Heterogeneity in Explicit Space

We have introduced a framework for understanding and synthesizing the role of organismal and physical agents of heterogeneity in ecological systems. The framework comprises flow chains that identify agents and controls on organismally and physically generated heterogeneity (Figure 2.3). Biologically generated heterogeneity layers on top of the heterogeneity of a physical template. The physical template results from tectonics, geology, climate, and geomorphology (Turner et al. 1991; Cadenasso et al., in prep.). How these layers interact as either top-down or bottom-up controls is an important question that Kruger is ideally suited to answer (Chapter 6, this volume). There are three kinds of ecological processes by which organisms add new layers of heterogeneity to the physical template: ecological engineering, boundary function, and patch dynamics. These broad processes provide the empirical content for the general models motivated by the framework and incorporate the interactions that ecologists traditionally study, such as competition, predation, demography, and succession. The three broad processes focus attention on the modification of system structure and the resultant pattern of heterogeneity.

Ecological Engineering

The term *ecological engineering* has been defined in ecology as the direct or indirect modulation, by one kind of organism, of the availability of resources to other organisms (Jones et al. 1994; Lawton and Jones 1995). Ecological engineering can operate via physical, chemical, or transport processes and is the basic mechanism by which organisms affect heterogeneity. Physical engineering will be our focus. It includes the structure that organisms build or modify or the structure of the organisms themselves (Pickett et al. 2000). It is these structures that directly or indirectly control the flow of resources to other organisms. The transformation of physical structure has ecological consequences because the structure directly controls or indirectly modulates the flow of consumable resources used by other species. The resources can be energy, materials, space, information, or other organisms.

In Kruger, engineering can take many forms. Grazers may accentuate the contrast between sodic sites with their sparse, low-statured woody vegetation and adjacent taller, denser savanna or riparian zones (Dye and Walker 1980). Consumption of plants and soil erosion are direct and indirect mechanisms, respectively, by which the accentuation might be accomplished (Scholes and Walker 1993). Reeds in sandy sections of rivers are also important engineers because they trap sediment, stabilize soil for other vegetation, and provide habitat for other organisms. In the Kruger rivers, reeds may affect not only habitat but also consumable resources. They may maintain sandbars, which retain backwater ponds that are important sources of water for some animals during drought. The trails engineered by large mammalian herbivores, or herds of herbivores, can be important for the movement of surface waters and nutrients and are locations of high erosion potential along stream banks. The breaking or uprooting of trees by elephants alters the canopy of the savanna and therefore changes species mixtures and biomass distribution (Chapters 8 and 11, this volume).

Engineers exert indirect control over other organisms in the system. Because the engineer builds or modifies a structure into a new state, and the structure controls or modulates the resource flux, direct material or energetic participation by the engineer is not necessary for engineering effects to exist. This is in marked contrast to biotic or abiotic resource exploitation, where energy and material exchange is essential.

Physical state change is central to the system model (Figure 2.3) because it is the physical change in the environment that enables differential control or modulation of resource flows. For example, elephants transform trees and undisturbed soil to woody debris with soil pits and mounds (Cumming 1992; Guy 1976). The ways in which the physical state change exerts direct resource control or indirect resource modulation depend on how the resource flows work in a particular system. This is the interface with the responder in the agent-substrate model (Figure 2.3). Direct control has two sources. First, the structure created can be a resource. The size of a baobab tree directly determines the amount of living space for other organisms. Second, the flow of the resources can be regulated by the structure. Canopy structure of large riparian trees may control throughfall and stemflow during rain events, which may directly control the distribution and abundance of water and nutrient resources for some understory plants. Indirect modulation also occurs when the structure acts on other forces in the system that control resource flow. Organisms downstream from a reed bed are affected by the hydrological control over sedimentation that is exerted by the reeds.

Although there is value in single-species engineering models illustrated in this chapter, particularly for systems in which there is a predominant engineer that builds major structure in the environment (e.g., river channel trees, elephants, or termites), almost all habitats are physically engineered by numerous organisms at different scales, in different ways, with different consequences (Lawton and Jones 1995). The existence of diverse engineers in Kruger is a unique feature of the system for advancing ecological understanding.

To translate a conceptual model with these characteristics into an operational, system-specific, scaled model entails integration with any boundaries present and with a patch dynamics framework.

Ecological Boundaries

The second category of processes by which organisms generate spatial heterogeneity is through boundary function. Boundaries exist as discontinuities between contrasting habitats. For example, boundaries may be compositional and architectural transitions between climatically or geologically maintained vegetation types (Holland et al. 1991), such as the savanna-forest transition in moister areas of Kruger. Many other gradational boundaries exist in response to transitions in soil, substrate, or other variables on many scales, such as the riparian-savanna boundaries in Kruger (Figure 2.5). However, not all boundaries are gradual. At the opposite extreme, boundaries may be distinct, having sharp structural or functional gradients. Most sharp boundaries are the result of human activity, ecological engineering, or intense physical disturbance. Biotically engineered boundaries include those around treefalls created by insects, buffalo wallows, or excavations by warthogs.

Boundaries can affect the functioning of populations, landscapes, and ecosystems by modulating fluxes (Pickett and Cadenasso 1995). The contrast between the patches may be in the potential to move materials, energy, organisms, or information or in the relative magnitudes of these ecological commodities. The boundary between the two contrasting patches has a structure that differs from the two patch types. In some cases the boundary combines features of the two types, and in other cases the boundary can be a unique element of heterogeneity. The flux between the two patch types may be affected by the structure of the boundary between them. There are many mechanisms by which such modulation may occur, but the net effect logically can be either enhancement of the flow, reduction in the flow, or no effect on the flow (Cadenasso et al. 2003).

Although the study of boundaries has been largely descriptive and static (Murcia 1995), there are a growing number of examples of boundary function

FIGURE 2.5. Three positions along a riparian-upland boundary on the Phugwane River in the northern section of Kruger. The three photos represent, from left to right, vegetation immediately on the stream channel bank, boundary vegetation, and upland savanna vegetation. The contrasting three-dimensional structures across this boundary zone may affect the fluxes of materials and organisms, or of processes such as fire, across the boundary. (Photos by K. Schwarz.)

based on their structural characteristics (Cadenasso and Pickett 2000, 2001; Weathers et al. 2001). The deposition of nitrogen into temperate forest edge zones is substantially higher than elsewhere in the forest (Weathers et al. 1995, 2001; Erisman and Draaijers 1995). This led us to hypothesize that in the arid environment of the Kruger, where dust may be a significant landscape flux, the riparian-savanna edge may play a filtering role for aerial deposition. Boundaries may also filter seed flux (Cadenasso and Pickett 2001) or mediate the impact of herbivores on tree seedlings (Cadenasso and Pickett 2000). These processes may act in the patchy environment of Kruger, where interactions between herbivores and riparian-savanna boundaries, or with edges of sodic and nonsodic sites, or between burned and unburned patches (Chapter 11, this volume) may be mediated by structural or compositional differences at those boundaries.

Boundaries have a functional role in the heterogeneity generated by organisms (Pickett and Cadenasso 1995): boundaries control the expression of engineering over space. This recognizes that all ecological systems are open to fluxes of matter, energy, and information (Wiens 1992; Pickett and Cadenasso 1995). Boundaries are important both because they are a result of organism engineering and because organisms, as mobile engineers, interact with boundaries (Cadenasso and Pickett 2000, 2001). Therefore, boundaries are controllers affecting the agent, the substrate, and the response to heterogeneity (Figure 2.2).

Some boundaries generate organismal response that itself transforms the heterogeneity of the environment. An example is the creation of sodic patches in riparian-upland boundaries in Kruger (Rogers 1995). This is an active area of research, and many questions remain. Generally, areas of sodic soil accumulate minerals and fine particles at the soil surface and support contrasting vegetation and herbivore activities that differ from those found in the normal savanna soil. However, it has been suggested that trees on islands in the Okavango Delta may increase salinity by pumping mineral-laden water to the soil surface (McCarthy and Ellery 1994). Engineering by trees of the soil properties in the boundary area enhances the boundary formation, ultimately leading to distinct patches. This example illustrates the feedbacks between physical factors at boundaries and amplification by engineering.

Patch Dynamics and Heterogeneity

Patch dynamics accounts for both organismal and physical causes of heterogeneity. The agents of patch formation can be arranged in a causal hierarchy (cf. Pickett et al. 1989). Slowly evolving agents of tectonics, geology, climate, and geomorphology act on the coarse scale (Chapter 5, this volume). At finer scales, physical disturbances such as wind, fire, flood, and landslides (Dale et al. 1999) contribute to a mosaic of heterogeneity. Fire and flood are medium-scale physical agents of disturbance that are especially active in savanna (Furley et al. 1992). Organism establishment, growth, and behavior are key agents of fine-scale patch origin and change. At the finest scales, we have noted the differences in selectivity and control of engineering by organisms and the complementary class of disturbance by physical forces. Humans are a major source and modifier of patches in landscapes worldwide, and they act from the fine to the coarse scale. The interaction of people in Kruger savanna is of exceedingly long duration (Chapters 11 and 19, this volume). Patches in real mosaics therefore reflect a complex of causes, ranging from the persistent physical template, the engineering effects of many kinds of organisms including humans, and the dynamics within patches subject to a variety of controllers.

Patches have complex structure, reflecting the hierarchy of causes that have acted in a particular place. Patches of a particular spatial scale may be resolved into constituent patches on finer scales (Kotliar and Wiens 1990). At the coarsest scale in Kruger, vast riparian and upland patches can be resolved, whereas at a much finer scale, sodic patches can appear as a distinct component of the boundary between the larger riparian and upland patches. Patches may also have either abrupt or gradual boundaries at a specified scale, as in the broad gradients bounding *vleis* on basalt compared with the steep architectural shifts from *spruits* to upland savanna on granite. Complexity within patches may result from the persistence of structural legacies or surviving propagules as a result of low or moderate levels of "diffuse disturbance" such as drought-induced mortality of trees (Chapter 11, this volume). The complexity of patch structure results from gradients in the underlying physical template, the differential performance of the organisms that engineer patches, or the characteristics of the organisms that form the substrate on which engineering or disturbances act. Alluvial and nonalluvial soil interactions with groundwater flow from rivers or from the adjacent upland, interacting with the behavior of herbivores, combine to influence the location and spread of sodic patches near rivers in Kruger. Any heterogeneity in the substrate that existed before the action of an engineer or disturbance can modify the subsequent pattern of heterogeneity. Here, then, is another node of interaction between coarse-scale drivers of tectonics, geology, climate, and geomorphology, with finer-scale organismal sensitivities and behaviors. Resolving coarse-scale versus fine-scale control is an open question for Kruger (Chapter 11, this volume).

Patch dynamics are spatially explicit. Rather than considering patches as a population characterized by aggregate or averaged properties, the configuration and geographical location of patches are analyzed. Granite block boundaries near to or distant from rivers are a case in point. Patches are functionally connected with the larger spatial context. The surroundings of a focal patch or system are themselves a complex mosaic, which can control the access of engineers to the patch (Chapter 8, this volume) or the availability of resources in it. Patch connectivity depends on the nature of the flux through the mosaic and the existence of boundaries at various scales and their impact on lateral flows in the landscape. An example of the function of configuration appears in the Kruger rivers (Chapter 9, this volume). After severe floods, the successional patterns in alluvial sections of the rivers differ depending on what kind of substrate exists upstream of the section. Hence, the configuration of bedrock, mixed anastomosing, and alluvial reaches upstream of a focal reach determine the successional patterns in the river channels.

Patch dynamics puts ecological engineering into a spatial context. Although engineering acts on specific locales, the spatial context is not explicit in the concept. Patch dynamics makes the spatial context explicit and relates the effects of engineering in different locations in a landscape to one another. Boundaries make the interactions and connections between patches explicit. They are a key mechanism by which spatial configuration affects landscape function. In general terms, patch dynamics has been shown to be a major contributor to species coexistence and the structuring of assemblages and ecosystems in many places (Pickett and White 1985a; Clark 1991; Fisher 1993; Hansson et al. 1995; Rogers 1997). Because of its functional significance, patch dynamics is also a major focus for ecosystem management (cf. Chapter 3, this volume). As one of the few large unconstrained landscape systems to have an intact fauna of megaherbivores, a complex geological template, a layer of fire dynamics, and extreme rainfall variation in time, Kruger can expose the complexities and dynamics of patches as few other systems can.

Conclusion

Heterogeneity is a major driver of the function of ecological systems and of their richness and productivity. Organisms make and maintain a vast store of the world's heterogeneity on fine to medium scales (Huston 1994; Pickett 1998) but do so in the context of coarse-scale, slower drivers that determine the physical template. This chapter has identified the most important elements that must be considered to understand the sources and roles of spatial heterogeneity in Kruger.

The framework outlined here casts the savanna landscape, including its embedded drainage network, as a physical and biological mosaic that is integrated by fluxes of materials, energy, and organisms across it. The framework can be viewed as a multidimensional space whose axes represent the causes of heterogeneity, the kinds of fluxes involved, and the potential outcomes of heterogeneity. As a tool, the framework presents the entities and processes that can be combined to build functional models of landscape heterogeneity. Such models are key to an understanding of the Kruger landscape and to its sustainable management.

References

Allen, T. F. H., and T. W. Hoekstra. 1992. *Towards a unified ecology.* New York: Columbia University Press.

Belsky, A. J. 1984. Smallscale pattern in grassland communities in the Serengeti National Park, Tanzania. *Vegetatio* 55:141–151.

Belsky, A. J., and C. D. Canham. 1994. Forest gaps and isolated savanna trees: an application of patch dynamics in two ecosystems. *BioScience* 44:77–84.

Breen, C. M., K. H. Rogers, and P. J. Ashton. 1988. Vegetation processes in swamps and flooded areas. Pages 223–247 in J. J. Symoens (ed.), *Vegetation of inland waters.* Dordrecht, The Netherlands: Kluwer.

Cadenasso, M. L., and S. T. A. Pickett. 2000. Linking forest edge structure to edge function: mediation of herbivore damage. *Journal of Ecology* 88:31–44.

Cadenasso, M. L., and S. T. A. Pickett. 2001. Effects of edge structure on the flux of species into forest interiors. *Conservation Biology* 15:91–97.

Cadenasso, M. L., S. T. A. Pickett, T. L. Benning, H. C. Biggs, A. Gaylard, C. James, R. J. Naiman, and K. H. Rogers. In prep. Integrating savanna landscapes: the link between river networks and terrestrial patches across multiple scales.

Cadenasso, M. L., S. T. A. Pickett, K. C. Weathers, and C. G. Jones. 2003. A framework for a theory of ecological boundaries. *Bioscience* (in press).

Clark, J. S. 1991. Disturbance and tree life history on the shifting mosaic landscape. *Journal of Ecology* 72:1102–1118.

Cumming, D. H. M. 1982. The influence of large herbivores on savanna structure in Africa. Pages 217–245 in B. J. Huntley (ed.), *The ecology of African savannas.* Berlin: Springer-Verlag.

Dale, V. H., A. E. Lugo, J. A. MacMahon, and S. T. A. Pickett. 1999. Ecosystem management in the context of large, infrequent disturbances. *Ecosystems* 1:546–557.

Dollar, E. S. J., K. H. Rogers, C. S. James, M. Thoms, and M. R. Rountree. In prep. A scaled hierarchical framework for understanding pattern and process in river systems. In preparation.

Dye, P. J., and B. H. Walker. 1980. Vegetation-environment relations on sodic soils of Zimbabwe, Rhodesia. *Journal of Ecology* 68:589–606.

Erisman, J. W., and G. P. J. Draaijers. 1995. *Atmospheric deposition in relation to acidification and eutrophication.* Amsterdam: Elsevier.

Fisher, S. G. 1993. Pattern, process, and scale in freshwater systems: some unifying thoughts. Pages 575–597 in P. S. Giller (ed.), *Aquatic ecology: scale, pattern, and process.* Oxford, UK: Blackwell Scientific Publishers.

Furley, P. A., J. Proctor, and J. A. Ratter. 1992. *Nature and dynamics of forest-savanna boundaries.* London: Chapman & Hall.

Griffioen, C., and T. G. O'Connor. 1990. The influence of trees and termite mounds on the soils and herbaceous composition of a savanna grassland. *South African Journal of Ecology* 1:18–26.

Guy, P. R. 1976. The feeding behaviour of elephant (*Loxodonta africana*) in the Sengwa area, Rhodesia. *South African Journal of Wildlife Research* 6:55–63.

Hansson, L., L. Fahrig, and G. Merriam. 1995. *Mosaic landscapes and ecological processes.* New York: Chapman & Hall.

Heritage, G. L., B. P. Moon, G. P. Jewitt, A. R. G. Large, and M. Rountree. 2001. The February 2000 floods on the Sabie River, South Africa: an examination of their magnitude and frequency. *Koedoe* 44:37–44.

Holland, M. M., P. G. Risser, and R. J. Naiman. 1991. *Ecotones: the role of landscape boundaries in the management and restoration of changing environments.* New York: Chapman & Hall.

Huston, M. A. 1994. *Biological diversity: the coexistence of species in changing landscapes.* New York: Cambridge University Press.

Jax, K., C. Jones, and S. T. A. Pickett. 1998. The self-identity of ecological units. *Oikos* 82:253–264.

Jones, C. G., J. H. Lawton, and M. Shachak. 1994. Organisms as ecosystem engineers. *Oikos* 69:373–386.

Kolasa, J., and C. D. Rollo. 1991. Introduction: the heterogeneity of heterogeneity: a glossary. Pages 1–23 in J. Kolasa (ed.), *Ecological heterogeneity.* New York: Springer-Verlag.

Kotliar, N. B., and J. A. Wiens. 1990. Multiple scales of patchiness and patch structure: a hierarchical framework for the study of heterogeneity. *Oikos* 59:253–260.

Laws, R. M. 1970. Elephants as agents of habitat and landscape change in East Africa. *Oikos* 21:1–15.

Lawton, J. H., and C. G. Jones. 1995. Linking species and ecosystems: organisms as ecosystem engineers. Pages 141–150 in C. G. Jones and J. H. Lawton (ed.), *Linking species and ecosystems.* New York: Chapman & Hall.

McCarthy, T. S., and W. N. Ellery. 1994. The effect of vegetation on soil and ground water chemistry and hydrology of islands in the seasonal swamps of the Okavango Fan, Botswana. *Journal of Hydrology* 154:169–193.

Murcia, C. 1995. Edge effects in fragmented forests: implications for conservation. *Trends in Ecology and Evolution* 10:58–62.

O'Neill, R. V., D. L. DeAngelis, J. B. Waide, and T. F. H. Allen. 1986. *A hierarchical concept of ecosystems.* Princeton, NJ: Princeton University Press.

Pickett, S. T. A. 1998. Natural processes. Pages 11–19 in M. J. Mac (ed.), *Status and trends of the nation's biological resources.* Reston, VA: U.S. Department of Interior, U.S. Geological Survey.

Pickett, S. T. A., and M. L. Cadenasso. 1995. Landscape ecology: spatial heterogeneity in ecological systems. *Science* 269:331–334.

Pickett, S. T. A., and M. L. Cadenasso. 2002. The ecosystem as a multidimensional concept: meaning, model, and metaphor. *Ecosystems* 5:1–10.

Pickett, S. T. A., M. L. Cadenasso, and C. G. Jones. 2000. Generation of heterogeneity by organisms: creation, maintenance, and transformation. Pages 33–52 in M. Hutchings (ed.), *Ecological consequences of habitat heterogeneity*. New York: Blackwell.

Pickett, S. T. A., S. L. Collins, and J. J. Armesto. 1987. A hierarchical consideration of causes and mechanisms of succession. *Vegetatio* 69:109–114.

Pickett, S. T. A., J. Kolasa, J. J. Armesto, and S. L. Collins. 1989. The ecological concept of disturbance and its expression at various hierarchical levels. *Oikos* 54:129–136.

Pickett, S. T. A., J. Kolasa, and C. G. Jones. 1994. *Ecological understanding: the nature of theory and the theory of nature*. San Diego, CA: Academic Press.

Pickett, S. T. A., and K. H. Rogers. 1997. Patch dynamics: the transformation of landscape structure and function. Pages 101–127 in J. A. Bissonette (ed.), *Wildlife and landscape ecology*. New York: Springer-Verlag.

Pickett, S. T. A., and P. S. White. 1985a. *The ecology of natural disturbance and patch dynamics*. Orlando, FL: Academic Press.

Pickett, S. T. A., and P. S. White. 1985b. Patch dynamics: a synthesis. Pages 371–384 in S. T. A. Pickett (ed.), *The ecology of natural disturbance and patch dynamics*. Orlando, FL: Academic Press.

Rogers, K. H. 1995. Riparian wetlands. Pages 41–52 in G. I. Cowan (ed.), *Wetlands of South Africa, their conservation and ecology*. Pretoria: Department of Environmental Affairs.

Rogers, K. H. 1997. Operationalizing ecology under a new paradigm: an African perspective. Pages 60–77 in S. T. A. Pickett (ed.), *The ecological basis of conservation: heterogeneity, ecosystems, and biodiversity*. New York: Chapman & Hall.

Rountree, M. W., K. H. Rogers, and G. L. Heritage. 2000. Landscape state change in the semi-arid Sabie River, Kruger National Park, in response to flood and drought. *South African Geographical Journal* 82:173–181.

Scholes, R., and B. Walker. 1993. *An African savanna: synthesis of the Nylsvlei study*. New York: Cambridge University Press.

Scholes, R. J. 1985. Drought related grass, tree, and herbivore mortality in a southern African savanna. Pages 350–353 in J. C. Tothill (ed.), *Ecology and management of the world's savannas*. Canberra: Australian Academy of Science.

Shachak, M., and S. Brand. 1991. Relations among spatiotemporal heterogeneity, population abundance, and variability in a desert. Pages 202–223 in J. Kolasa (ed.), *Ecological heterogeneity*. New York: Springer-Verlag.

Shachak, M., and C. G. Jones. 1995. Ecological flow chains and ecological systems: concepts for linking species and ecosystem perspectives. Pages 280–294 in C. G. Jones (ed.), *Linking species and ecosystems*. New York: Chapman & Hall.

Turner, M. G., R. H. Gardner, and R. V. O'Neill. 1991. Potential responses of landscape boundaries to global environmental change. Pages 52–75 in M. M. Holland (ed.), *Ecotones: the role of landscape boundaries in the management and restoration of changing environments*. New York: Chapman & Hall.

Weathers, K. C., M. L. Cadenasso, and S. T. A. Pickett. 2001. Forest edges as nutrient and pollutant concentrators: potential synergisms between fragmentation, forest canopies, and the atmosphere. *Conservation Biology* 15:1506–1514.

Weathers, K. C., G. M. Lovett, and G. E. Likens. 1995. Cloud deposition to a spruce forest edge. *Atmospheric Environment* 29:665–672.

Wiens, J. A. 1992. Ecological flows across landscape boundaries: a conceptual overview. Pages 216–235 in F. di Castri (ed.), *Landscape boundaries*. New York: Springer-Verlag.

Wiens, J. A. 2000. Ecological heterogeneity: an ontogeny of concepts and approaches. Pages 9–31 in M. Hutchings, E. A. John, and A. J. A. Stewart (eds.), *The ecological consequences of environmental heterogeneity*. Malden, MA: Blackwell Science.

Chapter 3

Adopting a Heterogeneity Paradigm: Implications for Management of Protected Savannas

KEVIN H. ROGERS

Conservation of protected savannas has been dominated by a focus on charismatic species populations, homogeneity of habitat, and equilibrium and carrying capacity perspectives of system state, without regard for scale (Rogers 1997; Stalmans et al. 2001), largely because historically savanna management has been heavily influenced by stable state concepts developed for commercial agriculture (Peel et al. 1999). However, the pervasiveness of spatial heterogeneity and flux in ecosystems means that incorporating these concepts into management is emerging as a major challenge for savanna conservation and national park management.

Biodiversity is slowly becoming an important theme in savanna conservation, but all too often it is viewed from the very limited perspective of maintaining species richness. In contrast, a heterogeneity paradigm emphasizes that ecological systems function across a full hierarchy of physical and biological components, processes, and scales in a dynamic space-time mosaic (Pickett et al. 1997). Reconciling these contrasting perspectives is not as daunting as it may seem. First, we need to recognize that biodiversity encompasses compositional, structural, and functional elements of ecosystems, each manifest at multiple levels of interconnected organization from genes to landscapes (Noss 1990; Figure 3.1). However, this definition lacks explicit scaling of the spatial and temporal pattern and interactions of ecosystem components (Chapter 2, this volume).

The theory of patch dynamics (Pickett and Rogers 1997) provides the scale perspective we need to add to this multiple-level perspective of biodiversity. Recognition that patches may be nested within one another leads one to the concept of hierarchical patch dynamics (Wu and Loucks 1996; Figure 3.2), which provides context for species distributions and interactions with all the components of the landscape. Together with the multilevel biodiversity con-

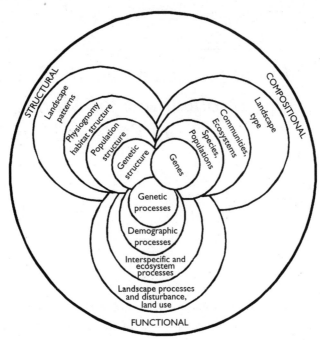

FIGURE 3.1. Structural, compositional, and functional biodiversity (after Noss 1990).

cept, it underpins the new thrust to emphasize heterogeneity in the management of ecological systems (Rogers and Bestbier 1997). The integration of these two concepts forms an appealing basis for incorporating the processes generating heterogeneity into management.

Heterogeneity and Management

The key determinants of heterogeneity (agents, substrates, controller, responder, and engineer, outlined in Chapter 2, this volume) can be easily related to this concept of hierarchical patch dynamics across multiple levels of organization (Figure 3.2) to provide a model useful to managers. At any time different organisms recognize and respond to patchiness across different ranges of scale. The finest scale of patch response is the grain, and the coarsest, usually the organism's home range, is the extent (Kotliar and Wiens 1990). Similarly, disturbance events and ecological engineers act as agents of change on biotic and physical components of the hierarchically nested patch mosaic (substrate) over particular scales. Controllers also act across particular ranges of scale.

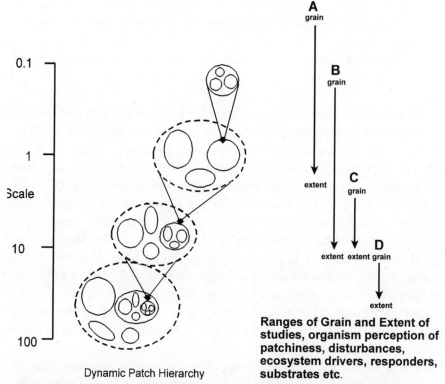

FIGURE 3.2. Relationships between the key components of heterogeneity (after Kotliar and Wiens 1990, using concepts from Chapter 2, this volume).

This model of heterogeneity allows the manager to view the landscape as a nested patch hierarchy generated by key formative processes rather than merely as the geographic arrangement of ecosystem components. The challenge faced by scientist and manager alike is to avoid becoming bogged down in the complex array of substrate and responder attributes (e.g., species richness, geological and geomorphic complexity, climatic variables) of heterogeneity and to extract the key agents of change (e.g., the dominant physical and ecological engineers) that managers can manipulate with the limited tools (e.g., fire, rifle, bulldozer) at their disposal. Scientists and managers therefore need to:

- Generate spatially and temporally explicit goals for heterogeneity management;
- Identify key agents and controllers of change that managers can manipulate if the need to influence the nature and direction of heterogeneity change arises;

- Design monitoring and modeling exercises to focus on spatially explicit relationships between change agents and attributes of heterogeneity at specific scales;
- Ensure that management is achievable within the limits of the tools and resources available.

There are two other important challenges for protected area and savanna managers that must be met if heterogeneity is to be managed effectively. The first is to embrace the uncertainties of nature's space-time complexity in the context of the regional, and even global, socioeconomic system in which the protected area is embedded. Kruger provides excellent examples (Chapter 1, this volume). The perennial rivers that arise west of the park boundary and flow into Mozambique cannot be conserved without extensive negotiations between a very wide range of local, national, and regional stakeholders (Chapter 21, this volume). At a broader scale, endangered species (elephant, rhino) management illustrates the need to consider international treaties and a range of international interest groups with widely differing worldviews and predilections for compromise. In general, managing protected areas in the broader socioeconomic context rests on the need to integrate the full range of stakeholder value systems.

The second is that because heterogeneous nature is in a continual state of flux and our understanding of ecosystem functioning is poor, dealing with uncertainty from an imperfect knowledge base must become central to effective management.

Adaptive management has emerged as the only widely recognized model for managing uncertainty in interactive social and ecological systems. However, it has not generally been implemented for savanna or biodiversity management and is currently facing an implementation credibility crisis (Box 3.1). Central issues in this crisis are a narrow focus on scientific experiments as primary means of learning while managing, overly ambitious monitoring and modeling programs, poor articulation of management goals, a failure to engage stakeholders, and a failure to manage concomitant change in implementing organizations and people.

A minimum set of conditions for successful adaptive management under a heterogeneity paradigm therefore consists of broader understanding of the learning process in adaptive management to include knowledge generation, transformation, and application; explicit and achievable goals of ecosystem heterogeneity and flux compatible with stakeholder needs and values; achievable monitoring and appropriate modeling that serve management goals, stakeholder needs, and the learning process; and appropriate institutional design, function, and reward systems for a socially responsible learning organization. Each of these conditions is discussed in turn.

> **BOX 3.1**
>
> ## An Analysis of the Implementation Crisis Facing Adaptive Management
>
> - The central tenet of adaptive management, "management by experiment," leads ecologists to seek manipulative, ecosystem-level experiments as solutions to management problems.
> - Managers are uncomfortable with the additional uncertainty and risk of experimental manipulation but comfortable with a trial-and-error approach that justifies the next generation of decisions when management actions fail to produce a desired result.
> - There is too much turf protection by managers and scientists who do not recognize the need for different roles in a science-management partnership.
> - Implementation becomes bogged down by the tyranny of modeling and modelers in pursuit of the ultimate model or technology.
> - The nature of ecosystem structure and functioning desired by management usually is not articulated and does not incorporate the range of stakeholders' value systems.
> - Monitoring programs are too ambitious and are abandoned because they are unachievable or become ends in themselves rather than means to a management end.
> - Too much unstructured information from scientists, modelers, and monitoring paralyzes decision-making processes.
> - The adaptive management process often is superimposed on bureaucratic institutional structures and processes. The learning-by-doing process flounders when the fundamental management axiom of "form must follow function" is ignored.
> - Individuals within organizations are unable to adapt to the new ways of thinking and functioning. When the new system undermines old comfort zones, staff become indecisive and unproductive and revert to familiar operating rules.
>
> Based on Walters 1997; Ludwig 2001; Rogers 1998; Rogers et al. 2000a; and Hanna 2001.

The Learning Process in Conservation

Manipulative experiments designed by expert scientists are central to the adaptive approach of learning by doing. Anything else is dismissed as trial-and-error management (Walters and Green 1997). But the limited applicability of this narrow perspective of learning is being exposed in both science (Pickett et al. 1994; Ludwig 2001) and management (Senge 1990; Allee 1997). It is time to broaden the concept of learning in adaptive management, especially in savanna conservation, where learning has not been made explicit and the focus has remained on charismatic fauna rather than on the processes of science and management.

There are three reasons for this:

- If learning is not explicit and carefully managed, informal mental models of scientists and managers come to dominate in a self-reinforcing communication circle. The consequence is that a body of pseudo-fact becomes the basis of decision making (Rogers and Biggs 1999). Such management has been more common than we might like to admit in savanna ecosystem conservation.
- Because ecosystems are so complex, scientists want to confront managers with the complexity they spend so much effort studying. It is sobering to contrast the many components and complex interactive processes ecology has unraveled with the small number and simplicity of the tools (e.g., fire, a gun or trap, food, water, earthmoving equipment, and occasionally money) available to the manager. We have a lot of learning to do if we are to convert the products of science into achievable goals and implementable solutions for conservation.
- Too many ecologists and their funding agencies indulge in what those in the technology transfer field call a strategy of hope. The hope is that good science will inevitably lead to information that someone will find useful (Rogers 1997). The problem is that report after report, publication after publication, sits on the shelf only to be read by other scientists and not implemented by managers.

Pseudo-fact, ecosystem complexity, the limited management toolbox, and the strategy of hope all tell us it is time to make the learning process more explicit and managed in ecosystem conservation.

Scientific Learning

Ecosystems are complex, nonlinear, dynamic, and self-organizing, permeated by uncertainty and discontinuities. Folke and Berkes (1998) point out that the kind of science this implies is a move away from the positivist emphasis on objectivity toward a recognition that fundamental uncertainty is large, yields are unpredictable, certain processes are irreversible, and qualitative judgments do matter. They suggest caution about assuming that the Western system of acquiring scientific knowledge is a universal epistemology and embrace the value of indigenous knowledge but go no further in proposing an alternative. The new philosophy of science explained by Pickett et al. (1994), on the other hand, provides excellent insight into the spectrum of scientific processes used to improve knowledge and thereby illustrates the important but limited role of experiments.

The new philosophy of science emphasizes two departures from the hypothesis-testing, inductive approach of adaptive management. First, science is a logical process of objectively relating a conceptual construct of the system under study to observable phenomena. Second, objectivity results not from the method of the individual experimenter but from the cancellation of individual bias by the active participation of a community of scientists in an open-ended interrogation of nature. Three tools of science are used in this interrogation by which scientists merge the conceptual construct with observed phenomena to incrementally improve understanding: **generalization** (construction of pattern in classification or synthesis), **causal explanation** (quantification of relationships), and **testing**. Testing is the comparison of an expectation or hypothesis derived from theory with observations and may be by either **confirmation** (most common) or **falsification** (most venerated). There are three modes of testing: **experiment, comparison,** and **correlation.** This understanding about the process of scientific investigation places experimentation in its proper context and give us a much broader basis for incorporating science into adaptive management.

The recognized success of the Kruger river research program (Breen 2000) can be attributed to adoption of this approach to learning (Chapter 21, this volume). Most importantly, the program began with a very clear set of conceptual constructs that guided research into the natural river systems and their management.

Treating policy or management action as an experiment presents a limited view of the processes involved in improving our understanding. It may be better to consider policy as a conceptual construct that we would relate to observable phenomena (natural or managed components of ecosystems) through a complex set of projects using the range of tools of science. This would require scientists, managers, and policymakers to collectively translate broad policy statements into conceptual constructs that are scientifically useful and compatible with management potential. In most cases links between management action, science, and policy are informal and tenuous at best. There is an urgent need for a theory of science, management, and policy integration.

Any consideration of linking policy, management action, and science in this way exposes the need for very careful management of the knowledge that will be generated by a large and diverse set of research projects in any one conservation area. For example, Kruger has about 300 research projects in progress at any one time. This number can be expected to grow, but the number of scientific staff who have to absorb the findings will not grow in proportion. Ensuring that science is used in management decision making is no trivial task and certainly cannot be left to the strategy of hope.

Managing Knowledge for Adaptive Management

A recurrent theme in modern business management is the understanding that sustainable operations need the capacity for ongoing learning and continuous transformation (Allee 1997). There is much emphasis on making ongoing learning an explicit part of management and on making knowledge management (as opposed to mere data or information management) an explicit and successful part of learning. Knowledge has come onto center stage and is seen as information, combined with experience, context, interpretation, and critical reflection (Reynolds 1998). It has been suggested that the emphasis on synthesis and theory testing in academia can be combined effectively with the action orientation of practitioners to generate creative learning organizations, even within government bureaucracies.

Broadly speaking, knowledge management should be responsible for generating wisdom, which teams use for decision making. Control of the quantity, quality, and form of information reaching appropriate people is imperative for effective decision making. Knowledge management is about strategically and creatively reducing the complexity of data and information into knowledge and wisdom to facilitate effective decision making. Thus, it moves an organization well beyond the confines of mere information management. The work of a knowledge manager has three foci: generation of knowledge, which entails its creation, acquisition, synthesis, and adaptation; concentration of knowledge centers on its capture, transformation, and representation; transfer of knowledge between locations and, most importantly, its absorption by the recipient (Meyers 1996).

Despite the imperative for these processes as precursors to effective decision making, knowledge management largely remains informal and ad hoc in conservation. Under severe resource constraints, an uncertain knowledge base, and a limited management toolkit, pragmatic learning by doing in a structured science-management partnership becomes an imperative. We have to learn how to manage knowledge better.

Setting Achievable Goals for Heterogeneity Management

An almost universal barrier to good management is poor translation of policy into achievable and scientifically defendable operational goals, supported by a process with which to audit their achievement. Consequently, management actions are not effectively tested against societal value systems or against scientific understanding, leading to a lack of direction, purpose, and transparency in management (Rogers and Biggs 1999).

An important obstacle to translating policy into goals is the need for a system that integrates stakeholder value systems with the scientific rigor needed to define the desired ecosystem endpoints of management actions. Very few management exercises have ever achieved this, and an effective trace of the policy implications of management actions usually is impossible. The situation is complicated by our poor understanding of ecological and social systems and their propensity for change (Folke and Berkes 1998).

The recognition that protected areas cannot operate in isolation from the socioeconomic system in which they are embedded means that incorporating societal values into conservation management has become an imperative. The problem is that values are too fuzzy for many ecologists and conservationists, who want hard facts or pragmatic solutions, respectively. However, it is important to recognize that values are the reference points people use for evaluating alternatives. Values range from ethical principles that must be upheld to guidelines that assist in making trade-offs or setting priorities. They represent the real needs of individuals and groups at any level of society.

Increasingly, effective decision-making methods incorporate a philosophical approach to understanding and articulating values and methods to use these values to create and select alternatives (Keeney 1992). Value-focused decision making is especially productive in situations involving multiple stakeholders. The best alternative for any one stakeholder is that which meets the values of others and makes them better off. Explicit inclusion of other stakeholders' values in alternatives you propose will lend support to your choice.

The process of developing a collective vision is central to incorporating values into adaptive management (Rogers et al. 2000b). Vision provides the first step in integrating social values, scientific knowledge, and management experience in a multiparty system. The first step is exposing disparate mental models and developing the common purpose and knowledge base on which consensus thrives.

Vision is a valuable tool for converging energies, but its value depends on how well it is translated into reality. This is a tricky task and requires a facilitator skilled at negotiating consensus rather than compromise. When everyone is focused on the common needs and values embodied in a broad consensus vision, the template for converging it with reality exists.

A good model for achieving this convergence is an objectives hierarchy (Keeney 1992), which divides the vision into achievable targets for ecosystem management (Rogers and Biggs 1999). The cascading linkages provide increasing detail of explanation of what the world should look like. Each step in the hierarchy is developed by the same negotiation process as the vision. When used in this way, a common vision preempts conflict.

Within an organization, a vision is a philosophical statement of intent that describes the core business of the organization. It largely reflects the societal values embedded in broad policy. It is the sort of statement that chief executive officers use in their dealings with the press or politicians and is readily understood by all stakeholders.

It is not enough to provide a broad vision and expect everyone to know what to do or to expect scientists to feel comfortable with its lack of detail and potential scientific ambiguities. The vision must be broken down into a hierarchical series of objectives of increasing focus, rigor, and achievability. Objectives are qualitative articulations of values defined in the vision and operating principles of the organization. They form a foundation on which to develop quantitative operational goals.

The finest level of a hierarchy is defined by such goals. These goals may be either institutional or conservation goals. Institutional goals define achievable targets for managing institutional structures and processes and are fundamental to defining the resources available to management. Conservation goals define targets for ecosystem management and therefore must be scientifically rigorous without compromising the value systems embodied in the vision.

Setting targets that clearly and unambiguously describe the complex, dynamic ecosystem to be conserved is a major challenge (Rogers and Bestbier 1997), but without them effective management is unlikely. These targets cannot be subjective and must integrate scientific understanding and professional judgment. Most importantly, they must effect change in ecosystem characteristics and must represent hypotheses or conceptual constructs that can be used to focus learning to improve understanding in an adaptive management cycle.

There have been few attempts to set such targets for ecosystem management. In Kruger they are called thresholds of potential concern (TPCs). TPCs are hypotheses of upper and lower levels of acceptable ecosystem change. Therefore, their validity and appropriateness are always open to challenge, and they must be adaptively modified as understanding and experience of the system being managed increase (Chapter 4, this volume).

Achievable Monitoring and Appropriate Modeling

Modeling of system response to potential management actions and subsequent monitoring of system response are central to adaptive management. Too often however, these exercises are designed more to serve science or scientists than to serve adaptive learning. Ensuring that they are appropriate to the purpose of the exercise and achievable within resource constraints becomes central to successful adaptive management.

Monitoring

Good management always incorporates a monitoring or assessment program (Noss 1990; Christensen 1997), which should provide a reality check on current system state, provide a cross-check on the accuracy of the problem definition and appropriateness of the selected solution, audit goal achievement, and, on a longer time scale, feed back to the evaluation of vision and policy. However, science and management, especially in savanna conservation, seldom formalize these operational rules. Monitoring has largely focused on censuses of large game and veld condition, more useful to farmers than to heterogeneity conservation. In such instances programs are not designed to monitor the consequences of specific management actions, let alone the changing perspectives of stakeholders. Rather, they aim to assess the general state of a limited set of ecosystem components. Seldom are results from different aspects of monitoring integrated and synthesized to improve understanding of the ecosystem as a whole. As a consequence, management loses direction and lacks a defensible scientific base, becoming reactive to the array of surprises nature, economics, and social adjustments deliver.

It is essential to specify the scientific and management questions monitoring is intended to answer and validate the relationships between indicators and components of heterogeneity they represent. Heterogeneity should be monitored and changes predicted, at multiple levels of organization over a range of spatial and temporal scales. Indicators should be identified for all levels of organization, but selecting an appropriate range for a particular issue is difficult. Indicators can be viewed as surrogates for the ecosystem structure and function the manager wants to achieve. Good indicators are sensitive enough to provide an early warning of change, are widely applicable or distributed over an appropriate geographic range, can provide continual assessment over a wide range of stress, are independent of sample size, are easy and cost-effective to measure, enable discrimination between natural fluxes and anthropogenic stress, and are relevant to ecologically significant phenomena and specified management goals (Noss 1990).

No single indicator has all of these attributes, and because heterogeneity monitoring must cover a wide range of scales relating to structure, composition, and function, a set of complementary indicators is needed (Rogers and Biggs 1999). Furthermore, it is time to incorporate flux into monitoring to ensure understanding of rates, frequency, magnitude, durations, and trajectories of change across organizational levels, recognizing the social complexity that continually interacts with this ecological complexity.

A growing school of thought (Holling 2001) holds that the diversity and complexity of interactive social and ecological systems can be traced to a small number of biotic and abiotic variables and processes with few species or groups of species running these processes.

Similarly, proponents of learning-by-doing business management (Senge 1990) distinguish detail or attribute complexity from dynamic complexity that arises from critical cause-and-effect relationships. Managing dynamic complexity is seen as providing greater leverage than managing detail complexity. We need to identify the few variables that lead to complexity and resilience in ecosystems and develop management and monitoring systems around them.

The research-management partnership in the Kruger has focused on identifying the main agents, natural and anthropogenic, of ecosystem change and on identifying indicators of these agents and of system response to them. The collective wisdom of the partnership, guided by a value-based objectives hierarchy, was used to define, in spatial and temporal terms, the upper and lower levels of this small number of variables with large effects on ecosystem state. These concepts are dealt with in more detail in Chapters 4 and 9.

Finally, monitoring programs must not become ends in themselves and must be fully integrated into adaptive management to explicitly serve the learning and decision-making processes. In Kruger an iterative management process ensures that research, predictive modeling, and management operations interact with monitoring to develop, test, and modify management goals (Chapter 4, this volume).

Appropriate Modeling

The complexity issue has dogged modeling just as it has monitoring. Debates on what constitutes appropriate modeling for management abound. In-depth discussions can be found in Starfield (1997) and Walters and Korman (1999). In general, the idea of single large, multipurpose models that describe all dynamics is inappropriate because the model needed is determined by the question asked.

There is a growing realization that one often learns more from the process of building a model than from the model output. Models should be viewed as experiments, as hypotheses, conceptual constructs, and problem-solving tools that guide the learning process. Therefore, the questions we ask and the processes by which we use models become as important as how we build them.

The move therefore is toward smaller, purpose-built models that can be developed rapidly to address key uncertainties and needs (Mackenzie et al. 1999). The models should embrace the small number of variables and processes that define dynamic complexity. In the longer term, model development is iterative, leading to improved model structure, performance, and learning as monitoring and research improve data availability and understanding.

The ultimate test of a model in adaptive management is not how accurate or truthful it is but how well it serves the learning-by-doing process. It is a means to an end, not an end in itself.

Appropriate Institutional Design

As the multidimensional complexity of savanna ecosystem conservation becomes more evident, so does the need to create different institutional structures with a new set of operational strategies and tactics. The old model of rangers and scientists operating independently to manipulate and understand the conserved area, in isolation from the surrounding landscape and communities, is over.

When new operating systems are developed, they are too often superimposed on old bureaucratic structures and processes with too much inertia to respond appropriately or sustainably (Rogers et al. 2000b). The structure and process of conservation management must both be reconfigured, with three main foci: managing knowledge to ensure its internal capture, transformation, and effective use, as discussed earlier; building partnerships with stakeholders and external knowledge generators; and creating learning institutions that strategically serve adaptive management.

Partnerships

The need to build partnerships with stakeholders in savanna conservation is not new, and a number of examples are found in Africa (Getz et al. 1999). It is new in South Africa, however, and we have many challenges to face, but unlike most other countries, we have highly supportive legislation. The Constitution, the Water Act, and the National Environmental Management Act require cooperative governance across all levels of society to provide equity, efficiency, and sustainability of access to resources, and are designed to enable citizens to control their own futures and participate in managing natural resources.

The days of conservation agencies making decisions for the people are gone, and the norm will become the formation of partnerships in which all parties (government, agencies, business, and the public) are stakeholders in resource management. Clearly, a central theme of conservation management must be securing the support and compliance of citizens and government in cooperative partnerships of consensus management (Rogers et al. 2002), hence the focus on constituency building in South African National Parks and Kruger (Chapter 1, this volume). The challenges will be enormous and will require a range of special skills and people not previously seen in this sphere of public service.

The need to build partnerships between scientists and managers to gather and synthesize new knowledge is equally paramount but just as seldom formalized. In most cases research is designed to meet the needs of scientists rather than managers, and a better balance is needed. More importantly, a broader spectrum of activities is needed to ensure that research results are transformed into wisdom for decision making (Rogers et al. 2000b).

It is rare for an ecological study to tell us both how an ecological system works and how to achieve a particular management goal. Good research and development organizations know that to avoid the strategy of hope and remain competitive, they must institute vigorous and structured technology transfer systems. If ecology and conservation are to break out of the strategy of hope, they too must have explicit processes for finding consensus, for product development, and for transfer to management (Rogers 1997).

Building Learning Institutions

A recurrent theme in business management is the understanding that sustainable business entails the capacity for ongoing learning and continuous transformation (Allee 1997). Over the last decade industry and businesses have become aware of the need to create learning institutions through a combination of adaptive and generative leadership (Senge 1990). Senge describes an adaptive process as one of coping and a generative process as one of creating. Bureaucratic institutions cope, learning institutions create and are therefore able to adapt. Adaptation is at the core of adaptive management, yet little attention has been paid to this imperative (Rogers et al. 2000b).

Four characteristics are central to learning organizations (Senge 1990): the leader as a designer, teacher, and steward; mental models; shared vision; and practice.

Leadership

Developing a learning culture entails a move away from regulatory, authoritarian line management and toward a new style of leadership. Generative leaders are defined as designers of common purpose and core values, of strategies and structures for guiding decisions, and of effective learning processes; teachers who help people achieve more accurate, insightful, and empowering views of reality; and stewards for both the people and the vision of the enterprise.

This kind of leadership is imperative if adaptive, learning-by-doing management is to be institutionalized in cooperative governance.

Mental Models

Mental models are our worldview: deeply ingrained assumptions and generalizations of how we see the world and our actions in it. They govern how we think and how we interact. They may be purely professional (a Poperian, holistic, reductionist, or synthetic approach to science), or they may be central to one's personal history. Their influence pervades partnerships. Individuals' mental models must surface in a way that strengthens both the individual and the partnership. This process is tricky because it confronts cherished believes and assumptions. But in the end a balance must be found between the individual's model and the partnership's mental model. Rather than trying to make an engineer a biologist, one should manage the creative tension of their interaction.

SHARED VISION

An accurate picture of current reality is just as important as a compelling picture of a desired future (Senge 1990). It is essential to have vision and root it in reality (Rogers et al. 2000a) while remaining aware of the present. Partnerships built on a vision of an uncertain but exciting future can scale back in time to identify and examine the more pressing problems constructively if they have a sense of where they are going.

PRACTICE

Even though we all know that sports teams improve with practice, we expect partnerships, prototypes, and new operational procedures to work from the start. They don't; they evolve. And we need practice, experimentation, and failures if we are to learn to make them work properly. Nowhere else should we be more prepared for this need than in adaptive management. We must build practice into reward systems, which bureaucracies and science do not do (Rogers 1997).

Conclusion

It is time for protected area managers and their agencies to change the way they do business. A few good examples of this change are emerging (Chapter 2, this volume; Salafsky et al. 2001), but much work remains if we are to develop both the theory and practice for doing so. This chapter has highlighted some of the essential ingredients of such an approach.

It is essential to begin with very clear conceptual constructs of the system to be understood and managed and of the management system to be used and tested. Different ecosystems require different research and management strategies, so it is essential to start with a good understanding of the system of focus. Research and management must become integrated learning processes, generating understanding in a structured knowledge management system. This is unlikely to be achieved unless there is a clear sense of purpose, or vision, that is explicitly translated into achievable and auditable operational goals. These goals must attract full buy-in from all stakeholders and fully describe the ecosystem heterogeneity to be achieved.

Managers have a very limited set of tools with which to build the desired spatiotemporal ecosystem heterogeneity and to preserve stakeholder values. Given that ecological understanding is tenuous at best, a pragmatic learning-by-doing or adaptive management partnership becomes imperative. Predictive modeling and monitoring of ecosystem response to the key agents of change should be integrated to ensure that the learning-by-doing process is strategic rather than reactive.

Implementing these measures will mean new operating rules and processes for scientists, managers, and their institutions. To think that these processes can simply be instituted in existing bureaucracies is naive. Protected area conservation agencies must be reconfigured to focus on knowledge management and partnerships with stakeholders. They must build adaptable, learning institutional cultures and operating systems in which new-style leaders institute reward systems to stimulate innovation and risk taking in strategic adaptive management.

References

Allee, V. 1997. Transformational learning. *Executive Excellence* 14:12.

Breen, C. M. 2000. *The Kruger National Park Rivers Research Programme final report. Incorporating the contract and review reports.* Report No. TT130/00. Water Research Commission, Pretoria, South Africa.

Christensen, N. L. 1997. Managing for heterogeneity and complexity in dynamic landscapes. Pages 67–186 in S. T. A. Pickett, R. S. Ostfeld, M. Shachak, and G. E. Likens (eds.), *Enhancing the ecological basis of conservation: heterogeneity, ecosystem function and biodiversity.* New York: Chapman & Hall.

Folke, C., and F. Berkes (eds.). 1998. *Linking social and ecological systems. Management practices and social mechanisms for building resilience.* Cambridge, UK: Cambridge University Press.

Getz, W. M., L. Fortmann, D. Cumming, J. du Toit, J. Hilty, R. Martin, M. Murphree, N. Owen-Smith, A. M. Starfield, and M. I. Westpal. 1999. Sustaining natural capital and human capital: villagers and scientists. *Science* 283:1855–1856.

Hanna, S. 2001. Managing the human-ecological interface: marine resources as example and laboratory. *Ecosystems* 4:736–741.

Holling, C. S. 2001. Understanding the complexity of economic, ecological and social systems. *Ecosystems* 4:390–405.

Keeney, R. L. 1992. *Value-focused thinking: a pathway to creative decision making.* Cambridge, MA: Harvard University Press.

Kotliar, N. B., and J. A. Wiens. 1990. Multiple scales of patchiness and patch structure: a hierarchical framework for the study of heterogeneity. *Oikos* 59:253–260.

Ludwig, D. 2001. The era of management is over. *Ecosystems* 4:758–764.

Mackenzie, J. A., A. L. van Coller, and K. H. Rogers. 1999. *Rule-based modelling for management of riparian systems.* Report No. 813/1/99. Water Research Commission, Pretoria, South Africa.

Meyers, P. S. (ed.). 1996. *Knowledge management and organizational design.* Boston: Butterworth-Heinemann.

Noss, R. F. 1990. Indicators for monitoring biodiversity: a hierarchical approach. *Conservation Biology* 4:355–364.

Peel, M. J. S., H. Biggs, and P. J. K. Zacharias. 1999. The evolving use of stocking rate indices currently based on animal number and type in semi-arid heterogeneous landscapes and complex land-use systems. *African Journal of Range and Forage Science* 15:117–127.

Pickett, S. T. A., J. Kolasa, and C. Jones. 1994. *Understanding in ecology: the theory of nature and nature of theory.* San Diego, CA: Academic Press.

Pickett, S. T. A., R. S. Ostfeld, M. Shachak, and G. E. Likens (eds.). 1997. *Enhancing the ecological basis of conservation: heterogeneity, ecosystem function and biodiversity.* New York: Chapman & Hall.

Pickett, S. T. A., and K. H. Rogers. 1997. Patch dynamics: the transformation of landscape structure and function. Pages 101–127 in J. A. Bissonette (ed.), *A primer in landscape ecology.* New York: Springer-Verlag.

Reynolds, M. 1998. Reflection and critical reflection in management learning. *Management Learning* 29:183–200.

Rogers, K. H. 1997. Operationalising ecology under a new paradigm. Pages 60–77 in S. T. A. Pickett, R. S. Ostfeld, M. Shachak, and G. E. Likens (eds.), *Enhancing the ecological basis of conservation: heterogeneity, ecosystem function and biodiversity.* New York: Chapman & Hall.

Rogers, K. H. 1998. Managing science/management partnerships: a challenge of adaptive management. *Conservation Ecology* 2(2):R1. Online: http://www.consecol.org/vol2/iss2/resp1.

Rogers, K. H., and R. Bestbier. 1997. *Development of a protocol for the definition of the desired state of riverine systems in South Africa.* Pretoria: Department of Environmental Affairs and Tourism.

Rogers, K. H., and H. Biggs. 1999. Integrating indicators, end points and value systems in the strategic management of the Kruger National Park. *Freshwater Biology* 41:439–451.

Rogers, K. H., C. M. Breen, J. Jaganyi, D. Roux, T. Sherwill, B. van Wilgen, E. van Wyk, and F. Venter. 2002. Fundamentals of co-operative governance for water management in South Africa: developing a collective rationality for managing a common property resource. Appendix 7 in B. van Wilgen et al. (eds.), *Principles and processes for supporting stakeholder participation in integrated river management.* Water Research Commission Project K5/1062 Final Report, Pretoria, South Africa.

Rogers, K. H., D. Roux, and H. Biggs. 2000a. Challenges for catchment management agencies: lessons from bureaucracies, business and resource management. *Water SA* 26:505–511.

Rogers, K. H., D. Roux, and H. Biggs. 2000b. The value of visions and art of visionaries. *Conservation Ecology* 4(1):R1. Online: http://www.consecol.org/vol4/iss1/resp1.

Salafsky, N., R. Margulis, and K. Redford. 2001. *Adaptive management: a tool for conservation practitioners.* Washington, DC: Biodiversity Support Program.

Senge, P. 1990. The leader's new work: building learning organizations. *Sloan Management Review* Fall 1990. Pages 440–463.

Stalmans, M., E. Witkowski, K. Balkwill, and K. H. Rogers. 2001. A landscape ecological approach to address scaling in conservation management and monitoring. *Environmental Management* 28:389–401.

Starfield, A. M. 1997. A pragmatic approach to modeling for wildlife management. *Journal of Wildlife Management* 61:261–270.

Walters, C. J. 1997. Challenges in adaptive management of riparian and coastal ecosystems. *Conservation Ecology* 1(2):1. Online: http://www.consecol.org/vol1/iss2/art1.

Walters, C., and J. Korman. 1999. Cross-scale modeling of riparian ecosystem responses to hydrologic management. *Ecosystems* 5:411–421.

Walters, C. J., and R. Green. 1997. Valuation of experimental management options for ecological systems. *Journal of Wildlife Management* 61:987–1006.

Wu, J., and O. L. Loucks. 1996. From balance of nature to hierarchical patch dynamics: a paradigm shift in ecology. *Quarterly Review of Biology* 70:439–466.

Chapter 4

An Adaptive System to Link Science, Monitoring, and Management in Practice

HARRY C. BIGGS AND KEVIN H. ROGERS

Kruger has implemented a unique version of adaptive ecosystem management (strategic adaptive management [SAM]) built on a base of recent developments in ecology and business management (Chapter 3, this volume). New paradigms in ecology stress complex adaptive systems and heterogeneity, and business management now emphasizes that organizations need to continually reinvent themselves through purposeful knowledge diffusion. Establishment of SAM was favored by an interaction between certain catalysts and an existing legacy in Kruger. It differs from conventional adaptive management in having a stronger emphasis on the forward-looking component, attempting to swing the bulk of decisions into proactive rather than reactive mode. It has a strong goal-setting component evidenced by a well-developed objectives hierarchy (Keeney 1992) and strongly articulated monitoring endpoints (called thresholds of potential concern [TPCs]). The objectives hierarchy and endpoints act as a nexus for connecting science, monitoring, and management in an innovative and motivating way. This chapter describes the new management system and the challenges it presents.

Our purpose is to demonstrate how Kruger has addressed a pervasive deficit in conservation: the effective integration of science and management in and around conserved areas (Chapter 3, this volume). Ideally, the science elements (stored knowledge and ongoing research), the monitoring elements (regular state-of-the-system measurements, often classed together with the science elements), and the management elements (direct action modifying or maintaining the system) will operate as a smoothly integrated three-part unit serving common objectives.

Background

Since the 1950s, scientists and managers have collaborated under various management circumstances in Kruger, with a steadily increasing knowledge base. When the legacy that arose is viewed with current understanding, we can see that Kruger has behaved as an adaptive institution (*sensu* Holling 2001). It went through decades of the conservative buildup of connectedness and potential, entering the release phase in the adaptive cycle in the early 1990s, thus opening up a range of alternative possible management trajectories. The reorganization phase of the evolving intersection between science and management is the central theme of this chapter. The background that gave rise to this new interface was crucial to its establishment (Box 4.1).

Core Elements of the New Kruger Management System

Biodiversity management initiatives, from grassroots (Salafsky et al. 2001, or http://www.fosonline.org) to theoretical (Chapter 3, this volume), stress a small set of generic needs for success: recognition that we are dealing with spatially and temporally complex adaptive systems, clear purpose and goals, participative learning by all stakeholders and not just by their advisors, monitoring to test assumptions, and adaptive organizational processes that promote institutional curiosity and the ability to capitalize on experience, new knowledge, and surprises. The new management system of Kruger embraces each of these needs through four core elements:

- A new vision statement, heavily influenced by repeated public participation, explicitly embraces spatiotemporal heterogeneity. It is based on the three pillars of biodiversity (composition, structure, and function) and the recognition that national parks should embrace the wilderness concept and provide benefits to the populace. The statement reads, "To maintain biodiversity in all its natural facets and fluxes and to provide human benefits in keeping with the mission of the South African National Parks in a manner which detracts as little as possible from the wilderness qualities of the Kruger National Park" (Braack 1997a).
- A hierarchy of objectives (Keeney 1992; Chapter 3, this volume), an inverted tree of goals, branching downward from value-laden vision statement with increasing explicitness to technically stated ecosystem and institutional goals. The objectives hierarchy fills in the middle ground between high-level vision statements and the explicit lower-level (what exactly, by whom, and when) statements needed to realize the vision. The full Kruger objectives hierarchy and a description of techniques used to derive it are available in Braack (1997a). Figure 4.1 illustrates a small section of the objectives hierarchy relating to the influence of the atmospheric system on biodiversity conservation, of cardinal importance in the decades ahead. Table 4.1 depicts an overall outline of themes covered by the objectives hierarchy.

BOX 4.1
A Favorable Legacy and the Right Catalysts at the Right Time

How did the Kruger National Park come to develop the SAM processes it now uses? For half a century, scientists and managers worked together in Kruger building a solid foundation of interaction:

- Since the 1950s Kruger has employed its own scientists, who could interact with external scientists and assimilate their findings into the broader Kruger experience.
- Managers and scientists have long been exposed to each others' products and demands (Chapter 1, this volume). They have benefited from regular joint decision making, mutual respect for each other's disciplines, sufficient capacity and resources, and sufficient time to learn jointly from the experience that resulted.
- There is a close-knit organizational culture and much similarity between individuals in beliefs and background.

In short, a community of practice with elements of the key processes of diffusion, communication, and adoption has existed for many years in Kruger, effecting coordination between scientists and managers. Although they do not always conform to modern expectations (e.g., pseudo-fact [Chapter 3, this volume] may have dominated certain decisions), numerous examples of science-management links exist:

- Decisions on elephant culling (1967–1994) were based on research in Kruger that suggested that numbers above one elephant per square mile might permanently jeopardize vegetation recovery (Whyte et al. 1999). Careful annual monitoring of elephant numbers guided culling quotas, which thus provided putative vegetation protection.
- Mammals were exported to establish viable populations in other parks. Offtake limits in Kruger, immobilization and sedation, and likelihood of survival under conditions of transport, arrival, and release (Novellie and Knight 1994) were all based on research.
- Lions were culled in the 1970s in an attempt to reduce pressure on wildebeest and zebra populations after research demonstrated that traditional migration routes had been cut off by the western boundary fence and population modeling predicted predation impacts (Joubert 1986).
- Similarly, the culling of species such as buffalo and the evolution of water provisioning and fire policies were guided by experienced scientists and managers, based on the best information available at the time.

The science-management interaction in Kruger evolved over 40 years to form a strong partnership, but by the 1980s Kruger was seen as becoming increasingly isolated and insular. The sociopolitical changes of the early 1990s challenged the homophily of culture, belief, and background in Kruger, exposing the park as part of a bigger dynamic. The new open environment catalyzed change, and the conservative partnership provided a firm base from which to capitalize on recent innovations in resource management (Chapter 3, this volume):

- The viewpoint of nature as in balance, linear, predictable, and controllable was challenged by one of flux and socioeconomic complexity (Chapter 3, this volume).
- The history of knowledge compartmentalization (Chapter 22, this volume) and the limited use of integrating tools such as models became a constraint to advancing scientific understanding. However, interdisciplinary integration characterizing local river initiatives influenced scientific activity in general.
- The insular and autocratic decision-making organization could no longer function unchanged under the new sociopolitical system. Significant changes in administration and external partnerships necessitated that choices be justified in terms of explored options, forcing reconsideration of entrenched policies.
- Fundamentals of many Kruger science-management activities, such as monitoring, came under the scrutiny of modern science and management practices (Chapter 3, this volume), prompting a reorientation toward biodiversity and heterogeneity. This important theme is continued through this chapter.

Kruger's ability to meet these substantive challenges was significantly aided by the existence and success of the Kruger National Park Rivers Research Program (RRP; Chapter 9, this volume). Until this time external researchers had operated independently within Kruger, each collaborating with perhaps one Kruger scientist in one field. The RRP organized a strong group of researchers into a cohesive program that enjoyed autonomy from but a great deal of interaction with Kruger scientists. The RRP offered a useful model for concept development and a broader application of adaptive management for Kruger.

Initially, diffusion of RRP innovations into the broader Kruger science and management community was limited because most management and almost all research historically had been focused on the terrestrial system. The catalyst for the full-scale revision of Kruger's objectives and management approach was the call for elephant conservation and management reforms in the mid-1990s (Whyte et al. 1999). With the globalization of South Africa came international pressure to justify the need for culling, and there was a growing recognition that elephant had to be understood and managed as part of a dynamic ecosystem. The imperative of elephant management and intellectual advances brought to the table by the RRP merged to catalyze a new management philosophy, a new vision, an objectives hierarchy, and explicit management targets (Rogers and Bestbier 1997; Rogers and Biggs 1999; Braack 1997a, 1997b).

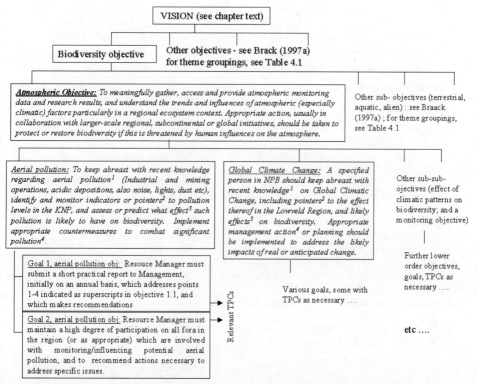

FIGURE 4.1. An example of a small portion of the Kruger objectives hierarchy. The section chosen shows the layers from the conceptual (top), just below the vision statement, to the detail (bottom). The top sections are value laden and involve public participation; the bottom sections are technical and constructed by managers and scientists.

- The concept of TPCs, a set of operational goals that together define the spatiotemporal heterogeneity conditions for which the Kruger ecosystem is managed. TPCs are defined as upper and lower levels along a continuum of change in selected environmental indicators. When this level is reached, or when modeling predicts it will be reached, it prompts an assessment of the causes of the extent of change. The assessment provides the basis for deciding whether management action is needed to moderate the change or to recalibrate the TPC. TPCs form the basis of an inductive approach to adaptive management because they are invariably hypotheses of limits of acceptable change in ecosystem structure, function, and composition. Therefore, their validity and appropriateness are always open to challenge, and they must be adaptively modified as understanding and experience of the

TABLE 4.1

An outline of topics covered by the objectives hierarchy of Kruger and the associated main monitoring (and hence threshold of potential concern [TPC]) themes. The actual objectives of the atmospheric component are shown in Figure 4.1 as an example of how each topic was broken down hierarchically.

MAIN GROUPS OF BIODIVERSITY MANAGEMENT OBJECTIVES IN KRUGER MANAGEMENT PLAN	MAIN THEME AREAS IN MONITORING PROGRAM
Atmospheric: global climate change, pollution, climate recording and networking	Woody vegetation, including aerial photos*
Aquatic: pans, rivers (public relations, legal, biodiversity, integrated catchment management, relationship with upland), surface water distribution (water provisioning policy over the landscape and its effects)	Herbaceous vegetation*
	Rare biota*
	Large mammal responders (aerial census)*
	Invertebrates
	Spatial patchiness
Terrestrial research: fire, predation, herbivory, disease, nutrient cycling, pollination	Alien biota (mainly invasive plants)*
Terrestrial management: fire, erosion, disease, fencing, land acquisition and consolidation, illegal exploitation, plant and animal population management, pollination	Birds, fish, amphibians, reptiles, and small mammals
	Pollination
Terrestrial monitoring: monitoring program, TPCs	Disease
	Climate (tourism staff also assist with monitoring)
Alien impact: strategy, prevention, eradication, prohibition, research, awareness	Fire*
	Erosion
	Landscape water
	Nutrient cycling
Other major objectives of Kruger:	River water quality and flow*
Human benefits: neighbor relations, staff, tourism and hospitality, problem biota	Wilderness qualities
Wilderness: awareness, zonation, policy and law, auditing, networking	
Integrated environmental management (the balancing objective): agreeing on a desired state, levels and mechanisms of trade-offs between objectives, and integrated environmental management and EIAs	

The management plan is subject to interim updating, audits, and major revision every 5 years. Each theme has TPCs, although most climate measurements are not thus linked.
*All programs in which rangers have more than just incidental involvement in at least one major component of the actual monitoring in that program.

system being managed increase (Rogers and Bestbier 1997). An illustrative set of TPCs is presented in Table 4.2.

The joint suite of TPCs represents an overall and multidimensional envelope within which flux or variation of the ecosystem is acceptable to both scientists and managers operating under the vision statement. The wider the TPCs are set, the bigger the envelope of possible system behaviors and patterns. Widely set TPCs imply that managers choose a risk-tolerant approach that avoids blatantly unsatisfactory trajectories but

TABLE 4.2

Three illustrative thresholds of potential concern (TPC) descriptions, given in sufficient detail to understand the background, definitions, scale descriptors, and rationale.

BIOPHYSICAL THEME, EXAMPLE OF TPC, AND BACKGROUND	SIMPLIFIED PARTIAL WORDING OF TPC	TIME AND SPACE SCALES FOR COLLECTION AND EVALUATION	RATIONALE AND COMMENT
Spatial heterogeneity of woody vegetation, measured as the percentage of woody cover Background: For biodiversity management, Kruger, a long, narrow park, has recently been zoned divided into four major blocks: two contiguous blocks in the center currently designated for high elephant impact (source) and two peripheral blocks at either end designated for low impacts (sink). These may swap later.	Inside any one of the four elephant management zones making up Kruger, woody cover should not drop below 80% of its highest-ever value; the mean drop parkwide should not exceed 30%.	Digital airborne remote sensing at 0.5-m resolution every 3 years, analyzed every 3 years using standardized algorithms, with calibrated corrections for scale and resolution changes in historical photography used for benchmarks.	It is believed desirable to eventually subject all large areas of the park to varying elephant impacts to ensure spatiotemporal flux in disturbance pressure. This may allow change between grass and woodland states as, for instance, recorded in East Africa (Dublin 1995)
Fire pattern, measured as long-term fire frequency Background: Assumption is that fire pattern parameters such as frequency, seasonality, intensity, annual extent, and size distribution are surrogates of biodiversity (van Wilgen et al. 1998).	Cumulative probability curve (proportion of area burnt vs. years since last fire) should not exceed stated limits specified at three points on an empirical Kruger curve typical of savanna systems: median (3.5–7.5 years), 80th percentile (5–10 years), and maximum postfire age (33 years)	Ongoing records of area burnt, at satellite image resolutions of 30m, 250m, and 1.1km, calculated annually at end of fire season and currently computed over past 30 years for coarse resolutions. Because the 30-year window moves on one year at a time, change develops slowly in this TPC.	Concern is that fires do not develop variable long-term frequency patterns. Because historical Kruger interfire period is deemed too frequent, the median TPC is acceptable only if greater than historical median of 3.5 years (lower limit).

(continued)

TABLE 4.2 (continued)

BIOPHYSICAL THEME, EXAMPLE OF TPC, AND BACKGROUND	SIMPLIFIED PARTIAL WORDING OF TPC	TIME AND SPACE SCALES FOR COLLECTION AND EVALUATION	RATIONALE AND COMMENT
Fish, measured as integrity of assemblage Background: This example is included as one of an index not derived specifically for Kruger's monitoring program but widely used in the broader region, in this case for river health assessments. Several such indices are incorporated in the Kruger program, also in the interest of meaningful regional results.	Fish assemblage integrity index (FAII), which takes into account intolerance index, expected frequency of occurrence, and health index per homogeneous stretch, drops below Class B (Kleynhans 1999).	Fish sampled by standardized techniques in slow, fast, shallow, and deep water and various habitat cover combinations every 2 years at six sites per Kruger stretch of river. Formula applied with expert judgment.	The assemblage should not deviate from the assemblage that would be expected in unmodified conditions (Class A) or at least in largely natural conditions with few modifications (Class B).

tolerates broadly varying conditions and patterns. On the other hand, a narrow set of TPCs implies a risk-adverse strategy that seeks to optimize for a narrow zone of variability and system behavior. Since the new vision statement was implemented, Kruger has opted for risk tolerance, believing this to allow development of greater resilience (Holling 2001).

- An adaptive decision-making process (Figures 4.2 and 4.3). The vision, objectives, and TPCs must be seen in their natural setting, the adaptive management cycle. There are many models of adaptive management (Chapter 3, this volume), all showing generically similar steps from participative visioning, through goal setting to multioption planning. The consequences of the options are thought through and tested for acceptability, and the best choice is then made and operationalized. Implementation is always accompanied by monitoring and by an evaluation and a conscious reflection step to feed back into another loop of this iterative process (Figure 4.2). We would like to stress the influence of the visioning and objective-setting step; in Kruger this affects generation of understanding (via research) and the identification of agents of change (Figure 4.3). The latter then spawns the TPCs and hence the monitoring program that audits their achievement.

Two indispensable steps govern the style and scope of feedback in the adaptive decision-making loop. The first is to determine whether the objectives and vision are being met once interventions are carried out (Figure 4.3, unpacked

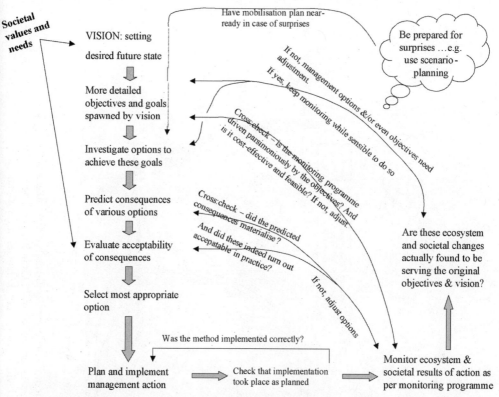

FIGURE 4.2. Generalized version of an adaptive management system, emphasizing feedbacks between objectives, actions, and monitoring.

in more detail in Figure 4.2). The second is that in Kruger, the very specific TPC step (Figure 4.3) dictates whether, when, and how management action will take place and elegantly ensures that subsequent steps check that the outcomes of management actions meet the objectives and vision.

The scope of audit thus includes ensuring that the physical implementation took place properly, that the system returned to within the TPC and that the consequences or side effects of the management actions were acceptable, that the monitoring system to achieve this is feasible and efficient, and, most important, that the objective was indeed served by the TPC returning to within limits. This comprehensive cross-checking system, if documented in the excellent tradition of Kruger's earlier managers (Chapter 1, this volume), should ensure that future managers are not left wondering why their predecessors made certain choices.

Knowledge sharing is crucial in adaptive management; we therefore stress the importance of role overlap (Figure 4.3) between scientists, managers, and other stakeholders. The central elements of knowledge sharing are discussed in more detail in Box 4.2.

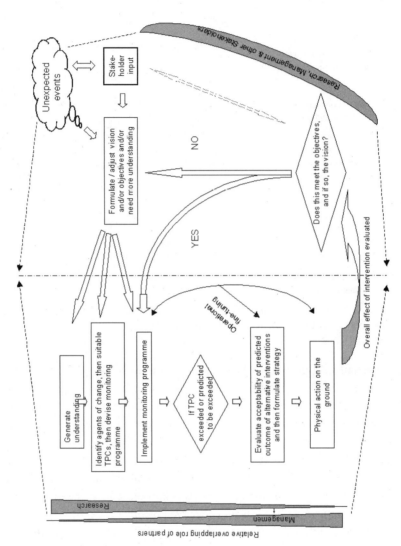

FIGURE 4.3. Strategic adaptive management processes aimed at illustrating the linkage of research, monitoring, and management processes via the use of TPCs in Kruger. These processes reflect how science and management interact with environmental changes and societal values.

BOX 4.2
Self-Evolving and Informal Communities as the Basis for the Transfer, Adoption, and Diffusion of Knowledge

DIRK ROUX

Sharing of perspectives and knowledge, adoption of new knowledge, and subsequent diffusion take place largely within informal and self-organizing communities. These communities of practice (COPs) provide structure for networking across organizational and disciplinary boundaries and provide connectivity between research (knowledge supply) and management (knowledge demand) fraternities. COPs serve two main purposes: to deemphasize organizational boundaries and to amplify individual knowledge at the group, organizational, and institutional levels. These communities are defined in terms of interest rather than geographic location.

Identification with and participation in a community is essential for developing good relationships between individuals and groups within a specific social system. Important relationship issues include degree of trust, ease of working together, capacity for joint problem solving, ability to resolve conflicts, openness and quality of communication, and capacity for considering what is best for the community. Through community interaction, researcher, manager, and stakeholder perspectives can coevolve in a process of mutually dependent learning.

There are three key knowledge processes within this community of practice: transfer, adoption, and diffusion.

Transfer

Knowledge transfer is essentially a communication process whereby personal experiences and tacit knowledge are shared based on common or overlapping interest. Even when knowledge predominantly moves in one direction, usually from researcher to end user, the two or more parties must participate in a series of communication exchanges as they seek to establish a mutual understanding about the meaning of the knowledge. If feasible, clear articulation of demand from the client side is very helpful. Early and ongoing interaction between researchers and managers of natural resources is the safest way to increase the degree of compatibility between knowledge innovations and the needs of resource management (ideally, the innovator should help to shape these needs). Scientists become aware of implementation realities, such as capability and resource constraints, at an early stage. Managers get to experience the new knowledge and technology in action during pilot applications and have time to develop ownership. The overall result is that a natural progression is fostered among all parties, from research to design to adoption and subsequent implementation.

Adoption

Scientists often mistake the handing over of a final research report for the transfer of knowledge or technology. Resource managers are left with a product in which they have little ownership that may not be suitable to their particular set of resource realities. True transfer ends with adoption, which implies both emotional and financial commitment to sustained or routine use. Once opinion leaders have adopted new knowledge, it is transferred to community and organizational levels where formal resource allocation takes place. This knowledge can be diffused and deployed so that, through practical application, its latent value can be realized.

Diffusion

Diffusion, as defined by Everet Rogers (1995), includes spontaneous or unplanned spreading of new ideas and innovations as well as planned, directed, or managed spreading (also called dissemination) of new ideas and innovations. The rate at which diffusion takes place determines to a large degree the rate at which individuals and organizations can expand and renew their knowledge and technological capabilities. When new knowledge is created, adopted, and diffused, social change occurs. Therefore, diffusion has a direct influence on the ability of people and organizations to respond to change. Two key necessary conditions for successful diffusion in resource management agencies are critical levels of people's capacity to absorb new knowledge and critical levels of organizational capability (combination of knowledgeable people, equipment, strategies, and logistical infrastructure) to deploy and apply new knowledge.

Derivation and Influence of the Objectives Hierarchy

The first step in deriving the Kruger objectives hierarchy was to identify the potential key elements (biodiversity, human benefits, wilderness, naturalness, and custodianship) of a vision statement, representing societal needs and values for Kruger. These were then debated at public meetings, and suggested changes were incorporated for the next iteration of public meetings. This publicly accepted vision was broken down at a series of (mainly internal) meetings into a hierarchical series of objectives of increasing focus, rigor, and achievability. This process was also guided (Rogers and Bestbier 1997) by context (international, national, and local realities, such as obligations under biodiversity agreements) and by a set of operating principles (such as least interference in the ecosystem to achieve a particular aim and use of the precautionary principle). Upon completion of the hierarchy, management (Braack 1997b) and monitoring policies were derived, which were compliant with these objectives.

The finest level of the hierarchy represents achievable and testable operational targets. The concept of these targets eventually crystallized as TPCs at a groundbreaking meeting of 30 South African scientists helping South African National Parks (SANParks) to derive a monitoring program to support the new objectives. The recommendations of the meeting determined the current spread of monitoring themes summarized in Table 4.1. The emphasis on heterogeneity in the objectives necessitated a broadening of the scope of the monitoring program to assess the consequences of agents of change for biodiversity management. To cement public acceptance, several of the new management policies, including the contentious proposal for elephant management, were taken back to public forums, at which the policy was traced back to the original vision.

The new objectives hierarchy resulted in several new or revised policies:

- A revised elephant management plan, with high- and low-impact zones in an ecosystem-level adaptive management experiment (Whyte et al. 1999)
- A plan for a new 20-year landscape-level fire management trial (Braack 1997b) designed to elucidate the effects of a range of fire management systems and overcome limitations of previous fine-scale experiments
- A revised water-provisioning policy (Braack 1997b) stressing surface water as a determinant of landscape heterogeneity (Chapter 8, this volume)
- A new recreational opportunity zonation plan with varying grades of wilderness purity (Braack 1997b), expanding opportunities in a heterogeneous land-use framework
- New liaison rules to elicit benefits for and greater participation by neighbors (Braack 1997b) who had been previously marginalized

Not all of these changes originated during this revision; indeed, several were evolving at the time. The new management plan with its explicit objectives gave each issue a firmer framework. Most were directly influenced by the heterogeneity paradigm (Chapter 3, this volume), a cornerstone principle in the review. The commitment of the park to reconsider the plan every 5 years builds in longer-range adaptive possibilities to the policy management life cycle. One internal midterm audit on achievement in terms of objectives has already been conducted.

Staff restructuring and reallocation took place in research and management components to better address objectives. Research regrouped into system ecology, species and communities, and human impact programs (see Figure 22.3, this volume), and a river manager was appointed for the first time. The new broad scope of the objectives hierarchy exposed many new research needs by an initiative that audited the existing and historical range of research products against information needs defined by the objectives. These were publicized

(Freitag and Biggs 1998) as opportunities for research collaboration and drew in a new range of participant scientists and research approaches.

TPCs as Mediators of the Science-Monitoring-Management Relationship

By its nature, monitoring should provide a common working ground for research and management. The use of TPCs mandates and further integrates participation of both research and management. Although monitoring in Kruger is a joint function of Scientific Services and Management components, we suggest that the Kruger experience with TPCs can also contribute to other settings, such as in agencies with separate monitoring sections. We discuss the role of TPCs in Kruger by following the sequence of steps outlined in Figure 4.3.

Under the vision and objectives of the Kruger, the Research Section is seen as being responsible for generating the understanding needed to set and test the TPCs. For research, this process represents the development of hypotheses (relevant to the objectives) in varying levels of confirmation. Indeed, research not substantively linked to this aim has come to be regarded as less necessary. On the other hand, much fundamental research ceased to appear remote to park needs, linking to management and conservation (via the objectives hierarchy and the TPCs derived under it) and giving it new justification. Useful TPCs have become sought-after deliverables in the organization, and research is judged according to the ability to deliver them. Research deemed at first sight unconnected to TPCs is considered important if it generates a better basis for the interpretation of existing TPCs or the development of new and ecologically more appropriate TPCs.

The Monitoring Initiative's task under the vision is to measure and assess the state of the Kruger ecosystem in a focused way. Since the advent of TPCs, researchers and others responsible for monitoring jointly develop and evaluate the monitoring program with TPCs as the central focus. Because prior agreement has been reached on endpoints (TPCs), they are taken seriously in formal organizational decision making. This provides powerful motivation for monitoring staff, who appreciate the context and constructive outcome of their efforts. In practice, the initial calibration or setting of TPC levels at which the stakeholders agree to become concerned about ecosystem change can be difficult, yet it is essential to initiate SAM. For certain TPCs, these levels are little more than educated initial guesses, but for others can already be closely linked to levels presaging well-understood ecosystem endpoints. The only monitoring themes we consider partially exempt from TPCs belong to the "significant surprise" category. Where important unexpected events cannot be predicted, their indicators must thus be measured continuously, even without a TPC framework; this includes routine weather measurements and aspects of hydrology.

The Management Section oversees and implements the activities that will support ecosystem integrity, such as taking physical action (e.g., closure of water, erection of fences, translocation of animals) or seeing that the system is left to recover on its own. Under TPC-based operating rules, they manage by exception and therefore should act only when a TPC is exceeded or, better still, once it is predicted to be exceeded. If managers or any other group are uncomfortable about the criteria and monitoring themes that have been selected (albeit with their initial input), they are free to challenge either the use of particular TPCs or their particular levels, via the Research Section, which is responsible for maintaining the list of TPCs. Once TPCs are tabled (formally reported as having been exceeded or likely to be exceeded)—and assuming they are not immediately recalibrated—management is activated via the rest of the SAM cycle. Options are generated, consequences predicted, and acceptability evaluated. The best strategy is chosen, implemented, and continuously reevaluated against the stated objectives and goals. Management's ability to return the situation to within TPCs thus becomes a visible process in the organization. This lends transparency to the process, which prevents misunderstandings and any perception of deception. There is an explicit additional check to ensure that returning the system to within the TPC concerned actually serves the intended objectives and the vision. This continual feedback and crosschecking adjusts the system in a truly adaptive way, and the more goal-oriented and predictive the system is, the more strategic it becomes, approaching the ideals of SAM.

In this way, the suite of TPCs acts as a central hub around which research, monitoring, and management activities can be sensibly unified. This works if there is committed and full buy-in to a common vision and objectives, an agreed decision-making process (in the Kruger case, SAM) is used consistently in a business environment conducive to its success, and the TPCs are viable.

TPCs are likely to be robust, defensible, acceptable, and practical if they are biologically and ecologically meaningful, statistically definable, robust, and defensible; logical and concise, unequivocally stated in exact detail; conceptually understandable and sufficiently intuitive to be manager-friendly; and technically and financially feasible to develop, implement, monitor, and maintain. Choice of indicator has a major influence on feasibility.

When TPCs are exceeded or predicted to be exceeded, a formal submission must be made to a joint decision-making committee tasked with decisions on ongoing ecosystem management. In the Kruger case, this committee contains senior science and management staff and meets bimonthly, although provision is made for shorter-term handling of emergencies. The typical headings or topics addressed in a submission for evaluation by such a body are as follows:

- Background: discuss relevant objectives, why the TPC was originally chosen, and what it states and means; identify events that led up to

exceedance or predicted exceedance and the negative implications for biodiversity conservation
- Exact statement of exceedance, usually repeating the exact wording of TPC, and measurements that show exceedance or predicted exceedance and, in the latter case, by when and with what confidence
- Additional supportive or collateral evidence to assist interpretation
- Alternative possible management responses, with pros, cons, and their anticipated results

When such a submission is tabled in this way, the joint decision-making committee can effectively evaluate the acceptability of predicted outcomes and formulate strategy, enabling specific actions to be launched (Figure 4.3). Sometimes the outcome is that more information is required before a final decision can be made, but it is extremely important to ensure that this does not become an excuse for inaction. Table 4.3 provides narrative examples of the life cycles of TPCs that have been tabled.

Outcomes of TPC Use

In more than 3 years of formalized operation, Kruger has tabled 21 different TPCs, many two or three times. The necessity of repeat notification is an issue Kruger is learning to deal with. Alien species invasions provide a good example (Chapter 19, this volume) because the mandate of a national park, ideally, is to preclude them entirely. The first-level TPC in Kruger therefore is exceeded on arrival or impending arrival. Because it is seldom possible to eradicate them once they are introduced, it is counterproductive for TPC alarm bells to ring every time a plant is sighted. The next level of alarm can be invoked on criteria relating to spread or densification (Chapter 19, this volume). In other cases, such as complex, chronic "outside" threats (a cardinal example is river sedimentation from upstream areas outside Kruger; Table 4.3), keeping the TPC on the list can motivate managers to not let the matter rest until the ecosystem is considered out of danger. Clearly, too many persistent alarms will dilute interest and effort, underlining the need to balance institutional response capacity with reality.

The breakdown of categories of these tabled TPCs is: river flow and quality, 5; alien plants, 8; alien fish, 2; *Varroa* mite (a serious alien parasite of bees), 1; alien birds, 1; rare antelope, 3; and fire, 1. Although they can be classified in slightly different ways, it is clear that management focus in Kruger has shifted since the introduction of SAM. This shift is away from traditional wildlife management topics such as water provision, population regulation, and fire management. Many other changes, such as paradigm shifts in ecological understanding and poverty relief campaigns to clear alien species (Chapter 19, this

TABLE 4.3

Four illustrative TPCs that have been tabled in Kruger, with a description of the lead-up and outcome.

EVENT OR LEAD-UP	TCP EXCEEDED OR EXPECTED TO BE EXCEEDED	DESCRIPTION AND OUTCOME
Silt release episode on Olifants River (from Phalaborwa barrage, just upstream of Kruger's western boundary)	Water quality: turbidity exceeded maximum as set for Olifants River. Resulted in fish kills.	Had happened several times in previous decade; meant to have been prevented each time thereafter; committee decided letter of complaint (as in previous years) insufficient; outcome was serial meetings with full environmental management plan developed by the company operating the barrage.
Unintended fire (cause of fire meant to be primarily lightning; fire started by transmigrants crossing park from Mozambique and by other anthropogenic causes regarded as undesirable)	Area burnt from anthropogenic causes exceeded allowable percentage in all years since 1997.	Early warning in van Wilgen et al. (1998) that cause of fire has no biological significance. Committee seemed initially unconvinced of seriousness of TPC despite having approved it originally. After several resubmissions with more detail and adjustment, recalibrated for evaluation over a composite 10-year period starting in 1992. Later seen to be clearly on trajectory to exceedance of this TPC, and original TPC deemed retrospectively to have been correct. Major policy revision followed; new policy implementation began in 2002.
Sabie River alluviation and loss of bedrock influence (as triggered by various models developed by Kruger National Park Rivers Research Programme)	Directional loss of bedrock influence over a predicted 20-year period, following monotonic change since earliest aerial photos in 1940s, with modeling support on both riparian species and geomorphic indicators (see Chapter 22, this volume).	A major seven-point strategy and longer-term integrated catchment management outline was prepared after second retabling, with good ownership of problem taken. February 2000 catastrophic floods were thought initially to have solved problem (which until then was beyond model's verifiable capability), but it soon became clear that sediment was only redistributed and that the major trend was still present.
New occurrence of alien plant *Chromolaena odorata* reported along rivers	New occurrence: alien with listed high index of potential threat. Multifocal, reported by alien plant removal.	Working for Water crews removed individuals found, but islands in rivers problematic. Surveillance sharpened. Subsequent relistings led to development of multitier TPC system, with confinement at low densities (with appropriate TPCs) now targeted.

volume) have also influenced this shift. In concert with these other changes, TPCs have been influential in making Kruger a functional adaptive management site.

The preemptive nature of Kruger's SAM system has resulted in more proactive behavior. For instance, river sedimentation patterns have been predicted to reach unacceptable levels many years hence (Chapter 9, this volume), but actions are now taking place to remedy this situation ahead of a crisis (Table 4.3). However, it is not always possible to predict the exceedance of a TPC. For example, in the past there have been frequent unexpected silt releases from a barrage operated by a mining company in the Olifants River west of Kruger's boundary, resulting in fish kills. When the most recent release was dealt with under the SAM system (Table 4.3), Kruger was able to negotiate an environmental management plan with the mining company, which is likely to provide a durable solution.

Rare antelope management issues have also benefited from SAM; in previous years, they were controversial and characterized by inaction (Grant and van der Walt 2000). The recognition that sable antelope populations had fallen below TPC limits led to rapid agreement that there was a problem. A concerted research effort resulted and is expected to provide the understanding needed to solve the problem. TPCs had thus facilitated agreement among holders of divergent opinions. Like all other TPCs, the rare antelope TPCs are contestable. What made the difference this time was the clear understanding that decisions should be made on a TPC and that this TPC could not simply be changed. The debate around actual levels of TPCs must take place outside the pressurized atmosphere of the joint decision-making meeting. When tabling of a TPC challenges vested interests, there may be a temptation to recalibrate during the meeting to simply avoid the problem.

This type of situation has happened only once in the 3 years of operation of the system. The TPC for unintended fire (Table 4.3) was changed in such a way that it would be evaluated after 10 years rather than annually. Long before this 10-year period had elapsed, managers experienced difficulties stopping unintended fires, and scientists predicted that the TPC would be exceeded regardless of management action. This led to a major fire policy revision (Chapter 7, this volume) that in retrospect could have been elicited sooner, after the first tabling of the TPC.

National and international counterparts have expressed much interest in this system; two formal independent evaluations bear mention. One study (Duff 2002) evaluates Kruger for progress in SAM; although much remains to be achieved, the design is described as promising. A recent strategic review by McKinsey and Company (2002) proposed this adaptive management approach for wider use in SANParks.

Challenges

The explicit articulation in the objectives hierarchy provides clarity for each stakeholder concerned directly or indirectly with the biodiversity and ecosystem briefs of the Kruger mission to formulate their approaches to meet these aims in a heterogeneous ecosystem. There is a common ethic of operation (SAM), and there are endpoints consistent with societal values and operating principles. There are also emerging challenges.

Knowledge Management

Maintaining and updating a bank of TPCs, especially when they are being actively used, criticized, and refined, entails the challenges of maintaining any volatile living document. Protocols allowing challenge and revision in an orderly but nonimpeding way, and the consequent updating and version control, are issues facing the Kruger Research Section.

Knowledge management (Allee 1997) is a central issue. After a TPC is tabled, the tendency is for several unpredictable threads of information flow to arise as implementation proceeds, especially in response to a novel or complex threat. These threads are documented at varying (not always appropriate) levels of quality. A protocol and environment are needed in Kruger for a continuous "roping together" so that the organization benefits appropriately from the overall experience. This is one of the more difficult aspects of knowledge management (Box 4.2). Shared learning is occurring in several ways; for instance, a core team of enthusiasts has formed a community of practice that continually reworks and improves the SAM system (Rogers et al. 2000). The ultimate aim is to achieve a SAM-TPC system that runs smoothly, with products leading to the next actions in a sustainable feedback loop.

Management Activities outside the TPC Framework

Certain classes of management activity continue to receive resources and attention without being evaluated against a TPC framework, such as bovine tuberculosis control. Here veterinary regulations and clinical-scale sentiments play a strong role. The pressure triggering such management actions thus is usually legal or sociopolitical (nontariff trade barrier laws or a scare that a certain disease may have disastrous effects) and not the result of the exceedance of any TPC. TPCs and these realities must converge: either managers should agree to be less concerned (if current TPCs are not likely to be exceeded), or appropriate but explicit sociopolitical TPCs should be developed. The latter might take place in the same way that environmental water requirements (used as Kruger

TPCs) are balanced in legislation against other water needs (Chapter 21, this volume). This would require a skilled SANParks negotiator to continuously facilitate trade-offs between veterinary control and biodiversity aims, much as the Kruger river manager does for water releases to meet biodiversity goals. This will prevent operation outside the framework of common agreement about the desired envelope of conditions for Kruger.

How Many TPCs?

Kruger was overambitious in its first attempt at SAM. The full planned suite of themes and TPCs has yet to be implemented. Even if this were feasible, some consider it undesirable because the large number would tend to limit ecosystem flux and reduce ecosystem resilience. The large initial set of TPCs is helping us to learn, and feedbacks within SAM should lead to downscaling to the most parsimonious, effective set. Scaling down may also be driven by cost. We should measure what is needed, not what the organization knows how to measure or what biologists like measuring, remembering that the wide scope of Kruger's objectives inherently leads to more rather than fewer TPCs.

How Early Is Too Early, Given the Risk of a False Alarm?

The challenge is to blow the whistle before exceedance, whenever trajectories are seen to be heading in the wrong direction. At the same time, one wants as few as possible false alarms. This is a challenge in a system where wide variation within the accepted envelope is desired. In our experience, scientists want to check the validity of their measurements rather than blow the whistle. To counter this caution, it may be advisable to encourage TPC submission even if subsequent confirmatory investigations or developments lead to withdrawal of the warning as false. In practice, no false warnings appear to have occurred in Kruger to date, although we feel that some TPCs that should been tabled have not been.

Basing management on monitoring may be necessary but not sufficient in that even the Kruger TPC system, tuned to its most sensitive possible level under the vision, may not successfully identify all threats early enough. A threat recognition and amelioration system (Margoluis and Salafsky 2001, or http://www.bsponline.org), sensibly combined with this SAM-TPC system, may prove productive.

Looking from Inside or Outside the Desired Envelope

Many management plans for conservation agencies set targets that they attempt to reach rather than defining a desirable envelope in which they want to stay, which is the strategic feature of SAM. This dichotomy may reflect the level of existing compliance to the desired set of ecosystem conditions. In agencies regulating areas that are mostly outside the desired envelope, targets may be a more practical formulation. The Kruger comes with a history, at least if judged by perceptions, of being mainly inside the envelope, so the TPC formulation expressed in this chapter is considered appropriate.

Conclusion

This chapter has discussed SAM, a strategic (forward-looking) application of adaptive management with compatible and well-articulated goals and endpoints. The nuances of a TPC are: a worry level to monitor, a hypothesis to examine, a traceback to a particular agent of ecosystem change, an achievable environmental goal, and one dimension of the composite desired envelope represented by objectives. These have proved of value to Kruger in integrating science, monitoring, and management in a system characterized by heterogeneity. We believe the greatest leverage for SAM can now be obtained via development of ongoing shared learning skills, particularly as ecosystem management widens its stakeholder base. Future evaluations will tell to what extent we have advanced constructively.

References

Allee, V. 1997. *The knowledge evolution: Expanding organisational intelligence.* Newton, MA: Butterworth-Heinemann.

Braack, Leo. 1997a. *A revision of parts of the management plan for the Kruger National Park. Vol. 7: An objectives hierarchy for the Kruger National Park.* South African National Parks, Skukuza, South Africa. Online: http://www.parks-sa.co.za.

Braack, Leo. 1997b. *A revision of parts of the management plan for the Kruger National Park. Vol. 8: Policy proposals regarding issues relating to biodiversity maintenance, maintenance of wilderness qualities, and provision of human benefits.* South African National Parks, Skukuza, South Africa. Online: http://www.parks-sa.co.za.

Dublin, H. T. 1995. Vegetation dynamics in the Serengeti-Mara ecosystem: the role of elephants, fire and other factors. Pages 71–90 in A. R. E. Sinclair and P. Arcese (eds.), *Serengeti II, dynamics, management and conservation of an ecosystem.* Chicago: University of Chicago Press.

Duff, J. E. 2002. *An assessment of adaptive management practices in South Africa National Parks and Ezemvelo KwaZulu-Natal Wildlife.* Unpublished M.Env.Dev. thesis, University of Natal, Pietermaritzburg, South Africa.

Freitag, S., and H. C. Biggs. 1998. *KNP management plan objectives hierarchy: published work, projects in progress, opportunities for participation.* Special Internal Report of the Kruger National Park, SANParks, Skukuza, South Africa.

Grant, C. C., and J. L. van der Walt. 2000. Towards an adaptive management approach for the conservation of rare antelope in the Kruger National Park: outcome of a workshop held in May 2000. *Koedoe* 43:103–112.

Holling, C. S. 2001. Understanding the complexity of economic, ecological and social systems. *Ecosystems* 4:390–405.

Joubert, S. C. J. 1986. The Kruger National Park: an introduction. *Koedoe* 29:1–11.

Keeney, R. L. 1992. *Value-focused thinking: A path to creative decision making*. Cambridge, MA: Harvard University Press.

Kleynhans, C. J. 1999. The development of a fish index to assess the biological integrity of South African rivers. *Water SA* 25:265–278.

Margoluis, R., and N. Salafsky. 2001. *Is our project succeeding? A guide to threat reduction assessment for conservation*. Washington, DC: Biodiversity Support Programme.

McKinsey and Company. 2002. *SANParks and McKinsey Final Meeting Report, following strategic review*. Report to SANParks, Skukuza, South Africa.

Novellie, P. A., and M. Knight. 1994. Repatriation and translocation of ungulates into South African national parks: an assessment of past attempts. *Koedoe* 37:115–119.

Rogers, E. M. 1995. *Diffusion of innovations*. New York: Free Press.

Rogers, K. H., and R. Bestbier. 1997. *Development of a protocol for the definition of a desired state of riverine systems in South Africa*. Department of Environmental Affairs and Tourism, Pretoria, South Africa.

Rogers, K. H., and H. Biggs. 1999. Integrating indicators, end-points and value systems in the strategic management of the Kruger National Park. *Freshwater Biology* 41:439–451.

Rogers, K. H., D. J. Roux, and H. C. Biggs. 2000. Challenges for catchment management agencies. Lessons from bureaucracies, business and resource management. *Water SA* 26:505–512. Online: http://www.wrc.org.za.

Salafsky, N., R. Margoluis, and K. Redford. 2001. *Adaptive management: a tool for conservation practitioners*. Washington, DC: Biodiversity Support Program.

van Wilgen, B. W., H. C. Biggs, and A. L. F. Potgieter. 1998. Fire management and research in the Kruger National Park, with suggestions on the detection of thresholds of potential concern. *Koedoe* 41:69–87.

Whyte, I. J., H. C. Biggs, A. Gaylard, and Leo Braack. 1999. A new policy for the management of Kruger National Park's elephant population. *Koedoe* 42:111–132.

PART II
A Template for Savanna Heterogeneity

The key components of ecological heterogeneity were outlined by Pickett et al. (Chapter 2) as agents, substrates, controllers, and responders. The chapters in Part II of this book concern the main agents, substrates, and controllers of Kruger's heterogeneity, setting up Part III to deal with a selected set of responders. Agents of heterogeneity covered here include fire, herbivory, and drainage, and substrates range in scale from the geological formations underlying the whole of Kruger to nutrient pools in particular soil types and plant communities. The availability of surface water is an example of a controller in that the dispersion of drinking sites influences the distribution and abundance of water-dependent grazing ungulates and therefore has a controlling influence on agents such as herbivory and fire. Another controller is land use in the upper catchments to the west of Kruger's boundaries, which strongly influences the flow of water and the deposition and movement of sediment along Kruger's main rivers. Frequent visits back to Chapter 2 (especially Figures 2.2 and 2.3) while reading through Part II will help the reader understand both the conceptual framework and the functional links between key components of savanna heterogeneity as they arise in each chapter. For an applied perspective, Rogers and O'Keeffe (Chapter 9) describe how heterogeneity thinking was first brought into the Kruger experience through the Rivers Research Programme, in which the conceptual framework proved useful for integrating science and management in the service of ecosystem conservation.

In simple terms the topics covered in this part of the book represent the main drivers of Kruger's heterogeneity. The spatial configuration and physicochemical properties of parent rock, interacting with a dynamic climatological pattern, give rise to the functional attributes of ecosystem processes such as erosion, deposition, production, decomposition, combustion, and consumption. All these attributes form a driving template for ecosystem heterogeneity, which is responded to and further embellished in multiple feedback loops by a variety of species. In this part of the book we deal with the template; the species come next.

Chapter 5

The Abiotic Template and Its Associated Vegetation Pattern

FREEK J. VENTER, ROBERT J. SCHOLES, AND HOLGER C. ECKHARDT

Over very long time scales the physical and biological components of habitats provide the template on which evolution forges the characteristics of species' life history strategies, the basis of biological heterogeneity. Over ecological time scales the spatial configuration and dynamics of resources and constraints presented by the physical substrate provide the context for biological heterogeneity (Chapter 2, this volume). Physical heterogeneity is the subject of this chapter.

In savannas, fire and herbivory have received much, perhaps even disproportionate, attention as top-down controllers of ecosystem structure and function. The potential influence of bottom-up control has received much less attention, but there is good evidence (Chapter 11, this volume) of fundamental bottom-up control from the physical template. This is especially true for savannas, such as Kruger, with intermediate rainfall (Chapter 10, this volume) where the soil acts as substrate and controller (*sensu* Chapter 2, this volume) of both biotic and abiotic components of the ecosystem. No two soil bodies are completely identical because every soil is a unique product of local parent material, climate, organisms, relief, and time (Jenny 1941). In areas of intermediate rainfall, water, acting as a vector of particulate and dissolved materials, becomes a primary agent of soil heterogeneity and consequent biotic responses. An understanding of the distribution and properties of soils therefore greatly enhances understanding of other savanna ecosystem features and processes.

This chapter includes an overview of topography, climate, geology, and the broad soil and vegetation patterns associated with them. Links between the abiotic template, ecosystem drivers such as fire, and vegetation are explored. We focus specifically on catenas of the two dominant geological formations (granitic and basaltic) as examples of interactions between lithology, landform, soil, water, and plants. In conclusion we propose a hierarchical patch mosaic of the physical template that will aid understanding and management of the Kruger system.

Topography

Kruger is part of the northeastern South African lowveld, the low-lying area between the footslopes of the Drakensberg Great Escarpment to the west and the extensive Mozambique coastal plain to the east. The lowveld, which consists mainly of plains with low to moderate relief, forms part of a broad landform pattern, the Eastern Plateau Slope (Kruger 1983). On average the lowveld lies at 300 m above sea level and has a gentle slope toward the east. The topography of Kruger (Figure 5.1, Table 5.1) largely reflects differences in resistance to weathering of the underlying rocks and in the intensity of dissection in areas that flank the major rivers.

Most of the geomorphic features of the lowveld, southern Africa, and indeed Africa as a whole, derive from the birth of Africa as a continent in the middle Cretaceous period (100 million years ago) (King 1978). The geomorphic features are the result of a complex combination of isostatic readjustments (uplifts and deformations) of the continental crust and glacio-eustatic oscillations in the sea level, followed by incision and planation processes (du Toit 1954; King 1978; Partridge and Maud 1987; Moon and Dardis 1988). These processes form erosion surfaces such as the highveld and lowveld on a macro scale. The dominant process of planation in Africa is that of backward retreat of scarps, formed by the incision of drainage lines, and the lateral spread of the newly formed valleys (footslopes) at the expense of former erosion surfaces (King 1978). The undulating landscape that is left behind is gradually flattened and smoothed out. Because new erosion cycles are initiated at the coastline and along drainage lines, whenever differential uplift takes place, the southern African geomorphology typically is multicyclic. Relicts of older erosion surfaces remain within younger surfaces, similar to terraces formed along a river (King 1978).

Several such erosion surfaces exist in the eastern parts of South Africa. The major geomorphologic events that led to the present situation in the lowveld are described in King (1967, 1978), van der Eyk et al. (1969), Venter and Bristow (1986), Partridge and Maud (1987), and Moon and Dardis (1988).

After the dismantling of Gondwanaland, the Great Escarpment retreated from east to west, leaving at its foot the lowveld as a new erosion surface and exposing underlying rock formations. The Indian Ocean lapped the eastern foot of the present-day Lebombo Mountains (du Toit 1954; Truswell 1977) and, as a consequence of isostatic readjustment, retreated eastward to form the coastal plain. The Great Escarpment therefore must have retreated some

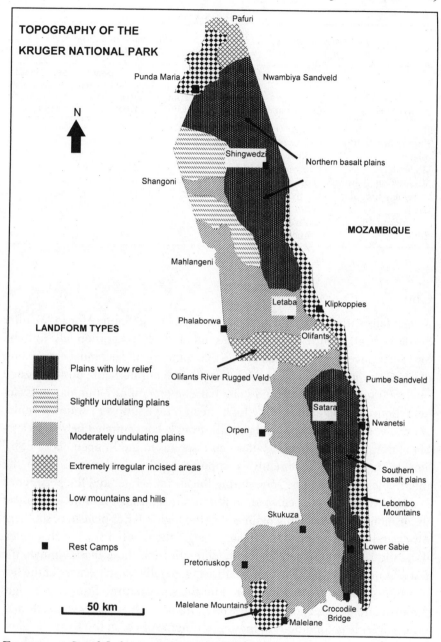

FIGURE 5.1. Simplified topographic map of Kruger.

60–100 km in approximately 100 million years. Subsequent downward erosion and flattening of the lowveld was influenced by the resistance to weathering of exposed rock formations, which led to the present topography of Kruger (Figure 5.1, Table 5.1).

TABLE 5.1

Major characteristics of the landform types illustrated in Figure 5.1 that occur in Kruger (adapted from Kruger 1983).

LANDFORM TYPES	CURVATURE	RELIEF (M)	STREAM FREQUENCY (N/KM^2)	PERCENTAGE OF AREA WITH SLOPES LESS THAN 2%	PERCENTAGE OF AREA WITH SLOPES MORE THAN 15%
Plains	Straight	10	0.4	97	0
Slightly undulating plains	Convex and concave	15	1.1	89	0
Moderately undulating plains	Convex and concave	23	2.3	16	2
Extremely irregular areas	Convex and concave	35	8.2	0	36
Low mountains and hills	Convex and concave	114	3.2	0	61

Climate

Kruger falls within two climate zones defined by the South African Weather Service (Weather Bureau 1986). The south and central portion falls into the lowveld bushveld zone (rainfall of 500–700 mm/year and potential evaporation of 6 mm/day in October), and the north is in the northern arid bushveld zone (300–500 mm/year rainfall and potential evaporation of 7 mm/day in October). Both are characterized by high mean temperatures in summer, mild, generally frost-free winters, and rainfall strongly concentrated between October and April. This climatic pattern corresponds to the Köppen class "BSh." The strong rainfall seasonality is imposed by the northward shift of the intertropical convergence zone during the austral winter and the persistence of a stable high-pressure cell over southern Africa during this time. Because of its location on the low-lying plain east of the high interior plateau of southern Africa and its proximity to the Indian Ocean, the weather systems that determine the Kruger climate are slightly different from those that dominate the rest of southern Africa. The temperature is generally warm because of the low elevation and subtropical location, and the summertime humidity is high. Approximately once a fortnight in winter, midlatitude westerly waves (fronts) bring cold, clear conditions to most of southern Africa. In the Kruger region, the effect is often to draw in moist maritime air from the Indian Ocean, leading to windy and overcast conditions and occasional light drizzle, followed by clear, calm conditions.

In summer, the combination of moisture and heat creates an unstable atmosphere that results in the powerful convective thunderstorms that deliver most of the rainfall. Occasionally a tropical cyclone penetrates from the warm waters of the Mozambique Channel and deposits up to half of the mean annual rainfall within a few days, leading to extensive flooding. High winds are rare.

The general climate pattern of hot, wet summers and mild, dry winters has been in place in the Kruger region for several million years (Tyson and Partridge 2000). During this time there have been long-term periodic variations in temperature and rainfall, known as the Pleistocene glacial and interglacial cycles. During glacial periods, the last of which ended 16,000 years ago, temperatures were about 5°C lower than at present, and rainfall is inferred to have been less than at present (Tyson and Partridge 2000). Over the past 3,000 years the climate is believed to have been broadly comparable to the current climate, with a medieval warm period ending at 1200 AD, followed by the Little Ice Age, 1 to 2°C cooler, until the seventeenth century and gradual warming thereafter (Tyson and Partridge 2000). The climate of Kruger is projected to become significantly warmer (by 2–6°C) during the twenty-first century because of human-induced climate change. Great uncertainty is associated with future rainfall trends, which could be up to 20 percent higher or lower.

Rainfall

Most of the precipitation in Kruger is in the form of rainfall. There is a trend of decreasing mean annual rainfall from south to north and east to west (Figure 5.2, Table 5.2), except in the extreme northwest and southwest, where topography influences rainfall. The rainfall totals for the hydrological year (July to June) are approximately normally distributed, with a standard deviation of around 120 mm for most sites. As a result, the interannual coefficient of variation of rainfall, which has a large impact on the variability of grass production, ranges from 25 percent in the south to 35 percent in the north (Schulze 1997). If a drought year has rainfall less than 50 percent of the long-term mean, such years can be expected to occur approximately 20 percent of the time. Since 1981 at least three extreme droughts have occurred (Figure 5.3), and there are indications that they may be on the increase (Preston-Whyte and Tyson 1988).

It is apparent from the time series of annual rainfall (Figure 5.3) that there is no overall trend in rainfall, but there are coherent periods of above- and below-average rainfall, which have been called cycles. There is an element of chance in these wet and dry runs, which would be expected even if rainfall in adjacent years were completely independent. There is also an element of nonrandomness, accounting for about one-third to one-half of the variance: the dry sequences generally are associated with the positive phase of the El Niño–Southern Oscillation (ENSO) events, which have been reported to have a periodicity of approximately 18 years in southern Africa (Tyson 1985) and per-

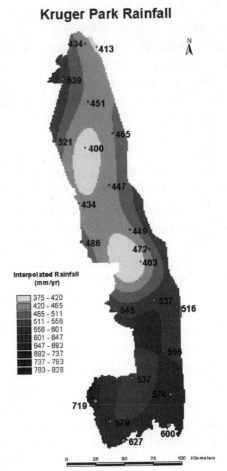

FIGURE 5.2. Distribution of annual rainfall across Kruger (Weather Bureau 1986).

sist for more than 1 year. The predictive capacity of this cycle is low (and possibly getting lower because of apparent changes in the dynamics of ENSO), so it is not possible to project wet and dry cycles more than approximately 1 year in advance.

The covariance between rainfall in Kruger is high on the annual scale and declines progressively on the monthly and daily scales. In other words, when Komatipoort, in the south, experiences a prolonged drought, it is likely that Punda Maria, in the north, also experiences below-average rainfall. This is because the climate over the entire lowveld is governed by the same system. However, because storm cells have elliptical footprints (minor axis 15 km, major axis 66 km; Dixon 1977), rain events can be geographically isolated. Therefore, there is little benefit for a migratory species to move in a north-south direction

TABLE 5.2

Climate data for sites from the south to the north of Kruger. Values are the monthly totals or averages over the period of record (data courtesy of the South African Weather Service).

	JAN.	FEB.	MAR.	APR.	MAY	JUN.	JUL.	AUG.	SEP.	OCT.	NOV.	DEC.
Komatipoort 24.4330° S, 31.9500° E, 140 m asl (1961–1995)												
Rain (mm)	114	95	63	39	15	12	8	9	18	46	66	85
Skukuza 24.98° S, 31.60° E, 263 m asl (1961–1990)												
Rain (mm)	93	87	73	33	14	10	10	6	26	35	76	84
Tmax	32.6	32	31.2	29.4	27.9	25.9	25.9	27.3	29.3	29.8	30.6	31.9
Tmin	20.6	20.4	19.1	15.4	10.0	5.6	5.7	8.7	12.8	16.0	18.1	19.7
Cloud (%)	46.2	41.8	42.3	33.0	25.8	22.5	23.9	30.8	38.6	50.0	52.9	50.0
Sun hours	7.4	7.7	7.4	7.7	8.0	8.1	8.2	8.2	8.0	7.4	6.9	7.1
Radiation (MJ/day)	20.9	21.2	19.1	17.4	15.1	14.2	14.8	15.8	17.4	18.3	19.1	21.3
Satara 24.40° S, 31.77° E, 275 m asl (1961–1990)												
Rain (mm)	79	64	73	26	8	11	13	14	13	50	58	93
Tmax	33.7	32.7	31.5	29.9	28.5	25.9	25.9	27.3	29.1	29.6	31.0	32.3
Tmin	21.1	20.8	19.6	16.8	12.9	9.4	10.0	11.8	14.1	16.5	18.2	20.0
Letaba 23.85° S, 31.58° E, 215 m asl (1961–1990)												
Rain (mm)	78	58	27	31	10	5	6	8	12	39	54	76
Tmax	34.1	33.5	32.6	30.5	28.5	26.1	26.4	28.1	30.3	31.1	32.0	33.3
Tmin	22.0	21.5	20.5	17.1	12.0	7.8	8.5	10.8	14.6	17.3	19.4	21.0
Shingwedzi 23.10° S, 31.43° E, 215 m asl (1961–1990)												
Rain (mm)	59	71	38	29	9	4	3	5	12	35	51	84
Tmax	34.1	33.1	32.6	30.7	28.5	26.1	26.5	28.1	30.5	31.2	32.7	33.4
Tmin	21.4	20.7	19.8	16.7	11.4	7.4	7.7	10.1	13.8	17.0	19.4	20.7
Punda Maria 22.68° S, 31.01° E, 462 m asl (1961–1990)												
Rain (mm)	64	84	23	32	15	3	6	5	28	48	77	65
Tmax	32.3	32.0	31.5	29.6	27.3	25.2	24.9	26.5	28.8	29.7	31.5	31.6
Tmin	20.8	20.7	19.9	17.8	14.8	12.2	12.3	13.5	15.7	17.7	19.0	20.2
Pafuri 22.4170° S, 31.2170° E (1988–2001)												
Rain (mm)	111	70	49	11	10	4	8	3	4	30	29	76

asl = above sea level.
Tmax = mean maximum temperature.
Tmin = mean minimum temperature.

to evade droughts, but there is benefit in short-term local movement in pursuit of greener pastures. There may historically have been benefit in migrating west during dry periods, up the rainfall gradient toward the escarpment. This option is now largely precluded by fences, roads, and incompatible land uses.

Rainfall in Kruger typically occurs as discrete events, often as thunderstorms. Daily rainfall at a site is statistically partly dependent on rainfall during the previous day (Zucchini and Adamson 1984; Scholes et al. 2001). The mean storm size (i.e., the mean annual rainfall divided by the mean number of rainy days per year) is around 9 mm (±0.5 mm), with little seasonal variation. The storms typically are of short duration (minutes to a few hours). As a result, the rainfall

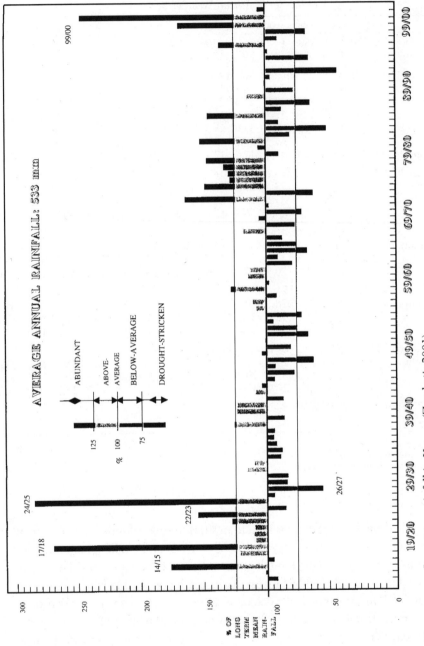

FIGURE 5.3. Long-term rainfall in Kruger (Zambatis 2001).

intensity often is high, leading to high erosion rates on unprotected soil, and flash floods in the generally ephemeral drainage lines are a common phenomenon. When tropical moisture is fed through the Intertropical Convergence Zone (ITCZ) into a cutoff low-pressure cell, intermittent rain may continue for several days. These events are seldom characterized by thunder and may cause flooding.

The stochastic nature of the rainfall, coupled with the high potential evaporation, imposes a pulsed nature on biological processes in the soils and plants of Kruger (Chapter 11, this volume). Their physiology is highly active for several days or weeks after rain, followed by a variable period of inactivity before the next rain.

Radiation

Kruger receives slightly less radiation than many areas of comparable latitude in southern Africa because of the cloudiness caused by its position between the escarpment and ocean. Table 5.2 presents the mean sunshine hours and atmospheric transmissivity through the seasonal cycle at Skukuza. Transmissivity is a function principally of the aerosol optical thickness, which is particularly affected by dust and smoke from wildfires during the dry winter.

Incoming shortwave radiation is calculated from records of sunshine duration using the Angstrom relation, calibrated for a radiometer at Nelspruit, 120 km to the southwest. The high solar radiation is the principal reason for the high mean temperatures and the high potential evapotranspiration recorded in Kruger. Shortwave albedo ranges from 25 percent in winter to 13 percent in summer, when the ground is covered in green vegetation, and drops to 6 percent for the few weeks after a fire. The isothermal evaporation therefore is on the order of 3 to 6 mm per day in winter and summer.

Temperature

There is a slight spatial trend in temperature from cooler in the south to hotter in the north, and there is the usual inverse relationship between temperature and altitude, with a lapse rate of about 0.66°C/100 m. Catabatic flow of cold, dense air into the drainage lines is a prominent nighttime feature in winter and can lead to near-zero temperatures in the valleys. True frosts occur occasionally in the hilly country of the south and rarely on the plains of the north. Nevertheless, every decade or so an exceptionally strong cold front does cause freezing temperatures throughout Kruger. If associated with rainfall, as sometimes happens in late winter or early spring (September), it can lead to mortality in plants and animals.

Humidity

The high daily range in temperature (i.e., the difference between the daily minimum and maximum screen temperatures) is caused by the low humidity of the air, particularly during the winter. This results in the mean monthly relative humidity at midday in summer ranging from 50 to 53 percent and in winter from 37 to 42 percent. Individual days can be much drier or wetter. The low winter humidity is an important factor controlling the rate of spread and intensity of fires: fires go out above a relative humidity of about 50 percent and are highly prone to "spotting" (ignition by burning embers ahead of the fire front) below about 15 percent.

Wind

The dominant wind directions are southeast and northwest, which correspond to synoptic conditions in which a low-pressure cell is located over the Mozambique Channel and a high-pressure cell over the interior, respectively. The average wind speed is around 2.5 m/second in most seasons, with windless conditions experienced about 50 percent of the time. Windy conditions usually are associated with fronts. Bergwind conditions (hot, dry winds blowing off the escarpment to the west) occur on a few occasions per year and create hazardous fire conditions.

The General Influence of Climate on Soils, Vegetation, and Animals

Soil profiles generally become shallower as rainfall decreases toward the north. This is particularly true for the coarse-grained soils derived from granitic materials, where soil depths decrease from approximately 150 cm in the Pretoriuskop area (rainfall 750 mm/year) to 30 cm north of Phalaborwa (rainfall 350 mm/year). There is also a decrease in the number of soil types that reflect wet soil conditions (Venter 1990), particularly plinthite or FeMn precipitations in parts of the soil that are subjected to regular alternative saturated and drier conditions. This means that the diversity of soil types also decreases with decreasing rainfall. On the other hand, the decreased leaching associated with lower rainfall leads to free $CaCO_3$ being more abundant in soil profiles of the drier north.

Vegetation reflects annual rainfall patterns clearly. The most striking of these is the change from the lowveld bushveld zone, where the rainfall is more than 500 mm/year, to the northern dry bushveld zone, where rainfall generally is below 500 mm/year and where mopane (*Colophospermum mopane*) usually dominates the woody vegetation on soils with more than 15 percent clay. Clear indications are also present in the Pretoriuskop area (rainfall 750 mm/year), where indicator species for wet conditions and drainage lines in drier parts (e.g., common cluster fig [*Ficus sycomorus*]) tend to occur on crests and elsewhere away from streams (Gertenbach 1983).

Another important aspect of rainfall patterns is that the so-called wet and dry cycles significantly influence grass cover, fire regime, animal population dynamics and movements, and the proliferation of certain animal diseases (Chapters 7, 10, and 17, this volume). Drought years may have up to 26 percent less rain than normal years (Gertenbach 1980; Venter and Gertenbach 1986). Grass cover during extended dry periods (e.g., 1982–1983, 1991–1992, and 1994–1995 in Figure 5.3) declines to very low levels over large areas of Kruger, reducing the potential for fires. During these dry periods plains-loving animals such as zebra and blue wildebeest increase in number, whereas long-grass feeders such as buffalo, roan antelope, sable antelope, reedbuck, and tsessebe show decreases in population numbers (Chapter 15, this volume). Anthrax is one of the animal diseases that is associated with climatic conditions. Outbreaks usually occur toward the end of extended dry periods and may have a significant influence on numbers of susceptible animals such as roan antelope (Chapter 17, this volume).

The climatic variations have a significant influence on both terrestrial (Chapter 15, this volume) and aquatic (Chapter 21, this volume) systems of Kruger, and any deviation (especially reduced rainfall) caused by global climate change may necessitate drastic changes to the way in which Kruger is managed in the future.

Geology

A diverse assemblage of igneous, sedimentary, and metamorphic rocks, as well as unconsolidated sediments, which covers a time span of more than 3.5 billion years, occurs within the borders of Kruger. The distribution and characteristics of these different rocks are summarized in Figure 5.4 and Table 5.3. More detailed descriptions can be found in Schutte (1986), Barton et al. (1986), Frick (1986), Bristow (1986), Walraven (1986), Bristow and Venter (1986), and Sweeney (1986).

The strike of the lithology usually is north-south, so that the geological succession changes from west to east, subdividing Kruger into roughly north-south bands of different geology (Figure 5.4). Granitic rocks in the west and basaltic rocks in the east underlie the majority of Kruger. A thin north-south strip of sedimentary rocks separates the granitic and basaltic rock formations. A diverse assemblage of rock formations in the extreme north results in this area being unique in Kruger.

This diversity in parent materials is manifested in a large variety of soils, which support different plant communities and animal populations.

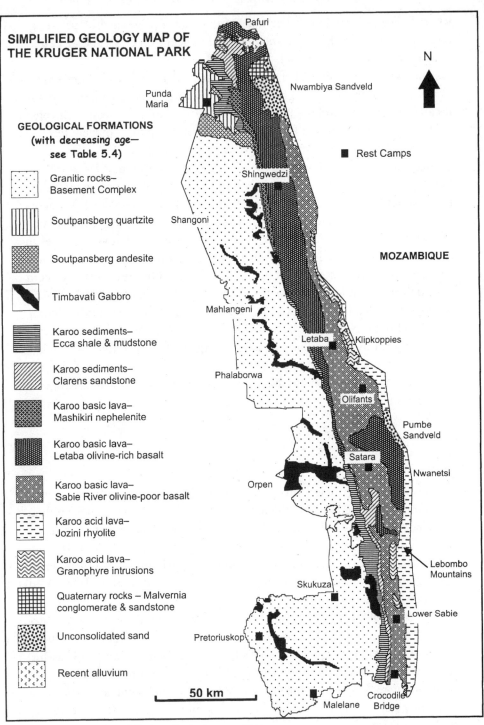

FIGURE 5.4. Simplified geological map of Kruger.

TABLE 5.3

Summary of the stratigraphy of the major rock formations in Kruger
(after Schutte 1986; Barton et al. 1986; Bristow and Venter 1986).

AGE (MA)	MAJOR UNITS	MAJOR SUBUNITS	DOMINANT ROCK TYPES
<130	Recent deposits	—	Alluvium, sand
		Malvernia formation	Conglomerate and sandstone
<175	Karoo sequence	Tshokwane granophyre	Granophyre
~175		Jozini rhyolite formation	Rhyolite, dacite
~190		Sabie River basalt formation	Olivine-poor basalt
~200		Letaba basalt formation	Olivine-rich basalt
~200		Mashikiri nephelinite formation	Nepheline lavas
~200–300		Clarens sandstone formation	Fine-grained sandstone
		Ecca group	Shale, mudstone, grit, conglomerate, coal
~1,000	Timbavati gabbro		Gabbro, quartz gabbro, olivine gabbro
~1,800	Soutpansberg group	Nzhelele formation	Quartzitic sandstone, shale, basalt
		Wyllies Poort quartzite formation	Quartzite, sandstone
		Fundudzi formation	Sandstone, quartzite
		Sibasa basalt formation	Basalt
~2,050	Basement complex	Phalaborwa igneous complex	Syenite
~2,200		Tsheri pegmatite	Muscovite-bearing pegmatite
~2,650		Baderukwe granite	Granodiorite, granite
~3,200		Nelspruit granite suite	Granite, gneiss, migmatite
~3,500		Orpen gneiss	Gneiss
>3,500		Makhutswi gneiss	Gneiss, migmatite, amphibolite
		Goudplaats gneiss	Gneiss, migmatite, amphibolite
		Murchison sequence	Amphibolite, schist
		Barberton sequence	Schist, amphibolite

Soils and Vegetation

The strong correlation between the geology and soils accentuates the geogenetic nature of the soils, which is to be expected of a young erosion surface such as the lowveld, and the moderate rainfall of the area (Venter 1986, 1990; Chapter 10, this volume). The response of the vegetation and animal populations to the template presented by the geology (which is reflected by the soils) and changes caused by the ecosystem drivers such as rainfall and fire have led to a complex patch mosaic.

Kruger is situated the Zambezian Domain of the Sudano-Zambezian Region (Menaut 1983). This domain occupies 3.77 million km^2, from the southern Congo to KwaZulu-Natal and from the Atlantic to Indian Oceans. In other words, most of the plant taxa (and, to a large extent, the animal taxa) of Kruger have a wide regional distribution or have close relatives with a wide regional distribution.

Seven biomes are recognized in southern Africa (Rutherford 1997), and Kruger falls within the savanna biome, which is defined as having a discontinuous overstory of woody plants and a herbaceous layer dominated by C_4 grasses. Savannas cover 60 percent of sub-Saharan Africa (Scholes and Walker 1993) and 12 percent of the global land surface (Scholes and Hall 1996). Clear large-scale species richness patterns (indexed as the number of taxa per 10,000 km^2) can be observed on the African continent. The highest richness in mainland Africa is in the Southern Hemisphere. Plant species richness of southern African savannas is high (5,788 species; White 1983), on both a global and a biome scale, and is second only to that of the fynbos biome. In Kruger, 1,998 indigenous plant taxa have been collected (Zambatis 2002). Unlike the fynbos, the savanna biome is also exceptionally rich in mammals, birds, reptiles, fish, amphibians, and insects. Low and Rebelo (1996) recognize seven vegetation types in the Kruger part of the lowveld.

The cover of woody plants in the Kruger savannas ranges from widely dispersed individuals with a total tree canopy cover of 5 percent to near-closed canopy woodlands with a cover of around 60 percent. Our use of the term *woodland* is reserved for areas with tree cover greater than 35 percent and height greater than 5 m. If the tree cover is less than 35 percent, it is called tree savanna. Shorter (>5 m), closed formations are thickets, and if the canopy is not closed it is called bushveld. More than 75 percent of Kruger is in the latter category: trees 2–5 m tall, tree cover 20–40 percent. The rest is tree savanna, woodlands, and thickets, with true forests occurring patchily along rivers.

The long, dry winter and the prevalence of dry grass fuel lead to a regime of frequent fires (Chapter 7, this volume). This fire regime is central to the persistence and structure of the Kruger savannas (Chapter 11, this volume). Where there is high grass biomass and fires are more intense, such as on the basalt plains, the savannas are more open, and where fires are less intense, such as on the granites, they are more closed.

The savannas of southern Africa can be broadly divided into an arid and a moist group (Huntley 1982). In Kruger they can be grouped as infertile or fertile based largely on soil properties. Savannas on nutrient-poor substrates derived from granite or sandstone tend to be dominated by trees in the families Combretaceae (especially *Combretum* and *Terminalia* spp.) and Caesalpinaceae. These trees are deciduous, with broad leaves and no thorns. The grasses such as herringbone grass (*Pogonarthria squarrosa*), *Eragrostis* spp., and *Aristida* spp. are wiry, unpalatable, and sparse. Savannas on more fertile substrates (clay soils) are dominated by the Mimosaceae (especially *Acacia* spp.), which are also deciduous, with fine compound leaves and many thorns. Because these plants have mycorrhizal associations in their root systems and high nitrogen concentrations, they are preferred by herbivores. The nutrient-rich savannas have nutritious, high-bulk grasses such as small buffalo grass (*Panicum coloratum*), red grass (*Themeda triandra*), and bushveld signal grass (*Urochloa mosambicensis*).

There are many other ecological correlates of this basic split and some important exceptions. Mopane (*Colophospermum mopane*) is one such exception: it is an unusual member of the Caesalpinaceae that dominates in

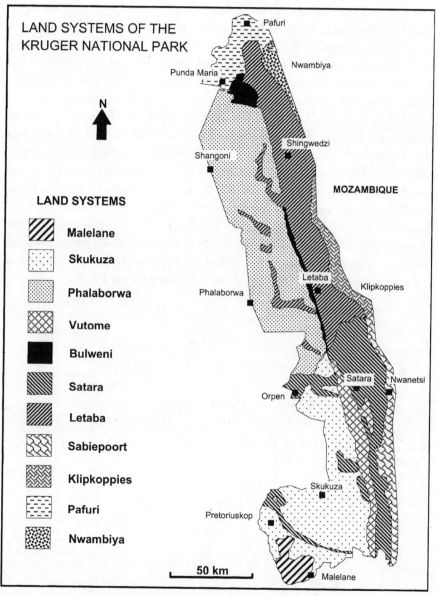

FIGURE 5.5. Distribution of land systems in Kruger (Venter 1990).

fertile but hot and dry environments because oils in its leaves protect it from desiccation.

The savannas of Kruger are split between the two main ecological types. Broad-leaved savannas occupy approximately 75 percent of Kruger (of which 50 percent are mopane), and the remaining 25 percent are made up of fine-leaved savanna. The Kruger savannas vary significantly in structure and composition at the local scale, and the factors that determine this heterogeneity are related mostly to soil, fire, and climate (Chapter 11, this volume). Savannas thus occur everywhere in Kruger except on soils that are periodically waterlogged, such as marshes (*vleis*) and seeplines, which are treeless and covered by sedges and moisture-loving grasses; on substrates poorly permeable to roots or water (e.g., some sodic sites or shallow soils over unfractured rock), which are treeless and may be almost bare of vegetation; and in areas where there is a perennial water source within the rooting zone (e.g., riparian forests and fire-protected gullies in mountains).

The vegetation and land classifications most widely used for management and research are based on the Gertenbach (1983) and Venter (1990) classifications. The Gertenbach system focuses on dominant woody vegetation, recognizing 35 landscapes. The Venter system is a hierarchy of land classes. It recognizes 11 land systems and 56 land types and focuses on the association between geology, terrain morphology, soils, and woody vegetation. A simplified overview of the broad soil, geomorphology, and vegetation pattern in relation to land systems is described by Venter (1990) (Figure 5.5, Tables 5.4–5.7). Soils are described in general terms, but where soil classification names are used, they are according to the South African soil classification system (Soil Classification Working Group 1991).

TABLE 5.4

Morphometric features of land systems in Kruger.

NAME	AREA HA	AREA % OF PARK	STREAM FREQUENCY (N/KM^2)	LOCAL RELIEF (M)	SLOPE CLASSES (% OF AREA)			
					0–2%	2–6%	6–15%	>15%
Malelane	41,326	2.1	3.15	114	0	9	30	61
Skukuza	382,045	19.6	2.33	23	16	69	13	2
Phalaborwa	518,266	26.6	2.96	18	47	40	9	4
Vutome	78,819	4.1	1.21	15	71	26	3	0
Bulweni	32,384	2.2	0.34	12	73	13	10	4
Satara	275,867	14.2	1.39	23	59	35	5	1
Letaba	356,664	18.3	1.19	14	59	30	7	4
Sabiepoort	84,122	4.3	1.42	34	11	43	21	25
Klipkoppies	45,733	2.4	2.06	72	7	20	21	53
Pafuri	80,424	4.1	1.95	40	17	32	25	26
Nwambiya	40,879	2.1	0.12	15	91	9	0	0
Total	1,936,529	100	—	—	—	—	—	—
Average	—	—	1.65	35	41	30	13	16

TABLE 5.5

Summary of the geology, soils, topography, vegetation, and rainfall of land systems in the Kruger.

LAND SYSTEM	GEOLOGY AND SOILS	TOPOGRAPHY	VEGETATION	RAINFALL (MM/YR)
Malelane	Granitic rocks. Rock outcrops and stony soils	Low mountains and hills	Broad-leaved bushveld	600–700
Skukuza	Granitic rocks. Uplands: sandy soils. Bottomlands: duplex sodic clay	Slightly undulating plains	Uplands: broad-leaved bushveld. Bottomlands: fine-leaved bushveld (acacias)	500–750
Phalaborwa	Granitic rocks, often migmatized. Uplands: sandy soils. Bottomlands: loam or duplex sodic clay	Slightly to strongly undulating plains	Uplands: broad-leaved bushveld. Bottomlands: broad-leaved bushveld (mopane)	450–600
Vutome	Karoo sedimentary rocks. Sandstone: fine sand. Shale: duplex sodic clay	Flat to slightly undulating plains	Sandstone: broad-leaved bushveld. Shale: fine-leaved woodland (acacias)	500–650
Bulweni	Karoo and Soutpansberg sedimentary rocks. Sandstone: rock outcrops and sand. Shale: paraduplex clay	Flat to moderately undulating plains	Sandstone: broad-leaved bushveld. Shale: broad-leaved woodland (mopane)	450–500
Satara	Mafic volcanic rocks (basalt). Red and dark clays	Flat plains	Fine-leaved tree savanna (acacias)	500–650
Letaba	As for Satara	As for Satara	Broad-leaved shrubveld (mopane)	450–500
Sabiepoort	Acid volcanic rocks (rhyolite). Rock outcrops and stony soils	Low mountains and hills	Broad-leaved deciduous bushveld (no mopane)	500–650
Klipkoppies	As for Sabiepoort	As for Sabiepoort	Broad-leaved bushveld (mopane present)	450–500
Pafuri	A wide range of sedimentary and volcanic rocks and associated soils, as well as alluvial floodplains	Slightly to strongly undulating plains	Broad-leaved dry bushveld (mopane present)	400–650
Nwambiya	Unconsolidated recent sand and conglomerate	Flat plains to intensely incised areas	Broad-leaved thickets	<450

The Malelane, Skukuza, and Phalaborwa Land Systems

The areas in the west are underlain by granite and gneiss (see geological map, Figure 5.4, and land system map, Figure 5.5) and characterized by very dis-

TABLE 5.6

A general description of the soils related to the different land systems in Kruger.

LAND SYSTEM	CREST	MIDSLOPE	FOOTSLOPE	VALLEY BOTTOM
Malelane	Rock outcrops and shallow, stony soils; very shallow, coarse sand and loam	Rock outcrops and stony soils	Shallow to deep coarse to medium sand and loam; areas with steep slopes dominated by stony (mainly talus) and rock outcrops	Rock outcrops and stony soils in association with shallow to deep gray and brown alluvial loam and clay
Skukuza	Moderately deep to shallow red and brown coarse sand and loam	Shallow gray coarse sand	Duplex sodic soils with shallow gray and brown sand or loam abruptly overlying calcareous, prismatic clay	Complex association of deep brown sand and calcareous clay and rock outcrops
Phalaborwa	Shallow to very shallow red and brown sand and loam	Shallow to very shallow gray sand or red loam	As for Skukuza land system, but shallow to moderately deep red loam to sandy clay also common	As for Skukuza
Vutome	Shale and mudstone: brown duplex soils: loam abruptly overlying calcareous and sodic prismatic clay Sandstone: deep red fine sand	Shale and mudstone: moderately deep brown, occasionally calcareous calcareous paraduplex clay (sand or loam not abruptly overlying clay) Sandstone: low ridges, koppies, and rock outcrops in association with deep red fine sand		As for Skukuza
Bulweni	Moderately deep to shallow red and brown structured clay	Shallow red, brown, and black weakly to strongly structured clay and loam	Deep to moderately deep black and brown calcareous clay, often displaying cracks	As for Skukuza
Satara				A complex association of black and brown calcareous alluvial clay and loam in various stages of profile development

Letaba	Shallow to moderately deep black calcareous clay	Moderately deep to deep black, expansive calcareous clay
Sabiepoort	Stony soils and rock outcrops in association with very shallow grayish brown loam	Stony soils (mainly talus) and rock outcrops in association with very shallow to moderately deep brown loam and clay
Klipkoppies	As for Sabiepoort	Grayish brown calcareous clay, shallow to moderately deep black calcareous clay
Pafuri	Sandstone: stony soils and rock outcrops; shallow to deep red and yellow-brown coarse to fine sand Basalt: stony soils and rock outcrops; very shallow brown and black calcareous clay and loam	Complex association of sand, clay, and stony and shallow calcareous loam
Nwambiya	Very deep red coarse sand	

	Shallow black calcareous clay	As for Satara
	Mainly rock outcrops and stony soils	Mainly rock outcrops and stony soils; occasional shallow to moderately deep brown clay
	As for Sabiepoort	Deep red alluvium
	Very deep to shallow yellow coarse sand; shallow gray sand and duplex soils in association with pans and depressions	

TABLE 5.7
Summary of the dominant woody vegetation of land systems in Kruger.

NAME	LAND UNIT			
	CREST	MIDSLOPE	FOOTSLOPE	VALLEY BOTTOM
Malelane	Broad-leaved, moderately dense *Combretum apiculatum* bushveld; several rare species		Broad-leaved, dense *C. apiculatum* bushveld on shallow soils; dense *C. apiculatum*, *Combretum collinum*, and *Combretum zeyheri* bushveld on deeper soils	Dense riverine bushveld, woodland, and forest with heterogeneous vegetation composition
Skukuza	Broad-leaved, moderately dense *C. apiculatum* and *C. zeyheri* bushveld	Broad-leaved, moderately dense *Terminalia sericea* and *C. zeyheri* bushveld	Fine-leaved, open *Acacia gerrardii* and *Euclea divinorum* shrubveld; *Combretum hereroense* and *Acacia nigrescens* often very prominent	As for Malelane
Phalaborwa	Broad-leaved, moderately dense bushveld: *C. apiculatum* on sandy soils and *Colophospermum mopane* on loam and clay	As for crest, but *T. sericea* occasionally dominant on gray sand	Open to moderately dense *C. mopane* bushveld	As for Malelane
Vutome	Shale and mudstone: fine-leaved, dense *Acacia welwitschii* and *E. divinorum* woodland; *Spirostachys africana* prominent near drainage channels Sandstone: broad-leaved, dense bushveld dominated by *T. sericea* and *C. zeyheri*			As for Malelane
Bulweni	Broad-leaved, moderately dense *T. sericea*, *C. zeyheri*, and *C. collinum* bushveld on deep sand; dense *C. mopane* and *E. divinorum* woodland and thicket on paraduplex clay; *S. africana* prominent near drainage channels			As for Malelane
Satara	Fine leaved, open *A. nigrescens* and *Sclerocarya birrea* tree savanna; *Dichrostachys cinerea* always a prominent shrub	Fine-leaved, moderately dense *A. nigrescens* and *D. cinerea* bushveld	Fine-leaved, open to moderately dense *C. hereroense*, *D. cinerea*, and *E. divinorum* shrubveld; *Acacia borleae* in localized dense shrub thickets	Dense riverine bushveld with heterogeneous vegetation composition; *Hyphaene natalensis*, *Lonchocarpus capassa*, and *Acacia xanthoploea* often the dominant woody plants

Letaba	Dense to moderately dense C. mopane shrubveld	As for crest but species preferring shallow soils occur as well	Open to dense C. mopane bushveld	As for Satara
Sabiepoort	Broad-leaved, dense to moderately dense C. apiculatum and Pterocarpus rotundifolius bushveld	Broad-leaved, moderately dense C. apiculatum and P. rotundifolius bushveld	Fine-leaved, dense C. hereroense and A. nigrescens bushveld; A. gerrardii and D. cinerea also prominent	Dense riverine bush with heterogeneous vegetation composition
Klipkoppies	Broad-leaved, open to moderately dense C. apiculatum and C. mopane bushveld		Moderately dense C. mopane woodland	As for Satara
Pafuri	Deep sand: broad-leaved, moderately dense Burkea africana, T. sericea, and Pseudolachnostylis maprouneifolia woodland Shallow, stony sand: broad-leaved, moderately dense C. apiculatum and Kirkia acuminata tree savanna Very shallow calcareous clay: moderately dense C. mopane and Commiphora glandulosa woodland and bushveld		Dense C. mopane and E. divinorum woodland or bushveld	As for Satara
Nwambiya	Broad-leaved Baphia massaiensis and Guibourtia conjugata thickets	Broad-leaved, dense Xeroderris stuhlmannii, C. apiculatum, and C. zeyheri bushveld	Complex association of a variety of communities that are associated with different soil types	

tinctive catenal sequences of soils from crest to valley bottom: reddish or yellowish brown sand, gray hydromorphic sand and clay (seasonally waterlogged band of soils along the contour of the slope, called a seepline), grayish brown sodic duplex soils (sand or loam abruptly overlying dispersed clay affected by the presence of sodium), and mixed alluvial soils. The ecological consequences of this pattern are described in more detail in a later section on the ecological significance of catenas in Kruger.

The vegetation of the sandy uplands in the southwest (i.e., the area underlain by granitic rocks) is characterized by dense deciduous broad-leaved bushveld, typically dominated by *Combretum* spp. (red bushwillow [*C. apiculatum*], large-fruited bushwillow [*C. zeyheri*], and variable bushwillow [*C. collinum* subsp. *suluense*]) and silver cluster-leaf (*Terminalia sericea*). Along footslopes where duplex soils are found, open, small-leaved shrubveld with thorny woody plants occurs. Dominant species are red thorn (*Acacia gerrardii*), knob thorn (*Acacia nigrescens*), common false-thorn (*Albizia harveyi*), sickle bush (*Dichrostachys cinerea*), magic guarri (*Euclea divinorum*), and russet bushwillow (*Combretum hereroense*). The grasses are nutritious bulk growers and, if the sandy A horizon is thick, may reach a much higher biomass than on the crests. Silver cluster-leaf (*Terminalia sericea*) and large-fruited bushwillow (*Combretum zeyheri*) often dominate the wetter seepline on the midslope and form a prominent band along the contour. Hydrophilic grasses such as gum grass (*Eragrostis gummiflua*) and sedges are dominant along seeplines. In general the grass cover is sparse in dry years and moderate to dense in wet years, becoming denser and taller as the average rainfall increases toward the southwest. In the Pretoriuskop area (annual rainfall 750 mm), tall thatch grasses are dominant.

The granitic areas in the northern half of Kruger consist mainly of dry *Combretum apiculatum* and *Colophospermum mopane* bushveld. Red bushwillow (*C. apiculatum*) is confined to the shallow gravel and sandy soils of crests. Mopane (*C. mopane*) dominates the more clayey soils (soil clay content >15 percent) derived from greenstone rocks and dolerite dikes and on the duplex soils along footslopes.

Certain granitic areas do not display the typical catena pattern. These areas are characterized by the presence of metamorphic rocks such as greenstones; mountainous areas with thin soils, as in some parts of the Malelane Land System; areas that have been intruded by numerous dolerite dikes; and areas that are intensely dissected. Metamorphic greenstone rocks such as amphibolite and schist, which are relicts of older greenstone belts in the granite-gneiss body, crop out more often in the northern areas. They form red, sandy clay loam soils characterized by *Colophospermum mopane* woodland or bushveld. These areas usually are less undulating.

Dolerite dikes of Karoo age often abound in the basement rocks. They occur as long narrow strips (usually 20–50 m wide), often in two distinct directions.

The dikes are basic rocks similar in composition to basalt because they were the feeders for the lavas that formed the basalt. They weather into clayey, structured, fertile soils, and their presence causes a marked change in soil patterns. In certain areas they occur in dense swarms and form complex checkered soil patterns, which differ from normal granite-gneiss soil patterns. Many soils in areas where dolerite dikes are common are of binary origin, created by colluviation of dolerite material over weathered granite-gneiss.

Vutome and Bulweni Land Systems

Sedimentary rocks of the Ecca group (i.e., gray mudstone and shale; see Figures 5.4 and 5.5) are easily weathered and form large areas of brown and gray, clayey, structured soils in the southern, central, and far northern regions. In the southern and central regions sodic duplex soils (sand or loam abruptly overlying dispersed clay) are dominant, whereas in the Punda Maria area paraduplex soils (sand or loam topsoil grading gradually into clay subsoil) are dominant. The reasons for this variation are not clearly understood.

Fine-leaved woodlands and thickets of Delagoa thorn (*Acacia welwitschii*) and magic guarri (*Euclea divinorum*) occur in the area underlain by Ecca sediments in the southern and central regions. In the north, where paraduplex soils occur, mopane (*Colophospermum mopane*) forms almost monospecific closed woodland.

Soils derived from sandstone of the Clarens formation (fine grained) and quartzite of the Soutpansberg group (medium to coarse grained) usually occur in small areas directly adjacent the Ecca soils and are included in these land systems. Areas underlain by these rocks usually are characterized by deep red and yellow sand. Where prominent hills or koppies occur, shallow skeletal soils and rock outcrops are dominant, and deep soils are limited to pockets between rock outcrops. This is especially true of the Punda Maria–Pafuri area, where koppies of the Clarens sandstone formation occur most extensively and give rise to outstanding scenery. The sandy soils derived from Clarens sandstone usually are favored by broad-leaved bushveld or tree savanna dominated by silver clusterleaf (*Terminalia sericea*) and large-fruited bushwillow (*Combretum zeyheri*) and, in the northern regions, wild seringa (*Burkea africana*).

The Satara and Letaba Land Systems

Weathering of basic igneous rocks (mostly basalt; Figure 5.4) forms clayey soils by virtue of their mineralogical composition and the prevailing climate of the lowveld. More detail about the basaltic soils is presented in the next section, which deals with the significance of the catena on basalt.

The area underlain by basic rocks south of the Olifants River (Satara Land System) consists mainly of fine-leaved tree savanna or bushveld, dominated by knob thorn (*Acacia nigrescens*), sickle bush (*Dichrostachys cinerea*) and marula (*Sclerocarya birrea*). North of this river the basic rock plains (Letaba Land System) are characterized by *Colophospermum mopane* shrubveld, of which leadwood (*Combretum imberbe*) is a conspicuous component. A lush grass cover is characteristic of the basaltic areas under high rainfall conditions.

Timbavati gabbro is an intrusive basic rock (Figure 5.4) that occurs sporadically as sills within the granites. It forms part of the Satara and Letaba land systems (Figure 5.5) because the soils associated with the gabbro sills are strong correlates of the soils derived from basalt. Along the contacts between gabbro intrusions and granite, extensive colluviation of gabbroic material over weathered granite has resulted in the formation of soils of binary origin (in which stone lines are common) on downslope positions.

The Sabiepoort and Klipkoppies Land Systems

Rhyolite and granophyre, which occur along the eastern boundary (Lebombo hills and low mountains, Figures 5.4 and 5.5), are extremely resistant to weathering, and hills consisting mainly of stony soils and rock outcrops are common. These areas are characterized by little or no saprolite (layer of partly weathered rock), whereas most of the other soils grade into saprolite up to several meters thick. The Lebombo mountains are characterized by deciduous bushveld in which red bushwillow (*Combretum apiculatum*) is the dominant species south of the Olifants River and *C. apiculatum* and *C. mopane* are the dominant species north of this river. Several other tree species are limited in distribution to the Lebombo Mountains.

The Nwambiya and Pumbe Areas

Deep to very deep, coarse, sandy soils are found along the eastern boundary at Pumbe and Nwambiya (Figure 5.4), the latter area being the larger of the two. The Pumbe area is small and has been classed with the Sabiepoort land system. The soils are predominantly red along crests and midslopes and grade into yellow-brown and gray soils along footslopes. The sand is several meters (up to 6 m) deep in some places. The Pumbe vegetation consists of moderately dense to open broad-leaved (red bushwillow [*Combretum apiculatum*], large-fruited bushwillow [*C. zeyheri*], silver cluster-leaf [*Terminalia sericea*], and kudu-berry [*Pseudolachnostylis maprouneifolia*]) bushveld. In the Nwambiya area broadleaved deciduous thickets dominate the deep red, sandy soils (sand camwood [*Baphia massaiensis*] and copal wood [*Guibourtia conjugata*]), grading into dense wing bean (*Xeroderris stuhlmannii*), red bushwillow (*Combretum apiculatum*), and large-fruited bushwillow (*C. zeyheri*) bushveld, and mopane (*Colophospermum mopane*) woodland.

The Pafuri Land System

The Punda Maria–Pafuri area is complex and consists of various geological formations (Figure 5.4), associated soils, and a large number of plant communities. It varies from rugged hills and gorges (Luvuvhu River and tributaries) to lowland floodplain, the only true floodplain in Kruger. Because of the complexity and uniqueness of the area, it was classified as one land system (Venter 1990). Van Rooyen (1978) and Gertenbach (1983) listed the important plant communities as *Colophospermum mopane* woodland and forest; *Burkea africana* and *Pseudolachnostylis maprouneifolia* broad-leaved bushveld; *Kirkia acuminata*, *Afzelia quanzensis*, and *Combretum apiculatum* broad-leaved tree savanna; *Androstachys johnsonii* and *Croton pseudopulchellus* dry thicket and woodland; and *Acacia albida* and *Ficus sycomorus* river forest.

Alluvial Soils

Alluvial soils, although not shown as a separate land system, occur along most drainage lines, and their extent increases with the size of the drainage line. Older river terraces and gravels also occur along the major rivers. The most extensive alluvial deposits are found along the Limpopo and Luvuvhu rivers. The sediments flanking the Limpopo River are mostly sandy, whereas those along the Luvuvhu consist of deep red silt. Significant alluvial deposits occur along the Shingwedzi River system.

The plant communities of the alluvial sediments are species-rich. Tree species such as jackal-berry (*Diospyros mespiliformis*), weeping boer-bean (*Scotia brachypetala*), leadwood (*Combretum imberbe*), matumi (*Breonadia salicina*), river bushwillow (*Combretum erythrophylum*), water elder (*Nuxia oppositifolia*), common cluster fig (*Ficus sycomorus*), sausage tree (*Kigelia africana*), ankle thorn (*Acacia robusta*), ana tree (*Faidherbia albida*), and nyala tree (*Xanthocercis zambesiaca*) are but a few examples of this community.

Ecological Significance of the Catena: Examples from Granitic and Basaltic Areas

Several studies have focused on the characteristics of the catena in granitic landscapes, the processes of its formation, and its ecological significance (Purves 1973; Dye 1977; Tinley 1979; Webber 1979; Olbrich 1984; Scholes 1986; Venter 1986, 1990; Fraser et al. 1987; Munnik et al. 1990; Chappel 1992;). Thrash et al. (1991) have studied the impact of artificial water on certain aspects of soils as they occur in the catena, and Webber (1979) investigated the influence of fire on different soils in the experimental burning plots, which include cate-

TABLE 5.8
Comparison of ecological aspects of granite- and basalt-derived soils in Kruger.

PARAMETER	GRANITE	BASALT
Dominant rock-forming minerals	• Quartz, K-feldspar (weather resistant).	• Plagioclase, pyroxene (easily weathered).
Soil formation	• In situ clay-forming potential is low. • Mostly coarse grains; finer clay minerals and soluble products wash out.	• In situ clay-forming potential is high. • Mostly fine clay minerals are formed.
Geomorphic consequences	• Coarse material is moved slowly by overland flow (usually shifted locally). • Undulating landscapes form.	• Fine material is easily taken into suspension by running water and transported away. • Flat plains landscapes form.
Leaching processes	• Fine materials are easily leached from sandy soils at hilltops and moved to bottomlands because of higher soil permeability and undulating landscape. • Crests become sandier and nutrient poor, whereas footslopes become more clayey and nutrient rich. • A distinctive catena is formed, with a seepline (subsurface water forced to the surface by clay layer during wet periods) separating uplands and bottomlands. • Clayey footslopes expand uphill over time as clay accumulates along the seepline zone.	• Slower movement of water through clay soils and flatter landscapes inhibit the large-scale movement of material downslope, as in the case of the granitic soils. • Catenas are not so distinctively defined, but $CaCO_3$ does accumulate in low-lying areas. • Expansive 2:1 clays often occur in bottomlands, forming deep cracks upon drying, usually with few woody plants because their roots tear.
Clay and soil characteristics	• Crests have low clay (>15%), footslopes have high clay (25–35%). • Crests are dominated by 1:1 clays and footslopes by 2:1 clays. • Monovalent cations (especially Na) cause clay in bottomlands to deflocculate as soon as they occupy 15–20% of exchange sites on clay. • Duplex soils form along bottomlands (sand or loam abruptly overlying clay). • Bottomlands are susceptible to erosion.	• Crests have high clay (35–40%), footslopes very high clay (50–60%). • 2:1 clays dominate (mainly smectite clays). • Divalent cations (Ca and Mg) dominate exchange sites, causing clay particles to be drawn to each other by van der Waals forces. They are thus flocculated and fairly stable and form blocky structures as a result of their potential to expand upon wetting and shrink upon drying.

Plant and animal consequences

- Sandy uplands
  - Low soil organic material
  - Low clay
  - Low cation exchange capacity
  - Low base status
  - Low water-holding capacity
  - *Combretum* trees
  - Low N availability
  - Unpalatable grasses
  - Low herbivore biomass
- Clay bottomlands
  - Moderate soil organic material
  - High clay
  - High cation exchange capacity
  - High base status
  - High water-holding capacity
  - *Acacia* shrubs
  - High N availability
  - Palatable grasses
  - Moderate herbivore biomass

- Clayey basaltic soils
  - High soil organic material
  - High clay
  - High cation exchange capacity
  - High base status
  - High water-holding capacity
  - *Acacia* trees
  - High N availability
  - Palatable grasses
  - High herbivore biomass

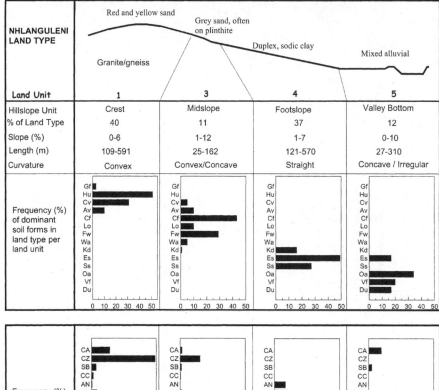

FIGURE 5.6. An example of a catena on granite, as found in the Nhlanguleni land type just north of Skukuza (after Venter 1990). This area has no dolerite dykes or greenstones and is ideal for the study of catenas on granite. The symbols used for the soil forms (Soil Classification Working Group 1991) and dominant woody plant species are explained in the table below.

Soil forms		Woody plant species
Gf - Griffin	} Red and yellow sand	CA - *Combretum apiculatum*
Hu - Hutton		CZ - *Combretum zeyheri*
Cv - Clovelly		SB - *Sclerocarya birrea*
Av - Avalon		CC - *Combretum collinum*
Cf - Cartref	} Grey sand, often on plinthite	AN - *Acacia nigrescens*
Lo - Longlands		PR - *Pterocarpus rotundifolius*
Fw - Fernwood		TS - *Terminalia sericea*
Wa - Wasbank		SM - *Strychnos madagascariensis*
Kd - Kroonstad	} Duplex, sodic clay	CH - *Combretum hereroense*
Es - Estcourt		AG - *Acacia gerrardii*
Ss - Sterkspruit		ED - *Euclea divinorum*
Vf - Vilafontes	} Mixed alluvial	AH - *Albizia harveyi*
Oa – Oakleaf		CI - *Combretum imberbe*
Du - Dundee		LC - *Lonchocarpus capassa*
		SA - *Spirostachys africana*
		MS - *Maytenus senegalensis*

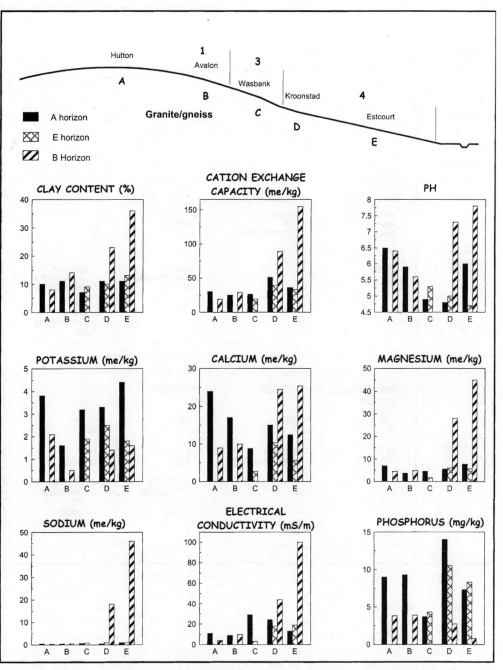

FIGURE 5.7. Typical physical and chemical properties of soils along a granitic hillslope in the Nhlanguleni area, Skukuza Land System.

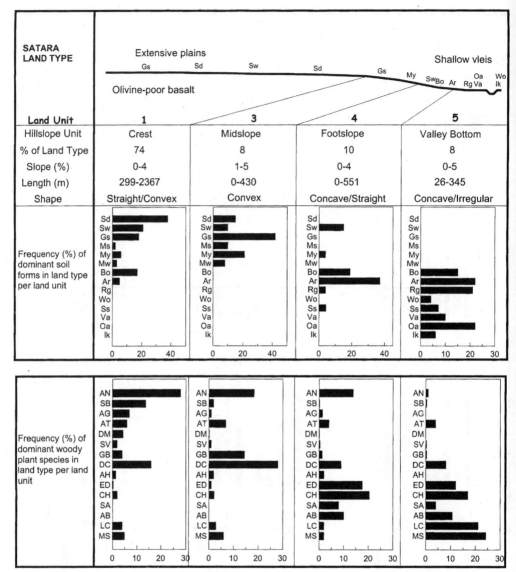

FIGURE 5.8. An example of a catena on basalt as found in the Satara land type (after Venter 1990). Symbols for dominant soil forms, are according to the South African soil classification system (Soil Classification Working Group 1991), are as follows: Sd (Shortlands) and Sw (Swartland) are red and brown structured clay loams; Gs (Glenrosa) and Ms (Mispah) are red and brown shallow loams; My (Mayo) and Mw (Milkwood) are dark shallow clays; Bo (Bonheim) is a dark structured clay; Ar (Arcadia), Rg (Rensburg) and Wo (Willowbrook) are dark expansive clays; Ss (Sterkspruit) is a duplex, sodic clay; Va (Valsrivier), Oa (Oakleaf) and Ik (Inhoek) are mixed alluvials. Dominant woody plant species are as follows: AN, *Acacia nigrescens*; SB, *Sclerocarya birrea*; AG, *Acacia gerrardii*; AT, *Acacia tortilis*; DM, *Dalbergia melanoxylon*; SV, *Securinega virosa*; GB, *Grewia bicolor*; DC, *Dichrostachys cinerea*; AH, *Albizia harveyi*; ED, *Euclea divinorum*; CH, *Combretum hereroense*; SA, *Spirostachys africana*; AB, *Acacia borleae*; LC, *Lonchocarpus capassa*; MS, *Maytenus senegalensis*.

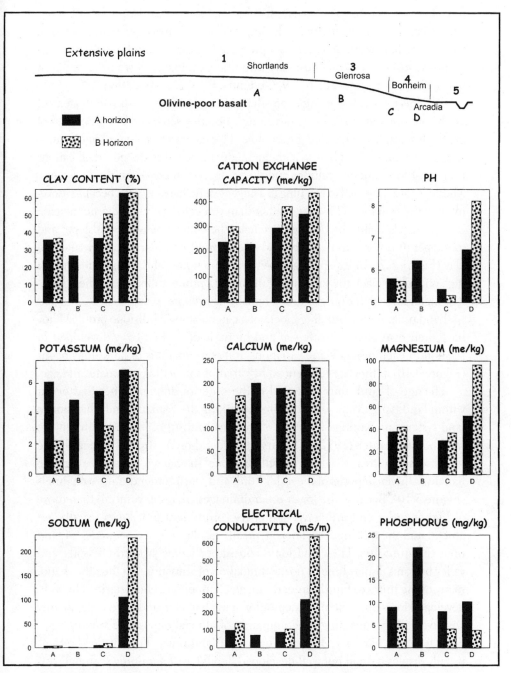

FIGURE 5.9. Typical physical and chemical properties of soils along a basaltic hillslope in the Satara Land System.

nas. However, very few studies have focused on the catenas associated with other geological formations, including basalt, or made comparisons between them. More than 80 percent of Kruger is represented by either granitic or basaltic formations, and catenas manifest very important ecological systems at different scales (Table 5.8, Figures 5.6–5.9; Gertenbach 1983; Venter 1986, 1990).

Granitic areas of Kruger often are characterized by a very distinctive catena of sandy soils on crests and clayey soils along footslopes, separated by a thin band of gray hydromorphic soils called the seepline. This has much ecological significance (Figures 5.6 and 5.7). There is a gradual reduction in clay and adsorbed cations from crest to midslope positions and then a marked increase on the footslopes. Midslopes are subjected to leaching to a larger extent than are crests because movement of water in the soil is lateral rather than vertical. An abrupt, drastic increase in both clay and absorbed cations occurs at the contact between midslopes and footslopes in B horizons that is caused by an abrupt transition between sand and clay (Figure 5.7). In exceptionally wet years the water table along midslopes is high enough to reach the soil surface, making it a prime wallow area. These localities therefore switch from dry to wet on fairly short times scales.

Land units are the building blocks of a catena along a hillslope profile. They have their own characteristic soil and biotic features that have developed in relation to each other. Many variations and combinations of land units occur (Figure 5.10) as they are influenced by variations in geology, climate, and age.

The age of land units is related to the order of drainage lines that occur within land types. Many drainage lines draining the Skukuza and Phalaborwa land systems are young. They are of the first to third order according to the method of Strahler (1952). The first- to second-order streams generally are small, seasonal streams, which are still erosional in character, and they lack the flanking alluvial deposits of the older third- and fourth-order drainage channels (Figure 5.10). Because the lower-order drainages are more youthful than those of higher order and are actively incisive, prominent footslopes usually are absent, and the hillslopes (catenas) typically display a 1, 3, 5 sequence of land units (Figure 5.10). Their midslopes consist of shallow or lithosolic soils, and valley bottoms display little or no accumulation of alluvium. On the other hand, many of the third- to fourth-order drainage channels are characterized by well-developed footslopes, often are depositional in character, and have wide, slightly concave valleys consisting predominantly of alluvial deposits of varying ages. They typically have a 1, 3, 4, 5 sequence of land units. Their valley bottoms usually display a small but well-developed floodplain, containing seasonal pans and animal wallows, and levees flanking the channel.

A land element is the lowest level in the patch hierarchy of land classes (Table 5.9). Land elements are subdivisions of land units, which means that land units are not necessarily homogeneous components of a hillslope. Land units may represent either fairly homogeneous soil bodies or assemblages of soil

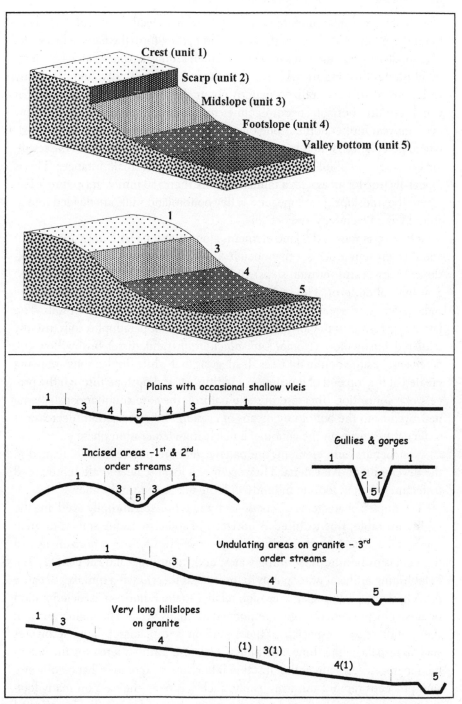

FIGURE 5.10. An illustration of the different terrain morphological units (top; Soil Classification Working Group 1991) and examples of different hillslope types (bottom) in the lowveld and Kruger (after Venter 1990).

types that have similar properties. Soil types, with a specific soil profile with distinguishing horizons, characteristics, and properties, form the basis of this scale. The subdivision of land units into land elements along a typical catena on granite illustrates this (Figure 5.11, Randspruit land type south of Skukuza). Many variations of this generalized pattern occur. For example, the frequency at which plinthic horizons occur declines markedly as the rainfall drops below 600 mm/year further north. Along midslopes, where water accumulates through lateral percolation, much larger variations in soil climate exist, with the resultant development of many more different soil types over small distances. This is especially true for areas with a rainfall of more than 550 mm/year (Figure 5.11), where the midslope, footslope, and valley bottomland units are divided into 2, 3, and 2 land elements, respectively.

Other variations of the land elements along footslope land units are brought about by erosion processes that operate along the bottom parts of certain footslopes. Mini-scarps (erosion steps) 20 to 50 cm high and erosion gullies up to 2 m high often form. The mini-scarps usually result in the formation of bare sodic patches, whereas the gullies may strip parts of the footslope of soils. The latter case results in the formation of shallow stony or paraduplex soils, usually with red bushwillow (*Combretum apiculatum*) and russet bushwillow (*C. hereroense*) as the dominant trees. It suggests that while duplex soils are being created at the top end of footslopes, and thus expand upslope through the process of solonization (increase in sodic nature), they are simultaneously being destroyed from the bottom by means of erosion and solodization (decrease of sodic nature) through the action of a newly initiated erosion phase.

Noncognate land elements are patches that are not necessarily related to the development of a catena. They occur at a local scale and are represented by features such as termite mounds, dolerite dikes, and mud wallows.

The duplex character (i.e., coarse-textured A horizon abruptly overlying the highly unstable, fine-textured B horizon) of sodic footslope soils is of great importance. The A horizon acts as a protective layer for the B horizon against the effects of raindrops, running water, and trampling (Thrash et al. 1991). Establishing artificial water points in these areas therefore is a mistake. Because the A horizon is stable against erosion relative to the B horizon, its deterioration by sheet erosion generally goes unnoticed for many years. The higher nutrient status of these soils, especially if the A horizon is thin, attracts large herbivores such as impala, white rhino, and blue wildebeest, further increasing the potential for erosion. As soon as the unstable B horizon is exposed, rill and gully erosion proceeds rapidly and irreversibly; such has been the case on many footslopes in the privately owned nature reserves west of Kruger. Reclamation of these soils is extremely difficult because conditions such as a high pH, toxic levels of certain minerals, and dry conditions caused by poor infiltration prevent vegetation establishment.

There is a distinct decrease in the concentration of phosphorus and the less mobile exchangeable cations (potassium, calcium, and to a lesser extent magnesium) from A to B or E horizons for sandy soils of crests and midslopes (Figure 5.7). This can probably be attributed to the larger amounts of organic material contained in A horizons. This increases cation exchange capacity and therefore their capacity to retain soluble minerals. Sandy soils of granitic crests generally are low in nutrients compared with soils derived from basalt (Figures 5.7 and 5.9). This is well reflected in the general distribution of large herbivores (Chapters 10 and 15, this volume). Very different processes operate in basaltic areas with clay soils, resulting in different soil and vegetation patterns than found on granite (Figures 5.6 and 5.8). An example from the Satara land type (which forms part of the Satara Land System) on the basaltic plains is used to illustrate this (Figures 5.8 and 5.9).

The soils of the Satara Land System generally are high in clay and nutrients (Figure 5.9) and are dominated by *Acacia* trees that enhance nitrogen availability and therefore attract herbivores. Large concentrations of game occur in this land system (Chapter 15, this volume). Along this basalt catena there is an increase in pH and most of the exchangeable cations, in a downslope direction (Figure 5.9). Soil reaction usually ranges from slightly acid to strongly alkaline. Calcium is the dominant exchangeable cation, but in the horizons of expansive soils, where maximum accumulation has occurred, there are also high concentrations of sodium and magnesium. Magnesium and sodium, which are the two most mobile cations, generally display slight to very pronounced increases with depth. Potassium and phosphorus display a decrease with increasing depth in most profiles.

The cation exchange capacity of both A and B horizons generally is greater than 200 me/kg soil and even exceeds 500 me/kg soil in some of the vertic horizons (Venter 1990), compared with 20 to 150 me/kg for granite soils. These high values indicate the dominance of 2:1 layer silicates (mostly smectite minerals) in the black expansive soils. The intermediate to low values that characterize the red paraduplex soils on the crest indicate a mixture of 1:1 (probably kaolinite) and 2:1 lattice clays. The base saturation of the majority of soils on basalt exceeds 80 percent (Venter 1990), and many A and B horizons that are derived from olivine-rich basalt or occur in bottomlands are calcareous.

A Patch Hierarchy of the Physical Template

The first intensive soil mapping surveys in Kruger (Harmse and van Wyk 1972; Harmse et al. 1974) generated detailed soil maps. They were found to be very useful when dealing with local issues but confusing when used to plan management for Kruger as a whole. Therefore, Venter (1982) and Gertenbach (1983) consolidated much of this information. Gertenbach (1983) delineated 35 landscapes in a system that has been used widely for management planning. Venter

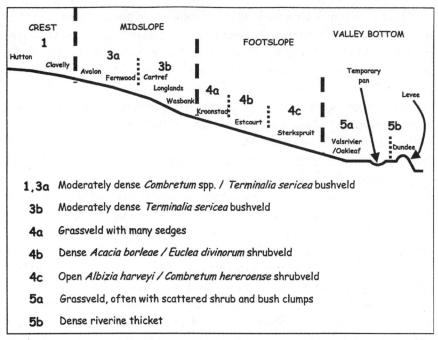

FIGURE 5.11. A typical hillslope profile on granite in the Randspruit land type in the southern part of Kruger, illustrating how land units can be further subdivided in land elements (after Venter 1990). Soil forms are according to the classification system for South Africa (Soil Classification Working Group 1991).

(1990) devised a four-level (four-scale) land classification system. Only two of these scales (land systems and land types) have been used to any extent, and it has proved more cost-effective to map small areas in detail at appropriate scales, when the need arises, than to map the whole park. Similar difficulties with scale were experienced in the Rivers Research Programme (Chapter 9, this volume).

Geomorphologic processes operate at different spatial and temporal scales, and they formed the basis for the present patch hierarchy classification (Figure 5.12, Table 5.9). The process of scarp retreat, for example, is noticeable at a regional scale (Drakensberg Great Escarpment of more than 1,000 m height is slowly retreating from the east coast), down to gullies or even small mini-scarps of 20 cm in height that progress quickly. Continental movements relative to the sea level initiated the Great Escarpment, whereas a road, game trail or trampled area may initiate a gully or mini-scarp. The terrain morphological units (TMUs; crest, scarp, midslope, footslope, valley bottom; Figure 5.10) that are used in the South African context to describe landscape morphology and development (Soil Classification Working Group 1991) are present at both of the above-mentioned spatial scales. Here we will use them as a reference for a patch hierarchy conceptual model for land, although all the units may not always be present.

This land class hierarchy model was developed for the lowveld landscape (Figure 5.12, Table 5.9), in the same way a patch hierarchy classification was developed for the Sabie River by Rogers and Bestbier (1997). This process helps to focus attention on the appropriate scales and processes at which particular ecological studies should be conducted or management actions taken, thereby promoting understanding of links between abiotic and biotic facets of ecosystems (Chapter 2, this volume). Heterogeneity in Kruger therefore can be described at different scales because it is often a direct consequence of the influence of either soil differences (patches linked to the geological template and the geomorphic development of the landscape) or climatic perturbations.

The patch hierarchy for land classes is a helpful tool for understanding the physical template and its influence on the biotic facets of an area the size of Kruger. However, it is necessary to discuss the abiotic aspects and their links to other ecosystem drivers.

Influence of the Abiotic Template on Savanna Heterogeneity

Soils exert an influence on terrestrial plants in three main ways: soil affects plants through its ability to absorb, store, and thereby provide water; soil provides plants with the nutrients needed for normal growth and metabolic processes; and physical or chemical conditions of soils may inhibit the penetration of plant roots and the volume of soils they use. These soil characteristics greatly influence the species composition and structure of plant communities and are therefore important factors in generating and maintaining heterogeneity. In terrestrial ecosystems with semiarid climates, soil and climate are major causes of heterogeneity.

Geology, soil, and fire interact to cause differences in vegetation structure and composition on granites and basalts (Figures 5.13–5.15). Grass production is significantly higher on the basaltic soils of the Satara area than on granitic soils of the Skukuza area (Trollope and Potgieter 1986), resulting in higher-intensity fires on basaltic soils. The higher grass biomass on clayey basaltic soils than on sandy granitic soils is a consequence of much higher production and differential species composition. The clayey basaltic soils usually are dominated by C_4, summer-growing perennial bunchgrasses such as red grass (*Themeda triandra*), stinking grass (*Bothriocloa radicans*), and small buffalo grass (*Panicum coloratum*), which often grow to 50 cm or higher. Wiry grasses, often annual pioneers, that produce low bulk dominate the sandy granitic soils. Dominant species include herringbone grass (*Pogonarthria squarrosa*), sand quick (*Schmidtia pappophoroides*), curly leaf (*Eragrostis rigidior*), blue-seed grass (*Tricholaena monachne*), small rolling grass (*Trichoneura grandiglumis*), and *Aristida* spp., which seldom grow to 50 cm.

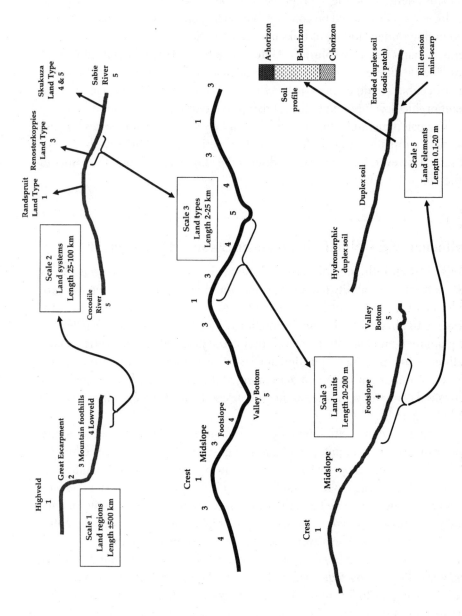

FIGURE 5.12. A proposal for a patch hierarchy for Kruger based on land classes; read in conjunction with Table 5.9 (after Rogers and Bestbier 1997).

TABLE 5.9

A proposed hierarchy of patches for the lowveld and Kruger.

	LAND REGION	LAND SYSTEM	PATCH SCALES LAND TYPE	LAND UNIT	LAND ELEMENT
Approximate map scale	Smaller than 1:1,000,000	1:1,000,000	1:250,000	1:30,000	1:10,000
Land classification scale	Major geographic region with similar macroclimate, containing one or more land systems	One or more land types with similar geology, geomorphology, and mesoclimate	Recurring soil pattern associated with land units (e.g., along catenas)	Association of soil types with similar properties	Single soil type or other feature such as sodic patch
Floristic classification scale	Biomes	Vegetation types	Vegetation communities	Vegetation associations	Taxa (species)
Drainage scale	One or more major river basins or parts thereof	Sections of major rivers (5th- to 6th-order catchments)	Low-order (2nd–3rd) streams	Overland flow, gullies, or low-order (1st–2nd) streams	Overland flow, rills, or mini-scarps onto sodic patches if present
Science and management application scale	Catchment studies, neighbor relations, transfrontier parks	Catchment studies, studies on wide-ranging animals, broad-scaled management planning	Studies on focused management planning, territorial animals	Autecological studies	Sodic patch studies, small animal studies
Possible terrain morphological units					
1: Crest	Highveld	Randspruit land type	Crests of hillslope (hilltop)	Land surface outside of gully	Land surface outside of rill or mini-scarp
2: Scarp	Great escarpment	—	—	Gully sides	Rill sides or mini-scarp
3: Midslope	Foothills	Renosterkoppies land type	Midslope of hillslope (seepline)	Rugged area at gully foot	Rugged area at rill or mini-scarp foot
4: Footslope	Lowveld	Skukuza land type	Footslope of hillslope (bottomland)	Gully floor	Sodic patch floor
5: Valley Bottom	Coastal plain	Sabie River valley	Valley bottom (stream channel and flanking alluvium)	Stream channel	Stream channel

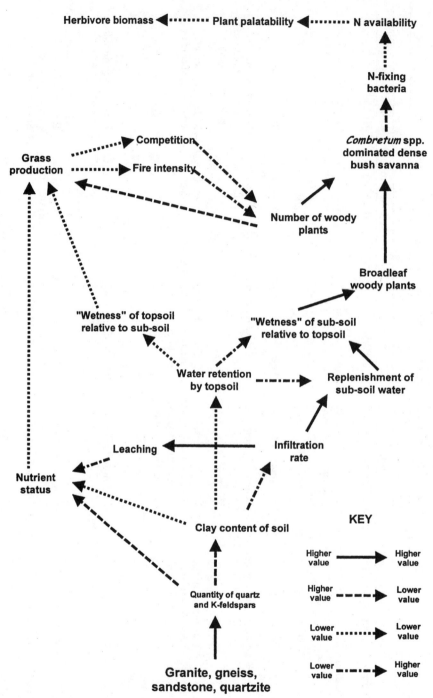

FIGURE 5.13. Ecological links on granitic sandy soils in the southern part of Kruger.

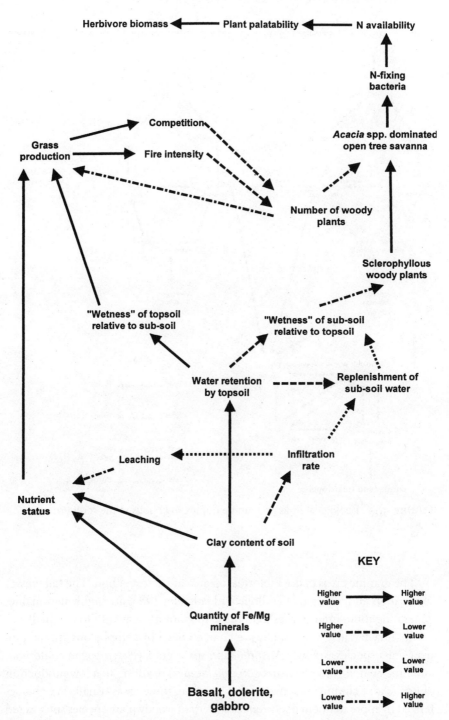

FIGURE 5.14. Ecological links on basaltic clay soils in southern Kruger.

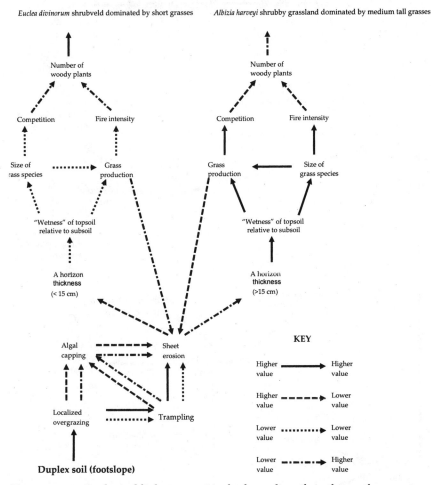

FIGURE 5.15. Ecological links on granitic duplex sodic soils in the southern part of Kruger.

The granitic areas of the Pretoriuskop area are an exception. The tall grasses in this high rainfall region (Trollope and Potgieter 1986) are highly flammable, generating more intense fires than those in granitic areas with low rainfall.

Shallow basaltic soils usually are characterized by xerophylous annual pioneer grass species, such as *Aristida* spp., spear grass (*Heteropogon contortus*), and nine-awned grass (*Enneapogon cenchroides*) resulting in a low production of forage and combustible material. Therefore, these areas usually are characterized by more frequent occurrences of certain woody plant species such as red bushwillow (*Combretum apiculatum*), flaky thorn (*Acacia exuvialis*), lowveld cluster-leaf (*Terminalia prunioides*), and white raisin (*Grewia bicolour*) than in adjacent areas with deeper soils. These phenomena demonstrate the importance of fire in maintaining certain structural features of the vegetation in areas with clayey soils.

The thickness of the A horizon usually is the critical factor determining vegetation composition and structure in duplex soils (Figure 5.15). The critical thickness of these A horizons is 15–20 cm. The species composition and physiognomy are very different on either side of this threshold. Where A horizons of duplex soils consist of deep apedal sand or loam and sodium, pH, and electrical conductivity values are low, a dense grass cover is established, fires are more intense, and woody plant stature is kept low. However, when A horizons are thin, grasses are exposed not only to a very dry soil climate but also to the unfavorable chemical conditions of B horizons. Grass species that are common on these duplex soils are short, creeping lawn grasses with broad leaves such as curly-leaved dropseed (*Sporobolus nitens*), pan dropseed (*S. ioclados*), carrot-seed grass (*Tragus berteronianus*), and common crowfeet (*Dactyloctenium aegyptium*). The low growth form and biomass of these grasses lead to infrequent low-intensity fires. Therefore, woody plant species that grow in sodic duplex soils (magic guarri [*Euclea divinorum*], tamboti [*S. africana*], Delagoa thorn [*Acacia welwitschii*], and horned thorn [*A. grandicornuta*]) usually form dense stands.

The harsh conditions of B horizons of sodic duplex soils are illustrated by the fact that several plant species that occur frequently on these soils also occasionally occur on rock outcrops and koppies (Gertenbach 1983). Examples include jacket-plus (*Pappea capensis*), tamboti (*S. africana*), lowveld cluster-leaf (*Terminalia prunioides*), Delagoa thorn (*Acacia welwitschii*), and the impala lily (*Adenium multiflora*). The shrubs *Acacia borleae* and *Euclea divinorum* often occur together on either duplex soils (southern region) or vertic soils with high values for sodium or electrical conductivity. Magic guarri (*E. divinorum*) also occurs on other soils (paraduplex and melanic clay) with high sodium and electrical conductivity values.

In the northern areas mopane (*Colophospermum mopane*) is the dominant woody plant species on soils with >15 percent clay. Van Rooyen (1978), Gertenbach and Potgieter (1979), Fraser et al. (1987), and Gertenbach (1983, 1987) have described the vegetation and soil types in some of these areas. The limited occurrence of *C. mopane* in the northern regions may be related to the dry climate of the area. However, because *C. mopane* also occurs in other parts of southern Africa where the rainfall exceeds 500 mm/year, the high evaporation rate of the northern regions probably also plays an important role.

Conclusion

This chapter has emphasized heterogeneity of the physical template as substrate for the interactive agent, controller, and responder processes (Chapter 2, this volume) that ultimately manifest themselves in the overall heterogeneity of the Kruger ecosystem. The dynamics of this complex patch mosaic and the causal mechanisms are considered in varying detail in subsequent chapters. However, so much research in Kruger has focused on the biotic interactions, especially those involving large herbivores and fire, that much remains to be learned about the role of the physical template as substrate for heterogeneity at multiple scales. We examined a range of correlations between template characteristics, vegetation, and animal distribution patterns that should form good hypotheses for further research. The highly heterogeneous nature of the template provides excellent opportunities for exploring the relative roles of abiotic (bottom-up) and biotic (top-down) controls of heterogeneity, which Robert Scholes et al. (Chapter 11, this volume) suggest is an important new frontier in savanna ecology.

The conservation and management of national parks is a task of great responsibility and often involves compromise between conflicting interests. Managers must find the golden mean between conserving ecological heterogeneity and wilderness qualities and providing for human benefits (e.g., ecotourism) (Chapter 4, this volume). Consequently, South African National Parks has long realized that management and development must have a sound knowledge base and that management strategies should be revised as that base expands. This chapter has emphasized the role of the abiotic template in generating heterogeneity and the complexity of abiotic-biotic relationships. The philosophy, management options, and research proposals we make are relevant to both Kruger, other parks under the auspices of South African National Parks, and other parks in Africa.

References

Barton, J. H., Jr., J. W. Bristow, and F. J. Venter. 1986. A summary of the Precambrian granitoid rocks of the Kruger National Park. *Koedoe* 29:39–44.

Bristow, J. W. 1986. An overview of the Soutpansberg sedimentary and volcanic rocks. *Koedoe* 29:59–67.

Bristow, J. W., and F. J. Venter. 1986. Notes on the Permian to recent geology of the Kruger National Park. *Koedoe* 29:85–104.

Chappel, C. 1992. *The ecology of sodic sites in the Eastern Transvaal lowveld.* Unpublished M.Sc. thesis, University of the Witwatersrand, Johannesburg, South Africa.

Dixon, M. J. 1977. *Proposed mathematical model for the estimation of the aerial properties of high-intensity, short duration storms.* Report TR 78. Department of Water Affairs, Pretoria, South Africa.

du Toit, A. L. 1954. The geology of South Africa. 3rd edition. Oliver and Boyd, Edinburgh, UK.

Dye, P. J. 1977. *Vegetation-environment relationships on Rhodesia sodic soils.* Unpublished M.Sc. thesis, University of the Witwatersrand, Johannesburg, South Africa.

Fraser, S. W., T. H. van Rooyen, and E. Verster. 1987. Soil-plant relationships in the central Kruger National Park. *Koedoe* 30:19–34.

Frick, C. 1986. The Phalaborwa syenite intrusions along the west-central boundary of the Kruger National Park. *Koedoe* 29:45–58.

Gertenbach, W. P. D. 1980. Rainfall patterns in the Kruger National Park. *Koedoe* 23:35–44.

Gertenbach, W. P. D. 1983. Landscapes of the Kruger National Park. *Koedoe* 26:9–121.

Gertenbach, W. P. D. 1987. *'n Ekologiese studie van die suidelikste mopanieveld in die Nasionale Krugerwildtuin.* Unpublished D.Sc. thesis, University of Pretoria, Pretoria, South Africa.

Gertenbach, W. P. D., and A. L. F. Potgieter. 1979. Veldbrandnavorsing in die struikmopanieveld van die Nasionale Krugerwildtuin. *Koedoe* 22:1–28.

Harmse, H. J. M. Von, and P. van Wyk. 1972. *Verkenningsgrondkaart van die Suidelike Distrik van die Nasionale Krugerwildtuin.* Unpublished map. Potchefstroom University, Potchefstroom, South Africa.

Harmse, H. J. M. Von, P. van Wyk, and W. P. D. Gertenbach. 1974. *Verkenningsgrondkaart van die noordelike gedeelte van die Nasionale Krugerwildtuin.* Unpublished map. Potchefstroom University, Potchefstroom, South Africa.

Huntley, B. J. 1982. Southern African savannas. Pages 101–119 in B. J. Huntley and B. H. Walker (eds.), *Ecology of tropical savannas.* Ecological Studies 42. Berlin: Springer-Verlag.

Jenny, H. 1941. *Factors of soil formation.* New York: McGraw-Hill.

King, L. C. 1967. *South African scenery.* 3rd edition. Edinburgh: Oliver and Boyd.

King, L. C. 1978. The geomorphology of central and southern Africa. In M. J. A. Werger and A. C. van Bruggen (eds.), *Biogeography and ecology of southern Africa.* The Hague: Dr. W. Junk.

Kruger, G. P. 1983. *Terrain morphological map of southern Africa.* Pretoria: Department of Agriculture.

Low, A. B., and A. G. Rebelo. 1996. *Vegetation of South Africa, Lesotho and Swaziland.* Pretoria: Department of Environmental Affairs and Tourism.

Menaut, J. C. 1983. The vegetation of African savannas. Pages 109–149 in F. Bourliére (ed.), *Ecosystems of the world 13: tropical savannas.* Amsterdam: Elsevier.

Moon, B. P., and G. F. Dardis. 1988. *The geomorphology of southern Africa.* Johannesburg: Southern Book Publishers.

Munnik, M. C., E. Verster, and T. H. van Rooyen. 1990. Spatial pattern and variability of soil and hill slope properties in a granitic landscape. 1. Pretoriuskop area. *South African Journal of Plant and Soil* 7(2):121–130.

Olbrich, B. H. 1984. *A study on the determinants on seepline grassland width in the Eastern Transvaal lowveld.* Unpublished B.Sc. honors project, University of the Witwatersrand, Johannesburg, South Africa.

Partridge, T. C., and R. R. Maud. 1987. Geomorphic evolution of southern Africa since the Mesozoic. *South African Journal of Geology* 90:179–208.

Preston-Whyte, R. A., and P. D. Tyson. 1988. *The atmosphere and weather of southern Africa.* Cape Town: Oxford University Press.

Purves, W. D. 1973. *The stability of the clay fraction of light-textured soils derived from granite.* Unpublished report. Soil Science Society of South Africa.

Rogers, K. H., and R. Bestbier 1997. *Development of a protocol for the definition of the desired state of riverine systems in South Africa.* Pretoria: Department of Environment Affairs and Tourism.

Rutherford, M. C. 1997. Categorization of biomes. Pages 91–98 in R. M. Cowling, D. M. Richardson, and S. M. Pierce (eds.), *Vegetation of southern Africa.* Cambridge, UK: Cambridge University Press.

Scholes, R. J. 1986. *A guide to bush clearing in the Eastern Transvaal lowveld.* 2nd edition. Johannesburg: Resource Ecology Group–Botany, University of the Witwatersrand.

Scholes, R. J., N. Gureja, M. Giannecchinni, D. Dovie, B. Wilson, N. Davidson, K. Piggott, C. McLoughlin, K. van der Velde, A. Freeman, S. Bradley, R. Smart, and S. Ndala. 2001. The environment and vegetation of the flux measurement site near Skukuza, Kruger National Park. *Koedoe* 44:73–83.

Scholes, R. J., and D. O. Hall. 1996. The carbon budget of tropical savannas, woodlands and grasslands. In J. M. Melillo and A. Breymeyer (eds.), *Global change: carbon cycle in coniferous forests and grasslands.* New York: Wiley.

Scholes, R. J., and B. H. Walker. 1993. *An African savanna: synthesis of the Nylsvley study.* Cambridge, UK: Cambridge University Press.

Schulze, R. E. 1997. *South African atlas of agrohydrology and climatology.* Water Research Commission Report TT82/96. Water Research Commission, Pretoria, South Africa.

Schutte, I. C. 1986. The general geology of the Kruger National Park. *Koedoe* 29:13–37.

Soil Classification Working Group. 1991. *Soil classification: a taxonomic system for South Africa.* Memoirs on the Natural Resources of South Africa, 15. Department of Agricultural Development, Pretoria, South Africa.

Strahler, A. M. 1952. Hypsometric (area-altitude) analysis of erosional topography. *Geological Society of America Bulletin* 63:1117–1142.

Sweeney, R. J. 1986. Geology of the Sabie River Basalt Formation in the southern Kruger National Park. *Koedoe* 29:105–116.

Thrash, I., P. J. Nel, G. K. Theron, and J. du P. Bothma. 1991. The impact of the provision of water for game on the woody vegetation around a dam in the Kruger National Park. *Koedoe* 34:131–148.

Tinley, K. L. 1979. *Management ecology of the Sabie-Sand Wildtuin.* Report commissioned by the Sabie Sand Reserve, South Africa.

Trollope, W. S. W., and A. L. F. Potgieter. 1986. Estimating grass fuel loads with a disc pasture meter in the Kruger National Park. *Journal of the Grassland Society of Southern Africa* 3:148–152.

Truswell, J. F. 1977. *The geological evolution of southern Africa.* Cape Town: Purnell and Sons.

Tyson, P. D. 1985. *Climatic change and variability in southern Africa.* Cape Town: Oxford University Press.

Tyson, P. D., and T. C. Partridge. 2000. Evolution of Cenozoic climates. Pages 371–387 in T. C. Partridge and R. R. Maud (eds.), *The Cenozoic of southern Africa.* Oxford Monographs on Geology and Geophysics 40. Cape Town: Oxford University Press.

van der Eyk, J. J., C. N. Macvicar, and J. M. de Villiers. 1969. *Soils of the Tugela Basin.* Pietermaritzburg, South Africa: Town and Regional Planning Commission.

van Rooyen, N. 1978. *'n Ekologiese studie van die plantgemeenskappe van die Punda Milia-Pafuri-Wambiyagebied in die Nasionale Krugerwildtuin.* Unpublished M.Sc. thesis, University of Pretoria, Pretoria, South Africa.

Venter, F. J. 1982. *Schematic soil map of the Kruger National Park.* Unpublished 1:500,000 map. National Parks Board, Skukuza, South Africa.

Venter, F. J. 1986. Soil patterns associated with the major geological units of the Kruger National Park. *Koedoe* 29:125–138.

Venter, F. J. 1990. *A classification of land for management planning in the Kruger National Park*. Unpublished Ph.D. thesis, University of South Africa, Pretoria, South Africa.

Venter, F. J., and J. W. Bristow. 1986. An account of the geomorphology and drainage of the Kruger National Park. *Koedoe* 29:117–124.

Venter, F. J., and W. P. D. Gertenbach. 1986. A cursory review of the climate and vegetation of the Kruger National Park. *Koedoe* 29:139–148.

Walraven, F. 1986. The Timbavati gabbro of the Kruger National Park. *Koedoe* 29:69–84.

Weather Bureau. 1986. *Climate of South Africa*. Weather Bureau Publication 40. Department of Environment Affairs, Pretoria, South Africa.

Webber, N. W. 1979. *The effects of fire on soil/plant relationships in the southern part of the Kruger National Park. A study in soil geography.* Unpublished M.Sc. thesis, University of South Africa, Pretoria, South Africa.

White, F. 1983. *The vegetation of Africa*. Natural Resources Research XX. Paris: UNESCO.

Zambatis, N. 2001. *Rainfall data for the Kruger National Park*. Unpublished report. Scientific Services, Kruger National Park Skukuza, South Africa.

Zambatis, N. 2002. Checklist of species of the Kruger National Park, Skukuza Herbarium, Skukuza.

Zucchini, W., and P. T. Adamson. 1984. *The occurrence and severity of droughts in South Africa*. Water Research Commission Report 91/1/84. Water Research Commission, Pretoria, South Africa.

Chapter 6

Biogeochemistry: The Cycling of Elements

MARY C. SCHOLES, ROBERT J. SCHOLES,
LUANNE B. OTTER, AND ANDREW J. WOGHIREN

The existence of fertile and infertile habitats in southern African savannas, as a result of edaphic factors, has long been recognized. These habitats can easily be recognized in the Kruger, where the basaltic geological substrate in the eastern part of the park gives rise to fertile soils and the granitic parent material found in the west gives rise to nutrient-poor, sandy soils. These habitats can also be found at the landscape scale, where granitic catenas exhibit a sandy, nutrient-poor crest and a nutrient-rich valley resulting from the accumulation of clays at the foot of the catena. The nutrient-poor areas are dominated by broad-leaved woody vegetation, protected chemically from herbivory, with the nutrient-rich areas exhibiting fine-leaved acacia vegetation, protected by physical defenses.

The understanding of the functioning of these savanna types is made more complex when one introduces fire, which burns with different intensity at different times of the year and under a range of weather conditions (Chapter 7, this volume). Hundreds of different gases and aerosol particles are produced when savanna vegetation burns. The relative proportions of these compounds vary in response to both fire intensity and plant species composition, which in turn is differentially distributed across the savanna landscape.

The role of biotic factors in creating spatial pattern and the implications for community and ecosystem dynamics is less well understood than the role of abiotic factors (i.e., topography and precipitation) in creating spatial pattern (Augustine and Frank 2001). Large mammalian herbivores are an important dictating element of savannas that not only respond to spatial heterogeneity in plant communities but also affect spatial heterogeneity in the landscape. The direct effects of wildlife herbivory on ecosystems are well studied in ecosystems such as Yellowstone (Frank and Groffman 1998) and the Serengeti (Caughley 1982; McNaughton 1985), where it has been shown that ungulates can modify conditions for themselves and for other organisms above and below ground.

Knowledge generated in other ecosystems, even other savannas, cannot simply be transferred to Kruger because very different biotic and abiotic factors are operative. This chapter emphasizes these differences and their relationship to the nutrient-poor status of the soils, the regular fires, and the prevalence of large herbivores.

A biogeochemistry perspective focuses on the fluxes of matter between pools or compartments (Figure 6.1). The movement of chemical substances between the compartments (e.g., atmosphere and terrestrial biosphere) and the transformations are called biogeochemical cycles. This chapter describes how the spatial heterogeneity of the fertile and infertile areas influences the size of the elemental pools and the transfer of elements between the terrestrial and atmospheric components. Seasonal rainfall patterns, their spatial and temporal variability, and the amount are primary determinants of the rates of biological transfers. Nutrient-poor and nutrient-rich southern African savannas are given context by reference to the biogeochemistry of other ecosystems. Limited data are available for Kruger, and the authors have included data from the Nylsvley Nature Reserve, which was the principal study site for the Savanna Biome Programme from the mid-1970s to mid-1980s and is located 200 km north of Johannesburg.

Elemental Pools in Savannas

The nutrient budgets of southern African savannas, like those of other ecosystems, have cycles within cycles, operating at different scales and rates and intersecting at certain points. For example, nitrogen cycling occurs within minutes when it happens through fire-enhanced mineralization, hours when through digestion in an animal gut, and years when through microbial decomposition in the litter layer or soil. Similarly, the difference between nutrient-rich and nutrient-poor savannas lies not only in the total quantity of nutrients present but also in the temporal heterogeneity at which they are turned over (Scholes and Walker 1993).

Plant Pools

Total plant nitrogen pools range from 182 to 300 kg $N \cdot ha^{-1}$ for the South African ecosystems; no comparable data are available for other ecosystems. Total plant phosphorus pools are estimated from Nylsvley to be 2 kg $P \cdot ha^{-1}$, with less than 1 percent of the system phosphorus being in the vegetation. No data are available for total plant phosphorus pools from Kruger, but the nutrient-poor savannas are expected to be similar to the Nylslvey estimates. The nutrient-rich areas on the basalts are expected to have much higher total plant phosphorus pools. These levels of high phosphorus fertility are able to sustain much higher animal numbers than the sandier soils.

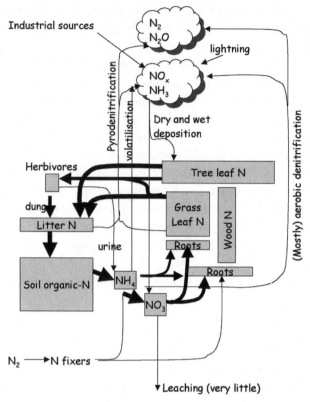

FIGURE 6.1. A simplified schematic representation of the Kruger terrestrial ecosystem, showing the main pools (boxes) and fluxes (arrows) for nitrogen.

Soil Pools

Total soil nitrogen values for southern African savannas range from 3,060 to 4,635 kg N·ha^{-1} (Scholes and Walker 1993; Woghiren 2002). Relative to other ecosystems, even the southern African nutrient-rich areas are very low in soil nitrogen, phosphorus, and carbon. Microbial carbon accounts for about 0.2 percent of the total, and the labile fraction is about 10 percent. This labile fraction was markedly affected by the frequency of burning, with annual burns significantly reducing the size of this pool (Otter 1992).

Exchanges between Terrestrial Pools

Equilibrium levels of soil organic matter are determined by the balance of production of biomass, stabilization of detritus, and mineralization of organic materials (including decomposition, herbivory, and fire). Nitrogen mineralization plays a significant role in transforming nitrogen in soils. The major factors that influence the rate of decomposition of litter, and thus the mineralization and immobilization balance of nitrogen, are environmental parameters (soil moisture, temperature, aeration, and pH) and the chemistry of the litter, particularly its nitrogen, lignin, and phenolic content (Swift et al. 1979). Because nitrogen often is the most limiting nutrient to terrestrial plants, it follows that differences in the rates of mineralization, immobilization, and nitrification can have profound effects on primary productivity.

Microbial immobilization of nitrogen is a transient phenomenon in savannas because of the short life span of microbes and their sensitivity to water stress. In temperate forests, large amounts of nitrogen may be immobilized early in the growing season, but this declines as the season progresses (Aber and Melillo 1979). In contrast, seasonal measurements of nitrogen mineralization rates in Kruger showed net mineralization in the early summer (September to December), and immobilization at all other times of the year (Table 6.1). The granitic soils have a smaller pool of nitrogen with a high turnover rate, whereas the basaltic soils have a much greater pool of nitrogen with an overall slower turnover rate (C. C. Grant 2000–2001). Mineralization rates peaked in the spring, with values of $0.2 \mu g \ N \cdot g^{-1}$ soil·day^{-1}, and immobilization rates peaked in the autumn at $-0.06 \mu g \ N \cdot g^{-1}$ soil·day^{-1} (Figure 6.2; Woghiren 2002). Immobilization rates were found to be much greater on the basaltic soils ($-0.09 \ \mu g \ N \cdot g^{-1}$ soil·day^{-1}) than on the granitic soils ($-0.0003 \ \mu g \ N \cdot g^{-1}$ soil·day^{-1}).

Ammonium is the dominant inorganic nitrogen ion in all the sites measured in Kruger, in contrast to Nylsvley, where nitrate was the dominant ion (Scholes and Walker 1993; Woghiren 2002; C. C. Grant 2000–2001). The dominance of ammonium could result in less leaching from the system and a lower potential for nitrogen trace gas emissions. The most striking feature of the nitrogen cycle is the pulsing of the mineralization rates immediately after rainfall events. Very slow rates of net immobilization and mineralization take place throughout the year, but the advent of the spring rains leads to an immediate response by the soil microbes, with short-lived bursts of mineralization at high rates. Nitrogen mineralization rates from a number of other ecosystems again highlight the low rates measured in Kruger (Table 6.1). Impacts of herbivores on the biogeochemistry of the ecosystem are profound. Ungulates affect the rate of nutrient cycling and thus nutrient availability to plants in different ways. Herbivores often increase (Holland and Detling 1990) but sometimes decrease (Ritchie et al. 1998) the rate of nutrient cycling (Table 6.1), and a range of hypotheses have been presented in the literature to explain the observed patterns.

Annual nitrogen mineralization has been estimated using a combination of field data and modeling that calculates the soil moisture and temperature conditions throughout the year. Estimates for the Skukuza flux site in Kruger are 58 and 78 kg $N \cdot ha^{-1} \cdot year^{-1}$ for the broad-leaved and fine-leaved sites, respec-

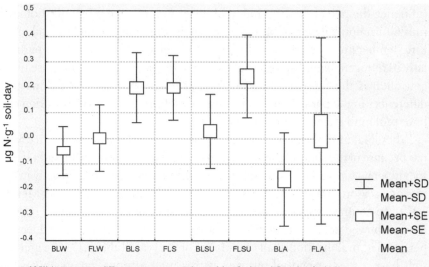

FIGURE 6.2. Seasonal variation of nitrogen mineralization on both sites using all data points ($n = 200$). BLA = broad-leaved autumn; BLS = broad-leaved spring; BLSU = broad-leaved summer; BLW = broad-leaved winter; FLA = fine-leaved autumn; FLS = fine-leaved spring; FLSU = fine-leaved summer; FLW = fine-leaved winter.

tively. These may be overestimated because of the scarcity of data across the seasons. Estimates for a nutrient-poor savanna at Nylsvley were 35 kg $N \cdot ha^{-1} \cdot year^{-1}$ (Scholes and Walker 1993).

Exchange between Land and Atmosphere

Several of the key element cycles in the Kruger ecosystem contain a loop through the atmosphere. The physical process of exchange between land and atmosphere is linked to the exchange of radiant energy and "momentum" (wind energy). The exchange of water determines the duration of the period for which plant and microbial activity can occur on the land.

Exhanges of Energy, Momentum, and Water

The semiarid tropics are among the sunniest ecosystems in the world: close enough to the tropics to get high insolation year-round but not so close as to have a permanent blanket of cloud. As a result, the annual shortwave radiation at Skukuza is about 6,526 MJ (ranging from 21.27 $MJ \cdot m^{-2} \cdot day^{-1}$ in November to 14.17 $MJ \cdot m^{-2} \cdot day^{-1}$ in June). Given that the rainfall at Skukuza averages 547 mm,

TABLE 6.1

Soil and plant elemental pool sizes and nitrogen mineralization rates from a range of ecosystems. Work was conducted in the nutrient-poor Burkea savannas of Nylsvley (Scholes and Walker 1993), on the crest and valley sites of a catena in Kruger (Woghiren 2002), and on the basaltic and granitic soils of Kruger (Otter 1992; C. C. Grant, unpublished data 2000–2001). The work of Ruess and McNaughton (1987) was based on the fertile soils of the Serengeti grasslands; Sirotnak and Huntly (2000) in riparian areas in Yellowstone National Park (exclosure without voles); Holland and Detling (1990) in the grasslands of Wind Cave National Park (colonized and uncolonized with prairie dogs), South Dakota; Pastor et al. (1993) in the boreal forests of Isle Royale, Michigan (exclosures without moose); and Tracy and Frank (1998) in Yellowstone National Park (exclosure without elk, bison, and pronghorn antelope).

	NYLSVLEY (SCHOLES AND WALKER 1993)	KRUGER (WOGHIREN 2002[1]; OTTER 1992[2]; C. C. GRANT, UNPUBLISHED DATA 2000–2001[3])		OTHER ECOSYSTEMS (RUESS AND MCNAUGHTON 1987[1]; SIROTNAK AND HUNTLY 2000[2]; HOLLAND AND DETLING 1990[3]; PASTOR ET AL. 1993[4]; TRACY AND FRANK 1998[5])
	NUTRIENT-POOR	NUTRIENT-POOR	NUTRIENT-RICH	
Plant nitrogen (kgN ha^{-1})	280	300[1]	182[1]	
Plant phosphorus (kgP ha^{-1})	2	NA	NA	
Soil nitrogen (kgN ha^{-1})	3,060	3,310[1]	4,635[1]	6,630–15,210[1] 6,630–9,750[3] 9,750–85,800[4] 3,510–37,050[5]
Soil phosphorus (kgP ha^{-1})	896	1,950–3,900[3]	1,560–5,070[3]	2,535–23,712[1]
Total soil carbon (%)	1.5	0.6[2]	2.2[2]	1.7–4.5[1] 5.2–34.9[4] 1.0–16.2[5]
Microbial biomass (mgC.kg^{-1})	100–1,780	20.0[2]	42.0[2]	470–3,160[1] 166–1,539[5]
Soil nitrogen mineralization rate (μgN.g^{-1}soil.day^{-1})	0.2–0.4	–0.15–0.2[1] 0.002–0.15[3] 0–0.24[1] –0.09 to –0.25[3]		0.05–3.17[1] –4.7–7.0[2] (presence of voles) 0.4–0.7[2] (control vole site and vole exclosure) 0.08–0.42[3] (colonized and uncolonized sites) 1.4–1.9[4] (control sites) 1.5–3.2[4] (exclosure) 0.162–0.282 (ungrazed – grazed sites)[5]

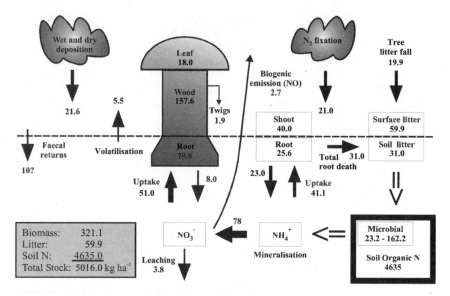

Fine-leafed savanna

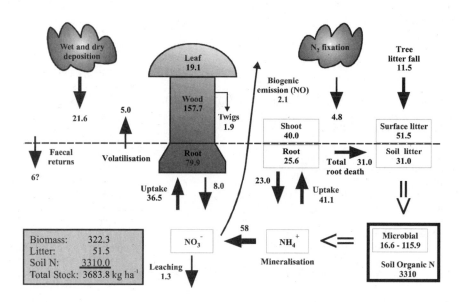

Broad-leafed savanna

FIGURE 6.3. Nitrogen cycles for the broad-leaved and fine-leaved savannas at the Skukuza flux site, showing the relative sizes of all pools and fluxes (kgN·ha^{-1}·year^{-1}). Arrow sizes are proportional to the size of each flux.

a maximum of 1,343 MJ·m^{-2} of the annual incoming energy can be converted to latent energy of evaporation. The shortwave albedo (fraction of incoming radiation that is reflected by the vegetated surface) is .13 in summer and .2 in winter (except briefly after fires, when it reaches .4 for a few days because of the blackened surface). Thus ~980 MJ of the incoming radiation is reflected. The daily mean surface temperature ranges from 27°C in summer to 16°C in winter, resulting in longwave radiation of about 364 MJ·m^{-2}. A tiny fraction of the incoming radiation (around 8 MJ·m^{-2}) is converted to chemical energy for plant production. The remaining energy (~3,830 MJ·m^{-2}, more than half of the budget) must be dissipated as sensible heat, or the convection of the warmed atmosphere. This is the ultimate driver behind the high temperatures in the lowveld and the development of a deep, turbulent mixed layer at the interface between the land and atmosphere.

The potential evaporation at Skukuza (i.e., the capacity of the net radiation to evaporate water and the atmosphere to absorb it) averages 5 mm·day^{-1}, ranging from 3.6 mm·day^{-1} in winter to 6.6 mm·day^{-1} in summer. Given that the median size of a rainfall event is less than 3 mm, and there are on average 65 rainy days in a year, of which 53 are in summer, the rainfall from one event often fully evaporates before the next event occurs. Most biological activity in plants and soils depends on water availability. It therefore occurs predominantly in summer, in intense pulses lasting a few days to a week or two, interrupted by low-activity periods of similar duration.

Emissions from Soils

Soils are an important source of NO_x (nitric oxide [NO] + nitrogen dioxide [NO_2]) and nitrous oxide (N_2O). Davidson and Kingerlee (1997) estimated the global NO budget to be 21 Tg N·year^{-1}, with tropical savannas and woodlands contributing 35 percent. Southern African savannas are estimated to produce 0.15×10^{-3} kg N·m^{-2}·year^{-1} as NO (Otter et al. 1999), whereas their contribution to the N_2O budget is uncertain because of the scarcity of data. However, nitrogen emissions from savannas are dominated by NO· N_2O emissions often are below the detection limit (>2 ng N·m^{-2}·s^{-1}) (Levine et al. 1996). Scholes et al. (1997) measured N_2O fluxes in a South African savanna that were on average 8 percent of the NO_x emissions.

A wide range of NO fluxes have been reported for African savanna systems (0.05–34 ng N·m^{-2}·s^{-1}) (Serca et al. 1998; Otter et al. 1999; Parsons et al. 1996; Scholes et al. 1997; Kirkman et al. 2001; Levine et al. 1996). The major controllers of NO fluxes are soil nitrogen, soil moisture, and soil temperature, so variation in these factors between sites causes emission rate variability. The first rains after a long dry season produce a pulse of NO. This can cause the NO flux and N_2O fluxes to increase up to 10-fold in savannas and woodlands (Meixner et al. 1997; Scholes et al. 1997). Fires in savannas also produce a 10-fold increase in NO emissions from the soil, with fluxes at Skukuza increasing to a maximum of 76.9 ng N·m^{-2}·s^{-1} within a day of burning (Levine et al. 1996). These pulsing effects decline after 3 to 4 days. Modeled annual estimates of NO emissions were 2.7 and 2.1 kg N·ha^{-1}·year^{-1} for the nutrient-rich and nutrient-poor savannas of Kruger, respectively (Figure 6.3). These values are not significantly different, with the temporal variability in the rainfall being the major driver and not the spatial variability in substrate quality and quantity (Woghiren 2002).

Emissions from Fires

The products of combustion of savanna vegetation include several hundred different gaseous compounds and a variety of particles (aerosols). The exact composition of the emissions depends primarily on the nature of the fuel and secondarily on the combustion conditions (smoldering versus flaming). The bulk of fires in the lowveld consume predominantly dry grass, with a small contribution of dead, fallen leaves and twigs of trees and shrubs, some live leaf material, and a lingering smoldering emission from dung and fallen logs (Shea et al. 1996). Therefore, the emission pattern from fires in Kruger is uniform. The average values given by Delmas et al. (1995) for African savannas are within the range measured for lowveld fires during the Southern African Fire-Atmosphere Research Initiative (SAFARI) 92 and SAFARI 2000 campaigns. Per kilogram of dry fuel consumed, the most important categories of emission are carbon dioxide (1,640 g; the mass is greater than the fuel because the oxygen in extracted from the air), carbon monoxide (65 g), hydrocarbons excluding methane (3.1 g), methane (2.4 g), and particles (10 g total, of which 7 g is carbon). The quantity of nitrogen-containing gases emitted depends on the nitrogen content of the fuel (which varies with site, species, and time of year) and the intensity of the fire. On average, around three-quarters of the nitrogen in the fuel is converted to gases (half as N_2, a quarter each NO_x and NH_3, and less than 1 percent as N_2O).

The absolute emissions from fires for a particular location are calculated by multiplying the mean fire frequency (1/return time, in years) by the mean fuel load and the aforementioned emission factors. For example, the flux site at Skukuza has experienced eight fires in 40 years (frequency = 0.2 year). This is higher than the Kruger average but not exceptional for the wetter, higher-production southeastern part of the park (van Wilgen et al. 2000). The fuel load averages around 330 g·m^{-2} (Trollope and Potgieter 1985; Shea et al. 1996). Thus the annualized mean emission of carbon through burning, assuming a 6-year mean return, is 25 g C·m^{-2}·year^{-1} and nitrogen is 0.3 g N·m^{-2}·year^{-1}. The mean methane emission is 0.13 g CH_4·m^{-2}·year^{-1}, nonmethane hydrocarbon 0.16 g C·m^{-2}·year^{-1}, N_2 is 0.16 g·m^{-2}·year^{-1}, NH_3 is 0.1 g·m^{-2}·year^{-1}, and NO_x is

0.17 g·m^{-2}·year^{-1}. Sites located on granitic soils in drier areas of the park have emissions 25 to 50 percent of these values (because of less frequent fires and lower fuel loads), whereas sites on basalts have up to twice the nitrogen emissions (because nitrogen content of the fuel is higher) for a similar rainfall regime and about the same nitrogen emissions but lower carbon emissions on drier sites.

Note that the carbon emissions are approximately balanced by regrowth of the fuel in subsequent years; therefore, savanna burning at a return frequency similar to the historical record has no net effect on atmospheric CO_2 levels. At the scale of the entire lowveld landscape, a large fraction of the N emitted as NO_x and NH_3 returns to the ecosystem as wet and dry deposition, and a portion of the N_2 emitted is balanced by biological nitrogen fixation. At the scale of the southern African subcontinent, even in preindustrial times there must have been a net transfer of nitrogen from land to ocean, partly as a result of the frequent fires. At the scale of a burned patch (a few hectares to a hundred thousand hectares) the emission-deposition loop through the atmosphere for carbon and nitrogen is not necessarily balanced. Thus, landscape facets burned more frequently than the median decline in available system nitrogen and carbon, and landscape facets protected from fire accumulate nitrogen and carbon (Jones et al. 1990).

Emissions from Vegetation

Vegetation emits a wide variety of volatile organic compounds (VOCs), such as methane, isoprene, monoterpenes, ethylene, methylbutanol, methanol, and ethanol, into the atmosphere. Vegetation emissions are the source of 90 percent of the global volatile organic carbon budget (1,150 Tg C·year^{-1}) (Guenther et al. 1995). In terrestrial systems the most important VOCs are isoprene and monoterpenes. On a global basis 175–503 Tg C in the form of isoprene and 127–480 Tg C as monoterpenes are produced annually (Fall 1999). Southern African savannas have been estimated to produce 18–74 Tg C·year^{-1} (Otter et al. 2002). This is substantially larger than the anthropogenic (0.61 Tg C·year^{-1}) and pyrogenic (0.55 Tg C·year^{-1}) emissions in the region (Otter et al. 2001).

Similar to photosynthesis, isoprene emissions are controlled mainly by light and temperature (Lerdau et al. 1997). High emitters typically convert 1–2 percent of the fixed CO_2 to isoprene (Harley et al. 1999). Unlike isoprene, monoterpenes can be stored by the plant in specialized structures such as resin ducts and glandular trichomes, and emission rates are controlled by the size of the stored pool, volatilization, and diffusion processes. Recent studies indicate that there are a number of light-dependent monoterpene-emitting species, such as *Colophospermum mopane* (Table 6.2).

VOC emission rates and compounds are species-specific; there does not appear to be any clear phylogenetic basis for these emissions (Harley et al. 1999). Emission capacities for a variety of South African savanna species are

TABLE 6.2

Emission rates for some of the dominant woody species found in South African savannas.

FAMILY	SPECIES	ISOPRENE ($\mu g\ C \cdot g^{-1}\ soil \cdot hr^{-1}$)	MONOTERPENES ($\mu g\ C \cdot g^{-1} \cdot hr^{-1}$)
Anacardiaceae	*Lannea schweinfurthii*	0.1[a]	
	Rhus leptodictya	54	1.1
	Sclerocarya birrea	<0.5	<0.5
Apocynaceae	*Carissa edulis*	0	
Balanitaceae	*Balanites maughamii*	1.7[a]	
Caesalpinaceae	*Burkea africana*	51 (36–62)	<0.5
	Colophospermum mopane	<0.5	34 (16–52)
	Schotia brachypetala	0.1[a]	
Combretaceae	*Combretum apiculatum*	<0.5	<0.5
	Terminalia sericea	<0.5	1.3
Ebenaceae	*Diospyros mespiliformis*	Low	
	Euclea natalensis	0	
Euphorbiaceae	*Securinegia virosa*	81	4.7
	Spirostachys africanum	67.5[a]	
Mimosaceae	*Acacia nigrescens*	110	0.7
	Acacia nilotica	<0.5	<0.5
	Acacia tortilis	<0.5	8.8
	Dichrostachys cinerea	0.1[a]	
Papilionaceae	*Lonchocarpus capassa*	28.3[a]	
	Peltophorum africanum	0	
Rhamnaceae	*Ziziphus mucronata*	0.3[a]	
Tiliaceae	*Grewia bicolour*	0.1[a]	
	Grewia flavescens	<0.5	0.5
	Grewia hexamita	0.1[a]	

[a]Emission rates are measured on small potted plants and therefore may be slightly underestimated, but they are a good indication of whether a species is an emitter. Monoterpene emission rates were not measured for these species.

shown in Table 6.2. Further emission capacities are given in Guenther et al. (1996) and Otter et al. (2002). The landscape average isoprene emission capacity for savannas is estimated at 9 mg $C \cdot m^{-2} \cdot hr^{-1}$, and for monoterpenes it is 0.04–3 mg $C \cdot m^{-2} \cdot hr^{-1}$ (Otter et al. 2002). The vegetation around the tower site at Skukuza is characterized as a *Combretum apiculatum* and *Acacia nigrescens* savanna. *Combretum* species display very low emission rates of both isoprene and monoterpenes, with a landscape average emission capacity of less than 1 mg $C \cdot m^{-2} \cdot hr^{-1}$ (Guenther et al. 1996). *A. nigrescens* savannas are dominated by isoprene-emitting species and are estimated to have a landscape isoprene emission capacity of almost 9 mg $C \cdot m^{-2} \cdot hr^{-1}$. Because emissions depend on light, temperature, and foliar density, there is a seasonal variation in VOC emissions, and this is more pronounced for isoprene. In South African savannas emissions are less than 5 mg $C \cdot m^{-2} \cdot day^{-1}$ in winter, with peak emissions occurring from December to March.

TABLE 6.3

Approximate emissions of carbon dioxide, methane, and ammonia resulting from the principal large herbivores in Kruger, expressed on a $m^{-2}\cdot year^{-1}$ basis. Methane emission was assumed to be 6 percent of the gross energy budget of the herbivores and nitrogen excretion in dung to be 8 g.kg dry matter consumed (IPCC*). Urine N was assumed to be equal to the difference between N intake and N excreted as dung plus N needed for growth. All urine N was assumed to be converted to NH_3 and no part of the dung N. N_2O production was assumed to consume 2 percent of total excreted N (IPCC).

SPECIES	MASS kg	NUMBER	FORAGE QUALITY %IVD**	%N	CO_2 $g\cdot m^{-2}\cdot year^{-1}$	CH_4 $g\cdot m^{-2}\cdot year^{-1}$	NH_3 $mg\cdot m^{-2}\cdot year^{-1}$	N_2O $mg\cdot m^{-2}\cdot year^{-1}$
Elephant	1,850	8,000	0.4	1	7.27	0.04	11.45	1.48
Buffalo	530	22,000	0.45	1.5	6.21	0.04	34.29	1.90
Zebra	290	32,000	0.45	1	5.45	0.03	8.58	1.11
Warthog	65	100,000	0.5	1.6	4.36	0.03	27.50	1.42
Impala	45	150,000	0.6	1.6	4.00	0.03	25.24	1.31
White rhino	1,500	2,000	0.4	1.3	1.52	0.01	6.00	0.40
Hippo	1,200	2,400	0.4	1.3	1.52	0.01	5.97	0.40
Giraffe	850	4,600	0.6	1.6	1.45	0.01	9.14	0.47
Wildebeest	115	14,000	0.55	1.6	0.90	0.01	5.65	0.29
Kudu	120	4,000	0.6	1.6	0.24	0.00	1.53	0.08
Waterbuck	120	2,000	0.45	1.2	0.16	0.00	0.51	0.04
Total					33.08	0.21	135.86	8.90

* Intergovernmental Panel for Climate Change
** *In vitro* digestion

Emissions from Animals

The principal gaseous emissions resulting from herbivory are the CO_2 respired by the animal, methane resulting from enteric fermentation, and ammonia volatilized from the urine (Table 6.3). Methane is also emitted if anaerobic decomposition continues in the dung, but because most wildlife dung pellets are small, dry, and widely distributed, the emissions from this source are considered small.

Wet and Dry Deposition of Nitrogen and Sulfur

Atmospheric deposition may be considered to occur via three pathways: wet deposition in rain, dry deposition as dust, and deposition from mist (Turner et al. 1995). The forms of nitrogen and sulfur may vary between and within the categories of dry and wet deposition, with ammonia usually accounting for 50 percent of dry nitrogen deposition and nitrate and ammonium contributing equal amounts in wet deposition (Hesterburg et al. 1996). South Africa generates its power from coal burning, and the majority of the power stations are located in the high-lying regions of Mpumalanga. Kruger is situated downwind of these emission sources and is affected by deposition estimated to be 21.6 kg $N \cdot ha^{-1} \cdot year^{-1}$. Dry deposition is 15 kg $N \cdot ha^{-1} \cdot year^{-1}$, of which 8 kg $N \cdot ha^{-1} \cdot year^{-1}$ is ammonium. Nitrate (3.2 kg $N \cdot ha^{-1} \cdot year^{-1}$) and ammonium (3.4 kg $N \cdot ha^{-1} \cdot year^{-1}$) are present in equal quantities in the wet deposition. Previous work carried out at Nylsvley and in Zimbabwe gives estimates of 1.66–4.01 kg $N \cdot ha^{-1} \cdot year^{-1}$ (Scholes and Walker 1993; Weinmann 1955). Nylsvley is upwind of the major industrial centers, so these values can be regarded as background levels (Scholes and Walker 1993). Wet and dry deposition of sulfur has been estimated to be 5.7 and 8.2 kg $S \cdot ha^{-1} \cdot year^{-1}$, respectively (Olbrich 1995a, 1995b; Scholes and Scholes 1999).

Biological Nitrogen Fixation

The other major input of nitrogen to the ecosystem is biological nitrogen fixation (BNF). Most studies on nitrogen fixation have involved searching for nodules on excavated root systems. This approach, applied locally, has resulted in minimal success, and many authors have concluded that nitrogen fixation is limited in southern African savannas. In general, members of the legume subfamily Caesalpinioideae, which dominate the tree layer of the moist, nutrient-poor savannas, were found not to be nitrogen-fixing using the acetylene technique. However, many of the leguminous herbaceous species, such as *Elephantorrhiza elephantine*, *Tephrosia* spp., and *Indigofera* spp., were nitrogen fixing. Estimates ranged from 0.63 to 0.85 kg $N \cdot ha^{-1} \cdot year^{-1}$ (Grobbelaar and Rösch 1981).

However, recent studies in Kruger using the natural abundance N^{15} technique have shown that many of the members of the woody Mimosoideae are also able to fix nitrogen. The percentage N_2 contributed by nitrogen fixation to the specific plant nitrogen content ranged from 0 to 46 percent in *Acacia nigrescens*, 22 to 93 percent in *Acacia nilotica*, 71 to 88 percent in *Stylosanthes fruticosa*, and 25 to 63 percent in *Macrotyloma maraguense*, with a mean value of 23 percent and 57 percent for each of the *Acacia* species, respectively, and a mean value of 84 percent and 43 percent for each of the herbaceous species, respectively. Initial estimates of N_2 fixation to the landscapes are 21.0 kg $N \cdot ha^{-1} \cdot year^{-1}$ for the nutrient-rich patches dominated by the acacias, with the herbaceous plants contributing 4.8 kg $N \cdot ha^{-1} \cdot year^{-1}$. Estimates for the nutrient-poor patches, where there is little contribution from the woody species, would peak at 4.8 kg $N \cdot ha^{-1} \cdot year^{-1}$ (Woghiren 2002). Variability in fixation potential is high between individual plants and plant clusters. The amounts of nitrogen fixed are affected by soil factors, especially phosphorus and water availability.

Nitrogen Cycles in Fine-Leaved (Nutrient-Rich) and Broad-Leaved (Nutrient-Poor) Savannas

Figure 6.3 is a synthesis of the pool sizes and fluxes of nitrogen in the two savanna types in Kruger. The key points are the significant differences in the BNF rates of the two sites and the large deposition rates. Figure 6.3 does not adequately address the losses caused by fire and herbivory. An analysis must be conducted at the landscape scale and across geologies to determine where nitrogen is being gained or lost.

Carbon Uptake and Loss

Gross primary production (GPP) is equal to the sum of all photosynthesis within a given area and period. It can be expressed either in dry matter (DM) terms or in carbon terms: the carbon content of DM in savannas is around 0.42 g C in each 1.0g DM (42 percent). Almost exactly half of the GPP is returned to the atmosphere within minutes to days of its uptake, through the process of plant respiration. That which remains is equal to the growth in mass of the plant plus any non-CO_2 carbon losses from the plant (litterfall, root turnover, root exudates, and hydrocarbon emissions) and is known as the net primary production (NPP).

NPP is an important variable for describing ecosystems; it is a measure of the total amount of energy the ecosystem has to support biotic and human activity, and it is highly correlated with animal carrying capacity. Until recently, the NPP of Kruger has never been directly measured, although it has been estimated from models and from remote sensing. Some of the main components (aboveground grass, wood, twig, leaf, grass root, tree fine root, and tree coarse root) have been measured or estimated for adjacent or comparable ecosystems. For instance, Scholes (1988) recorded aboveground grass production at Klaserie to range from 100 $g \cdot m^{-2}$ on sandy, infertile sites with 35 percent tree cover and 500 mm annual rainfall to 150 $g \cdot m^{-2}$ on clayey soils with 20 percent tree cover and the same rainfall. After trees had been removed, aboveground grass production increased to 150 and 300 $g \cdot m^{-2}$, respectively.

Grass production is highly dependent on rainfall. Shackleton (1998) monitored tree wood increment at many savanna sites, including some in and adjacent to Kruger. He found that wood production averaged 3 percent of wood standing crop. Because the wood standing crop in the Kruger region ranges from 1,000 to 4,000 $g \cdot m^{-2}$ (averaging around 2,000 $g \cdot m^{-2}$), this translates to wood production of 60 $g \cdot m^{-2}$. The majority of lowveld trees are deciduous; therefore, annual leaf production is approximately equal to the peak tree and shrub leaf standing crop, which is 50–150 $g \cdot m^{-2}$ (Dayton 1978; Scholes 1988;

Shackleton 1998). Twig production is about a quarter of leaf production (Scholes and Walker 1993).

Together, these relationships suggest that total NPP at a site receiving 550 mm rainfall on granite-derived soils in Kruger, with a tree basal area around 7 $m^2 \cdot ha^{-1}$, would be about 419 g DM, 63 percent produced by trees and 37 percent by grasses. The NPP is almost equally split above and below ground.

Some Biogeochemical Issues in the Lowveld

The Kruger National Park is located in the eastern part of South Africa and is subjected to significant amounts of acid rain and other precursors of ozone formation from biomass burning.

Nitrogen Saturation and Acidification of Ecosystems

Nitrogen saturation may be defined as the availability of ammonium and nitrate in excess of total combined plant and microbial nutritional demand, or a situation in which outputs of nitrogen are equal to or greater than nitrogen inputs (Aber et al. 1989). N saturation may be associated with adverse effects on tree and grass physiology, leaching of nitrate from soils, increases in N trace gas emissions, and degraded water quality. Acid rain is a common phenomenon in Mpumalanga, with the pH of the rainfall being in the region of 4.2. Rain contains both sulfuric and nitric acid at a ratio of approximately 2:1.

Comparative studies between high-altitude grasslands and plantation forests on the Mpumalanga escarpment west of Kruger show that a decrease in soil pH of 0.3–1.0 units has occurred in the plantations. This decrease is a consequence of increased base cation uptake, the acidic nature of the tree litter, and acid deposition (Scholes and Scholes 1999). Nitrate leaching has been measured from afforested catchments, indicating that some perturbation has occurred in the nitrogen cycle with changed land use. This nitrate is entering the rivers that flow into Kruger, adding to the high levels of nitrogen and sulfur deposition. This loading initially may be beneficial because it acts as a fertilizer and will increase foliage palatability. However, the ecosystem may become saturated, causing changes in ecosystem nutrient rententivity, changes in species composition, and ultimately loss of productivity. These effects are thought to be several decades away in the park.

Formation of Tropospheric Ozone

Tropospheric ozone (O_3) is formed through the interaction of VOCs and NO_x in the presence of sunlight. A photoequilibrium exists between NO, NO_2, and

O_3, but in the presence of VOCs the equilibrium is broken and O_3 is produced. The ratio of NO_x to VOCs is also an important factor in determining whether O_3 is produced or destroyed (Meixner 1994). In urban areas the anthropogenic NO_x and VOCs dominate, whereas in rural areas biogenic emissions are more important. High tropospheric O_3 concentrations have negative impacts on vegetation and human health. On a regional scale, both biogenic and pyrogenic emissions are important for the formation of regional tropospheric O_3.

Large areas of Kruger burn annually and together with the biogenic emissions contribute to ozone levels. Ozone concentrations over the south Atlantic off the west coast of Africa peak during spring (August–October) (Fishman et al. 1991; Thompson et al. 1996). Biomass burning produces O_3 precursors, such as NO_x, CH_4, VOCs, and CO. Peak biomass burning in southern Africa occurs at the end of the dry season (July–August; Scholes et al. 1996), so it was initially thought that the high O_3 concentrations were caused by biomass burning.

The Southern African Fire-Atmosphere Research Initiative (SAFARI 92) concentrated on investigating trace gases produced by savanna fires. The study indicated that emissions from biomass burning could not be the sole contributor to the high O_3 concentrations. Both the numbers and the timing (biomass burning peak was earlier than O_3 peak) indicated this. The combination of the temporal heterogeneity of biogenic emission pulsing that occurs at the beginning of spring and the pyrogenic emissions are unique for this system and lead to an increase in O_3 off the coast of Africa from August to October.

Conclusion

Understanding the patterns and causes of spatial and temporal heterogeneity in ecosystem function—the processes associated with energy and nutrient flow in ecosystems—remains at the frontier of ecosystem and landscape ecology. Despite tremendous advances in understanding ecosystem processes over small spatial extents, there exists very little theory for predicting variability in ecosystem processes across heterogeneous landscapes. The library of empirical data is expanding. A more synthetic understanding of heterogeneity in ecosystem processes remains an important research need.

References

Aber, J. D., and J. M. Melillo. 1979. Litter decomposition: measuring relative contributions of organic matter and nitrogen in forest soils. *Canadian Journal of Botany* 58:416–421.

Aber, J. D., K. J. Nadelhoffer, P. Steudler, and J. M. Melillo. 1989. Nitrogen saturation in northern forest ecosystems. *BioScience* 39:378–386.

Augustine, D. J., and D. A. Frank. 2001. Effects of migratory grazers on spatial heterogeneity of soil nitrogen properties in a grassland ecosystem. *Ecology* 82:3149–3162.

Caughley, G. 1982. Vegetation and the dynamics of modelled grazing systems. *Oecologia* 54:309–312.

Davidson, E., and W. Kingerlee. 1997. A global inventory of nitric oxide emissions from soils. *Nutrients Cycling in Agroecosystems* 48:37–50.

Dayton, B. F. 1978. Standing crops of dominant *Combretum* species at three browsing levels in the Kruger National Park. *Koedoe* 21:67–76.

Delmas, R., L. P. LaCaux, and D. Brocard. 1995. Biomass burning emission factors: methods and results. *Environmental Monitoring and Assessment* 38:181–204.

Fall, R. 1999. Biogenic emissions of volatile organic compounds from higher plants. Pages 42–96 in C. N. Hewitt (ed.), *Reactive hydrocarbons in the atmosphere*. San Diego, CA: Academic Press.

Fishman, J., K. Fakhruzzaman, B. Cros, and D. Nganga. 1991. Identification of widespread pollution in the southern hemisphere deduced from satellite analyses. *Science* 252:1693–1696.

Frank, D. A., and P. M. Groffman. 1998. Ungulate vs. landscape control of soil C and N processes in grasslands of Yellowstone National Park. *Ecology* 79:2229–2241.

Grobbelaar, N., and M. W. Rösch. 1981. Biological nitrogen fixation in a northern Transvaal savanna. *Journal of South African Botany* 47:493–506.

Guenther, A., C. N. Hewitt, D. Erickson, R. Fall, C. Geron, T. Graedel, P. Harley, L. Klinger, M. Lerdau, W. McKay, T. Pierce, R. Scholes, R. Steinbretcher, R. Tallamraju, J. Taylor, and P. Zimmerman. 1995. A global model of natural volatile organic compound emissions. *Journal of Geophysical Research* 100:8873–8892.

Guenther, A., L. Otter, P. Zimmerman, J. Greenberg, R. Scholes, and M. Scholes. 1996. Biogenic hydrocarbon emissions from southern African savannas. *Journal of Geophysical Research* 101:25859–25865.

Harley, P., M. Lerdau, and R. Monson. 1999. Ecological and evolutionary aspects of isoprene emission. *Oecologia* 118:109–123.

Hesterburg, R., A. Blatter, M. Fahrni, M. Rosset, A. Neftel, W. Eugster, and H. Wanner. 1996. Deposition of nitrogen-containing compounds to an extensively managed grassland in central Switzerland. *Environmental Pollution* 91:21–34.

Holland, E. A., and J. K. Detling. 1990. Plant response to herbivory and belowground nitrogen cycling. *Ecology* 71:1040–1049.

Jones, C. L., N. L. Smithers, M. C. Scholes, and R. J. Scholes. 1990. The effect of fire frequency on the organic components of a basaltic soil in the Kruger National Park. *South African Journal of Plant and Soil* 7:236–238.

Kirkman, G. A., W. X. Xang, and F. X. Meixner. 2001. Biogenic nitric oxide emissions upscaling: an approach for Zimbabwe. *Global Biogeochemical Cycles* 15:1005–1020.

Lerdau, M., A. Guenther, and R. Monson. 1997. Plant production and emission of volatile organic compounds. *BioScience* 47:373–383.

Levine, J. S., E. L. Winstead, D. A. B. Parsons, M. C. Scholes, R. J. Scholes, W. R. Coffer III, D. R. Cahoon Jr., and D. I. Sebacher. 1996. Biogenic soil emissions of nitric oxide (NO) and nitrous oxide (N_2O) from savannas in South Africa: the impact of wetting and burning. *Journal of Geophysical Research* 101:23689–23697.

McNaughton, S. J. 1985. Ecology of a grazing ecosystem: the Serengeti. *Ecological Monographs* 55:259–294.

Meixner, F. X. 1994. Surface exchange of odd nitrogen oxides. *Nova Acta Leopoldina* NF 70:299–348.

Meixner, F. X., T. H. Fickinger, L. Marufu, D. Serca, F. J. Nathaus, E. Makina, L. Mukurumbira, and M. O. Andreae. 1997. Preliminary results on nitric oxide emissions from a southern African savanna ecosystem. *Nutrient Cycling in Agroecosystems* 48:123–138.

Olbrich, K. 1995a. *Air pollution impacts on South African commercial forests: a research synthesis: 1990–1995 and strategy for the future.* Unpublished contract CSIR Report, FOR DEA-874, CSIR Division of Forest Science and Technology, Pretoria, South Africa.

Olbrich, K. A. 1995b. *Research on impacts of atmospheric pollution and environmental stress on plantations.* Unpublished contract report, FOR-DEA 874, to the Department of Water Affairs and Forestry. CSIR Division of Forest Science and Technology, Pretoria, South Africa.

Otter, L., A. Guenther, and J. Greenberg. 2002. Seasonal and spatial variations in biogenic hydrocarbon emissions from southern African savannas and woodlands. *Atmospheric Environment.* 36:4265–4275.

Otter, L. B. 1992. *Soil carbon fractionation of sand and clay soils under different burning regimes.* B.Sc. honors thesis, University of the Witwatersrand, Johannesburg, South Africa.

Otter, L. B., L. Marufu, and M. C. Scholes. 2001. Biogenic, biomass and biofuel sources of trace gases in southern Africa. *South African Journal of Science* 97:131–138.

Otter, L. B., W. X. Yang, M. C. Scholes, and F. X. Meixner. 1999. Nitric oxide emissions from a southern African savanna. *Journal of Geophysical Research* 104:18471–18485.

Parsons, D. A. B., M. C. Scholes, R. J. Scholes, and J. S. Levine. 1996. Biogenic NO emissions from savanna soils as a function of fire regime, soil type, soil nitrogen, and water status. *Journal of Geophysical Research* 101:23683–23688.

Pastor, J., B. Dewey, R. J. Naiman, P. F. McInnes, and Y. Cohen. 1993. Moose browsing and soil fertility of Isle Royale National Park. *Ecology* 74(2):467–480.

Ritchie, M. E., D. Tilman, and M. H. Knops. 1998. Herbivore effects on plant and nitrogen dynamics in oak savanna. *Ecology* 79:165–177.

Ruess, R. W., and S. J. McNaughton. 1987. Grazing and the dynamics of nutrient and energy regulated microbial processes in the Serengeti grasslands. *Oikos* 49:101–110.

Scholes, M. C., R. Martin, R. J. Scholes, D. Parsons, and E. Winstead. 1997. NO and N_2O emissions from savanna soils following the first simulated rains of the season. *Nutrient Cycling in Agroecosystems* 48:115–122.

Scholes, R. J. 1988. *Response of three semi-arid savannas on contrasting soils to the removal of the woody component.* Ph.D. thesis, University of the Witwatersrand, Johannesburg.

Scholes, R. J., J. Kendall, and C. O. Justice. 1996. The quantity of biomass burned in southern Africa. *Journal of Geophysical Research* 101:23667–23676.

Scholes, R. J., and M. C. Scholes. 1999. Nutrient imbalances following conversion of grasslands to plantation forests in South Africa. Pages 175–189 in E. M. A. Smaling, O. Oenema, and L. O. Fresco (eds.), *Nutrient disequilibria in agroecosystems: concepts and case studies*. Wallingford, UK: CABI.

Scholes, R. J., and B. H. Walker. 1993. *An African savanna, synthesis of the Nylsvley study*. Cambridge, UK: Cambridge University Press.

Serca, D., R. Delmas, X. Le Roux, D. A. B. Parsons, M. C. Scholes, L. Abbadie, R. Lensi, O. Ronce, and L. Labroue. 1998. Comparison of nitrogen monoxide emissions from several African tropical ecosystems and influence of season and fire. *Global Biogeochemical Cycles* 12:637–651.

Shackleton, C. M. 1998. Annual production of harvestable deadwood in semi-arid savannas, South Africa. *Journal of Forest Ecology and Management* 112:139–144.

Shea, R. W., B. W. Shea, J. B. Kauffman, D. E. Ward, C. I. Haskins, and M. C. Scholes. 1996. Fuel biomass and combustion factors associated with fires in savanna ecosystems of South Africa and Zambia. *Journal of Geophysical Research* 101:23551–23568.

Sirotnak, J. M., and N. J. Huntly. 2000. Direct and indirect effects of herbivores on nitrogen dynamics: voles in riparian areas. *Ecology* 8:78–87.

Swift, M. J., O. W. Heal, and J. M. Anderson. 1979. *Decomposition in terrestrial ecosystems*. Studies in Ecology 5. Berkeley: University of California Press.

Thompson, A. M. 1996. Evaluation of biomass burning effects on ozone during SAFARI/TRACE-A: examples from process models. Pages 333–349 in J. S. Levine (ed.), *Biomass burning and global change*. Cambridge, MA: MIT Press.

Tracy, B. F., and D. A. Frank. 1998. Herbivore influence on soil microbial biomass and nitrogen mineralization in another grassland ecosystem: Yellowstone National Park. *Oecologia* 114:556–562.

Trollope, W. S. W., and A. L. F. Potgieter. 1985. Fire behaviour in the Kruger National Park. *Journal of the Grassland Society of Southern Africa* 2:17–23.

Turner, C. R., R. B. Wells, and K. A. Olbrich. 1995. Deposition chemistry in South Africa. Pages 80–85 in G. Held, B. J. Gore, A. D. Surridge, G. R. Tosen, G. R. Turner, and R. D. Walmsley (eds.), *Air pollution and its impacts on the South African highveld*. Cleveland, OH: Environmental Scientific Association.

van Wilgen, B. W., H. C. Biggs, S. P. O'Reagan, and N. Mare. 2000. A fire history of the savanna ecosystems in the Kruger National Park, South Africa, between 1941 and 1996. *South African Journal of Science* 96:167–178.

Weinmann, H. 1955. The nitrogen content of rainfall in southern Rhodesia. *South African Journal of Science* 51:82–84.

Woghiren, A. J. 2002. *Nitrogen characterization of the savanna flux site at Skukuza, Kruger National Park*. M.Sc. thesis, University of the Witwatersrand, Johannesburg.

Chapter 7

Fire as a Driver of Ecosystem Variability

BRIAN W. VAN WILGEN, WINSTON S. W. TROLLOPE,
HARRY C. BIGGS, ANDRÉ L. F. POTGIETER,
AND BRUCE H. BROCKETT

Fire is a common disturbance in many terrestrial ecosystems, where it consumes enormous quantities of plant biomass. In African savannas, fire, climatic variability, soils, and large herbivores interact to shape the vegetation, driving spatial and temporal heterogeneity across the landscape. Because fires are variable in frequency, season, and intensity, no two fires are the same, and their sequencing on a site has significant effects on the vegetation. Managers in the Kruger National Park have used fire for decades, yet it is still controversial despite a long history of fire research. This chapter reviews the history of fire management in Kruger and the evidence of its effects. The options for fire management are described and compared with approaches adopted in other important conservation areas of the world. We have focused on the highly variable and heterogeneous nature of fire regimes and their contribution to the dynamic character of savanna ecosystems.

Sources of Fire Variability in Kruger

Fire is one of a suite of drivers that influence ecosystem dynamics and heterogeneity. Because fires themselves depend on fuel and weather that vary over space and time, they cannot be considered in isolation. The most important sources of variability include soil fertility, rainfall, levels of herbivory, and variability in the conditions under which fires burn. Soil fertility influences the growth of grasses and trees, and rainfall influences the biomass of grasses that provide the fuel for fires. Climatic variability will influence the occurrence of conditions conducive to the initiation and spread of fires, and herbivory can reduce the fuel loads and fire intensity or even prevent fires.

In Kruger, the main source of variation in soils is found between granite-derived soils (nutrient poor) and basalt-derived soils (nutrient rich). Total annual rainfall can vary fourfold (from <200 to >800 mm in a given year), and more than twofold spatially (mean annual rainfall ranges from 350 to 750 mm at different locations). Superimposed on different soil types, these sources of variation produce varying quantities of grass fuels. Regular assessments of grass biomass in Kruger have shown that these fuel loads can vary at any given time between zero and 10,000 kg/ha^{-1}. Fires will not carry when fuel loads are below 2,000 kg/ha^{-1}, which means that fuel loads can vary fivefold (from 2,000 to 10,000 kg/ha^{-1}) in any year in which fires occur. The climatic conditions conducive to large fires also vary, with only half of the years providing the prolonged hot and dry conditions necessary for large fires to occur (van Wilgen et al. 2000). These conditions must coincide with fuel loads greater than 3,500 kg/ha^{-1} for extensive fires to occur.

The numbers of important grazing mammals fluctuate in Kruger mainly in response to rainfall (Mills et al. 1995). For example, the numbers of buffalo were 32,000 in 1981 and 14,000 in 1993. Grazers and fires are direct competitors for grass fuels, and at times when herbivore numbers are low, fires can burn more area at a higher intensity; conversely, when herbivore numbers are high, fire occurrence and intensity are less (see Chapter 17, this volume, for the effects of rinderpest).

The most important source of variability in fires is fire intensity. It depends on the amount of fuel consumed but also, to a large degree, on the conditions under which fires burn. Byram's (1959) fireline intensity (the most widely used measure of fire intensity) is strongly correlated with the ecological impacts of fire. It is calculated as the product of the heat yield of fuels, the amount of fuel consumed, and the rate of spread of the fire. Heat yields are measured in J/g^{-1}, fuel loads in g/m^{-2}, and rates of spread in m/s^{-1}, providing units of kW/m^{-1}. Of these factors, rate of spread has the greatest range in vegetation fires, varying from 0.1 to 100 m/min^{-1}. The value for fuel consumed in savannas can vary from about 10 to 1,000 g/m^{-2}. Heat yields vary so little (by about 10 percent) that they can be considered as constant. Fire intensity thus has a practical range of 10 to more than 20,000 kW/m^{-1}, primarily because of the large variation possible in spread rates (Stocks et al. 1997).

Fire intensity in savannas has significant effects on the trees and shrubs. Higher-intensity fires kill the aerial portions of trees, and they can resprout only from the base, whereas less intense fires allow aerial tissues to survive, and the height growth of trees is not affected (Trollope 1999). Thus a headfire, which burns at high intensity, can have very different effects from a backfire in the same area under the same conditions of fuel, season, and fire return period. Fires that burn at night rather than during the day or fires that originate from point ignition sources rather than from perimeter ignitions have substantially lower intensities. It is feasible that active management in Kruger would have had a significant impact on the spatial distribution of fire intensities and thus fire effects.

Fire as an Ecosystem Driver in African Savannas

Fire affects plant communities through the large-scale, episodic removal of biomass and the delivery, in some cases, of lethal temperatures to plant material. It has been described as a large generalist herbivore (Bond 1997) that sometimes competes with, sometimes replaces, and sometimes facilitates herbivory. There are conflicting views of the necessity of fire in the maintenance of savannas (Scholes and Walker 1993). Many studies have shown that the exclusion of fire from savannas leads to an increase in woody biomass, and this has been interpreted as evidence that fire is essential for the persistence of savannas. However, in dry environments, fire exclusion does not eventually lead to a forest but rather to a denser form of savanna (van Wilgen and Scholes 1997). Almost all plant species in arid savannas (such as those found in Kruger) are able to survive repeated, frequent burning largely by resprouting. Pockets dominated by fire-intolerant species are not a feature of the landscape. Debates on the role of fire in such savannas therefore focus not on the conservation of fire-sensitive species (as in many other ecosystems) but rather on their role in promoting dominance by certain species over others. Not all South African savannas fall into this category, however. For example, in the Hluhluwe-Umfolozi Park (mean annual rainfall 1,000 mm), the exclusion of fire and the eradication of elephants led to the development of fire-sensitive forest vegetation (Whateley and Wills 1996). Such patches also are fire resistant (van Wilgen et al. 1990b), suggesting that a stable forest state can be reached in savannas where rainfall is high and disturbances by fire and elephants are kept out.

Grass-tree coexistence may be made possible by separation of the rooting niche, with trees having sole access to water in deeper soil horizons and grasses having preferential access to, and being superior competitors for, water in the surface soil horizons (Walter 1971). In this model, climatic variability does not allow for dominance by either life form, and coexistence is possible in a variety of states (Walker and Noy-Meir 1982). Higgins et al. (2000) have proposed an alternative model in which interactions between life history characteristics of trees (sprouting ability, fire survival at different life stages, and mortality) and the occurrence of fires (which prevent recruitment of trees into adult life classes) could explain coexistence.

A History of Fire Management in Kruger

Fire management in Kruger over the past 75 years has reflected the evolution of an understanding of the role of fire. Early ideas were formed around equilibrium theory: fire was regarded first as something to be avoided and was later applied at a fixed return period when it was realized that fire was an integral part of the system. The development of nonequilibrium theories of savanna dynamics (Mentis and Bailey 1990; Walker 1989) saw a shift in policy toward burning under diverse rather than fixed conditions. This section reviews the management approaches and resultant fire regimes in Kruger.

Fire Management

The fire management of Kruger was characterized by prescribed burning for a long period: 36 years between 1956 and 1992 (Table 7.1). In the late 1980s, a number of concerns were raised about the putative effects of this policy (van Wilgen et al. 1998; Biggs and Potgieter 1999). These included the following:

- The observation that grass species characteristic of poorly managed pastures and overgrazing became dominant as a result of excessively frequent burning in combination with the artificial provision of water (Trollope et al. 1995).
- The observation that the numbers of large trees had declined (Trollope et al. 1998; Eckhardt et al. 2000).
- The practice of "ringburning," in which fires are ignited around the periphery of management blocks and allowed to burn toward the middle. Natural fires, such as those associated with lightning strikes, would spread out in all directions from a point, allowing the fire to develop a range of intensities as it spread. Ringburning can trap animals and leads to a disproportionately large area burning as a high-intensity headfire (an effect magnified by these fires being carried out during the day and seldom at night).
- The lack of variation associated with burning on a fixed cycle over a long period.

These concerns led to a debate on the role of lightning ignitions in producing a more variable fire regime based on point ignitions. Originally, it was suggested that lightning should play a more important role in the fire regime of Kruger, not necessarily that it should dominate or completely replace prescribed burning. However, a policy of natural fires was adopted when a majority of managers supported the notion of wilderness ecosystem management (Biggs and Potgieter 1999). This policy called for all lightning-ignited fires to burn freely, whereas all other fires were prevented, suppressed, or contained. The philosophy was that lightning fires should produce the same patterns of frequency, season, and intensity that characterized the regime under which Kruger's biota evolved. Therefore, they should be in line with Kruger's mission statement, which calls for the maintenance of biodiversity "in all its natural

TABLE 7.1

Different fire management policies applied in the Kruger National Park since its proclamation in 1926 (summarized from van Wilgen et al. 1998, 1990a).

DATES	APPROACH TO FIRE MANAGEMENT	RATIONALE FOR APPROACH
1926–1947	Occasional and limited burning.	Provision of green grazing for wildlife.
1948–1956	All prescribed burning stopped and firebreaks established to assist in the control of wildfires.	Concern about the perceived negative effects of fire.
1957–1980	Formal system of prescribed burning once every 3 years in spring on fixed management areas ("burning blocks").	Necessity of fire in maintaining ecosystem health recognized.
1981–1992	Rainfall, fire age, and specific objectives in local landscape taken into account, intended to result in variable periods between fires, with season varied between late winter, midsummer, and autumn.	Application of fires over a longer period spread the workload more evenly and ensured better use of postfire grazing.
1992–2001	Lightning fires combined with suppression of anthropogenic fires.	Simulation of natural conditions under which the biota evolved.
2002–present	A combination of patch burns in areas where fire is considered necessary according to ecological criteria until a target area (based on rainfall) is reached. Suppression of fires in areas where fire is not considered necessary and tolerance of all lightning fires.	Adaptive management and the realization that command-and-control approaches are not feasible. Integrated fire management system formulated to promote heterogeneity of effects on the biota, thereby promoting and maintaining biodiversity.

facets and fluxes." The decision to change to a lightning-driven policy was made in the face of an available alternative range condition burning system that had been developed for Kruger (Biggs and Potgieter 1999). This system (Trollope et al. 1995) is based on the botanical composition and biomass of the grass sward that would produce a high species diversity of perennial grasses when burned; however, it was considered by managers to be too oriented to agricultural practice and therefore unsuitable for adoption in a conservation area.

Fire History

The fire regimes that characterized Kruger before the early twentieth century are not known. Before the proclamation of the Sabi Game Reserve in 1898, the area was settled, first by San hunter-gatherers and later by migrating tribes from the north. The area was settled since at least the fourth century AD by Iron Age communities (Plug 1989). The environment was not suitable for agriculture or herding, and it appears that the main form of subsistence was from hunting (Plug 1987). Undoubtedly, indigenous peoples used fire, and this must have had an influence on the area.

The fire history of Kruger has been analyzed for the second half of the twentieth century (van Wilgen et al. 2000). This analysis showed that fires covering 16.79 million ha occurred between 1941 and 1996 (16 percent of the area burning each year on average). Of this area, 5.15 million ha was burned between 1941 and 1957, when limited prescribed burning and protection from fire took place. Between 1957 and 1991, 2,213 prescribed burns covering 5.1 million ha (46.3 percent of the 10.98 million ha burned during that period) were carried out. Lightning fires burned 2.5 million ha between 1957 and 1996, or 21.6 percent of the area burned during that period. The mean fire return period was 4.5 years, with intervals between fires from 1 to 34 years (Figure 7.1). The distribution around the mean was not symmetrical, and the median interval was 3.1 years. Some areas burned more often than others. Mean annual rainfall had a significant effect on fire return periods, with mean return periods of 3.5 years in areas receiving >700 mm and 5.0 years in areas receiving <700 mm.

The fire history of Kruger is a product of fire management efforts combined with other factors (Table 7.2). For example, despite management intentions, some areas burn in unintentional fires, and at other times scheduled prescribed burns could not be carried out (e.g., after years of low rainfall when grass biomass would have been low). Therefore, only about half of the area that burned between 1957 and 1992 (52 percent including firebreak burns) resulted from fires set intentionally. The occurrence of fire was strongly correlated with grass biomass (Figure 7.2), which in turn resulted from the amount of rainfall in the previous season. This relationship may have been as important in determining whether an area will burn and how much of it will burn as the reigning fire management policy.

Understanding Fires at a Landscape Scale

An understanding of the role of fire in Kruger has been built on research, observation at a broader scale, and modeling. In this section, we review the understanding that has resulted from the combined interpretation of the aforementioned approaches.

Fire Research

Fire research in Kruger was initiated after fires burned 25 percent of Kruger in 1953 (van Wyk 1971). An experiment was established to test the effects of season and frequency of fire on a series of plots (Box 7.1, Figure 7.3). An important task of the first researchers was to conduct this experiment, making them

Fire as a Driver of Ecosystem Variability 155

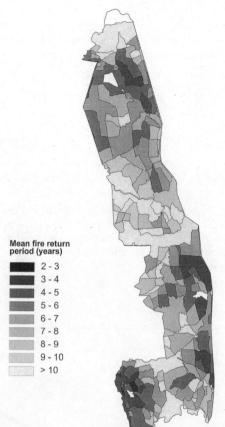

FIGURE 7.1. Map of Kruger showing mean fire return periods calculated from fire records between 1957 and 1996 (from van Wilgen et al. 2000).

the pioneers of the modern era of scientific management. The experiment has been maintained for almost half a century. Although it has been variously considered as having been conducted at too small a scale, based on a regular regime, and confounded by herbivory (Biggs and Potgieter 1999), it is a remarkable and rare asset, providing a unique resource for testing models of fire effects.

Interpretation from Observation

The aforementioned concerns about the role of fire in Kruger's management provided an impetus for analyzing its effects at scales larger than the experimental plots. In the absence of rigorous vegetation monitoring data before 1985, these studies have concentrated on fire's impacts on woody plants because they can be quantified from aerial and ground-based photography. A preliminary study (Trollope et al. 1998) using aerial photographs was unable to detect

TABLE 7.2

Features of the fire regime in the Kruger National Park between 1941 and 1996 (data from van Wilgen et al. 2000).

FEATURE OF FIRE REGIME	WET CLIMATIC CYCLE (1971–1980)	DRY CLIMATIC CYCLE (1983–1992)	ALL YEARS
Mean annual rainfall for period (mm)	638	450	534
Extent of fires (km²)	44,149	20,834	167,947
Frequency (mean fire return period in years)	4.3	9.1	4.5
Season	Fires concentrated in the late dry season (September)	Fewer fires overall, with more fires occurring in the early dry season	Fires in all months, but most (80%) from June to November, peak in September
Intensity	No data	No data	Intensities ranging from <100 to >10,000 kW/m^{-1}; relative distribution of intensities not known

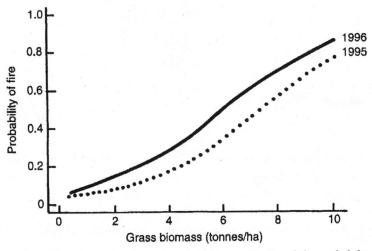

FIGURE 7.2. The relationship between grass biomass (fuel) and the probability of fire in two individual years in Kruger. Severe fire weather conditions occurred in 1996 but not in 1995 (from van Wilgen et al. 2000).

changes in the density of large trees between 1940 and 1960 but noted a dramatic decline between 1960 and 1989. A later study (Eckhardt et al. 2000) reported a decrease in cover of woody plants (all trees and shrubs) from 11.9 percent in 1940 to 4.3 percent on basalts but a small increase (from 19.7 to 22.1

BOX 7.1
Kruger National Park Experimental Burn Plots

Long-term experiments to test the application of fire treatments on fixed areas over extended periods are rare. The Kruger established a series of such trials in 1954. The trials covered fires at annual, biennial, and triennial (and later quadrennial and sextennial) intervals and in different seasons, as well as protection from fire. These trials have been maintained for almost half a century and have provided valuable information on the role of fire in Kruger. The following schematic shows the layout of one replication. Treatments were replicated four times in each of four major landscapes in Kruger. Each plot covers approximately 7 ha.

Feb	Oct	Aug	Apr	Dec	Aug	Apr	Aug	Feb	Oct	Dec	
3 yr	3 yr	3 yr	3 yr	3 yr	2 yr	2 yr	1 yr	2 yr	2 yr	2 yr	No fire

Principal Results to Date

- Fire has a marked effect of the structure but not the composition of woody vegetation (except in the high rainfall "sourveld" area). Trees and shrubs survive and mature in nonfire treatments but are significantly smaller where regularly burned.
- Frequently repeated fire increases grazing intensity, leading to dominance by pioneer or unpalatable perennial grass species (the plots were open to grazing).
- Seasonal effects manifest themselves through different fire behavior, and an understanding of resultant patterns has started to develop. For instance, in one landscape, a continuum exists from the least multistem coppicing in low-intensity February fires to the most in intense August fires, with interactions between season and frequency. Effects on the marula (*Sclerocary birrea*) have been documented.
- Fire behavior data collected during plot burns have been used to establish the importance of high-intensity winter fires in reducing the structural diversity of trees and shrubs.

Recent and current activities include work on soil properties and nutrients, mycorrhizal colonization, ant assemblages, and small mammals. Several major research programs involving aspects of ecosystem functioning are beginning or are being designed.

References

van der Schijff 1958; van Wyk 1971; Trollope 1999; Trollope et al. 1999; Enslin et al. 2000; Jacobs and Biggs 2001.

FIGURE 7.3. View of the experimental burning plots at Nwanetsi in the central Kruger in 1982. Fire was excluded from the plot on the right for 26 years, whereas the plot on the left was burned every 2 years in August. Frequent fire has eliminated large trees, which survive as stunted individuals, whereas exclusion of fire has clearly allowed large trees to survive and develop.

percent) on granites. However, Eckhardt et al.'s study showed that there was an overall decrease in the density of large (>5 m) trees on both substrates between 1984 and 1996. Both studies attributed the changes to interactions between fires and herbivores, particularly elephants.

The decline in large trees must be interpreted in relation to interactions between fire and use by elephants. Eckhardt et al. (2000) found a weak relationship between the cover of woody vegetation and fire frequency, with increased fire frequencies corresponding to decreases in cover. The weakness of the relationship suggests that fire was not the only factor causing the decline, which was most marked in large trees. Although mature trees can survive frequent fires, use by elephants and other animals can damage trees and allow fires to burn exposed areas of wood (Yeaton 1988). These scars tend to become larger with successive fires, and the trees eventually become structurally weakened and collapse. Frequent, regular fires (as practiced in Kruger for decades) would also have prevented smaller trees from developing into larger ones. Therefore, increased mortality and declining recruitment of juvenile to adult trees as a result of a combination of increasing numbers of elephants and frequent, regular burning appear to have resulted in a decline in large trees. Trollope et al. (1998) also argued that large trees could persist in areas where elephants, but not fire, were excluded (Figure 7.4).

FIGURE 7.4. View of a fenceline contrast in 1996 in northern Kruger. Elephants were excluded from the area on the left for 29 years but not on the right. Fire return periods were similar for both sides of the fence. Elimination of elephant damage has allowed large trees to persist in the presence of fire.

Modeling of Impacts

Spatial variability in fire, even at the scale of small plots, is a significant factor in allowing the coexistence of species in ecosystems (Hobbs and Atkins 1988; Yeaton and Bond 1991). However, the need to understand the complexity that arises from interspecific and intraspecific interactions between individuals, interactions across trophic levels, and interactions of organisms with the abiotic environment over space and time calls for complex modeling (Hartvigsen et al. 1998), and this has not yet been attempted in Kruger.

To date, only one conceptual model has rigorously examined the role of fire in determining tree-grass coexistence in African savannas (Higgins et al. 2000). This model, based in part on work done in Kruger, provides a framework for understanding the conditions that allow trees and grasses to coexist in a fire-prone and climatically variable environment. The model predicts that wet season droughts are an important factor limiting tree seedling establishment, and fires prevent the recruitment of trees into adult size classes. The model shows that the effects of variations in fire intensity on tree demography account for tree-grass coexistence and that removal of this variation can lead to the exclusion of trees. Unlike in Walker and Noy-Meir's (1982) equilibrium model, variance is central to coexistence. Variable fire intensities provide opportunities for trees to escape the flame zone, where they are most susceptible to fire, and to recruit into the more fire-resistant size classes. Declines in tree numbers in the model simulation result from continuing but low-level adult mortality. The rate of adult mortality caused by factors other than fire strongly influenced tree persistence in the model. The model suggests that low adult mortality rates (>.05) would be necessary for tree persistence, and the authors point to estimates of elephant-induced tree mortality rates of .18, reported by Thompson (1975) for Zimbabwean savannas, which suggest that the role of elephants as ecosystem modifiers should not be disregarded. This model supports many of the speculative explanations for the decline of large trees in Kruger and provides a useful framework for guiding a monitoring and evaluation program by pointing to the key parameters that influence tree-grass coexistence.

Are Large Fires Different?

Although the role of fire is well recognized in ecology today (Bond and van Wilgen 1996; Whelan 1995), the effects of very large and infrequent fires are not well understood. Fires sometimes qualify as large, infrequent disturbances (LIDs; Turner and Dale 1998), which are much larger in spatial extent or duration than those that typically affect ecosystems. For example, 1–3 percent of fires account for 97–99 percent of the landscape burned in some ecosystems (Bessie and Johnson 1995). The literature focuses on fires as LIDs in boreal forests (e.g., Foster et al. 1998), and this aspect of fires in savannas is less well understood. The type of response to an LID can be continuous (independent of the size of the disturbance), scale dependent, or of a threshold type. Threshold responses are those in which the response curve shows a discontinuity or a sudden change in the slope along the axis of increasing disturbance extent, intensity, or duration (Romme et al. 1998). Such a point would be reached when the size or intensity of the disturbance exceeds the capacity of the internal mechanisms to resist disturbance or where new mechanisms become involved. Understanding the consequences of LIDs is important for management. We need to know whether the substitution of smaller, more controllable disturbances (such as prescribed fires) for dangerous, uncontrollable fires could achieve management objectives. LIDs differ from smaller-scale disturbances in some but not all ecosystems by exhibiting threshold qualities that could cause the ecosystem processes to bifurcate (Romme et al. 1998). This could happen when species that would normally survive smaller fires are unable to survive or recolonize after larger fires. Evidence of thresholds reached after large fires is lacking for savanna ecosystems. Fires that could be regarded as LIDs occurred in 1953 and 1996 in Kruger, covering 25 and 22 percent of Kruger, respectively. The fires do not appear to have had any dramatic effects on the ecosystems in the medium term, indicating that they fit into the category of scale-independent responses. This supports the idea that substitution of occasional large fires by smaller prescribed burns would be acceptable.

Options for Managers

After deciding to abandon regular prescribed burning, managers in Kruger adopted a lightning fire approach in 1992. However, most of the area that burned in the first 9 years the policy was in force was burned by fires not started by lightning (lightning fires burned only 672,977 out of 3,160,680 ha, or 21.34 percent). This means that the goals of allowing a lightning-dominated fire regime to develop were not being met. This also meant that managers had to spend a lot of time containing fires. In response, in April 2002 Kruger's managers adopted an integrated approach that combines elements of patch mosaic, range condition, and lightning burns (Table 7.3). This means that many of the nonplanned ignitions can be left to burn (negating the need to contain all of the fires), but managers can exert some control by applying prescribed burns where the condition of the grass layer necessitates fire.

Management in Kruger aims to conserve biodiversity through the application of an appropriate fire regime. Because of the difficulties inherent in predicting the effects of fire on all facets of biodiversity, van Wilgen et al. (1998) have suggested that fire patterns be used as surrogate measures of biodiversity. For argument's sake, if it was postulated that each of the management systems proposed (Table 7.3) would be able to conserve biodiversity equally well, this could be tested by monitoring both the fire patterns that result (the surrogate measures) and the responses of biota. Under the new management policy, each of the biotic elements that are monitored would exist within an acceptable range of states that, if exceeded, would prompt an assessment of the causal factors. If fire patterns are also monitored, the changes can be interpreted against the background of a known fire history. This could lead to a change in fire management or the initiation of research (a response research framework; Rogers 1997). In the meantime, the fire patterns themselves can form goals against which managers can assess progress toward diversity conservation (van Wilgen et al. 1998).

Comparisons with Other Areas

Fire ecology is a developing field (Bond and van Wilgen 1996), with ideas and theories being formulated by few people. In many cases, ideas developed in

TABLE 7.3

Features of five approaches that have been considered as candidates for a fire management policy in Kruger National Park.

APPROACH	PRESCRIBED BURNING ON A FIXED CYCLE	LIGHTNING FIRES	PATCH MOSAIC BURNING	RANGE CONDITION BURNING	INTEGRATED FIRE APPROACH
Basic philosophy of approach	Regular fire is necessary to improve the quality of grass forage.	Lightning fires should produce the same fire regime as the one under which the park's biota evolved.	Application should result in a heterogeneous vegetation structure at a fine scale and thereby maximize biodiversity.	Given that the desired composition, structure, and dynamics of the vegetation are known, a fire regime can be selected to produce that vegetation.	Combining patch mosaic burning and lightning fire approaches will overcome problems associated with the latter.
Method of application	Fires are ignited in fixed areas on a fixed cycle.	Any fire ignited by lightning is allowed to burn freely, and all other fires are extinguished or contained.	Random point ignitions are spread over the fire season until a target area (based on early dry season grass biomass) is achieved.	Areas are burned when sufficient fuel is present and when grass species composition meets certain criteria.	Random point ignitions are combined with unplanned fires until a target area is reached, after which only lightning ignitions are allowed to spread.
Problems associated with approach	Lack of variation, negative effects on the vegetation, and problems associated with "ringburning."	In practice, most fires are ignited by nonlightning sources, and managers have to put a great deal of effort into fire control.	Safety concerns.	Perceptions that the approach is based on anthropogenic principles and is inappropriate for a conservation area.	Untested as yet.
Source	van Wilgen et al. 1990a.	Trollope et al. 1995.	Brockett et al. 2001.	Trollope et al. 1995.	Unpublished reports.

one part of the world are used in others. A comparison of the fire management approaches in large conservation areas elsewhere therefore could provide useful insights. In this section, we review the situation with regard to fire in three of the world's largest fire-prone conservation areas (Table 7.4). We chose these areas because good accounts of their fire ecology are available and because they are similar to Kruger in size and overall management goals.

Masai-Mara Game Reserve and Serengeti National Park

The area has experienced major vegetation changes in the past century, alternating between open grassland and dense woodland in response to changes in fire and herbivory (Dublin 1995). At the turn of the century the Serengeti-Mara was open grassland with lightly wooded patches, much as it is today. After the rinderpest epidemic of 1890, human and animal populations were dramatically reduced. Fires became infrequent because of low human populations, and elephant numbers were low after heavy hunting over several decades. Over the next 30–50 years, dense woodlands and thickets developed and persisted until the late 1950s.

The woodlands declined between the early 1960s and 1980 in response to increasing human populations that increased fires and drove the remaining elephants into the protected areas. Elephant numbers declined again as a result of poaching after 1980. These changes were accompanied by an increase in wildebeest numbers, from 250,000 in the 1960s to 1.5 million currently. Increased grazing pressure lowered the fuel loads and therefore the frequency and intensity of fires. Currently, woodland vegetation is again increasing in response to low elephant densities and reduced fire frequency and intensity.

Kakadu National Park

Kakadu's fire regime was dominated by aboriginal burning until the early twentieth century, when European settlers enforced fire protection, resulting in less frequent, late dry season burns (intense and extensive). This changed to early (low-intensity, patchy) dry season fires between 1980 and 1994, when attempts began to reconstruct an aboriginal fire regime (Russel-Smith et al. 1997). However, although 40 percent of the area burned at frequencies of at least 1 in 3 years, such frequencies cannot be sustained without substantial loss of obligate seeder species, which made up 54 percent of the sampled shrubby heath flora (Russel-Smith et al. 1998). The data also supported other observations concerning the catastrophic impact of contemporary fire regimes on fire-sensitive vegetation types in other sandstone regions of northern Australia. Based on these insights, Russel-Smith and his colleagues argue for the development of a program of strategic fire management, using low-intensity, early dry season burns

TABLE 7.4

Salient features of four large fire-prone conservation areas.

	KRUGER NATIONAL PARK, SOUTH AFRICA	MASAI-MARA GAME RESERVE AND SERENGETI NATIONAL PARK, EAST AFRICA	KAKADU NATIONAL PARK, NORTHERN AUSTRALIA	YELLOWSTONE NATIONAL PARK, NORTHERN UNITED STATES
Important features	All major species of African wildlife; perennial rivers.	Famous for its annual migration of wildebeest.	Australia's largest national park, known mainly for its wetlands and aboriginal rock art.	Established to protect thermal features such as geysers and hot springs but also known for its wildlife.
Approximate area (km²)	20,000	25,000	20,000	8,900
Dominant vegetation	Savanna woodlands.	Open grassland with lightly wooded patches.	Low, open woodland with scattered emergent trees; extensive floodplains.	About 80% forested and 13% meadow or grassland.
Fire climate	Wet summers and warm, dry winters.	Equatorial, with a marked dry season.	Monsoonal summer rains, with warm, dry winters.	Boreal, with very cold winters and warm, dry summers.
Fire regime	Winter fires at 3- to 6-year intervals.	Dry season fires at varying intervals.	Winter fires at approximately 3-year intervals.	Summer fires of low intensity every 20–25 years, with occasional (every 200 years?) high-intensity crown fires.
Issues relating to fire ecology and management	The relative merits of fixed and variable fire regimes. The role of fire-herbivory interactions in shaping vegetation structure.	Long-term (>100-year) changes in vegetation structure resulting from fire-herbivory interactions.	Reconstruction of an aboriginal fire regime to reduce the impact of fires on fire-sensitive species.	The role of infrequent, stand-replacing crown fires compared with frequent surface fires; the role of prescribed burning.
Similarities to Kruger National Park	—	The importance of fire-herbivory interactions.	The structure of the dominant vegetation.	Ecologically, there are few similarities; philosophically, the debate on the role of "natural" vs. prescribed fire has been similar.
Differences from Kruger National Park	—	A longer and more marked cycle between vegetation dominated alternately by trees or grasses.	The presence of fire-sensitive species. The absence of any large herbivores.	The phenomenon of stand-replacing fires and a much longer postfire succession. The absence of megaherbivores.

to break up continuous fine fuels, as was practiced in the past under intensive, traditional fire management by aboriginal people.

The role of aboriginal fire features prominently in debates around appropriate fire regimes for biodiversity conservation in Australia. There is little doubt that aboriginal burning played a central role in maintaining the landscapes subsequently colonized by Europeans, but the question of the original impact of humans is speculative (Bowman 1998). However, evidence has emerged that a fire regime of early dry season, low-intensity fires, as practiced uninterrupted by aboriginal people in areas adjacent to Kakadu, has resulted in a far more diverse ecosystem than exists in the Kakadu itself (Yibarbuk et al. 2001). The study attributed the ecological integrity of the site to continued human occupation and the maintenance of traditional fire management practices, which suppressed otherwise abundant annual grasses and limited the accumulation of fuels in perennial grasses or other litter. By contrast, fire management in Kakadu had failed to emulate the same type of fire regime, with resultant loss of diversity.

Yellowstone National Park

Yellowstone's fire history has been inferred from fire scars on long-lived trees, from sediment deposits dating back to 7,000 BP, and from charcoal deposits in lake sediments dating back to 17,000 BP. Low-intensity fires occurred at 20- to 25-year intervals, but high-intensity, stand-replacing fires occurred at intervals of 200 to 400 years. They were more frequent in warmer, drought-prone periods, such as AD 900–1300, than in cooler, wetter periods, such as 1550–1850 (Meyer and Wells 1997). The largest fires observed by managers in Yellowstone between 1931 and 1987 burned less than 15,000 ha. However, large fires occurred in Yellowstone in the summer of 1988; they burned 36 percent of Yellowstone and lasted for almost 2 months, until the onset of winter rains. The summer of 1988 was the driest in the 112 years, and many people were surprised at the extent of the 1988 fires (Christensen et al. 1989). However, recently completed fire history studies (Romme and Despain 1989; Millspaugh and Whitlock 1995) indicate that the 1988 fires were within the normal prehistoric fire sizes.

In the first century of Yellowstone's history, fire was regarded as a destructive force that should be fought. However, the policy changed in 1972 to one of allowing some lightning-ignited fires to run their course. Under this "natural" fire management, human-caused fires were suppressed, but lightning fires were allowed to burn unless they jeopardized human life, property, or endangered species (Franke 2000). Of the 368 lightning-caused fires that occurred from 1972 to 1987, 235 were allowed to burn; 208 of these covered 0.5 ha. Even in 1981, when Yellowstone had its busiest fire season during this period, only about 1 percent (8,500 ha) burned. Then in 1988, a combination of conditions never before seen in Yellowstone's history led to the burning of 327,000 ha. Under the existing policy, the fires initially were allowed to burn, but when they reached alarming proportions, the fires were regarded as "outside prescription," and huge sums of money were spent in a vain attempt to contain the fires. Despite changes to the fire policy after 1988 (Franke 2000), there is still controversy. Proponents of prescribed burning argue for setting prescribed fires to control the situation. This is a philosophy of "keeping nature in its proper place" and managing the area as a "safe and attractive" destination (Franke 2000, p. 45). It contrasts with the view that any interventions to suit the convenience and preferences of humans have no place in large national parks. Yellowstone illustrates the dilemma facing managers of natural areas elsewhere, but it is magnified by the infrequent occurrence of large and potentially damaging fires after many decades of fuel buildup.

Understanding the Role of Fire in Large Conservation Areas

The issues relating to fire ecology and management have been similar in the two African examples considered here but quite different in the Australian and American examples (Table 7.4). One of the striking differences between African savannas, and other fire-prone areas is the presence of megaherbivores (elephant, black and white rhinoceros, and hippopotamus). Megaherbivores often are considered a prime controlling influence on the tree-shrub-grass balance (Owen-Smith 1988; Owen-Smith and Danckwerts 1997), but we believe that fire is also a related, significant factor. Fire-megaherbivore interactions clearly are not significant in Australia and America. Africa remains a special case, where megaherbivores are extant and still have a major impact on the dynamics of vegetation, in combination with fire, in large conservation areas.

One question that is clearly highlighted in Australia is the importance of human fire practices on the ecology, survival, and conservation of the biota. The demonstration that aboriginal fire practices have resulted in diverse ecosystems and that practices that stopped or changed the traditional patterns had negative results must be considered in African conservation areas. Humans would have had a longer influence on fire regimes in Africa than anywhere else. Although archaeologists have championed the role of humans in influencing vegetation change, their arguments have not been critically examined by ecologists (Bond 1997). In Kruger, therefore, there is no basis for unraveling the questions of whether a "natural" fire regime (eliminating all human influence) should or can be established, what such a regime would look like, and whether such a regime is desirable. What has become clear is that a fire regime that relies solely on lightning ignitions is not practical and that a combination of human ignitions with some proportion of lightning-ignited fires may be the best compromise.

Conclusion

The experience gained over several decades of fire management in Kruger suggests that the lack of variability associated with prescribed burning at a fixed return period has been detrimental. The area that burns each year is strongly related to grass biomass, and it seems that management interventions may not be able to reduce this area significantly. However, it may well be possible for managers to have a significant influence on the variability of interfire periods by carefully choosing the areas where inevitable fires will occur. Evidence seems to suggest that by increasing this variability, and especially by ensuring that occasional longer intervals between fires occur, managers can ensure that trees are recruited into fire-resistant size classes. A detailed understanding of the responses of important individual species (currently lacking) would help in choosing the correct degree of variability and should form an important component of new research to assist managers.

References

Bessie, W. C., and E. A. Johnson. 1995. The relative importance of fuels and weather on fire behaviour in subalpine forests. *Ecology* 76:747–762.

Biggs, H. C., and A. L. F. Potgieter. 1999. Overview of the fire management policy of the Kruger National Park. *Koedoe* 42:101–110.

Bond, W. J. 1997. Fire. Pages 421–446 in R. M. Cowling, D. M. Richardson, and S. M. Pierce (eds.), *The vegetation of southern Africa*. Cambridge, UK: Cambridge University Press.

Bond, W. J., and B. W. van Wilgen. 1996. *Plants and fire*. London: Chapman & Hall.

Bowman, D. M. J. S. 1998. The impact of aboriginal landscape burning on the Australian biota. *New Phytologist* 140:385–410.

Brockett, B. H., H. C. Biggs, and B. W. van Wilgen. 2001. A patch mosaic burning system for conservation areas in southern Africa. *International Journal of Wildland Fire* 10:169–183.

Byram, G. M. 1959. Combustion of forest fuels. Pages 61–89 in K. P. Davis (ed.), *Forest fire: control and use*. New York: McGraw-Hill.

Christensen, N. L., J. K. Agee, P. F. Brussard, J. Hughes, D. H. Knight, G. W. Minshall, J. M. Peek, S. J. Pyne, F. J. Swanson, J. W. Thomas, and others. 1989. Interpreting the Yellowstone fires of 1988. *BioScience* 39:678–685.

Dublin, H. T. 1995. Vegetation dynamics in the Serengeti-Mara ecosystem: the role of elephants, fire and other factors. Pages 71–90 in A. R. E. Sinclair and P. Arcese (eds.), *Serengeti II: dynamics, management and conservation of an ecosystem*. Chicago: University of Chicago Press.

Eckhardt, H. C., B. W. van Wilgen, and H. C. Biggs. 2000. Trends in woody vegetation cover in Kruger, South Africa, between 1940 and 1998. *African Journal of Ecology* 38:108–115.

Enslin, B. W., A. L. F. Potgieter, H. C. Biggs, and R. Biggs. 2000. Long term effects of fire frequency and season on the woody vegetation dynamics of the *Sclerocarya birrea/Acacia nigrescens* savanna of the Kruger National Park. *Koedoe* 43:27–37.

Foster, D. R., D. H. Knight, and J. F. Franklin. 1998. Landscape patterns and legacies resulting from large, infrequent forest disturbances. *Ecosystems* 1:497–510.

Franke, M. A. 2000. *Yellowstone in the afterglow: lessons from the fires.* Mammoth Hot Springs, WY: U.S. National Park Service.

Hartvigsen, G., A. Kinzig, and G. Peterson. 1998. Use and analysis of complex adaptive systems in ecosystem science: overview of special section. *Ecosystems* 1:27–430.

Higgins, S. I., W. J. Bond, and W. S. W. Trollope. 2000. Fire, resprouting and variability: a recipe for grass-tree coexistence in savanna. *Journal of Ecology* 88:213–229.

Hobbs, R. J., and L. Atkins. 1988. Spatial variability of experimental fires in southwest Western Australia. *Australian Journal of Ecology* 13:295–299.

Jacobs, O. S., and R. Biggs. 2001. The effect of different fire treatments on the population structure and density of the Marula (*Sclerocarya birrea* (A. Rich.) Hochst. subsp. *caffra* (Sond.) *kokwaro* (Kokwaro and Gillet 1980) in the Kruger National Park. *African Journal of Range and Forage Science* 18:13–24.

Mentis, M. T., and A. W. Bailey. 1990. Changing perceptions of fire management in savanna parks. *Journal of the Grassland Society of Southern Africa* 7:81–85.

Meyer, G. A., and S. G. Wells. 1997. Fire-related sedimentation on alluvial fans, Yellowstone National Park. *Journal of Sedimentary Research* A67:776–791.

Mills, M. G. L., H. C. Biggs, and I. J. Whyte. 1995. The relationship between rainfall, lion predation and population trends in African herbivores. *Wildlife Research* 22:75–88.

Millspaugh, S. H., and C. Whitlock. 1995. A 750-year fire history based on lake sediment records in central Yellowstone National Park, USA. *Holocene* 5:283–292.

Owen-Smith, N. 1988. *Megaherbivores: the influence of a very large body size on ecology.* Cambridge, UK: Cambridge University Press.

Owen-Smith, N., and J. E. Danckwerts. 1997. Herbivory. Pages 397–420 in R. M. Cowling, D. M. Richardson, and S. M. Pierce (eds.), *The vegetation of southern Africa.* Cambridge, UK: Cambridge University Press.

Plug, I. 1987. Iron Age subsistence strategies in the Kruger National Park, South Africa. *Archaeozoologia* 1:117–125.

Plug, I. 1989. Aspects of life in the Kruger National Park during the early Iron Age. *South African Archeological Society Goodwin Series* 6:62–68.

Rogers, K. H. 1997. Operationalizing ecology under a new paradigm: an African perspective. Pages 60–77 in S. T. A. Pickett (ed.), *The ecological basis of conservation.* London: Chapman & Hall.

Romme, W. H., and D. G. Despain. 1989. Historical perspective on the Yellowstone fires of 1988. *BioScience* 39:695–699.

Romme, W. H., E. H. Everham, L. E. Frelich, M. A. Moritz, and R. E. Sparks. 1998. Are large, infrequent disturbances qualitatively different from small, frequent disturbances? *Ecosystems* 1:524–534.

Russel-Smith, J., P. G. Ryan, and R. Durieu. 1997. A LANDSAT MSS–derived fire history of Kakadu National Park, monsoonal northern Australia, 1980–94: seasonal extent, frequency and patchiness. *Journal of Applied Ecology* 34:748–766.

Russel-Smith, J., P. G. Ryan, D. Klessa, G. Waight, and R. Harwood. 1998. Fire regimes, fire-sensitive vegetation and fire management of the sandstone Arnhem Plateau, monsoonal northern Australia. *Journal of Applied Ecology* 35:829–846.

Scholes, R. J., and B. H. Walker. 1993. *An African savanna: synthesis of the Nylsvley study*. Cambridge, UK: Cambridge University Press.

Stocks, B. J., B. W. van Wilgen, and W. S. W. Trollope. 1997. Fire behaviour and the dynamics of convection columns in African savannas. Pages 47–55 in B. W. van Wilgen, M. O. Andreae, J. G. Goldammer, and J. A. Lindesay (eds.), *Fire in southern African savannas: ecological and atmospheric perspectives*. Johannesburg: Witwatersrand University Press.

Thompson, P. J. 1975. The role of elephants, fire and other agents in the decline of a *Brachystegia boehmii* woodland. *Journal of the Southern African Wildlife Management Association* 5:11–18.

Trollope, W. S. W. 1999. Veld burning: savanna. Pages 236–242 in N. M. Tainton (ed.), *Veld management in southern Africa*. Pietermaritzburg: Natal University Press.

Trollope, W. S. W., H. C. Biggs, A. L. F. Potgieter, and N. Zambatis. 1995. A structured vs. a wilderness approach to burning in Kruger in South Africa. Pages 574–575 in *Proceedings of the Vth International Rangeland Congress, 1995*. Salt Lake City, UT. Denver, CO: Society for Rangeland Management.

Trollope, W. S. W., A. L. F. Potgieter, H. C. Biggs, and L. A. Trollope. 1999. *Report on the Experimental Burning Plots (EBP) trial in the major vegetation types of the Kruger National Park*. Unpublished report, Scientific Services, South African National Parks, Skukuza, South Africa.

Trollope, W. S. W., L. A. Trollope, H. C. Biggs, D. Pienaar, and A. L. F. Potgieter. 1998. Long-term changes in the woody vegetation of the Kruger National Park, with special reference to the effects of elephants and fire. *Koedoe* 41:103–112.

Turner, M. G., and V. H. Dale, 1998. Comparing large, infrequent disturbances: what have we learned? *Ecosystems* 1:493–496.

van der Schijff, H. P. 1958. Inleidende verslag oor veldbrandnavorsing in die Nasionale Kruger-wildtuin. *Koedoe* 1:60–92.

van Wilgen, B. W., H. C. Biggs, S. O'Regan, and N. Mare. 2000. A fire history of the savanna ecosystems in the Kruger National Park, South Africa between 1941 and 1996. *South African Journal of Science* 96:167–178.

van Wilgen, B. W., H. C. Biggs, and A. L. F. Potgieter. 1998. Fire management and research in the Kruger National Park, with suggestions on the detection of thresholds of potential concern. *Koedoe* 41:69–87.

van Wilgen, B. W., C. S. Everson, and W. S. W. Trollope. 1990a. Fire management in southern Africa: some examples of current objectives, practices and problems. Pages 179–209 in J. G. Goldhammer (ed.), *Fire in the tropical biota: ecosystem processes and global challenges*. Berlin: Springer-Verlag.

van Wilgen, B. W., K. B. Higgins, and D. U. Bellstedt. 1990b. The role of vegetation structure and fuel chemistry in excluding fire from forest patches in the fire-prone fynbos shrublands of South Africa. *Journal of Ecology* 78:210–222.

van Wilgen, B. W., and R. J. Scholes. 1997. The vegetation and fire regimes of southern hemisphere Africa. Pages 27–46 in B. W. van Wilgen, M. O. Andreae, J. G. Goldammer, and J. A. Lindesay, *Fire in southern African savannas: ecological and atmospheric perspectives*. Johannesburg: Witwatersrand University Press.

van Wyk, P. 1971. Veld burning in the Kruger National Park. *Proceedings of the Annual Tall Timbers Fire Ecology Conference* 11:9–31.

Walker, B. H. 1989. Diversity and stability in ecosystem conservation. Pages 121–130 in D. Western and M. C. Pearl (eds.), *Conservation for the twenty-first century*. New York: Oxford University Press.

Walker, B. H., and I. Noy-Meir. 1982. Aspects of stability and resilience in savanna ecosystems. Pages 556–590 in B. J. Huntley and B. H. Walker (eds.), *Ecology of tropical savannas*. Berlin: Springer-Verlag.

Walter, H. 1971. *Ecology of tropical and subtropical vegetation*. Edinburgh: Oliver & Boyd.

Whateley, A. M., and A. J. Wills. 1996. Colonization of a sub-tropical woodland by forest trees in South Africa. *Lammergeyer* 44:19–30.

Whelan, R. J. 1995. *The ecology of fire*. Cambridge, UK: Cambridge University Press.

Yeaton, R. I. 1988. Porcupines, fires and the dynamics of the tree layer of the *Burkea africana* savanna. *Journal of Ecology* 76:1017–1029.

Yeaton, R. I., and W. J. Bond. 1991. Competition between two shrub species: dispersal differences and fire promote co-existence. *American Naturalist* 138:328–341.

Yibarbuk, D., P. J. Whitehead, J. Russell-Smith, D. Jackson, C. Godjuwa, A. Fisher, P. Cooke, D. Choquenot, and D. M. J. S. Bowman. 2001. Fire ecology and Aboriginal land management in central Arnhem Land, northern Australia: a tradition of ecosystem management. *Journal of Biogeography* 28:325–344.

Chapter 8

Surface Water Availability: Implications for Heterogeneity and Ecosystem Processes

ANGELA GAYLARD, NORMAN OWEN-SMITH,
AND JESSICA REDFERN

Water provision ranks with fire and culling policies as the main interventions available to managers of African savanna ecosystems (Owen-Smith 1996). However, our understanding of the relationship between biodiversity and surface water availability and distribution is limited because this relationship has not been examined within a hierarchy of scales. For example, Noss (1990) proposes that compositional, structural, and functional attributes of biodiversity should be examined at landscape to individual levels of organization. The availability and distribution of water sources can influence ecosystem structure and function at a range of scales and organizational levels through its influence on various processes and feedbacks affecting both animals and plants. For example, at a landscape level, the distributions of herbivore home ranges and vegetation communities may be influenced by the mosaic created by water sources. At the ecosystem and community levels, ecosystem processes such as nutrient cycling, predator-prey interactions, and interspecific competition may be influenced by the location of water sources. At a population level, surface water availability and distribution may influence herbivore survivorship through processes such as droughts and disease. Finally, individual herbivore drinking needs provide the mechanisms that produce the patterns observed at higher organizational levels.

In Kruger, management of surface water availability has been a priority since a water provision program was established in the 1930s (Joubert 1986). Although Kruger is traversed by five perennial rivers and a number of seasonal rivers, few natural sources of perennial water existed between these rivers when it was established in 1898 (Pienaar 1970). Given the influence of surface water availability on the distribution of most species of large mammalian herbivores (Western 1975; Owen-Smith 1996), large sections of Kruger supported very few herbivores, especially during the dry season, when animals concentrated along rivers and around the few long-lasting pools (Pienaar 1970). Furthermore, completion of the western boundary fence in 1961 prevented animal movements that may have taken place westward, possibly even as far as the wetter foothills of the interior escarpment.

The policy of establishing artificial water-points was adopted to support herbivore populations during droughts. Specifically, the aim was to provide an adequate network of reliable water-points through the construction of boreholes and earth dams (Pienaar 1970, 1983). Additionally, weirs and sluices were constructed to ensure adequate flow of perennial rivers during droughts, provide surface water for ungulates, and protect aquatic habitats, thus alleviating concerns that upstream demands on water would reduce the reliability of perennial rivers.

The water provisioning program was also designed to minimize the influence of temporal variability in rainfall (Gertenbach 1980) on herbivore populations (Pienaar 1983). In particular, it was suggested that increases in population sizes during periods of high rainfall could lead to population crashes during droughts if the area that populations could occupy under drought conditions was too limited (Pienaar 1983). Furthermore, there was a concern that without such a water stabilization program populations of many rare species, notably roan, tsessebe, and eland, could be lost during severe droughts, at least over a large section of the lower rainfall areas in the north (Pienaar 1983). At that time it was believed that an interventionist policy with regard to water would ensure "a healthy and productive environment which can accommodate long-term natural changes, is conducive to relative stability and which ensures population fluctuations of manageable proportions" (Pienaar 1983, p. 29).

Reasonable though this policy may have seemed at that time, the anticipated benefits did not materialize. Populations of rare antelope species such as roan, tsessebe, sable, and reedbuck fell to dangerously low levels during the extended drought period of 1982–1987, despite the widespread provision of water-points (Harrington et al. 1999; Chapter 15, this volume). Additionally, most dams did not hold water through extended dry periods. Indeed, research suggests that the provision of artificial water-points may have reduced herbivore diversity by expanding the distribution of common water-dependent species such as zebra and wildebeest, and concomitantly predatory species such as lions (Smuts 1978), at the expense of less common herbivores (Owen-Smith 1996). Moreover, recent evidence suggests that the widespread distribution of water-points may have homogenized woody vegetation structure across Kruger by spreading elephant impacts evenly across the landscape (Gaylard et al. in prep.).

Consequently, the former policy is now being reversed by closing a large proportion of Kruger's artificial water-points (Pienaar et al. 1997). This removal signals a shift in Kruger's management from the past paradigm of suppressing variability toward maintaining or even promoting the spatial and temporal fluxes believed to be necessary for sustaining biodiversity and crucial ecosystem processes (Chapters 2 and 3, this volume). In this chapter, we present findings from water-related research in Kruger that address the availability and distribution of surface water and its influence on ecosystem structure and function. Specifically, we describe the variability in the availability and distribution of surface water in Kruger at a range of spatial and temporal scales. We also explore the influence of surface water on terrestrial organisms, considering effects from the landscape to individual levels. We then consider the vegetation impacts created by the relationships between terrestrial organisms and water sources as well as the consequences of these impacts on heterogeneity and ecosystem processes. Finally, we discuss what the Kruger water provisioning program has taught us about the management of surface water.

Patterns of Surface Water Availability and Distribution

Surface water sources in Kruger comprise both natural and artificial sources.

Natural Sources of Surface Water

At the broadest scale, heterogeneity in the availability and distribution of natural surface water in Kruger is created by geology (Gaylard et al. in prep.). Specifically, the granitic areas on the western side of Kruger have a higher stream density than the basaltic plains in the east. Within each of these landscapes, surface water can be further differentiated into riparian and savanna water sources. Riparian water sources include perennial rivers that provide linear and temporally constant water sources. Five perennial rivers traverse Kruger from west to east (Chapters 9 and 21, this volume). Distances separating these rivers range from 35 km in the south to 180 km in the north.

Many ephemeral rivers also traverse Kruger. The larger ephemeral rivers retain surface water in numerous pools that are usually persistent throughout the dry season (Figure 8.1). Additionally, almost all of the larger ephemeral rivers, as well as several of the smaller tributaries, have dams or weirs constructed along their courses. These pools and dams form spatially and temporally variable point sources of surface water in the riparian landscape.

Natural savanna water sources are also point water sources and include springs, pans, and seasonally flooded drainage depressions known as *vleis*. The majority of the springs occur in the two higher-rainfall sections of Kruger, specifically the southwest and the northwest (Figure 8.1). Water availability in the numerous pans varies over the course of the dry season. The wallowing of animals such as buffalo plays an important role in the development and maintenance of pans (Young 1970; Chapter 5, this volume). The *vleis*

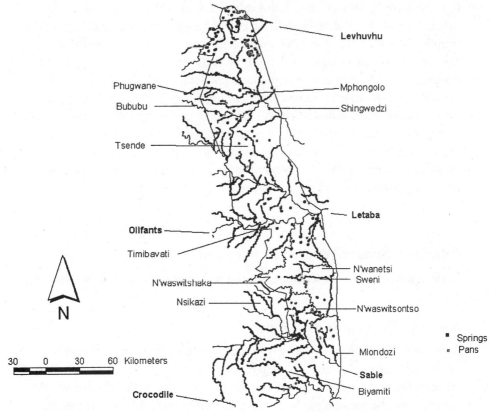

FIGURE 8.1. Location of major rivers, natural springs, and fountains in the Kruger National Park (perennial river names in bold).

are characterized by alluvial soils and tall grass swards (Joubert 1986) and occur mainly in the eastern basaltic regions that have poor infiltration rates.

Temporal variability in surface water availability and distribution is created by rainfall, evaporation rates, and geology (Gaylard et al. in prep.). Rainfall varies on an annual basis, with rain occurring mainly between October and March (Chapter 5, this volume). At the decadal time scale, an 18-year rainfall cycle consists of 9 years of above-average and 9 years of below-average rainfall (Gertenbach 1980). Additionally, the El Niño–Southern Oscillation produces less predictable fluctuations in rainfall (Chapter 5, this volume; Scholes and Walker 1993). Whereas rainfall provides the source of surface water, evaporation rates determine the persistence of water sources. The high radiation received by southern African savannas produces high temperatures and saturation deficits, and they consequently have high potential rates of evaporation

(Scholes and Walker 1993). Finally, the persistence of water sources is also influenced by geology. Where scouring has exposed bedrock, often in the bends of ephemeral rivers, surface water persists as pools along the river course. Surface water also persists for longer periods in pans occurring on the clayey basaltic soils in the east than on the more permeable granitic soils in the west.

Seasonal rivers typically flow once or twice during the rainy season. As the dry season progresses, surface water evaporates or drains into the riverbed, leaving a small number of persisting pools. Above-average rainfall years result in a larger number of persisting pools or rivers continuing to flow throughout the dry season. Conversely, when rainfall is below average, only a few of the deeper pools persist through the end of the dry season. Rainfall in previous years also has an important cumulative influence on how much surface water persists in seasonal rivers through the current year's dry season. Specifically, a series of dry years results in few persisting pools even after good rains, and more pools persist after a series of wet years even when the current rainy season is relatively dry (Table 8.1).

Artificial Sources of Surface Water

The water provisioning program superimposed spatially and temporally constant sources of surface water (more than 300 boreholes, earth dams, sluices, and weirs) on the heterogeneous availability and distribution of natural surface water. According to Pienaar (1970), the principles underlying the water provisioning program were to stabilize existing natural supplies, to provide additional artificial supplies in areas where natural supplies had previously existed, and to construct dams where storage capacity was adequate to guarantee water through periods of severe drought. Pienaar (1970) also envisaged the option of shutting down particular boreholes in rotation to minimize local overgrazing, but there is no record that this option was ever used. The initial boreholes and dams were constructed in 1933 after a persistent dry period. Other major projects were completed in the 1950s and 1960s (Figure 8.2). By 1995, 365 boreholes and about 50 earth dams had been constructed.

On a broad spatial scale, these artificial water sources have reduced the proportion of Kruger that is more than 5 km from surface water. Specifically, less than 20 percent is estimated to be further than 5 km from surface water during extreme drought conditions (Figure 8.3). On a finer spatial scale, water quality varies between boreholes and may influence their usage by different species. Sixty percent of the boreholes examined had pH values above 8, electrical conductivity ranged from 125 to 27,000 mS/m, and total dissolved solids ranged from 80 to 17,418 mg/L. Nitrite values were detectable for less than 5 percent of the samples taken. Nitrates, chlorine, sulfates, carbonates, fluo-

TABLE 8.1

Variability in surface water abundance (measured as the maximum percentage of the riverbed with surface water between seasonal periods), based on observations along three seasonal rivers in northern Kruger over three wet/dry seasons between 1997 and 1999 and 4 additional years for which records were available between 1942 and 1989 (from Gaylard et al. in prep).

YEAR	EARLY DRY (MAY)	MID-DRY (JUNE–JULY)	LATE DRY (AUG.–SEPT.)	WET (OCT.–APR.)	PREVIOUS ANNUAL RAINFALL (IN MM; JULY–JUNE)
1942			39		352
1971			5		464
1977			66		642
1989			2		293
1997	2.26	2.26	1.46	1.41	434
1998	0.71	0.71	0.56	95.25	205
1999	2.26	2.26	2.26	100	982

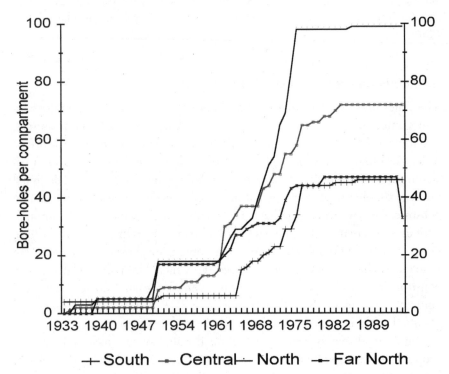

FIGURE 8.2. Number of boreholes supplied in different regions of Kruger during the Water for Game Program (1933–1994).

ride, potassium, sodium, magnesium, calcium, sodium bicarbonate, and sodium carbonate seldom reached concentrations that have been designated as unacceptable in drinking water for livestock (Young 1970). However, bicarbonate concentrations were almost always above acceptable levels for domestic drinking water.

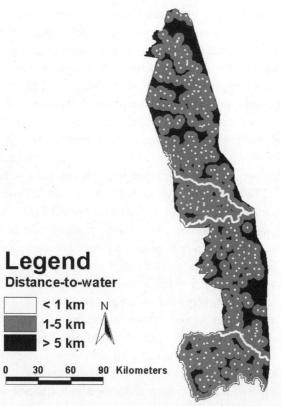

FIGURE 8.3. A projected worst-case scenario of the proportion of Kruger at varying distances from surface water under extreme drought conditions. (This includes all currently open boreholes and perennial rivers and springs that maintain surface water through extended dry periods. All dams and seasonal rivers are excluded.)

Faunal Patterns Influenced by Surface Water Use

The use of surface water by animals influences faunal distribution patterns across a range of spatial scales.

Landscape-Level Patterns

On the broadest scale, the mosaic created by the availability and distribution of surface water can influence species distribution patterns across the landscape. For example, wildebeest, buffalo, and giraffe generally are found within 0.5–1 km of water when feeding, and zebra and impala are found within 1–2 km of water (Wentzel 1990). Waterbuck prefer areas closer than 1 km from water regardless of activity (Engelbrecht 1986; Wentzel 1990). Less water-dependent species such as kudu and steenbok generally are found feeding more than 3 km from water (Cohen 1987; Wentzel 1990). However, large herbivore distributions relative to water also are influenced by geology and rainfall (Redfern 2002). In general, the dry season distribution of large herbivores relative to surface water (using ecological aerial census data collected during the dry seasons from 1981 to 1993; Viljoen 1996) shows a decline in herd density with increasing distance from water that is greatest for waterbuck, buffalo, and elephant, weakest for zebra, wildebeest, giraffe, and kudu, and intermediate for impala (Redfern 2002). Buffalo, zebra, and waterbuck occur further from water in dry years, when there is a broad-scale reduction in forage quantity. In contrast, wildebeest, elephant, and impala occur further from water in the western granitic landscapes, in which access to high-quality forage at broad scales is low, than in the eastern basaltic landscapes. These analyses suggest that in Kruger large herbivores, particularly grazers, alter their distributions such that they occur further from water to mitigate poor forage conditions. However, these findings must be interpreted within a context in which a maximum of only 8 percent of the available area was more than 5 km from water at the time of each census. In contrast, zebra on the northern plains selected areas 1–5 km from water but did not show a preference for landscapes defined by vegetation during average rainfall years. The closure of artificial water sources on the northern plains has resulted in a large proportion of this region occurring more than 5 km from water sources and hence greater water rather than forage limitations for some zebra (Davidson 2002).

Temporal variability in surface water availability also influences herbivore distribution patterns. In other parts of Africa, where artificial waterholes have not been introduced, herbivores show strong responses to the seasonal availability of surface water and green vegetation (Western 1975). When surface water and forage are scarce in the dry season, herbivores feed in areas close to surface water. With the first rains of the wet season, natural surface water and green, moisture-rich plants are abundant, and herbivores use these grazing areas that were avoided during the dry season when there was little surface water (Young 1970). However, the numerous boreholes provided in Kruger's previously waterless savanna areas have changed the temporal distribution patterns of water-dependent herbivores. Specifically, the frequency and duration of elephant visits to the riparian zone (the primary source of dry season surface water in areas without artificial sources) do not increase with changing surface water availability through the course of the dry season, even during very dry periods (Gaylard et al., in prep.). This is despite the fact that elephants traverse the riparian zone close to pools of surface water and focus their feeding activities around riparian surface water throughout the year.

Other species also exhibit a relationship between the mosaic created by surface water and faunal distribution patterns across the landscape. For example, the territories occupied by lion prides tend to be associated with the presence of perennial water-points, where sedentary water-dependent herbivores such as impala and waterbuck remain available as prey year-round (Smuts 1978). Additionally, the water provisioning program has affected the distribution of fish species. Construction of dams, such as the Kanniedood on the Shingwedzi River, has prevented the passage of anadromous fish and reduced their distribution (Olivier 2003). Construction of fishways has partly rectified the problem, as has the removal of some dams.

Community-Level Patterns

Compositional diversity of various faunal communities has been affected by the changes in surface water availability and distribution brought about by the water provisioning program. Diverse habitats are necessary for the coexistence of the spectrum of carnivores (Chapter 18, this volume); for example, cheetah prefer more open tree savanna, whereas wild dogs prefer more densely wooded areas. The compositional diversity of the carnivore community therefore is threatened by the homogenizing influence of herbivores on landscape vegetation structure when water provision is extensive.

The water provisioning program has also influenced the diversity of aquatic and avifaunal communities through the creation of habitats such as dams. For example, the diversity and number of fish species were reduced by the building of the Kanniedood Dam along the Shingwedzi River (Olivier 2003). The subsequent completion of a fishway at this dam has led to a slow recovery of the original fish community.

The decline in the roan antelope population has been attributed to increased competition for grazing with the escalating zebra population after the drilling of numerous boreholes on the northern basaltic plains (Grant 1999; Harrington et al. 1999). Finally, water-related changes in lion distributions and population sizes have altered the composition of the carnivore community. Increased scavenging opportunities provided by larger lion populations have caused the elimination of brown hyenas in Kruger through competition with concurrently increasing numbers of spotted hyena (Chapter 18, this volume).

Community-level patterns influenced by surface water availability and distribution therefore include direct changes to faunal compositional diversity and changes brought about indirectly by changing the frequency or intensity of ecosystem processes such as competition.

Population-Level Patterns

The availability and distribution of surface water also influence population growth rates via changes in immigration rates and animal survivorship. For example, zebra and wildebeest population growth over the past 20 years is positively correlated with increasing numbers of boreholes on the northern basaltic plains, whereas rare antelope (particularly roan) exhibit a negative correlation (Grant 1999). Similarly, the lion population in the central region doubled between 1933 and 1975 (Smuts 1978), coinciding with the period of augmentation and stabilization of water resources in this area. Survivorship within herbivore populations may be affected by the relationship between surface water distribution and forage availability. Specifically, a ubiquitous distribution of surface water results in spatially and temporally homogenous herbivore impacts across the landscape, leaving no areas of reserve forage during droughts (Walker et al. 1987). Whereas the central region of Kruger incurred moderate herbivore mortality (20–30 percent declines by sensitive species) during the severe 1982–1983 drought, the adjoining Klaserie Private Nature Reserve, where water-points were more closely spaced, incurred losses amounting to 70–90 percent of buffalo, waterbuck, warthog, wildebeest, zebra, sable, and impala populations. The catastrophic mortality in Klaserie was attributed to elevated population densities of species such as wildebeest compared with Kruger densities and the coalescence of heavy grazing areas resulting from the close spacing of water points (Walker et al. 1987).

Population growth rates and survivorship are also affected by the availability and distribution of surface water because large concentrations of animals around waterholes provide ideal opportunities for the spread of parasites and contagious diseases from infectious to susceptible animals (Young 1970). Parasitic organisms breeding in the water or mud surrounding waterholes include mosquitoes, fleas, and schistosomes (present in bilharzia-hosting snails). The large amounts of dung in the vicinity of waterholes also serve as breeding sites for several fly species. Diseases commonly spread at waterholes include mange (spread by skin contact), blackleg, anthrax, and foot-and-mouth disease (Young 1970; Chapter 17, this volume). Changed population sizes brought about by changes in the availability and distribution of surface water ultimately alter faunal diversity on larger scales by changing the relative abundance of animal species.

Individual-Level Patterns

The water needs of large herbivores differ between species. The frequency of visits to water sources has been found to be similar between seasons for elephant and buffalo. However, zebra, wildebeest, and impala visit water sources twice as frequently during the winter dry season as in the summer wet season (Young 1970). Seasonal differences in water source use by the less common ungulate species have not been studied.

Differences in species' water needs also exist at a finer temporal scale. Water intake per visit measured at artificial waterholes varies with the size of the animal (Young 1970; Table 8.2). The most water-dependent species (zebra, buffalo, and elephant) generally drink every 1–2 days during the dry season, although occasionally more than 3 days lapse between waterhole visits. Wildebeest and impala have somewhat longer intervals of 2–3 days between drinking (Young

TABLE 8.2

Water consumption and drinking intervals of the eight species consuming the largest amounts of water from artificial water sources in the Kruger National Park (from Young 1970).

SPECIES	MEAN WATER CONSUMPTION PER ANIMAL PER VISIT (LITERS; STANDARD DEVIATIONS IN PARENTHESES)	N (HERDS)	MEAN DRINKING INTERVAL (HOURS)	MAXIMUM DRINKING INTERVAL (HOURS)	MEAN DAILY WATER CONSUMPTION PER ANIMAL (LITERS)
Elephant (herds)	35 (6.9)	9			
Elephant (bulls)	77 (26.9)	23	43	>72	89
Zebra	4.7 (1.2)	26	35	>72	14
Buffalo	7.5 (2.3)	11	38	>47	21
Blue wildebeest	3.6 (0.4)	17	47	>76	8
Impala	0.9 (0.2)	27	68	>120	1
Giraffe	10.6 (2.7)	16			
Sable antelope	4.6 (0.7)	8			
Kudu	5.0 (1.3)	5			

1970). No information is available for the less water-dependent browsers, such as kudu and giraffe, or for the more rare antelope species.

The peak drinking time for most ungulate species is in the late morning (Young 1970). Wildebeest tend to arrive at water somewhat earlier than zebra. Buffalo commonly drink during the day in the winter dry season, but in summer their drinking time shifts to the late afternoon or early evening. Elephants drink both during the day and at night. Ambient temperatures above 35°C appear to inhibit animal movements to water, and few animals drink when the temperature is below 19°C (Young 1970).

Species also differ in their water source preferences. For example, elephants prefer drinking directly from cement reservoirs, perhaps preferring the cleaner water (Young 1970). Buffalo, zebra, and wildebeest tend to use earth dams rather than cement troughs, and buffalo also commonly drink from rivers (Young 1970). Additionally, roan antelope are reported to drink most often from natural pans or springs (Funston 1997). Individual herbivore drinking needs, drinking times, and water source preferences therefore provide the mechanisms that produce the patterns observed between herbivore populations and water sources at larger scales.

Vegetation Patterns Influenced by Surface Water

Surface water influences vegetation patterns across a range of spatial scales through its influence on faunal distribution.

Landscape-Level Patterns

Water sources can focus herbivore feeding activities, and hence vegetation impacts, in certain parts of the landscape. The availability and distribution of surface water therefore influence the landscape mosaic of vegetation condition and type. For example, the extent and degree of elephant impact on riparian woody vegetation in the north of Kruger are influenced by the particular spatial distribution of water sources (Gaylard et al. in prep.). In particular, the intensity of elephant impacts is patchy in seasonal riparian areas where water sources are isolated from other water; significantly heavier impacts occur in areas of the landscape that are close to surface water than in those that are remote from surface water. In contrast, homogeneous impacts occur in riparian areas with closely spaced water sources; the intensity of impacts is equal throughout the landscape. Thus, surface water distribution influences the structural heterogeneity of riparian woody vegetation at the landscape level. Patchy herbivore impacts also occur along linear sources of surface water such as perennial rivers. For example, heavier impacts arise on the herbaceous layers adjacent to pools sufficiently deep to support a hippo group than along other stretches of the river (McDonald 1992). These impacts are exacerbated where large concentrations of hippos occur at dams.

Ecosystem- or Community-Level and Population-Level Patterns

The radial movements of herbivores to and from individual water sources result in an increasing gradient in their grazing, browsing, and trampling impacts near waterpoints, known as "piospheres" (Andrew 1988; Owen-Smith 1996). Such piosphere effects have been documented for a range of vegetation variables in Kruger. A logistic equation (Thrash and Derry 1999) provides a good model of the relationship between increasing distance from water and herbivore-mediated changes to the herbaceous community, including a decreased abundance of annual plants (Thrash et al. 1993; Thrash 1998b), increased forage and fuel production potential (Thrash 1998b), increased basal cover (Thrash et al. 1991a; Thrash 1998b), and increased standing crop (Thrash 1998b). The sacrifice zone, where the herbaceous cover is severely depressed, typically occurs within 200–300 m of a water source (Thrash 1998b), whether natural or artificial (Thrash 1998a). However, projecting this sacrifice zone around water sources in Kruger indicates that no more than 4 percent of its area occurs within this zone (Harrington 1995; Thrash 1998b). The changes observed after the closure of boreholes on the northern plains illustrate the influence of surface water availability on herbaceous community composition. Borehole closure led to an increase in the relative abundance of decreased grass species, herbaceous cover, and grass height and a decrease in forbs and the abundance of bare patches (Davidson 1996; Zambatis 1996). Piosphere effects on herbaceous community composition therefore are localized and reversible.

Browsing at water sources similarly creates piosphere effects on woody vegetation. These piosphere effects include a decrease in tree density and an increase in scrub cover over a range of 2–5 km (observed at Wik-en-Weeg dam along the Phugwane river in the northwest; Thrash et al. 1991b). However, the presence of the dam did not lead to a change in the local woody community composition. Piosphere effects on woody vegetation at other types of waterpoints include a decrease in the density of mopane (*Colophospermum mopane*) shrubs and extremely high elephant damage to mopane and sickle bush (*Dichrostachys cinerea*) shrubs within 100–200 m of water (Fruhauf 1997). Moderate elephant damage to a range of tree species has been observed within 800 m of water sources, and lower levels of elephant damage to mopane shrubs occur up to 1,600 m from water (Fruhauf 1997). Piospheres are manifested in woody vegetation primarily through changes to local structural heterogeneity.

Piospheres traditionally have been viewed as undesirable components of the landscape because of the changes they bring about to local vegetation diversity. However, a multiscale approach emphasizes that because piospheres represent patches of contrasting vegetation, they contribute to heterogeneity at the landscape level. Increasing the number of water-points can negate this contribution if they are so close that the piospheres coalesce.

Other Ecosystem Processes Influenced by Surface Water Distribution

Large herbivores can influence ecosystem processes, such as soil, water, and wind erosion around water sources by trampling, digging, and reducing plant cover and litter (Cumming 1982). Establishing artificial water sources in previously waterless areas ultimately promotes soil erosion if large concentrations of herbivores denude soils through overgrazing (Venter 1990; Chapter 5, this volume). Cloven-hoofed animals and zebras, in particular, trample soil around waterholes and on the walls of earth dams (Young 1970). Severe trampling of topsoil can destroy the soil structure in clayey soils. In soils with a degraded structure, reduced infiltration and increased runoff accelerate water erosion on sloping areas and wind erosion on level areas (Venter 1990). In particular, large concentrations of game near the Tihongonyeni windmill were linked to the loss of 15–20 cm of topsoil (amounting to about 2.55 million tons of soil; Venter 1990). Within a 300-m radius of water, the infiltration rate is directly proportional to the distance from water sources (Thrash 1997).

Finally, concentrations of herbivores around waterholes can alter ecosystem function by changing nutrient distributions. For example, increased dung and urine in the vicinity of waterholes increases nutritive input to soils and grasses, leading to greater local productivity (James et al. 1999). Studies elsewhere in

Africa indicate that soils within 50 m of boreholes have significantly lower concentrations of aluminum and higher concentrations of nitrogen, calcium, magnesium, manganese, molybdenum, phosphorus, potassium, iron, and sodium than surrounding communities (Tolsma et al. 1987; Georgiadis and McNaughton 1990). Water-points therefore may be hotspots of elevated soil nutrients, although the effects of nutrient increases may not be expressed because of increased herbivore impacts and trampling (Chapter 10, this volume).

What Has the Kruger Experience Taught Us about Patterns of Surface Water Distribution and Ecosystem Heterogeneity?

The availability and distribution of surface water in semiarid savannas such as Kruger are spatially and temporally variable at multiple scales. Broad-scale heterogeneity in surface water availability and distribution is caused by geology (e.g., granite vs. basalt soils). Fine-scale heterogeneity in surface water availability and distribution is caused by rainfall patterns that vary over decades (wet and dry periods), years (El Niño oscillations), and seasons (summer rains and winter droughts) and over a range of spatial scales dependent on variability in storm size (Chapter 5, this volume). Evaporation and geology further contribute to heterogeneity in water source persistence.

The water provisioning policy aimed to stabilize the natural variability of the spatial extent and temporal persistence of water sources. It was envisaged that this stability would prevent the loss of rare game species and mitigate other long-term herbivore population fluctuations. These goals have not been realized: spatial and temporal heterogeneity in the availability of surface water persisted (although the ubiquitous boreholes have increased the availability and distribution of permanent water sources), populations of rare species declined, and populations of common herbivore species continued to fluctuate.

In view of this, management has reversed the former water provision policy and initiated closure of a number of boreholes (Pienaar et al. 1997). However, opinions regarding the effects of changed surface water provision on biodiversity in Kruger remain controversial because the issue has not been considered in a scaled fashion. By definition, biodiversity management in Kruger necessitates assessment of the consequences of various management strategies on diversity over a range of scales. This chapter integrates previous scale-neutral studies to improve our understanding of how variability in the availability and distribution of surface water influences animals and plants at landscape, ecosystem or community, population, and individual levels. This emphasizes how negative biodiversity consequences perceived at one scale may represent positive contributions to biodiversity at other scales.

Herbivore attraction to water sources is driven by processes operating at the individual level, such as drinking needs, drinking times, and water source preferences. In turn, the availability and distribution of water sources influence survivorship and immigration of herbivore and carnivore populations through ecosystem processes such as competition and predation. For example, studies of the herbivore community demonstrate that abundant water sources may exacerbate competition and predation, consequently benefiting common water-dependent herbivore species (e.g., zebra) at the expense of less water-dependent, rare species (e.g., roan). Studies of the carnivore community indicate that abundant water sources result in similar negative consequences for community diversity through competitive exclusion of less dominant species such as the brown hyena. Vegetation communities are influenced locally by surface water through the concentration of large herbivore impacts near water (piospheres). Local piosphere effects on vegetation communities include changes in herbaceous vegetation composition, as well as in woody vegetation structure, and reductions in the standing crop and basal cover of herbaceous vegetation. At a landscape level, studies have shown that herbivore distributions are correlated with distance to water and that the intensity of herbivore impacts is related to the spatial distribution of waterholes. Abundant water provision, though decreasing piosphere effects locally, therefore spreads these effects homogeneously across the landscape.

The evidence from Kruger studies presented in this chapter provides a crucial first step in developing an explicit understanding of how biodiversity is affected by the influence of surface water on plants and animals at multiple scales. However, a challenge for future research will be to explore the effect of surface water availability on a greater range of species (particularly smaller herbivores and carnivores, birds, and reptiles) to make explicit the linkages between surface water availability and biodiversity. As suggested by the studies presented in this chapter, such a synthesis must necessarily include a multiscale approach that addresses the impact of spatial and temporal variability in surface water availability and distribution on animals and plants at individual to landscape levels. Kruger currently contains an extensive network of artificial water sources, so a majority of the area lies within 5 km of permanent water sources. To evaluate the relationship between surface water availability and biodiversity in Kruger, managers must close a large number of these artificial water sources (Pienaar et al. 1997). Decisions regarding the number and location of boreholes to be closed should be made within Kruger's strategic adaptive management process (Chapters 3 and 4, this volume). This approach explicitly treats management actions as experiments (or learning by doing) and therefore incorporates the development of programs to monitor the outcome of management actions. Data collected during monitoring allow us to expand our understanding through modeling and predictive exercises, thereby improving future management decisions. Finally, analysis of these models provides an early warning system (Chapter 3, this volume) that can help us avoid problems caused by overabundant permanent water supplies or the possibility of thirst-induced drought mortalities.

References

Andrew, M. H. 1988. Grazing impact in relation to livestock watering points. *Trends in Ecology and Evolution* 3:336–339.
Cohen, M. 1987. *Aspects of the biology and behaviour of the steenbok* Raphicerus campestris *(Thunberg, 1811) in the Kruger National Park*. Unpublished D.Sc. thesis, University of Pretoria, Pretoria, South Africa.
Cumming, D. H. M. 1982. The influence of large herbivores on savanna structure in Africa. Pages 217–245 in B. J. Huntley and B. H. Walker (eds.), *Ecology of tropical savannas*. Berlin: Springer-Verlag.
Davidson, T. M. 1996. *Recovery of the herbaceous vegetation around three closed water holes on the northern plains of the KNP*. Unpublished B.Sc. honors thesis, University of the Witwatersrand, Johannesburg, South Africa.
Davidson, T. M. 2002. *Patch selection by zebra* (Equus burchelli) *on the northern plains landscape mosaic of the Kruger National Park: a hierarchical perspective*. Unpublished M.Sc. thesis, University of the Witwatersrand, Johannesburg, South Africa.
Engelbrecht, J. S. 1986. *Ekologiese skeiding van die rooibok* Aepyceros melampus Lichtenstein, *waterbok* Kobus ellipsiprymnus *(Ogilby) en die koedoe* Tragelaphus strepsiceros *(Pallas) in die sentrale gebied van die Nasionale Krugerwildtuin*. Unpublished M.Sc. thesis, University of Pretoria, Pretoria, South Africa.
Fruhauf, N. 1997. *Pattern of elephant impacts around artificial waterpoints in the northern region of the Kruger National Park*. Unpublished B.Sc. honors thesis, University of the Witwatersrand, Johannesburg, South Africa.
Funston, P. J. 1997. *Review and progress report on the Northern Plains Project May 1997*. South African National Parks Scientific Report 5/97. South African National Parks, Skukuza, South Africa.
Gaylard, A., R. N. Owen-Smith, and K. H. Rogers. In prep. a. Heterogeneity of surface water distribution: implications for large herbivore impacts and biodiversity management in savanna ecosystems. *BioScience*.
Gaylard, A., R. N. Owen-Smith, K. H. Rogers, and R. Biggs. In prep. b. Spatial and temporal variability in riparian habitat utilization by elephants in a southern African savanna. *Oecologia*.
Gertenbach, W. P. D. 1980. Rainfall patterns in the Kruger National Park. *Koedoe* 23:35–43.
Georgiadis, N. J., and S. J. McNaughton. 1990. Elemental and fiber contents of savanna grasses: variation with grazing, soil type, season and species. *Journal of Applied Ecology* 27:623–634.
Grant, R. 1999. *Status report on the Northern Plains Projects*. South African National Parks Scientific Report 01/99. South African National Parks, Skukuza, South Africa.

Harrington, R. 1995. *Herbivore and habitat changes associated with the roan antelope decline in the northern Kruger National Park.* Unpublished M.Sc. thesis, University of the Witwatersrand, Johannesburg, South Africa.

Harrington, R., N. Owen-Smith, P. C. Viljoen, H. C. Biggs, D. R. Mason, and P. Funston. 1999. Establishing the causes of the roan antelope decline in the Kruger National Park, South Africa. *Biological Conservation* 90:69–78.

James, C. D., J. Landsberg, and S. R. Morton. 1999. Provision of watering points in the Australian arid zone: a review of effects on biota. *Journal of Arid Environments* 41:87–121.

Joubert, S. C. J. 1986. *Masterplan for the management of the Kruger National Park.* Unpublished annual report, Vol. 1. National Parks Board, Skukuza, South Africa.

McDonald, I. R. 1992. *Impacts of large herbivores on the vegetation and landforms of the riparian zone of the Letaba River, Kruger National Park.* Unpublished report, University of the Witwatersrand, Johannesburg, South Africa.

Noss, R. F. 1990. Indicators for monitoring biodiversity: a hierarchical approach. *Conservation Biology* 4:355–364.

Olivier, L. 2003. *A study to determine the effectiveness and efficiency of the fishway in Kanniedood Dam in the Shingwedzi River, Kruger National Park.* Unpublished M.Tech. dissertation, Technikon Pretoria, Pretoria, South Africa.

Owen-Smith, N. 1996. Ecological guidelines for waterpoints in extensive protected areas. *South Africa Journal of Wildlife Research* 26:107–112.

Pienaar, D., H. Biggs, A. Deacon, W. Gertenbach, S. Joubert, F. Nel, L. van Rooyen, and F. Venter. 1997. *A revised water-distribution policy for biodiversity maintenance in the KNP.* Kruger Park Management Plan, Vol. 8. South African National Parks, Skukuza, South Africa.

Pienaar, U. de V. 1970. Water resources of the Kruger Park. *African Wildlife* 24:180–191.

Pienaar, U. de V. 1983. Management by intervention: the pragmatic/economic option. Pages 23–26 in R. N. Owen-Smith (ed.), *Management of large mammals in African conservation areas.* Pretoria: Haum Boekhandel.

Redfern, J. V. 2002. *Manipulating surface water availability to manage herbivore distributions in the Kruger National Park, South Africa.* Unpublished Ph.D. dissertation, University of California, Berkeley.

Scholes, R. J., and B. H. Walker. 1993. *An African savanna: synthesis of the Nylsvley study.* Cambridge, UK: Cambridge University Press.

Smuts, G. L. 1978. Interrelations between predators, prey and their environment. *BioScience* 28:316–320.

Thrash, I. 1997. Infiltration rate around drinking troughs in the Kruger National Park. *Journal of Arid Environments* 35:617–625.

Thrash, I. 1998a. Impact of large herbivores at artificial watering points compared to that at natural water points in Kruger National Park, South Africa. *Journal of Arid Environments* 38:315–324.

Thrash, I. 1998b. Impact of water provision on herbaceous vegetation in Kruger National Park, South Africa. *Journal of Arid Environments* 38:437–450.

Thrash, I., and J. F. Derry. 1999. The nature and modelling of biospheres: a review. *Koedoe* 42:73–94.

Thrash, I., P. J. Nel, G. K. Theron, and J. du P. Bothma. 1991a. The impact of water provision for game on the herbaceous basal cover around a dam in the Kruger National Park. *Koedoe* 34:121–130.

Thrash, I., P. J. Nel, G. K. Theron, and J. du P. Bothma. 1991b. The impact of water provision for game on the woody vegetation around a dam in the Kruger National Park. *Koedoe* 34:131–148.

Thrash, I., G. K. Theron, and J. du P. Bothma. 1993. Impact of water provision on herbaceous plant community composition in the Kruger National Park. *African Journal of Range and Forage Science* 10:31–35.

Tolsma, D. J., W. H. O. Ernst, and R. A. Verwey. 1987. Nutrients in soil and vegetation around two artificial waterpoints in eastern Botswana. *Journal of Applied Ecology* 24:991–1000.

Venter, F. J. 1990. *A classification of land for management planning in the Kruger National Park.* Unpublished Ph.D. dissertation, University of South Africa, Pretoria, South Africa.

Viljoen, P. C. 1996. *Ecological aerial surveys (EAS) in the Kruger National Park: summary of current methodology 1996 update. South African National Parks Scientific Report.* South African National Parks, Skukuza, South Africa.

Walker, B. H., R. H. Emslie, R. N. Owen-Smith, and R. J. Scholes. 1987. To cull or not to cull: lessons from a southern African drought. *Journal of Applied Ecology* 24:381–401.

Wentzel, J. J. 1990. *Ekologiese skeiding van geselekteerde herbivoorspesies in die benede Sabie-Krokodilriviergebied van die Nasionale Krugerwildtuin.* Unpublished M.Sc. thesis, University of Pretoria, Pretoria, South Africa.

Western, D. 1975. Water availability and its influence on the structure and dynamics of a savannah large mammal community. *East African Wildlife Journal* 13:265–286.

Young, E. 1970. *Water as faktor in die ekologie van wild in die Nasionale Krugerwildtuin.* Unpublished Ph.D. thesis, University of Pretoria, Pretoria, South Africa.

Zambatis, N. 1996. *Veld condition assessment in the KNP.* South African National Parks Scientific Report 21/96. South African National Parks, Skukuza, South Africa.

Chapter 9

River Heterogeneity: Ecosystem Structure, Function, and Management

KEVIN H. ROGERS AND JAY O'KEEFFE

Internationally, most river science has followed a unidisciplinary approach, and management has had the narrow focus of managing flow for single species. Increasingly there are calls for truly interdisciplinary studies that provide the basis for effective biodiversity management at the ecosystem level (Poff et al. 1997; Richter et al. 1997). Developing interdisciplinary understanding and implementable management systems for rivers was a dominant component of the Kruger experience in the 1990s. The main vehicle was the Kruger National Park Rivers Research Programme (RRP), which became a major learning experience for scientists, policymakers, and conservation managers. The success of the program is well recognized (Breen et al. 2000), but the lessons learned remain to be broadly assimilated.

Some fundamental challenges faced scientists in their efforts to inform management on Kruger's rivers. First, the scientific community needed to develop and articulate a common understanding of the heterogeneity and dynamics of the river systems, built from multiple scientific disciplines and a number of research projects. Second, scientists needed to adequately convey this understanding to managers so that they could form a hierarchy of visions, objectives, and goals (Chapter 3, this volume) that was consistent with scientific understanding and reflected the heterogeneity and dynamism inherent in these systems. This is probably the greatest challenge ecologists face today: adequately communicating their knowledge of spatial and temporal heterogeneity and dynamism in a way that enables managers to develop appropriate approaches to management.

This chapter deals with the way in which these challenges were met by the RRP and focuses on the sections of river within the park for which Kruger managers are directly responsible. The nature of the rivers beyond Kruger borders and the complexity of catchment management are discussed in Chapter 21.

Understanding River Heterogeneity

One of the aims of the RRP was to translate catchment-level controls on river form, governed by hydrology and sediment supply, into local hydraulic conditions and geomorphic character. Understanding the biotic responses to hydraulic and geomorphic structure and process became the basis for predicting biodiversity responses to changing catchment conditions. The program was guided by a series of conceptual models that illustrated the links between these elements of river structure and function (Breen et al. 1994). A summary (Figure 9.1) is provided as an aid to readers as they engage the text of this chapter.

Hydrology and Sediment Supply

Rivers in South Africa generally have highly variable flow regimes, with the Coefficient of Variation (CV) of mean annual runoff (0.78) being the highest for any country (Dollar 2000). The Kruger rivers are no exception. For example, the Sabie River, the river with the least variable natural flow regime, has a winter dry season discharge of 3–5 m^3/s, summer wet season base flows of 15–20 m^3/s, and a mean annual maximum of 289 m^3/s (SD 221 m^3/s). The CV of mean annual runoff of the Sabie River is best illustrated in terms of the main subcatchments. Upstream of the Sand River confluence the CV is 0.57, but below the confluence it is 1.2 because the Sand River has a CV of 2.1.

It is important to understand this flow variability in terms of its ecosystem consequences. Resh et al. (1988) defined floods with a discharge exceeding the long term mean by more than 2 SD as extreme events or large, infrequent disturbances (LIDs). Over the last 60 years there have been four flood events of 600 to 700 m^3/s in the Sabie that exceeded the mean by 2 SD. In February 2000 a flood with an estimated return period of 1:100 years reached some 6,000 m^3/s at Skukuza and 7,450 at Lower Sabie (Heritage et al. 2000). The flood exceeded the annual mean plus 2 SD by more than an order of magnitude. Park rangers' diaries suggest that the last flood of a similar magnitude was in 1925. The 600–700 m^3/s floods had small effects on the river compared with that of the February 2000 flood, raising the question of what constitutes an LID in these highly variable rivers.

Categorizing droughts in LID terms poses interesting problems. Over the last 60 years, seven years of low flow in which maximum flow was <1 SD of the mean (i.e., >68 m^3/s) could be considered drought years. In 1992 the rivers experienced the worst dry season flow in living memory when flow ceased at the Mozambique border, and many riparian trees died on both the Sabie and Luvuvhu rivers. There has been very little research into the effects of drought on these rivers (Pollard et al. 1996), and, given the growing demands for upstream

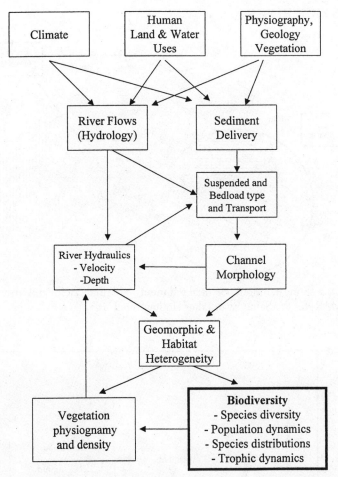

FIGURE 9.1. A summary of the central elements of river structure and function. More detailed conceptual models of each element guided research in the Kruger National Park Rivers Research Programme of the 1990s.

water uses, future research must explore the consequences of extreme low-flow events more closely.

Most of the mean annual runoff (71 percent) to the Sabie River is generated in the upper Sabie, Marite, and Noord Sand subcatchments, but most of the sediment (76 percent) is produced in the middle Sabie and Sand subcatchments. Only 2 percent of the mean annual runoff and 15 percent of the sediment are generated within Kruger (Figure 9.2; Heritage et al. 1997).

This spatial separation of water and sediment production has very important implications for management. Most water resource development and abstraction occur upstream of most land degradation and sediment production.

SABIE RIVER

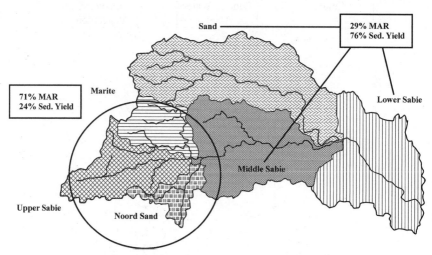

FIGURE 9.2. Subcatchment potential sediment yield (Sed. Yield) and mean annual runoff (MAR) for the Sabie River (after Heritage et al. 1997).

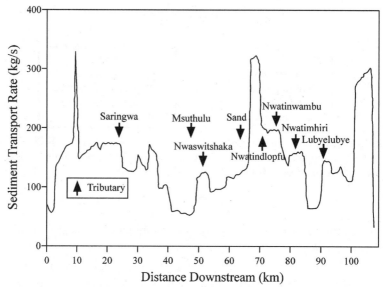

FIGURE 9.3. Regional in-channel sediment accumulation areas in the Sabie River (after Heritage et al. 1997).

This is likely to reduce sediment transport capacity and increase sediment storage in the rivers within Kruger.

A study of channel competence (a measure of a channels ability to transport sediment downstream; Heritage et al. 1997) identified a complex pattern of high- and low-sediment storage areas along the length of the Sabie and three zones of particularly high sediment storage (Figure 9.3): a braided section 10 km downstream of the western border, the section between the Sand and Nwatindlopfu tributaries (65–70 km downstream), and a mixed anastomosing section between Lower Sabie and the Lebombo Gorge (100–108 km). This understanding was particularly important in formulating thresholds of potential concern (TPCs) for river physical heterogeneity.

Geology

At the catchment scale Kruger's rivers flow across a very complex geological template (Cheshire 1994), with rocks (Figure 9.4) of different ages, origin, weathering potential, and weathering process, each of which has been intruded by at least three ages of diabase and dolerite dikes and sills. On the Sabie the youngest rocks (280–190 Ma) are the sedimentary and volcanic Karoo Sequence, followed by the Timbavati Gabbro intrusion (1,450 Ma), the sedimentary Transvaal sequence (2,300–2,220 Ma), and finally the oldest (3,500–2,800 Ma) Archean Basement complex of metamorphic gneiss and migmatites and intrusive granites. This differential lithology and structural variability has produced a geological template characterized by variable hardness and weathering. Therefore, the Sabie River has a concave but highly irregular, long profile punctuated by a series of knickpoints that define a number of subprofiles along the river (Figure 9.5). In the Lowveld Zone the dolerites, diabase, and gabbro are differentially affected by a combination of chemical and physical weathering processes that play a significant role in defining the macrochannel and active channel morphology. The consequence is a highly heterogeneous bedrock template with complex longitudinal and lateral changes in slope, roughness, scour zones (pools), and deposition zones (fluvial features). This heterogeneity generally is highest in the upper granitic areas, where regional dissection gradients are steeper, and is lower on the downstream basalts.

The floods of 1996 in the Olifants and 2000 in the Sabie stripped sediment from the channels and exposed this bedrock template. Rountree et al. (2000, 2001) demonstrated the extremely important role it plays in determining spatial and temporal patterns of sediment deposition and hence fluvial geomorphology. The influence of bedrock on flow and sediment deposition and movement patterns is evident at a number of scales and contributes greatly to structural heterogeneity:

- Kilometer-long sections determined by longitudinal subprofiles that are a response to variability in lithology, which form different channel types
- Tens to hundreds of meters in which dykes running across the river form obstructions to flow or depressions that hold water, depending on the weathering characteristics of different rock types (Cheshire 1994)

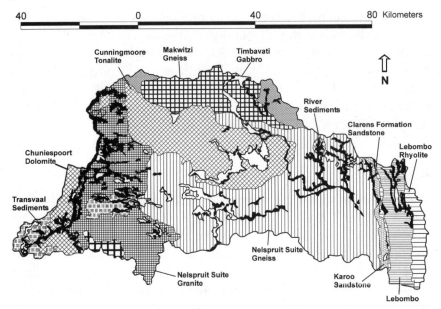

FIGURE 9.4. Coarse-scale geology of the Sabie River catchment (after Heritage et al. 1997).

- A meter or less behind individual rock outcrops or at sites of small-scale differential weathering in the same rock type

The bedrock template therefore influences fluvial processes from small-scale hydraulics to large-scale channel features. However, different channel types respond differentially to different flows, illustrating that the bedrock template is not the exclusive driver of structural characteristics or change pathways. Local hydraulic conditions, determined by sediment and especially vegetation, and variable sediment delivery controls (e.g., tributaries) all interact with the bedrock template to affect channel type change at particular sites on the rivers (Rountree et al. 2000; Kotschy et al. 2000).

Geomorphic Heterogeneity

Studies of fluvial geomorphology in Kruger began in 1990 and have been innovative and groundbreaking. Conventional wisdom at the time tended to link biotic response in rivers directly to hydrological parameters. RRP scientists were quick to realize that this would be inappropriate in these dynamic rivers, where changes in hydrology would generate rapid change in channel form and hence the template for biotic response. They understood that geomorphological pattern in space and time is exceptionally complex in these rivers and did not conform to textbook fluvial geomorphology. This complexity arises from the interaction

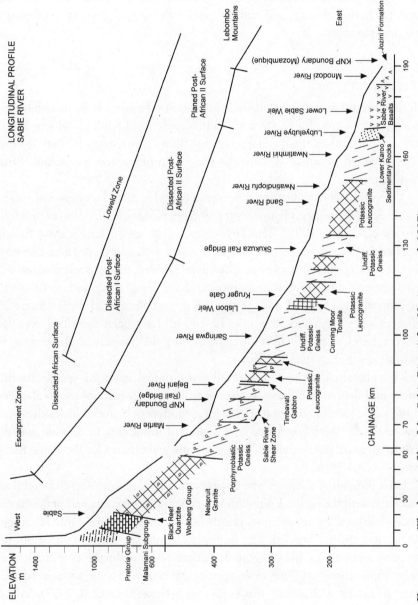

FIGURE 9.5. The long profile of the Sabie River (after Heritage et al. 1997).

of spatially and temporally variable sediment supply from the catchment, highly variable hydrology, a complex long profile, and complex local hydraulics generated by a heterogeneous bedrock template (Heritage et al. 1997). From the outset geomorphological studies were geared to understanding these interactions, predicting change, and making explicit the implications for biodiversity and river management.

Spatial Pattern

All the perennial Kruger rivers have incised into bedrock to form a channel-in-channel morphology with a steep macrochannel bank (10–15 m high) and a wide (200–600 m) macrochannel floor. Consequently, they do not exhibit typical floodplains, with the exception of the lower reaches of the Luvuvhu.

A hierarchical classification describes a complex nested sequence of zones, macroreaches, reaches, channel types, and geomorphologic units (Figure 9.6; van Niekerk et al. 1995). The largest-scale (100 m–10 km) feature that is repeated in the Kruger section of the river is the channel type. Five different channel types (alluvial single thread, alluvial braided, mixed pool-rapid, mixed anastomosing, and bedrock anastomosing) have been recognized according to the proportion of bedrock and alluvial influence (Moon et al. 1997). Some 24 different types of geomorphic unit (10–100 m) have been described over the 100-km stretch of river within the park, making the Sabie River the most geomorphically diverse river in Kruger.

This classification system formed the physical basis for predicting the nature and degree of change in river geomorphology in response to changes in the main catchment control variables. Studies of channel hydraulics and flow resistance were combined with sediment input data to develop a semi-quantitative dynamic model of geomorphological change for the Sabie River (Moon et al. 1997).

Each individual channel type displays unique hydraulic and channel flow character, with the bedrock anastomosing channels showing greatest flow resistance, followed by mixed pool-rapid, mixed anastomosing, alluvial single thread, and alluvial braided. The multiple channels at different elevations on the cross-sectional profile of bedrock sections presented particular challenges for hydraulic modelers. They developed a nonhorizontal water surface channel overspill model to deal with this complexity (Broadhurst et al. 1997). Mixed pool-rapid channel types showed a greater tendency to accumulate sediment, whereas anastomosing channels displayed greater potential sediment transport capacity. Morphological units are continually being created and destroyed along the river as a consequence of sediment deposition and erosion, leading to a number of channel type evolutionary pathways.

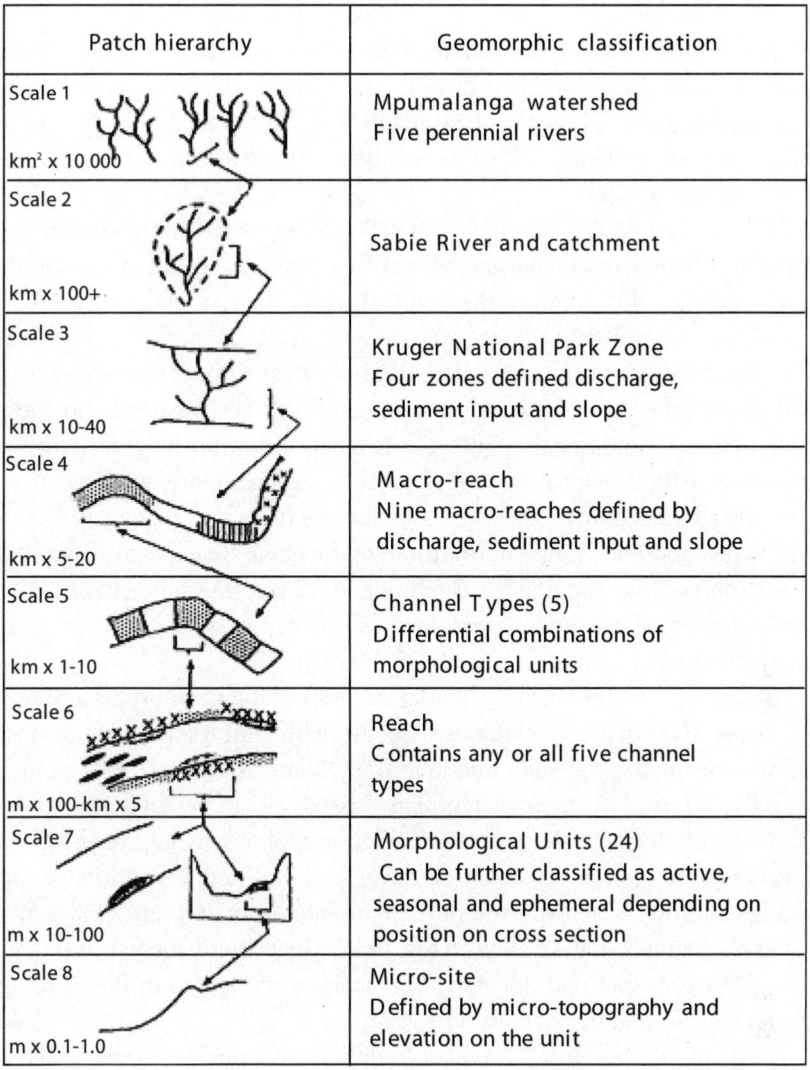

Patch hierarchy	Geomorphic classification
Scale 1 km² x 10 000	Mpumalanga watershed Five perennial rivers
Scale 2 km x 100+	Sabie River and catchment
Scale 3 km x 10-40	Kruger National Park Zone Four zones defined discharge, sediment input and slope
Scale 4 km x 5-20	Macro-reach Nine macro-reaches defined by discharge, sediment input and slope
Scale 5 km x 1-10	Channel Types (5) Differential combinations of morphological units
Scale 6 m x 100-km x 5	Reach Contains any or all five channel types
Scale 7 m x 10-100	Morphological Units (24) Can be further classified as active, seasonal and ephemeral depending on position on cross section
Scale 8 m x 0.1-1.0	Micro-site Defined by micro-topography and elevation on the unit

FIGURE 9.6. An illustrated hierarchical classification of the geomorphology of the rivers flowing through Kruger.

Temporal Pattern

Traditionally, bank-full discharge is considered to be the main channel-forming flow in alluvial temperate rivers. In more complex bedrock and alluvial rivers of semiarid areas, such as the Sabie, different flows inundate different geomorphic units depending on their size, height, and position on the cross-sectional elevation profile. Defining the flows that are effective in the construction and destruction of geomorphic features therefore is complex, but a broad categorization of geomorphic units as active, seasonal, and ephemeral proved useful in modeling exercises (Mackenzie et al. 1999). The active and seasonal features are likely to be modified by anthropogenic modification of the flow and sediment regimes. The ephemeral features are influenced only by very high-magnitude, low-frequency flood events that will not be affected by catchment development activities.

The general predictions, before the 2000 floods, were that a combination of water abstraction and increased sediment supply would result in increased sediment storage, alluviation of the macrochannel floor, a loss of bedrock influence, and general decline in heterogeneity of the geomorphic template in Kruger's rivers (Heritage et al. 1997; Birkhead et al. 2000). This was consistent with studies of past changes in river characteristics based on aerial photographic analysis (Carter and Rogers 1989). The loss of geomorphic heterogeneity would reduce structural biodiversity with probable negative consequences for biotic diversity (Rogers and Biggs 1999). Aquatic scientists and managers were very concerned about the longer-term future of the Sabie River because population growth and development in the catchment were certainly going to result in further reduction of flow and increase in sediment supply, which would only exacerbate the loss of bedrock influence in the river.

At the time of this writing, detailed analysis of the geomorphic response of the Sabie River to the 2000 floods is not available, but useful comparisons can be made using the equivalent flood in the Olifants River in 1996 (Rountree et al. 2001). More of the bedrock template was exposed in the Olifants than in the Sabie, suggesting a greater proportional removal of sediment, probably because vegetation succession was less well developed and had less influence on the change pathway. The exposure of the template on the Olifants and a 50-year aerial photograph analysis allowed researchers to examine the template and the fluvial features that had developed on it since the last such flood. Rangers' diaries suggest that this occurred in 1925.

This unique opportunity revealed 13 different channel type states and a complex, generally nonlinear and spatially variable pattern of change between them (Figure 9.7; Rountree et al. 2001). Mixed anastomosing channels were the most stable and remained so even after the large 1996 flood. The pool-rapid channels were most dynamic, with frequent flips between more alluvial and more bedrock-dominated states. Other channel types generally became more alluvial, but individual channels responded at different rates, and response was seldom linear. A generalized perspective of the multiple pathways is illustrated by the relationship between the degree of alluviation of a channel and gradient of the template. However, the study demonstrated that the control variables were even more complex, with local hydraulics, vegetation, and sediment delivery playing important roles.

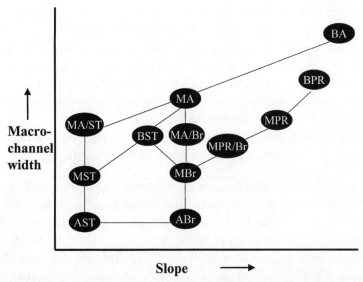

FIGURE 9.7. The complex pattern of potential changes in channel type observed in the Olifants River between 1947 and 2000 (after Rountree et al. 2002). Abr = alluvial braided; AST = alluvial single thread; BA = bedrock anastomosing; BPR = bedrock pool-rapid; Br = braided; BST = bedrock single thread; MA = mixed anastomosing; MBr = mixed braided; MPR = mixed pool-rapid; MST = mixed single thread; ST = single thread.

Vegetation Heterogeneity

Interactions between hydrology and fluvial geomorphology are critical determinants of vegetation heterogeneity in all river systems. Strong environmental gradients (vertical, lateral, and longitudinal) in flooding frequency, depth and duration, water availability from the water table, and substratum type all interact with a highly patchy geomorphological template to generate a diverse and dynamic environment for plant species colonization. Woody vegetation heterogeneity is recognizable at a range of scales. The herbaceous species composition has been less well studied, but an account can be found in Leroy (1997).

Spatial Pattern

Understanding species relationships to elevation at the cross-section scale facilitates understanding vegetation pattern. Species distribution along the elevation gradient (Figure 9.8; van Coller 1993) is unimodal and begins at the lowest elevations of the active channel features. A continuum of species turnover along the gradient is demonstrated by individual species peaks of highest abundance, ranging from matumi (*Breonadia salicina*) at lowest elevation (about 0.5 m) to tamboti (*Spirostachys africana*) at highest elevation (about 6 m). The upper limits of distribution show some pattern in that species are lost at three different elevations that correspond to the limits of active, seasonally flooded, and ephemeral features. A combination of these elevational distribution patterns and substratum characteristics of different geomorphic units leads to the vegetation communities recognized across the macrochannel.

The macrochannel bank is geomorphologically stable, with low rates of sediment deposition and erosion. It is characterized by two vegetation communities:

- The *Spirostachys africana* community, consisting of terrestrial species that occur in greater abundance in the riparian zone, is found at highest elevations and forms the boundary with terrestrial vegetation. Other species include knobthorn (*Acacia nigrescens*), russet bushwillow (*Combretum hereroense*), weeping wattle (*Peltaforum africanum*), white raisin (*Grewia bicolour*), and sickle bush (*Dichrostachys cinerea*).
- The *Diospyros mespiliformis* community occurs lower on the elevation profile and forms the transition to macrochannel floor communities. Other species include ankle thorn (*Acacia robusta*), Natal mahogany (*Trichelia emetica*), apple leaf (*Lonchocarpus capassa*), and the shrubs red spike-thorn (*Gymnosporia senegalensis*) and common wild currant (*Rhus pyroides*).

Four vegetation types are found along the macrochannel floor, which experiences frequent flooding, sedimentation, and erosion to generate a highly irregular topography. The amount of bedrock outcropping has important influences on vegetation of the macrochannel floor.

- The *Combretum erythrophyllum* community is found at the highest elevations on fine-textured substrata of alluvial islands. The majestic common cluster fig (*Ficus sycomorus*) the invasive lantana (*Lantana camara*) shrub, and weeping brides' bush (*Pavetta lanceolata*) are the other main woody species.
- The *Phylanthus reticulatus* community is found on lower-elevation, coarser sediment islands. Species composition is similar to that of the next community but with higher abundance of species such as *G. senegalensis* and white-berry bush (*Securinega virosa*).
- The *Phagmites mauritianus* reed dominates low-level, loose, coarse alluvium and exposed bedrock close to the active channel. Sandpaper fig (*Ficus capreifolia*), water pear (*Syzygium guineense*) and white-berry bush (*S. virosa*) are also common.
- The evergreen *Breonadia salicina* community also occurs at low elevations on bedrock sections of the active channel. Water elder

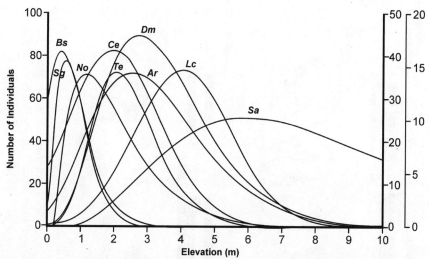

FIGURE 9.8. Distribution curves of the abundance of individuals (stem diameter >3 cm) of tree species in relation to elevation on the Sabie River. Ar = *Acacia robusta*; Bs = *Breonadia salicina*; Ce = *Combretum erythrophyllum*; Dm = *Diospyros mespiliformis*; Lc = *Lonchocarpus capassa*; No = *Nuxia oppsitifolia*; Sa = *Spirostachys africana*; Sg = *Syzygium guineense*; Te = *Trichelia emetica*. The three y-axis scales correspond: 0–100 for Bs and Sg; 0–50 for No, Ce, and Sa; 0–20 for Te, Ar, Dm, and Lc (after van Coller 1993).

(*Nuxia oppsitifolia*), water pear (*S. guineense*), water berry (*S. cordatum*), and reeds (*P. mauritianus*) are all common species.

A combination of differential species distribution across the elevation profile and the patchy distribution of geomorphic units leads to a highly complex and unique pattern of species community distribution across channel types (Figure 9.9; van Coller and Rogers 1996) in the Sabie River.

Temporal Patterns of Vegetation Change

Temporal patterns of vegetation change have been highly complex in all rivers (Carter and Rogers 1989; Rountree et al. 2000), and although they differ somewhat in each river, an explanation of changes based on the Sabie satisfies our need to examine temporal aspects of river heterogeneity in some detail.

Four patterns of long-term change have been identified using analysis of the aerial photographic record that stretches from 1940 to after the 2000 floods (Carter and Rogers 1989; Rountree et al. 2000; unpublished results). The probabilities of river landscape state change were assessed by recording changes in seven states (rock, reed, sand, herbaceous vegetation, terrestrial shrubs, woody vegetation, and trees), recognized in grid squares of 20 × 20 m on the ground.

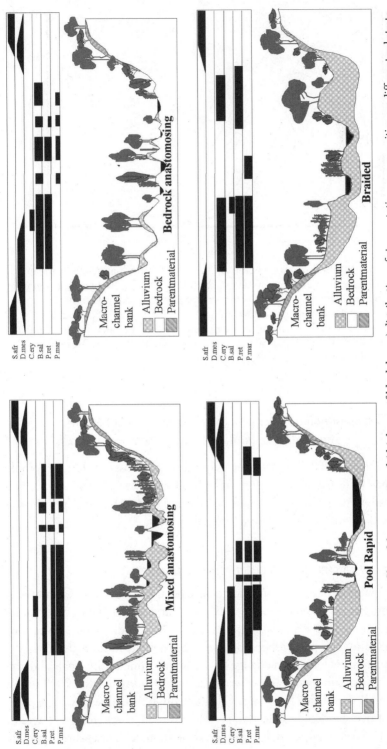

FIGURE 9.9. Diagrammatic profiles of the proportional (thickness of black boxes) distribution of six vegetation communities on different substrata of four different channel types on the Sabie River. B. sal = *Breonadia salicina*; C. ery = *Combretum erythrophyllum*; D. mes = *Diospyros mespiliformis*; P. mar = *Phragmites mauritianus*; P. ret = *Phylanthus reticulatus*; S. afr = *Spirostachys africana* (after van Coller and Rogers 1996).

In general, changes in state were complex, with no simple linear patterns, and the pattern of change changed over time (Figure 9.10).

- In 1940 the river tended to have a wide, sandy and rocky bed and probably was recovering from the large floods of 1925. The period 1940–1965 was one of dynamic change in state, with a small overall shift toward more herbaceous and reed cover with wider channels.
- From 1965 to 1985, state change was much more frequent and directional, with major reductions in area of physical substratum (rock, sand, and water) as it was colonized by reeds. Reed beds were in turn colonized by trees to form large patches of closed-canopy forest.
- Landscape state changes between 1986 and 1996 were even more complex and less directional than those in either of the two preceding periods. There was still an overall trend of vegetation establishment despite a severe drought in 1991–1992 that killed some trees and a 1:50-year flood in 1996 that removed some vegetation and caused small-scale erosion and deposition patches along the macrochannel floor. Changes as a consequence of both the drought and flood were most evident in bedrock anastomosing channels (Rountree et al. 2000). Tree roots could not easily track the drop in water level through rock during drought, and the steep slopes resulted in higher flow velocities during the flood.
- Little change in riverbed state was evident between 1996 and early 2000, but the 1:100-year flood caused a large reversal in the vegetation establishment trend of the last 60 years and set it back to something similar to that seen in 1940. Large areas of riparian forest, shrubs, and reed stands were removed and replaced by a complex mosaic of rock, sand, and water.

Although state change was different in all four periods, a general trend consistent with an episodic stripping model of geomorphic change emerges. Progressive sediment storage in the years between the two big floods of 1925 and 2000 reduced the surface area of water and provided an additional substratum for vegetation establishment. The general colonization sequence was reeds establishing on both rocky and sandy substrata, followed by trees that formed gallery forest. The other general trend, from a somewhat stochastic pattern of change in the early years after the 1925 flood to more directional change and then stabilization again in the last 20 years, suggests a succession sequence in which the potential for resetting by floods becomes less and less likely until a very large and infrequent disturbance resets the system.

The February 2000 floods were much bigger than any other event studied in the history of these rivers, and their impact and legacy are still largely unknown. They will provide many opportunities to increase our understanding of LIDs, to predict their role in ecosystem structure and functioning, and to incorporate the understanding and predictions into effective management strategies. These opportunities are discussed in more detail in Chapter 21.

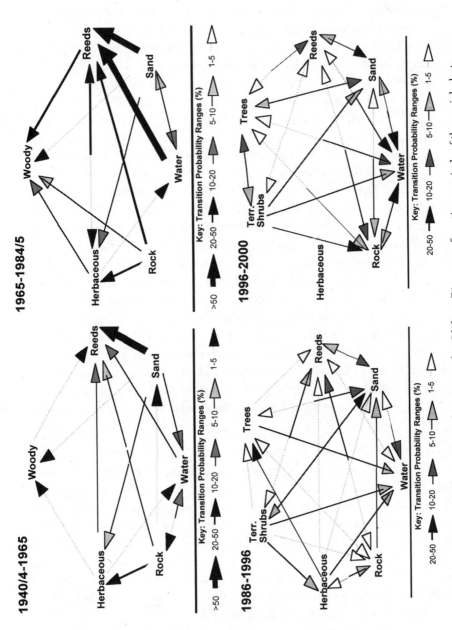

FIGURE 9.10. River-landscape state changes on the Olifants River over four time periods of the aerial photographic record.

Faunal Heterogeneity

The mobile vertebrate fauna of the rivers is much less patchily distributed than the geomorphology or vegetation. Russell (1997) examined fish species distribution in relation to a wide range of in-channel geomorphic features but was able to distinguish only three habitat-species associations, with much overlap between them (Figure 9.11). This study was conducted before detailed geomorphological studies of the rivers, and results are complicated by the coarse scale and selectivity of fish sampling techniques:

- A very distinctive rapids community of predominantly bottom-feeding periphytivores and insectivores such as *Chiloglanus paratus* and *C. pretorie*, in fast-flowing water over stony or bedrock substratum without aquatic plant cover.
- A pool community of 12 different species of detritivore, piscivore, and omnivore in deep pools with gravel or sandy substratum without plant cover.
- Marginal area communities of nine species of multiple-level insectivore that occur in shallow pools, backwaters, and bank margins frequently associated with rooted aquatic and overhanging vegetation.

A further 20 species of fish were variously found in two or more of these habitat types (Figure 9.11). The 38 fish species in the Sabie River showed very little change between 1960 and 1989, unlike those in some of the other rivers (Chapter 21, this volume). It therefore seems to have a stable and diverse fish fauna, although small spatial scale and short-term variability do occur in relation to flood and drought cycles (Weeks et al. 1996). Detailed studies of fish species distributions enabled Weeks et al. (1996) to describe hydraulic (velocity, depth) and substrate use and preference curves for the adult and juvenile stages of many of the more abundant species (see Figure 9.12 for examples). These provided essential understanding for assessment of environmental flow needs (Chapter 21, this volume).

Managing for River Heterogeneity

Four main problems faced scientists and managers concerned with managing river heterogeneity:

- Achieve the interdisciplinary understanding that will promote sensible interaction among scientists and managers.
- Articulate a vision and objectives hierarchy for river management and test it against public opinion.
- Propose a flow regime that would ensure that the objectives are met.

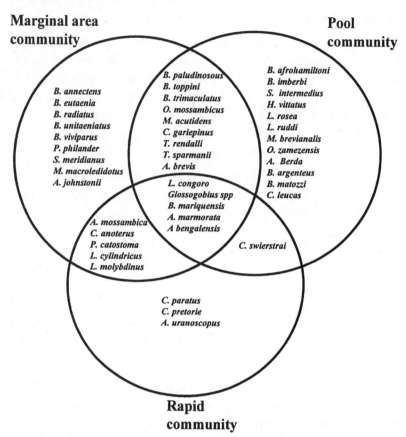

FIGURE 9.11. Fish community composition in different habitats of Kruger's rivers (after Russell 1997). *Barbus annectens, Barbus eutaenia, Barbus radiatus, Barbus unitaeniatus, Barbus viviparus, Pseudocrenilabrus philander, Serranochromis meridianus, Marcusenius macroledidotus, Aplocheilichthys johnstonii, Barbus paludinosous, Barbus toppini, Barbus trimaculatus, Oreochromis mossambicus, Micraleses acutidens, Clarius gariepinus, Tilapia rendalli, Tilapia sparmanii, Atherina brevis, Barbus afrohamiltoni, Brycinus imberbi, Schilbe intermedius, Hydrocynus vittatus, Labeo rosea, Labeo ruddi, Mesobola brevianalis, Opsaridium zamezensis, Acanthopagrus berda, Barbus argenteus, Barbus matozzi, Carcharhinus leucas, Labeo congoro, Glossogobius* spp., *Barbus mariquensis, Anguilla marmorata, Anguilla bengalensis, Chiloglanis anoterus, Peterocephalus catostoma, Labeo cylindricus, Labeo molybdinus, Chiloglanis paratus, Chiloglanis pretorie, Amphilius uranoscopus, Chiloglanis swierstrai.*

- Develop the potential to predict and monitor the response of river biodiversity to the proposed flow regime and audit that response against management objectives.

The development of an objectives hierarchy that integrates management of terrestrial and aquatic ecosystems of Kruger is described in Chapter 4 and the process of determining a flow regime in Chapter 21. This section deals with

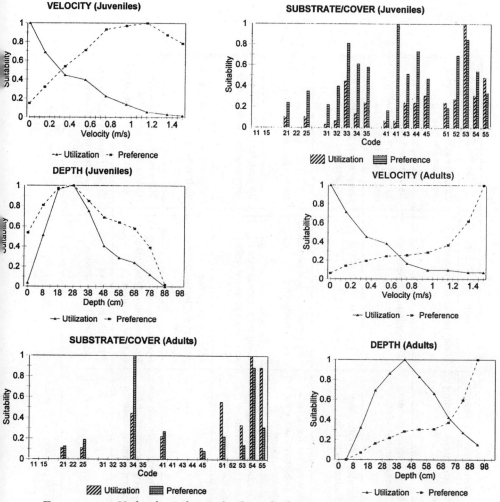

FIGURE 9.12. Hydraulic (velocity, depth) and substrate use and preference curves for the adult and juvenile *Labeo molybdinus* as an example of the fine-scale habitat information collected for fish species of the different communities shown in Figure 9.11. Code refers to a channel index code for cover (the tens) and substrate type (the units). 10 = no cover, 20 = offstream overhead cover, 30 = an instream object, 40 = instream overhead cover, 50 = a combination of cover types. The units 1–5 refer to fines and sand, gravel, cobble, boulder, and bedrock respectively (after Weeks et al. 1996).

achieving interdisciplinary understanding and setting the river TPCs against which response of river biodiversity to the proposed flow regime is audited.

Achieving Interdisciplinary Understanding

Any attempt to understand, let alone predict and communicate, the causal links between catchment processes and downstream biotic response is fraught with

Geomorphic Hierarchy	Management Scales	Prediction Scales	Biotic Response
Mpumalanga Watershed	Water resources planning scale	Water resources modelling	Catchment vegetation and land use change
Sable River and Catchment	Water resources planning and management scale	Rainfall/runoff modeling	Fish response to temperature. Vegetation response to rainfall
Kruger National Park Zone	Conservation policy set and success judged		Large scale pollution problems
Macro - Reach	Conservation manager's first subjective scale for decision making	Hydrological and sediment yield modelling	Non-repeating mosaic
Reach	River conservation goals set and representative mosaics monitored	Long term vegetation response to geomorphic change	Differential plant "community" distribution
Channel Type		Primary scale for predicting geomorphic change	Differential plant species distribution
Morphological Unit	Species management scale, e.g. Hippopotamus or invasive plant species	Hydrodynamic modelling. Fish community response to flow changes	Differential plant species distribution. Grain for established plant response
Micro-Site		Prediction of plant regeneration response	Grain- plant establishment response, invertebrates, fish fry, etc

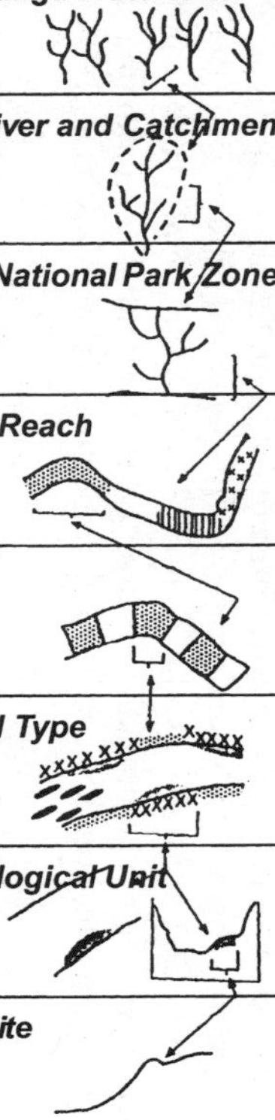

FIGURE 9.13. An illustration of the importance of a scaled understanding of rivers in science-management interaction (after Pickett and Rogers 1997).

problems of scale and interdisciplinarity (Rogers 1997; Bella and Williamson 1997). It is also complicated by many feedback mechanisms and contingencies that determine organism response and its manifestation.

The pragmatic approach was to focus attention on the downstream consequences of change in flow and sediment on the contemporary geological and geomorphological template. The RRP therefore aimed to develop the potential to predict local hydraulic and geomorphological response to catchment generated changes in hydrology and sediment yield (Moon et al. 1997; Jewitt et al. 1998; Mackenzie et al. 1999). A first step was to achieve integration between scientific disciplines and between science and management, from which would emerge appropriate insight and models. This was greatly facilitated by the scaled, system-level understanding generated during the RRP. Central to this understanding were a set of integrating frameworks that guided the research (Breen et al. 1994) and the hierarchical description of the physical template (Figure 9.6).

Each of the seven scales of the nested hierarchy has particular geomorphological correlates, focus for managers and planners, utility and target of prediction for conservation and resource management, type of mosaic, and biological response (Figure 9.13). Communicating between disciplines was very difficult until individuals and disciplines (e.g., water resource engineers, conservation officials, geomorphologists, and many biologists and ecologists) understood these scales, correlates, and links between them. This understanding allowed them to recognize the role their insights might play in an integrated effort of ecosystem management. The process of integrating the apparently divergent understanding of different disciplines also spawned a second wave of research and modeling focused specifically on integration, predictive potential, and decision making (Jewitt and Gorgens 2000).

The RRP provided a sound basic understanding of the river ecosystems that emphasized a heterogeneous habitat mosaic in space and time and recognized that the main drivers of this heterogeneity were a variable flow regime and sediment supply interacting on a complex geological template. The main threats to river biodiversity are reduction in flow and increased sediment supply and storage, which reduce bedrock influence on habitat diversity. Decreasing water quality is an issue in the Olifants and Crocodile Rivers (Chapter 21, this volume).

Monitoring and Management: Agents of Change, Indicators, and TPCs

Management of terrestrial and aquatic ecosystems of Kruger is integrated under the objectives hierarchy (Chapter 4, this volume). The vision requires managers to "maintain biodiversity in all its natural facets and fluxes and to provide human benefits, in keeping with the mission of the National Parks Board, in a manner which detracts as little as possible from the wilderness qualities of the

Kruger National Park" and "to maintain the intrinsic biodiversity (hydrological, geomorphic and biotic) of the aquatic ecosystems as an integral component of the landscape, and where necessary restore or simulate natural structure, function and composition" (Rogers and Bestbier 1997).

The objectives hierarchy requires management to set TPCs and identify agents and indicators of change in structural, functional, and compositional diversity (Chapters 3 and 4, this volume) for each of the rivers. TPCs were set for fluvial geomorphology, vegetation, fish communities, invertebrates, avifauna, the role of the riparian corridor as an altitudinal migration route into the catchments, water quality, and flow regime (Rogers and Bestbier 1997). The geomorphology and vegetation TPCs are illustrated here.

Two primary sets of change agents were identified in the five perennial rivers that run through the park. The first is an array of catchment activities affecting water quality that differ from river to river. The second is increased sediment storage in the rivers resulting from increasing sediment supply from poorly managed catchments and decreasing sediment transport capacity as a consequence of reduced flows (Heritage et al. 1997). This will have particularly serious consequences in most rivers because it is the mixed bedrock-alluvial nature of the riverbeds that generates a high habitat and species diversity. Alluviation as a consequence of increased sediment storage will reduce bedrock influence and homogenize the system.

The broad strategy of research and management was to accept, at least initially, that if management can achieve a specified spatial and temporal range of bedrock and alluvial conditions in the contemporary geomorphic template, biotic diversity will be promoted in this large protected area. The overall management objectives and TPCs therefore are defined by the structure of different types of river reach, with particular reference to the relative contributions of bedrock and alluvial features. Reaches are described on the basis of the proportions of different morphologic units, and differential proportions of these units distinguish different reach types. Management is aided by the designation and description of a few site-specific reaches as representative reaches, which form the focus of monitoring activities (Heritage et al. 2000).

Auditing Geomorphological Response on the Sabie River

A combination of criteria (the measured object or process), units of measurement, and scales of measurement were used to define geomorphological indicators of ecosystem response to the main agents of change (Table 9.1). Geomorphology represents a structural component of biodiversity of the river landscape. The geomorphic criteria were the five (of the total of nine) channel types characterized by bedrock influence. The 4 (of 23) bedrock geomorphic units susceptible to sedimentation (Heritage et al. 1997) are monitored by map-

TABLE 9.1

Monitoring the geomorphological (structural, landscape) component of biodiversity of the Sabie River.

Agents of change and rationale: Increasing sediment supply and decreased sediment transport capacity will result in increased sediment storage, alluviation, and loss of diversity of bedrock influence.

MEASUREMENT CRITERIA	MEASUREMENT UNITS	MEASUREMENT SCALE	THRESHOLDS OF POTENTIAL CONCERN
Selected channel types (5 of 9) in designated representative reaches: Bedrock anastomosing Alluvial anastomosing Mixed anastomosing Bedrock pool-rapid Mixed pool-rapid	Area of selected geomorphic units (4 of 14) on aerial photographs: Bedrock core bar Anastomosing bar Bedrock pavement Bedrock pool	Temporal: Every 5 yr and events (floods, droughts) greater than 1:25-yr return interval Spatial: 10^2–10^3 m per representative reach 20- × 20-m grid square	Directional loss of bedrock influence and water surface area at winter low flow (20-yr prediction) Anastomosing channel types: Bedrock core bars 50% cover or more Three units must be 2–10% of total area Pool-rapid channel types: Lateral and point bars 20% and pools 15% of total area

ping change on aerial photographs every 5 years and after droughts or floods of greater than 1:25-year return period. Monitoring thereby fills the dual function of auditing system response and establishing system variability.

Change in the proportion of bedrock defines the lower limits of TPCs (Table 9.1). At the scale of the Sabie River within Kruger (zone), the TPC would be reached if the indicators showed a directional loss of bedrock area, at winter low flow, over a prediction period of 20 years. A change over 20 years would demonstrate that characteristics of the main agent of change exceed natural variation (Rogers and Biggs 1999). Upper TPC limits may be the increase in proportions of alluvial features, but their measurement would be superfluous.

Auditing Vegetation Response to Geomorphic Change

The main premise behind vegetation TPCs is that alluviation will reduce the diversity of regeneration niches and flow reduction will result in the encroachment of terrestrial species into the riparian zone. Because vegetation distribution patterns are strongly correlated with geomorphology (van Coller et al. 2000), changes in spatial and temporal patterns of regeneration and mortality in representative reaches form the basis of vegetation TPCs and monitoring the vegetation component of biodiversity (Table 9.2).

The criterion of measurement is the population structure of indicator species of each of five of the six vegetation communities. Areal cover of *Phragmites mauritianus* Kunth is the unit of measurement for the sixth community (Rogers and Bestbier 1997). The important problem of alluviation allows us to focus monitoring on *Breonadia salicina* Ridsdale of pool-rapid channels types because these are predicted to be the first channel types to respond to increased sediment storage. The TPC for *B. salicina* populations is a loss of a negative J-curve in pool-rapid channel types because this is the primary site of recruitment of this species. A directional increase in areal extent of *P. mauritianus* cover over a 20-year prediction period would indicate a change in natural successional sequences. TPCs for other tree species are based on the assumption that at least one substantial recruitment event every 10 years would be enough to sustain populations, provided there was no unusual mortality.

The same process as that described for geomorphology was used to develop the vegetation TPCs and the associated monitoring program. A highly focused model predicting vegetation response to changing flow and geomorphology (Mackenzie et al. 1999) is discussed later in this chapter. The way in which continued research and modeling serve and are directed by TPC development can be illustrated by this vegetation example. A central assumption underlying the TPCs is that certain channel types are more susceptible to increased sediment storage and should become the focus of the monitoring program. A flume-based study of sediment transport will use different templates, which will simulate the main characteristics of different channel types, to determine their differential response to a range of sediment and flow characteristics. This will test the hypothesis that pool-rapid channel types will be the first of the bedrock types to become alluviated and support or refute the assumption that monitoring should be focused on them. Should the hypothesis be refuted, the science-management partnership will reassess the indicators used to define TPCs.

Integrated Modeling for Improved Science-Management Links

Integrative modeling has been central to making adaptive management in Kruger strategic. A small set of pragmatic models (Jewitt et al. 1998; Heritage et al. 1997; Mackenzie et al. 1999; Birkhead et al. 2000) assisted in the development, use, and auditing of river TPCs. Modeling was used to generate scenarios that aided in setting hypotheses of acceptable change, to define future monitoring needs, to provide managers with tools for predicting future trajectories of indicators of agents of change, and to audit TPCs.

TABLE 9.2

Monitoring the riparian vegetation component of biodiversity of the Sabie River.

Agents of change and rationale: Alluviation will reduce diversity of regeneration niches. Flow reduction will result in a terrestrialization of riparian zone. Vegetation distribution is strongly correlated with geomorphic diversity. Change in spatial and temporal patterns of regeneration and mortality in representative reaches forms the basis of thresholds of potential concern (TPCs).

MEASUREMENT CRITERIA	MEASUREMENT UNITS	MEASUREMENT SCALE	TPCs
Population structure of key species from each of 6 vegetation units: *Breonadia salicina* *Ficus sycomorus* *Phragmites mauritianus* *Trichelia emetica* *Combretum erythrophyllum* *Spirostachys africana*	Size class frequency distribution	Temporal: Every 3 yr and events greater than 1:25 yr Spatial: All representative reaches except: *B. salicina*, pool-rapid reaches only *C. erythrophyllum*, alluvial reaches only *S. africana*, macrochannel bank only	*B. salicina*: Negative J-curve population structure in pool-rapid channel types Other trees: Recruitment at least once in 10 years Mortality threshold uncertain *P. mauritianus*: Directional increase in areal extent (20-yr prediction)

A pragmatic approach to modeling was adopted whereby small objective-driven (often TPC-driven) models were developed as single-purpose management tools. The Breonadia model (Mackenzie et al. 1999) that was built to serve the geomorphology and vegetation TPCs is discussed here. The model is a standard, size-class population matrix model that is rule based and deterministic. Inputs include rainfall (in the form of daily values and states), hydrology (in the form of hydrological states), geomorphology (in the form of substrate types), and growth rates (in the form of size class longevity). Inputs and feedbacks determine the values of fecundity, survival, and the probability of staying in a size class or going to the next size class in the matrix. Feedback mechanisms include adult abundance effects on fecundity, density dependence effects on survival and fecundity for each size class, and population structure at the next time interval.

Rules took the form of IF-THEN or ELSE statements, which generate certain responses in the population structure depending on which environmental and vegetation conditions have been met. Rules had either quantitative or qualitative elements, depending on the level of confidence in data analyses. Qualitative rules were especially useful when data were lacking and expert opinion had to be used.

The model therefore was tested using expert opinion and knowledge to verify acceptable model behavior. A parameter sensitivity analysis was used to analyze uncertainty in the model (Mackenzie et al. 1999). Results from the sensitivity analyses were used to further validate the model, design research strategies, indicate potential system controls, test theory, and develop a monitoring program. In particular, they helped prioritize monitoring efforts because they highlighted the most sensitive parameters in the model and the most important in auditing the TPC.

The model is as much a heuristic tool as it is a predictive tool and should be frequently subjected to refinement by researchers, users, developers, and decision makers. Researchers use the model to highlight sensitive parameters and direct research efforts to improve the accuracy and reliability of model outputs and assumptions by improving the estimation of sensitive parameters. In so doing they improve understanding of the river systems. Users can improve parameter estimates by updating model defaults using data from monitoring and research. Their feedback will help developers refine rules and rule preferences and incorporate additional rules to address assumptions. Decision makers can also refine the model in that they can improve the critical values of the thresholds that define the TPCs. Managers who use the model to run TPC audits and scenarios of potential management actions can more easily articulate their needs for improved understanding and influence policy development.

The Breonadia model, though specialized in its application, can empower conservation managers around a stakeholder bargaining table. The model is essentially a river-section-scale tool (quantitative solicitation of a causal chain of our assumptions) that can be and has been applied to catchment-scale decisions, actions, or policy, by explicit definition, justification, and consideration of ecological flow needs for the Sabie River.

Integrated River Research, Modeling, Monitoring, and Decision Making

At the beginning of this chapter we suggested that the greatest challenge faced by ecologists today is adequately communicating their knowledge of spatial and temporal heterogeneity in a way that enables managers to respond to management problems. We can conclude this chapter by presenting a conceptual model of how research, predictive modeling, and operations interact with the monitoring of ecosystem response and TPCs (Figure 9.14). This process is essential in ensuring that the validity of TPCs is continuously tested, that they are based on current information, and that their use ensures that management is strategically adaptive rather than reactive.

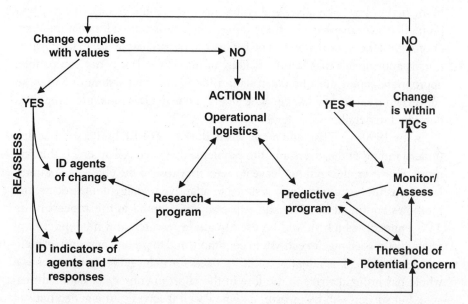

FIGURE 9.14. A conceptual model of how research, predictive modeling, and operations use thresholds of potential concern (TPCs) to interact with the monitoring of ecosystem response in strategic adaptive management (after Rogers and Biggs 1999).

Future research in Kruger must continue to identify the main agents, natural and anthropogenic, of change in river characteristics and identify indicators of these agents and of system response to them (Figure 9.14). The collective wisdom of science and management, guided by the objectives hierarchy, should be used to define, in spatial and temporal terms, the upper and lower levels of these indicators that will lead to concern about the degree of ecosystem change. These levels are the TPCs.

A structured monitoring program must assess change in these indicators and data fed back into models, such as the Breonadia model, in the predictive framework. Most monitoring exercises only alert management to changes that have already occurred. When monitoring is based on TPCs and coupled to predictive models, management becomes strategic. This is probably the single most important element of the strategic adaptive management process used in Kruger, one that sets it apart from most others and should be nurtured as we learn how to use it most effectively.

A change in ecosystem condition, measured during monitoring or predicted by a model, should prompt an assessment of the causes of change and a determination of whether it falls within the appropriate TPC (Figure 9.14). If it does, monitoring can continue as defined. If it falls beyond the TPC, the cause, degree, and nature of change must be assessed relative to the values embodied in the higher-level objectives and vision statement. If the change in ecosystem properties is compatible with these values (e.g., it was caused by a natural flood or drought), the information can be fed back to researchers to use in their experiments and models that test the validity of the TPCs. If it is not compatible, appropriate action must be taken within the operational system of Kruger to address the cause of change. The operational decision-making process is described in detail in Chapter 4.

In the 1990s scientists and managers in the Kruger RRP built a solid, scaled understanding of the rivers and the problems they face (Chapter 21, this volume). However, they did not leave it there; they rose to the challenge of translating this understanding into a strategic adaptive management process that promotes continual learning and reassessment. Central to this process is the TPC concept, which should be used to guide research and model development, to assess change in ecosystem condition, and to prompt appropriate management action. The adoption of this process for management of Kruger as a whole has brought rivers to the fore in the effort to conserve a national treasure, but success will be measured by how well it serves future generations of scientists and managers as they accumulate new understanding and face new problems.

References

Bella, D. A., and K. J. Williamson. 1997. Conflicts in interdisciplinary research. *Journal of Environmental Systems* 62:105–124.

Birkhead, A. L., G. L. Heritage, C. S. James, K. H. Rogers, and A. W. van Niekerk. 2000. *Geomorphological change models for the Sabie River in the Kruger National Park*. Water Research Commission Report No. 782/1/00. Water Research Commission, Pretoria, South Africa.

Breen, C. M., M. Dent, J. Jaganyi, B. Madikizela, J. Maganbeharie, A. Ndlovu, J. O'Keefe, K. H. Rogers, M. Uys, and F. Venter. 2000. *The Kruger National Park Rivers Research Programme: final report*. Water Research Commission, Pretoria, South Africa.

Breen, C. M., N. Quinn, and A. Deacon. 1994. *A description of the Kruger National Park Rivers Research Programme (second phase)*. Pretoria: Foundation for Research Development, Pretoria.

Broadhurst, L. J., G. L. Heritage, A. W. van Niekerk, K. H. Rogers, and C. S. James. 1997. *Translating local discharge into hydraulic conditions on the Sabie River: an assessment of channel flow resistance*. Water Research Commission Report 474/2/96. Water Research Commission, Pretoria, South Africa.

Carter, A. J., and K. H. Rogers. 1989. Phragmites reedbeds in the Kruger National Park: the complexity of change in riverbed state. Pages 339–340 in *Proceedings of the Fourth South African National Hydrological Symposium*. Vol. 4. Water Research Commission, Pretoria, South Africa.

Cheshire, P. 1994. *Geology and geomorphology of the Sabie River, Kruger National Park and its catchment area.* Centre for Water in the Environment Report 1/1994. University of the Witwatersrand, Johannesburg, South Africa.

Dollar, E. S. J. 2000. *The determination of geomorphologically effective flows for selected eastern sea-board rivers in South Africa.* Unpublished Ph.D. thesis, Rhodes University, Grahamstown, South Africa.

Heritage, G. L., L. J. Broadhurst, A. W. van Niekerk, K. H. Rogers, and B. P. Moon. 2000. *The definition and characterisation of representative river reaches.* Water Research Commission Report 376/2/00. Water Research Commission, Pretoria, South Africa.

Heritage, G. L., A. W. van Niekerk, B. P. Moon, L. J. Broadhurst, K. H. Rogers, and C. S. James. 1997. *The geomorphological response to changing flow regimes of the Sabie and Letaba river systems.* Water Research Commission Report 376/1/97. Water Research Commission, Pretoria, South Africa.

Jewitt, G. P. W., and A. H. M. Gorgens. 2000. Facilitation of interdisciplinary collaboration in research: lessons from a Kruger National Park Rivers Research Programme project. *South African Journal of Science* 96:410–414.

Jewitt, G. P. W., G. L. Heritage, D. C. Weeks, J. A. Mackenzie, A. van Niekerk, A. H. M. Gorgens, J. O'Keefe, K. Rogers, and M. Horn. 1998. *Modeling abiotic-biotic links in the Sabie River.* Water Research Commission Report 777/1/98. Water Research Commission, Pretoria, South Africa.

Kotschy, K. A., K. H. Rogers, and A. J. Carter. 2000. Patterns of change in reed cover and distribution in a seasonal riverine wetland in South Africa. *Folio Geobotanica* 35:363–373.

Leroy, M. R. 1997. *Relationship between exotic species distribution and geomorphology on the Sabie River in the Kruger National Park: implications for conservation.* Unpublished B.Sc. honors dissertation, University of the Witwatersrand, Johannesburg, South Africa.

Mackenzie, J. A., A. L. van Coller, and K. H. Rogers. 1999. *Rule-based modelling for management of riparian systems.* Water Research Commission Report 813/1/99. Water Research Commission, Pretoria, South Africa.

Moon, B. P., A. W. van Niekerk, G. L. Heritage, K. H. Rogers, and C. S. James. 1997. A geomorphological approach to the ecological management of rivers in the Kruger National Park. *Transactions of the Institute of British Geographers NS* 22:31–48.

Pickett, S. T. A., and K. H. Rogers. 1997. Patch dynamics: the transformation of landscape structure and function. Pages 101–127 in J. A. Bissonette (ed.), *Wildlife and landscape ecology*. New York: Springer-Verlag.

Poff, N. J., J. D. Allan, M. B. Bain, J. R. Karr, K. I. Prestegaard, B. D. Richter, R. E. Sparks, and J. C. Stromberg. 1997. The natural flow regime: a paradigm for river conservation and restoration. *BioScience* 47:769–784.

Pollard, S. R., D. C. Weeks, and A. C. Fourie. 1996. *A pre-impoundment study of the Sabie-Sand river system, Mpumalanga with special reference to predicted impacts on the Kruger National Park. Vol. 2. Effects of the 1992 drought on the fish and macro-invertebrate fauna.* Water Research Commission Report 294/2/96. Water Research Commission, Pretoria, South Africa.

Resh, V. H., A. V. Brown, A. P. Civich, M. E. Gurtz, H. W. Li, G. W. Minshall, S. R. Reice, A. L. Sheldon, J. B. Wallace, and R. C. Wissamar. 1988. The role of disturbance in stream ecology. *Journal of the North American Benthological Society* 7:433–455.

Richter, B. D., J. V. Baumgartner, R. Wigington, and D. P. Braun. 1997. How much water does a river need? *Freshwater Biology* 37:231–249.

Rogers, K. H. 1997. Operationalizing ecology under a new paradigm. Pages 60–77 in S. T. A. Pickett, R. S. Ostfeld, M. Shachak, and G. E. Likens (eds.), *Enhancing the ecological basis of conservation: heterogeneity, ecosystem function and biodiversity*. New York: Chapman & Hall.

Rogers, K. H., and R. Bestbier. 1997. *Development of a protocol for the definition of the desired state of riverine systems in South Africa*. Pretoria: Department of Environmental Affairs and Tourism. Online: http://www.ccwr.ac.za/knprrp/index.html.

Rogers, K., and H. Biggs. 1999. Integrating indicators, endpoints and value systems in strategic management of the rivers of the Kruger National Park South Africa. *Freshwater Biology* 41:439–452.

Rountree, M. W., G. L. Heritage, and K. H. Rogers. 2002. In-channel metamorphosis in a semi-arid, mixed bedrock/alluvial river system: implications for instream flow requirements. Pages 113–123 in *Hydroecology: riverine ecological response to changes in hydrological regime, sediment transport and nutrient loading*. Wallingford, UK: International Association of Hydrological Sciences (IAHS) Press. IAHS Publication 266.

Rountree, M. W., K. H. Rogers, and G. L. Heritage. 2000. Landscape state change in the semi-arid Sabie River, Kruger National Park, in response to flood and drought. *South African Geographical Journal* 82:173–181.

Russell, I. S. 1997. *Monitoring the conservation status and diversity of fish assemblages in the major rivers of the Kruger National Park*. Unpublished Ph.D. thesis, University of the Witwatersrand, Johannesburg, South Africa.

van Coller, A. L. 1993. *Riparian vegetation of the Sabie River: relating spatial distribution patterns to characteristics of the physical environment*. Unpublished M.Sc. dissertation, University of the Witwatersrand, Johannesburg, South Africa.

van Coller, A. L., and K. H. Rogers. 1996. *A basis for determining the instream flow requirements of the riparian vegetation along the Sabie River within the Kruger National Park*. Report 2/96. Centre for Water in the Environment, University of the Witwatersrand. Johannesburg, South Africa.

van Coller, A. L., K. H. Rogers, and G. L. Heritage. 2000. Riparian vegetation-environment relationships: complementarity of gradients versus patch hierarchy approaches. *Journal of Vegetation Science* 11:337–350.

van Niekerk, A. W., G. L. Heritage, and B. P. Moon. 1995. River classification for management. The geomorphology of the Sabie River. *South African Geographical Journal* 77:68–76.

Weeks, D. C., J. H. O'Keeffe, A. Fourie, and B. R. Davies. 1996. *A pre-impoundment study of the Sabie-Sand river system, Mpumalanga with special reference to predicted impacts on the Kruger National Park. Vol. 1. The ecological status of the Sabie-Sand river system*. Water Research Commission Report 294/1/96. Water Research Commission, Pretoria, South Africa.

PART III
Interactions between Biotic Components

It is through the study of species and assemblages that ecosystem heterogeneity takes on operational meaning because the behavior of these responders reveals the identities and effects of the drivers in the heterogeneity framework. As will become apparent in the chapters that follow, however, responders can also be drivers. For example, ungulate speciation is driven by spatial and temporal heterogeneity at the savanna biome level, and the diversity of species and body sizes within an ungulate guild has a feedback effect on vegetation diversity at the community level. Driving effects can also operate through intermediary responders, as when climatic variation causes fluctuations in the abundance of insects and therefore in the availability of food for insectivorous birds. The permutations are endless, but for a specific research problem, such as predicting the effects of a severe drought on the ungulate community, a functional awareness of ecosystem heterogeneity is helpful in tracing links from the responders of interest back to their drivers.

The contributors to this part of the book have adopted a heterogeneity mindset and emphasized interactions between, instead of descriptions of, the biotic components of Kruger's landscapes. The alternative, which we have made every effort to avoid, would have been a series of taxon-specific accounts of past research projects. The main trophic groups are covered (plants, herbivores, carnivores), and the best-studied taxa feature most prominently, so there are obvious omissions. Reptiles, amphibians, and fishes have had to be left out, and even some of the functionally important mammalian assemblages such as bats and rodents could not be given their own space in this book. Research opportunities await, and the aim of this book will have been achieved if future researchers use the information in these chapters to design and execute further research to fill in the missing pieces of the jigsaw puzzle.

Chapter 10

Interactions between Species and Ecosystem Characteristics

ROBERT J. NAIMAN, LEO BRAACK, RINA GRANT,
ALAN C. KEMP, JOHAN T. DU TOIT, AND FREEK J. VENTER

Discovering and quantifying species effects on ecosystem characteristics requires strong integration between biological and physical sciences. Although conceptually easy, it remains difficult in practice because of the enormous biocomplexity in natural systems, so as yet there is no unifying conceptual framework to explain the variety and the emergent features of these relationships (Brown 1995). Nevertheless, ecologists and managers recognize strong feedbacks between organisms and their environments, as indicated by the terms *keystone species* and *ecosystem engineer*, which are deeply integrated into the ecological literature (Jones et al. 1994; Power et al. 1996). Keystone species are ones whose impact on their community or ecosystem is disproportionately large relative to their abundance. Ecosystem engineers are organisms that directly or indirectly modulate the availability of resources to other species by causing physical state changes in biotic or abiotic materials. In doing so they modify, maintain, or create habitats. In either case, links between species and the environment—and often the consequences of those links—are expressed at the landscape scale for decades to centuries (Naiman 1988; Huntly 1995; Naiman and Rogers 1997).

In reality, animals affect ecosystems in two primary ways: by trophic and physical modifications of the environment (Figure 10.1). Existing vegetative communities consist of plants and plant parts that have not been eaten, soils contain biotic materials that have not been metabolized, and it is those uneaten and unmetabolized components that are measured and reported by ecologists. Processes hidden by the temporal and spatial perspectives of investigators often result in serious misjudgments about the environment and how it should be managed (Magnuson 1990). Additionally, further complications are introduced because links between species and ecosystems vary widely in time and space (Elton 1930; Turner et al. 1995). Population sizes often fluctuate over years to

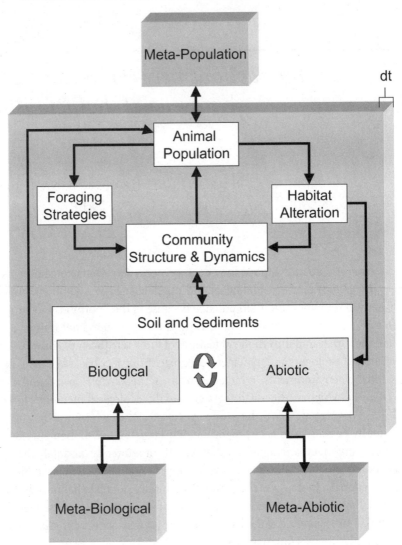

FIGURE 10.1. Species affect ecosystem characteristics in two primary ways: by foraging strategies or by physical modifications of the environment. The impacts are transmitted throughout biotic and abiotic components of the ecosystem, ultimately having important feedbacks to the species initiating the alterations. The strengths of the interactions vary over time (modified from Naiman 1988).

decades, animal distributions often track vegetation change (and vice versa) from decades to centuries, and the relative numbers of species in the community respond to competition, predation, disease, and other environmental influences. Consequently, Naiman and Rogers (1997) and others have argued for improving resource management by shifting from management of individual species for stability to managing them for population variability and for their interactive roles in the ecosystem.

In this chapter we examine recent theoretical, conceptual, and empirical developments related to links between species and ecosystem characteristics. We then address issues specifically relevant to Kruger that include the collective relationships of invertebrates, birds, mammals, and soil-plant communities that characterize African savannas. Finally, we provide a summary of general ecological principles to assist in guiding effective management prescriptions.

Theoretical, Conceptual, and Empirical Developments

A major barrier to linking species and ecosystems is the use of two fundamentally different currencies (Brown 1995). Most population and community ecologists use a dN/dt currency, which can be equated to the biological concept of fitness. Fitness is measured in terms of rates of change in numbers of individuals (N), species, heritable traits, or units of genetic information. In contrast, ecosystem ecologists use a dE/dt currency, which can be equated with fluxes and transformations of energy (E) and matter. Physical laws, such as those related to thermodynamics, chemical stoichiometry, and the periodic table, govern these transformations. Hall et al. (1992) and Brown (1995) correctly suggest that this issue can be resolved by recognizing that ultimately dN/dt or fitness can be expressed as an equivalent of dE/dt energetic or thermodynamic currency. This issue is fundamental to effectively linking species to ecosystem characteristics. If accomplished it would, at a minimum, encourage population biologists to quantify relationships between the biological components of fitness and their physical bases, and it would offer new insights into physiological constraints associated with organism-environment interactions. A general conceptual framework for effectively linking species to ecosystem characteristics requires an understanding of how fitness and population dynamics affect and are affected by fluxes and transformations of energy and matter. More simply stated, it means quantifying how organisms interact with their biogeochemical environment.

Brown (1995) and others are skeptical of formulating general predictive statements about the impacts of individual species on ecosystems with the current knowledge base. The reason is because the work performed on the ecosystem by each species in meeting its unique niche needs also is unique: unique in its type, magnitude, spatial and temporal distributions, and impact on the abiotic environment and other organisms. The impact depends on the specific environmental setting. Furthermore, the environment is a complex system full of nonlinearity, thresholds, amplifications, and damping effects, collectively making prediction difficult. This is not to say that the formulation of general principles is impossible; it is just a major challenge for ecologists to accomplish. Examining case studies, identifying coarse patterns in species-ecosystem links, and applying emerging mathematical tools are effective ways to begin.

Consider plant-herbivore interactions. They have a long history (e.g., Elton 1930), as do other ecological links, and provide a good example for formulating predictive ecological statements. The best evidence that herbivores are closely linked to ecosystem characteristics comes from experimental manipulations of herbivore populations either by direct exclusion or by manipulating higher-level consumers that control the size or composition of the herbivore trophic level (Huntly 1991, 1995). It is not well appreciated that the common effects of consumers on ecosystem processes are via feedbacks from trophic resources or from the abiotic environment (Naiman 1988). Most simply, the primary productivity and biogeochemical composition of ecosystems set maximum limits on the density or biomass of consumers that can be supported. Additionally, variations in primary productivity, the biogeochemistry of organic matter, and the stoichiometry of elements on small spatial scales also affect herbivore behavior and thus determine the pattern of their impacts within an area.

Organic materials not assimilated by consumers, and the metabolic excretions of consumers, eventually become incorporated into soils and help shape their characteristics. This is a multifaceted and dynamic process involving the chemical stoichiometry of the organic matter, consumer physiology, decomposition dynamics, and fire (Pastor and Naiman 1992; Wedin 1995). The chemistry of the organic matter is key because it can determine via which pathway the bulk of the primary production is shunted toward consumers (Figure 10.2). Species-level differences in tissue chemistry have important consequences for plant community dynamics, plant-herbivore interactions, nutrient-decomposition relationships, and the nature of disturbance regimes. Nevertheless, regardless of the pathway, the material not metabolized is eventually incorporated into the soil.

The following illustrations examine links between animal species and ecosystem characteristics. The topics addressed here are conceptually simple but enormously complex to quantify and manage against the backdrop of natural heterogeneity. Yet the lessons contained herein are of basic importance in maintaining the diversity and ecological vitality of Kruger for the long term.

Animal Effects on Plant Community Structure through Pollination and Seed Dispersal

Pollinators are as critical as light, water, and nutrients for plant persistence (Kearns and Inouye 1997) and hence vegetation structure, yet the pollination needs of most wild plants remain unknown. Additionally, widespread loss of pollinators is occurring, and this affects plants in subtle but meaningful ways (reduced seed set, less vigorous offspring, less pollen competition, and declin-

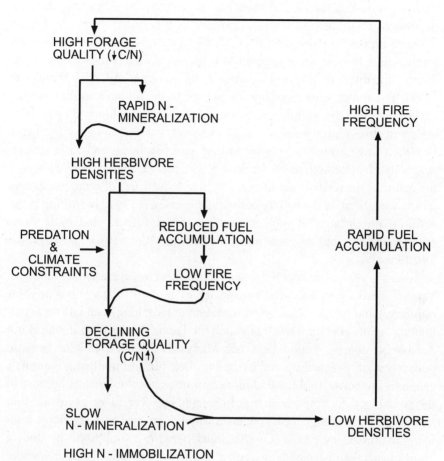

FIGURE 10.2. The biogeochemistry of plants as forage has strong influences on animal populations and on some physical disturbance regimes such as fire. C/N = carbon-nitrogen ratio.

ing fruit or seed production), and the loss may not be evident until long after the pollinator is extinct.

At least 67 percent of the world's plants depend on insects for pollination. Two key pollinator groups in Kruger are honeybees (Apidae) and fig wasps (Agaonidae). They have very different pollination strategies, both of which are important in determining vegetative patterns at the landscape scale. Although it is not yet clear to what extent the dominant plant species depend on honeybees for pollination—because beetles and many other insects species also frequent the same flowers but mostly in lesser numbers—there can be little doubt that honeybees are key species without which many of Kruger's plants would be dramatically reduced in abundance and distribution, if not extinct.

In a contrasting strategy, fig wasps have an extraordinarily narrow and interdependent relationship with figs. Survival of the 19 fig species in Kruger depends on near-microscopic sized (1- to 10-mm) wasps. Coevolution has resulted in only a few but usually only one species of wasp being capable of pollinating the internal flowers of a particular fig species (Galil 1977; Ware et al. 1993). The minute flowers of figs are located in the hollow central chamber of the fig "fruit": the syconium. The chamber entrance is blocked by an elaborate and closely interlocked maze of tiny bracts or scales, and the wasps negotiate their way through this obstacle using spadelike heads and other specialized body adaptations to reach the flowers (Chapter 12, this volume). However, the nature of the relationship is such that in any given area of occurrence there must always be at least a fig tree of the right species, at the right fruiting stage, within flying distance of another tree from which tiny fragile female wasps emerge, and the right wasp species for the trees (and the wasp) to persist for the long term.

Birds are also pollinators (Chapter 13, this volume), as are some mammals. Various small flying and arboreal species such as fruit bats (*Epomophorus wahlbergi*) and galagos (*Galago crassicaudatus*) have long been known as pollinating agents of certain trees, of which the baobab (*Adansonia digitata*) is a notable example (Coe and Isaac 1965). Mammals are effective vectors because pollen is readily picked up and carried in their fur, but nonflying mammals generally are considered to be of limited value as cross-pollinators because of their inability to fly from tree to tree. In Kruger, however, there is evidence that knobthorn (*Acacia nigrescens*) pollen is widely dispersed by a very large nonflying mammal, the giraffe (*Giraffa camelopardalis*). Knobthorns produce a profusion of creamy-colored, unscented, spicate ("bottle-brush") inflorescences in the late dry season after leaf-fall (August and September), and giraffes feed extensively on them (du Toit 1990a). Although flower consumption represents a cost to knobthorn plants, giraffes consume only a small fraction of available flowers, and they collect visible amounts of pollen on their heads and necks when feeding (du Toit 1992). In this way pollen is distributed to other flowering knobthorns throughout each giraffe's home range, which can cover >250 km^2 (du Toit 1990b). Despite the need for experimental evidence to test whether giraffes are effective pollinating agents of knobthorns, there is little doubt that other ungulates in Kruger (e.g., kudu [*Tragelaphus strepsiceros*]) transport pollen of various woody plants (e.g., red spike-thorn [*Gymnosporia senegalensis*], shaving-brush bushwillow [*Combretum mossambicense*]) that produce pale, unscented brushlike inflorescences in the dry season when insect activity is low (du Toit, pers. obs.).

Birds also are important in the population dynamics of some of the larger fruit-bearing trees, playing a role in seed dispersal, especially along major river drainages. Trees such as Natal mahogany (*Trichilia emetica*), sycamore fig (*Ficus sycamorus*), Cape ash (*Ekebergia capensis*), jackal-berry (*Diospyros mespiliformis*), and nyala tree (*Xanthocercis zambesiaca*) attract a variety of frugivores. These include many birds that cause little or no damage to the seeds and widely disperse them, even after rapid digestion of the fruity covering. Several prefer the more densely wooded habitats provided by watercourses, such as trumpeter and crowned hornbills (*Ceratogymna bucinator* and *Tockus alboterminatus*), black-collared barbets (*Lybius torquatus*), purple-crested lories (*Tauraco porphyreolophus*), and black-headed orioles (*Oriolus larvatus*). Their preference for riparian habitats also ensures that bird movement, and therefore their propensity for seed dispersion, is directed largely at some of the best sites for germination and plant growth. Furthermore, some frugivorous birds such as trumpeter hornbills usually retreat into dense tangles after feeding, where they regurgitate or drop seeds into thickets that subsequently afford protection to seedlings from browsing ungulates (Kemp 1995). Other frugivorous birds traverse the surrounding drier savanna as they move between watercourses or visit their roost and nest sites (e.g., green pigeon [*Treron calva*], black-eyed bulbul [*Pycnonotus barbatus*], crested barbet [*Trachyphonus vaillantii*], and brown-headed parrot [*Poicephalus cryptoxanthus*]). The effects of such itinerant dispersers are likely to be less obvious and more diffuse, given the range of directions in which they might choose to move. However, the effects of their dispersion are likely to be channeled to perches en route, such as tall trees in open bush and grassland or clumps of trees growing in association with termite mounds. The resulting seed rain is higher under such obvious perches, and the influence of birds on the dispersion of fruiting trees is suggested by the diversity of trees and thickets associated with termite mounds, in comparison to the surrounding savanna, and by the frequency of strangler fig (*Ficus natalensis*) seedlings growing on large, isolated, and often dead trees.

The overall effectiveness of birds as dispersers of seeds is best illustrated by the spread of certain alien plants, such as *Solanum*, *Melia*, and *Lantana* species. These species spread particularly rapidly along watercourses, suggesting that at least some of the patterns of plant distribution may be driven by avian dispersers attracted preferentially to such habitats, together with provision of good sites for germination.

Mammals are important agents of seed dispersal, carrying seeds in their fur and digestive systems (Lamprey 1967; Coe and Coe 1987; Miller 1994; Slater and du Toit 2002). Indeed, an important conservation issue in African savannas is the widespread replacement of diverse indigenous mammal assemblages by a few species of domestic livestock (usually cattle and goats), resulting in the loss of coevolved seed dispersal mechanisms (du Toit and Cumming 1999).

Animal Population Effects on Plant Community Structure through Foraging Strategies

Trophic diversification is a catalyst for further diversification. The very presence of certain organisms creates opportunities for other organisms to feed on those already present or use the products made available by them, thereby expanding the resource base even further. For example, the consumption of tree seedlings by various mammals, including rodents, lagomorphs, and ungulates, has a profound influence on the balance between woody and herbaceous vegetation in African savannas (Belsky 1984; Prins and van der Jeugd 1993). Another example is granivory by the red-billed quelea (*Quelea quelea*), which is perhaps the most important bird species in Kruger in terms of numbers and biomass, especially when breeding in colonies of millions of birds in the austral summer (Kemp et al. 2001). The breeding colonies offer the highest density of birds recorded for any species in the world (~16,000 nests/ha). Even though the ecological consequences of consumption by these colonies on seed production and recruitment of annual grasses remain to be established, the effects are visually strong. Nesting queleas forage within a radius of about 6 km (range 2–11 km, approximate area 13–380 km^2) around breeding colonies (Bruggers and Elliott 1989). Assuming queleas consume 20 percent of their body weight daily as seeds and insects, a nesting colony has the potential to consume 480 kg/ha each day. In fact, grass cover and recruitment are affected over a large radius around each colony, and the effects may be evident for more than 1 year. Given that the location, surface area, density, and effects of nesting red-billed queleas vary from summer to summer, such effects are of such a scale and duration that they are of broad ecological relevance.

The mere presence of some species also allows other species to prey, parasitize, or scavenge on those already present or to take advantage of their physical actions. For example, when blowfly maggots (*Chrysomyia albiceps* and *C. marginalis*) open a carcass to expose the nutrient-rich fluids inside, a wide range of small Diptera (Milichiidae and others) use the fluids too. Likewise, honey badgers (*Mellivora capensis*) use honey made by bees, and unused termite mounds provide shelter for dwarf mongoose (*Helogale parvula*) and a variety of reptile species. The diversity of invertebrates is amplified even further by the fact that for some there are different life stages, such as an egg stage, larval stage (e.g., caterpillar), pupal stage, and adult (e.g., moth), each of which attracts a different opportunistic or specialized predators and parasites. Each stage influences ecosystem characteristics in specific ways.

That canopy insects have nearly the same biomass as medium and larger mammalian herbivores (~2,400 kg/km^2; Braack, unpublished data) is particularly revealing of their ecosystem-scale importance. Given the extraordinary diversity and abundance of invertebrates, it is no surprise that insects serve as a major food resource for a wide range of other organisms. Within Kruger there are a large number of birds (e.g., bee-eaters, rollers, shrikes), lizards and snakes, frogs, and even mammals (e.g., aardvark, pangolin, and 39 species of insectivorous bats), which feed nearly exclusively on invertebrates. Invertebrates directly contribute to heterogeneity not simply through their own species richness but also as an abundant food resource, which allows other organisms to diversify.

The diet of the largest proportion of birds in Kruger by species, numbers, and biomass is invertebrates (once the omnivorous red-billed queleas have been excluded, although queleas also need insects to feed their young; Kemp et al. 2001). A large proportion of the insectivorous species, by numbers and body size, feed on larger arthropods whenever possible. These include hornbills, rollers, storks, plovers, and the larger shrikes and starlings. Most of the larger arthropods eaten have not been described or studied in detail but include Lepidoptera larvae, Orthoptera (especially grasshoppers, locusts, and mantids), Phasmida, Coleoptera and their larvae, and Arachnida such as solifugids and scorpions whenever these become available on the surface. There is no information on the population sizes or dynamics of these arthropods, but their large size suggests that their numbers and rates of recruitment are lower than for smaller species. This also suggests that birds, probably in conjunction with other predators, may impose important selection pressures on them.

Kemp et al. (Chapter 13, this volume) address the role of avian predators in shaping the evolution of cooperative breeding groups and the defense systems they use. This same predation potential also may affect a wide range of other small vertebrates. Habitats adjacent to Kruger support some of the highest densities of diurnal and nocturnal avian predators and scavengers recorded anywhere in Africa (Tarboton and Allan 1984; Simmons 1994; Davison 1998). The recorded upper limits suggest densities of 32 pairs of eagles per 100 km^2, 45 pairs of all diurnal raptors per 100 km^2, and 51 pairs of all diurnal raptors and scavengers per 100 km^2 (Simmons 1994). The implication is that, collectively, this top-down predation could cascade throughout the ecosystem in a manner similar to the better-understood ecological cascades described for predaceous fish in northern temperate lakes and tropical oceans (Carpenter and Kitchell 1993; Kitchell et al. 1999).

No figures are available for small mammalian predators in Kruger (the mongooses, genets, and civets), that would complement and extend the predation pressures provided by birds. However, the Serengeti ecosystem, of similar size (25,000 km^2) and with the same species diversity of birds (517 species) and small carnivores (20+ species) as Kruger, supports no less than 668 mongooses per 100 km^2 for only three species: ~376 per 100 km^2 dwarf mongoose, ~172 per 100 km^2 banded mongoose (*Mungos mungo*), and ~120 per 100 km^2 slender mongoose (*Galerella sanguinea*) (Sinclair and Arcese 1995). This suggests that the predation pressure from small carnivorous mammals is likely to be an important selective force shaping Kruger's small vertebrate communities.

Even in relatively natural ecosystems such as Kruger, death due to disease is a regular and natural event and represents yet another trophic-related species-ecosystem linkage (Chapter 12 and 17, this volume). Invertebrates are the single most important group of disease vectors in the animal kingdom, in terms of both the number of plant and animal diseases they transmit and the frequency by which they do so. Oxpeckers (*Buphagus* spp.) and vultures (family Accipitridae) also seem to have critical roles in disease dynamics, but few specific data are available (Chapter 13, this volume). Oxpeckers eat ticks, which are intermediate hosts for several diseases, and thereby control disease development in their mammalian hosts through tick consumption and cleaning of open wounds. Vultures and other scavengers remove, recycle, and transport carrion and associated diseases and parasites and are therefore ecologically important at a broader scale. The buffalo population in Kruger is the reservoir host for bovine tuberculosis, which has recently invaded through cattle from Mozambique, and although there is no evidence as yet that the buffalo population is being significantly affected (Rodwell et al. 2001), the disease is being passed on to their predators. Lions are susceptible to bovine tuberculosis infection (Keet et al. 1996), and there are potentially serious implications as this top predator begins to dwindle in Kruger. Although little is known about the ecology of diseases, the implications are so large that this topic has been identified as one of the grand challenges for ecology in the coming decade (National Research Council 2001).

Animal Population Effects on Soil Nutrients and Soil Moisture

Soils, together with their nutrient properties and vegetative communities, are exceptionally diverse in Kruger. Geological formations are oriented mainly north to south, causing a general east-west diversity of soil types. Additionally, the climate becomes drier to the north, causing a north-south diversity of climatic regimes. This is in contrast to very wet or very dry areas, where the importance of soil characteristics is less pronounced (Figure 10.3). Collectively, these soil characteristics greatly influence the composition and structure of plant communities and are therefore important factors in creating and maintaining habitat diversity in Kruger. In concert with the activities of animals, this has created a mosaic of ecological regions where different assemblages of plants and animals occur (Chapter 5, this volume). This is especially so for terrestrial ecosystems with semiarid climates, such as those in Kruger, where soil and rainfall are major causes of ecological diversity.

Although soils determine the composition of plant communities, individual plants influence soil characteristics at the local scale. A patchy distribution of grasses and trees leads to patchiness in water and nutrient use efficiencies on both spatial (trees can obtain water form deeper levels) and temporal (annuals

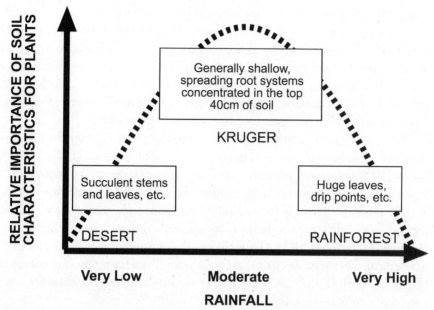

FIGURE 10.3. Conceptual model of the relative importance of soil characteristics for plants as a function of rainfall.

and perennials and early and late growers) scales (Baruch et al. 1995). Furthermore, the occurrence of nitrogen-fixing organisms, both free-living and those associated with plant roots, causes spatial heterogeneity in soil-available nutrients. Nitrogen and phosphorus flux is higher in fine-leaved savannas dominated by *Acacia* than in broad-leaved savannas (Scholes and Walker 1993). The rapidly decaying fine leaves are more favored by browsers than slowly decaying broad leaves. Leaf litter is three times higher below *Acacia* canopies than between them, and decomposition rates are three to five times higher for grass litter than for tree litter in broad-leaved savannas (Scholes and Walker 1993). This is reflected by the distribution of the palatable grass *Panicum maximum*, which prefers the moist and shaded habitats under trees (van Oudtshoorn et al. 2002).

Decomposition and the attendant cycling of nutrients is a function through which insects have a profound impact on savanna ecosystems. Three groups have crucial roles in Kruger: termites, dung beetles, and wood-boring beetles. Termites (Isoptera) have global importance, and their impact is particularly high in savanna ecosystems. Most species forage at night on dead and decaying plant material, such as woody debris, leaf litter, and even dry dung, consuming up to 55 percent of all surface plant litter. Digesting lignin and cellulose presents a major challenge to most animals, but some termite species have cellulose-degrading enzymes, and others have symbiotic microorganisms within the gut, making termites among the most efficient plant decomposers (Speight et al. 1999). A single species of common termite, *Macrotermes natalensis*, annually consumes ~20 kg/ha (Meyer 2001), suggesting that the collective litter and other woody material consumed by all species (in 22 genera) within Kruger must be considerable. That the majority of termites live inconspicuously below ground, where the food is brought, implies a major role in nutrient cycling, especially in making nutrients readily available to plant roots.

Where termites have built mounds on granite-derived soil, significantly elevated concentrations of almost all the mineral nutrients are evident (Table 10.1). Meyer (1997) calculated a density of 111 active mounds per square kilometer for all mound-building termites in the northern half of the Kruger on granitic soils. Termite mounds effectively form islands of nutrients, and grasses growing on the mounds are heavily used by herbivores, especially during dry seasons (see also Hesse 1955; Penzhorn 1982). This is often only along northern side of the mound, whereas palatable grass species as little as 5 m away from the mound are not eaten (Figure 10.4). Highly palatable grass species such as *Cenchrus ciliaris* also are more common on termite mounds (Trollope et al. 1989).

Wood-boring beetles, particularly the larval stages, which spend long periods tunneling and feeding on wood in dead or dying trees, are highly diverse and numerous in Kruger. Wood-boring beetles contribute significantly to the breakdown of wood by creating conditions conducive to fire. They colonize trees injured by elephant feeding, tunneling into and deepening the damage site, thereby weakening the tree, which often subsequently dies. This is especially evident in marula trees (*Sclerocarya birrea*).

The >120 dung beetle species in Kruger perform critical roles in removing the bulk of daily deposited dung in a short time. Most dung is buried underground as food for adult and larval beetles. In the process, nutrient-rich dung moisture and some dung itself enriches the soil for plant uptake. Dung beetles are critical in the warm, wet months, when they clear space for grass and shrubs to grow and deposit nutrients below ground that are used for the rapid flush of plant growth at this time. Extended drought seasons have a serious and long-lasting impact on general dung beetle populations. The severe drought of 1992 dramatically reduced dung beetle abundance, and consequently their activities, with full recovery not evident until several years later (Chapter 12, this volume). In Australia, the loss of native dung beetles has been estimated to potentially render millions of hectares useless for pasture production each year in addition to encouraging dung-breeding flies (Waterhouse 1977).

Mammals in African savannas have been best studied with regard to the influence of grazers on nutrient cycling (McNaughton et al. 1988) and browsers on woodland dynamics (Laws 1970; Pellew 1983; Dublin et al. 1990; Prins and van der Jeugd 1993; Cumming et al. 1997). The additional physical impacts of hoof action, dust bathing, and mud wallowing by large mammals produce

TABLE 10.1

Influence of termites on the chemical properties of surface soils. Mean value refers to soils taken from the termite mound. Mean change is relative to soils in the vicinity of the termite mound but not directly affected by termite activity. Data are condensed from Venter (1990).

VARIABLE	MEAN VALUE	MEAN CHANGE	RANGE
P (mg/kg soil)	20.0	151%	−35 to 377%
K (me/kg soil)	7.1	69%	−65 to 367%
Ca (me/kg soil)	179.9	543%	−4 to 2,371%
Mg (me/kg soil)	16.7	21%	−80 to 173%
Na (me/kg soil)	1.7	58%	−90 to 392%
pH	7.63	1.57 units	0.22 to 3.20 units

Soils evaluated included Kroonstad, Hutton, Clovelly, Glenrosa, Shortlands, Sterkspruit, Avalon, Mayo, Mispah, and Oakleaf forms from 14 sites in Kruger National Park.

FIGURE 10.4. Termite mounds illustrating the grazing halos where palatable plants have been consumed. Soils associated with termite mounds often are elevated in nutrients sought by grazing animals (see Table 10.1).

highly visible and characteristic features in the African savanna landscape (Cumming 1982; Naiman and Rogers 1997).

Distance to surface water is a powerful determinant of the distribution of herbivore biomass and density in African savannas in the dry season (Western 1975). Water distribution (Chapter 8, this volume) therefore is an equally powerful determinant of the distribution of herbivore dung. This is evidenced in the

TABLE 10.2

Soil nutrient data (mean ± SE per variable per site) from soil collected beneath Acacia tortilis *tree canopies at two sites in central Kruger National Park: near the Tshokwane ranger station and around a seasonal waterhole 4.8 km northeast of the ranger station, across the Nwaswitsontso River (du Toit et al. 1990).*

	CONCENTRATIONS IN DRY SOIL (MG/KG)		
SOIL NUTRIENTS	RANGER STATION	DIFFERENCE[a]	WATERHOLE
Total N	1,174 ± 123	<	1,946 ± 303
Nitrate N	0.81 ± 0.13	<	3.97 ± 0.60
Ammonium N	4.67 ± 0.18	<	12.61 ± 1.66
Total P	355 ± 16	<	1,303 ± 51
Soluble P	Trace	<	10.29 ± 2.63

[a]Differences ($p < .05$) determined a posteriori to analysis of variance using the Student-Newman-Keuls multiple range test.

Tshokwane region of central Kruger by a comparison of soil nitrogen and phosphorus concentrations beneath *Acacia* tree canopies around a seasonal waterhole and adjacent to a ranger station from which large herbivores had been excluded by fencing and human disturbance for about 50 years (du Toit et al. 1990). Both sites occupy the same position on the upper catena (the waterhole had formed from a dust bath that had become a mud wallow and then a pan), are on the same basaltic clay soil, and are >5 km apart. The waterhole site, with its high herbivore density, has significantly enriched soil compared with the ranger station site (Table 10.2). Therefore, waterholes may be hotspots of elevated soil nutrients despite the piospheres of degraded vegetation that surround them (Thrash 2000).

Although congregations of grazing and browsing ungulates (around waterholes, salt licks, and grazing lawns) create nutrient deposition sites, it does not necessarily follow that browsers and grazers influence the cycling of nutrients in the same manner. It is well established that intense grazing throughout the growing season maintains grazing lawns and fast nutrient cycles, and nutrients are retained close to the soil surface (McNaughton 1979, 1984; McNaughton et al. 1988). A process analogous to the maintenance of grazing lawns is found in a positive feedback loop (browsing-regrowth-rebrowsing) involving browsing ungulates and *Acacia* trees in central Kruger (Figure 10.5). The sustainability of the feedback loop remains to be tested, however, and evidence from other biomes indicates that intense browsing influences nutrient cycling in an opposite manner to grazing. In North American boreal forests the typical plant community response to intense selective browsing by moose (*Alces alces*) is

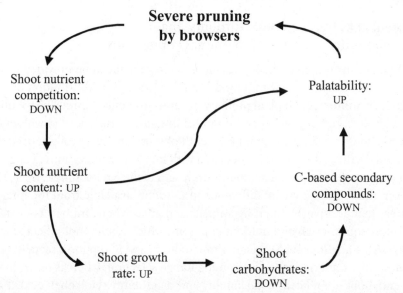

FIGURE 10.5. Positive feedbacks between browsers and *Acacia* in savannas (adapted from du Toit et al. 1990).

increased dominance by slow-growing, chemically defended woody plants that produce slow-decomposing litter (Bryant et al. 1991; Pastor et al. 1993). These slow cycles, which are prone to nutrient leaching, are maintained through positive feedback by virtue of slow-growing, unpalatable woody plants being adapted to nutrient deficient soils (Chapin et al. 1986). Therefore, patches of intense browsing ultimately could become unattractive to browsers, and the recovery of such patches to their previous states is questionable.

There is a dearth of information on the less visible effects of small or shy mammals. Burrowing mammals such as springhares (*Pedetes capensis*) and mole-rats (e.g., *Cryptomys hottentotus*) affect soil turnover, nutrient cycling, and seed dispersal in savannas, and these actions probably are significant at the ecosystem scale (Inouye et al. 1987). Nocturnal bark-gnawing porcupines (*Hystrix africaeaustralis*) exert subtle but ultimately impressive impacts on savanna woodland dynamics. Successive fires cause incremental dieback of tree stems behind bark scars so that wind action eventually snaps the weakened stems (Yeaton 1988). Another potentially important mammal is the nocturnal and solitary aardvark (*Orycteropus afer*), which although large (40–60 kg) is seldom seen and has been little studied (Taylor et al. 2002). By digging into the centers of termitaria and preying on termites (e.g., *Trinervitermes trinervoides*), aardvarks redistribute enriched soil and cause new termitaria to form in sites other than those that are destroyed, thus driving nutrient patch dynamics.

Feedbacks between Soil Nutrients, Plant Community Structure, and Animal Populations

Wild herbivores select the highest-quality forage that can be provided in sufficient quantities (Owen-Smith and Novellie 1982). The distribution of nutrients, via influences on plant productivity, thus influences animal distributions. The concentrations of mineral nutrients in sandy soils associated with granite formations generally are about one-tenth of those on clayey soils derived from basalt (Venter 1990). This is associated with higher concentrations of large herbivores on the clay soils than on the sandy soils of the Kruger (Table 10.3). However, there is a significant difference in nutrients available from the forage in areas that are sparsely used by animals and areas where animals concentrate within nutrient-rich and nutrient-poor soils, which often is not reflected in the soil (McNaughton 1988). Magnesium, sodium, and phosphorus seem to be the most important minerals determining forage selection, whereas nitrogen concentrations are related to palatability and digestibility (Holechek et al. 1982).

As discussed earlier, the high-nutrient islands created by termites attract large mammalian herbivores (Table 10.1). In March, the end of the growth season, the use rates of grasses on termite mounds are significantly higher ($p < .01$) on termite mounds (30 ± 5 percent) than on hillcrests (8 ± 4 percent). Grasses growing on termite mounds also have significantly higher nitrogen concentrations ($p < .01$), with 1.35 ± 0.07 percent on mounds and 0.92 ± 0.05 percent N in grasses growing on hillcrests. Phosphorus concentrations average 0.23 ± 0.02 percent in grasses growing on termite mounds and 0.20 ± 0.01 percent in grasses growing on hillcrests ($p = .28$), and sodium concentrations average 0.34 ± 0.1 percent in grasses growing on termite mounds and 0.27 ± 0.07 percent in grasses on hillcrests ($p < .01$).

Conclusion

The trophic and physical modifications of the environment illustrated in this chapter, and their interrelationships, suggest that there are many keystone species and ecosystem engineers in Kruger. Certainly the modifications are widespread, but for seminatural systems, the patterns and intensities are not unusual. Likewise, there may appear to be an extraordinary biocomplexity in Kruger, but other large natural areas have equally complex structures and functions across most spatial and temporal scales. Understanding and managing dynamic resources and learning how to predict the ecological consequences of increases or reductions in one or more species are key issues for Kruger.

Species-focused management in the past was dominated by concepts such as carrying capacity, which estimates an upper bound to the density of large

TABLE 10.3

Mean large herbivore (>11 kg) biomass between 1980 and 1991 in the four major land systems of the Kruger (see Chapter 5, this volume). The land systems represent the higher-rainfall zone (south) and the lower-rainfall zone (north) in the granites and basalts. The estimated herbivore biomass, the average number of hectares available for an individual herbivore, the available forage biomass for 1991, and the palatability of grass species (increasers and decreasers) are compared between the land systems.

LAND SYSTEM (LOCATION)	AREA (HA)	ESTIMATED ANIMAL BIOMASS (KG/HA)[a]	AREA/ HERBIVORE (HA), 1980–1991	FORAGE BIOMASS (KG/HA), 1991	DECREASERS[b] (%), 1991	INCREASERS[c] (%), 1991
Phalaborwa (northern granites)	499,457	32	11.1	2,028	23	69
Letaba (northern basalts)	341,886	29	11.4	2,970	25	68
Skukuza (southern granites)	372,214	27	7.2	3,099	35	40
Satara (southern basalts)	258,564	37	6.2	4,032	43	49

[a] Calculated from the aerial census by Kruger and from individual animal biomass data from Owen-Smith (1988).
[b] Decreasers are mostly palatable, perennial species and act as indicators of a moderate intensity of forage use.
[c] Increasers are mostly annual, less palatable species and are indicators of intense herbivore use.

animals and implies a balance-of-nature viewpoint. By contrast, ecosystem management recognizes spatial and temporal variability, or a flux-of-nature concept (Naiman 1992; Naiman and Rogers 1997). The former approach dampens extreme population and community changes and thereby reduces the strength and heterogeneity of links between species and ecosystems. We suggest that there are several guiding ecological principles for applying species-ecosystem links to effective resource management:

- Develop scientific perspectives and management strategies that emphasize decades rather than years.
- Manage both plants and animals for variability in time and space and for variable ratios between species.
- Allow contagion for the movement of information, nutrients, propagules, and organisms.
- Invest the effort and resources to discover, understand, and use the important species-ecosystem links, especially for invertebrates and the ecological consequences of disease.

As for being able to predict the consequences of species' increases or reductions, the scientific basis for forecasting, although making rapid progress, is not at a stage where precise predictions are possible. Kruger can make strong contributions to this issue and thereby be an example for the management of other large national parks around the world.

Acknowledgments

We thank the Andrew W. Mellon Foundation and the South Africa National Parks Board for generous support and encouragement. M. Coughenour, C. Jones, S. Bechtold, T. O'Keefe, D. Saah, and E. W. Seabloom graciously provided suggestions and comments for improvement of the manuscript.

References

Baruch, Z., A. J. Belsky, L. Bulla, C. A. Franco, I. Garay, M. Haridasa, P. Lavelle, E. Medina, and G. Sarmiento. 1995. Biodiversity as regulator of energy flow, water use and nutrient cycling in savannas. Pages 175–194 in O. T. Solbrig, E. Medina, and J. F. Silva (eds.), *Biodiversity and savanna ecosystem processes: a global perspective*. Berlin: Springer-Verlag.

Belsky, A. J. 1984. Role of small browsing mammals in preventing woodland regeneration in the Serengeti National Park, Tanzania. *African Journal of Ecology* 22:271–279.

Brown, J. H. 1995. Organisms and species as complex adaptive systems: linking the biology of populations with the physics of ecosystems. Pages 16–24 in C. G. Jones and J. H. Lawton (eds.), *Linking species and ecosystems*. New York: Chapman & Hall.

Bruggers, R. L., and C. C. H. Elliott (eds.). 1989. Quelea quelea. *Africa's bird pest*. Oxford, UK: Oxford University Press.

Bryant, J. P., F. D. Provenza, J. Pastor, P. B. Reichardt, T. P. Clausen, and J. T. du Toit. 1991. Interactions between woody plants and browsing mammals mediated by secondary metabolites. *Annual Review of Ecology and Systematics* 22:431–436.

Carpenter, S. R., and J. F. Kitchell (eds.). 1993. *The trophic cascade in lakes*. Cambridge, UK: Cambridge University Press.

Chapin, F. S. III, P. M. Vitousek, and K. Van Cleve. 1986. The nature of nutrient limitation in plant communities. *American Naturalist* 127:48–58.

Coe, M., and C. Coe. 1987. Large herbivores, acacia trees and bruchid beetles. *South African Journal of Science* 83:624–635.

Coe, M. J., and F. M. Isaac. 1965. Pollination in the baobab (*Adansonia digitata* L.) by the lesser bush baby (*Galago crassicaudatus* E. Geoffroy). *East African Wildlife Journal* 3:123–124.

Cumming, D. H. M. 1982. The influence of large mammals on savanna structure in Africa. Pages 217–245 in B. J. Huntley and B. H. Walker (eds.), *The ecology of tropical savannas*. Berlin: Springer-Verlag.

Cumming, D. H. M., et al. 1997. Elephants, woodlands and biodiversity in southern Africa. *South African Journal of Science* 93:231–236.

Davison, B. 1998. *Raptor communities in hill habitats in south-east Zimbabwe.* Unpublished M.Sc. thesis, University of Natal, Pietermaritzburg, South Africa.

Dublin, H. T., A. R. E. Sinclair, and J. McGlade. 1990. Elephants and fire as causes of multiple stable states in the Serengeti-Mara woodlands. *Journal of Animal Ecology* 59:1147–1164.

du Toit, J. T. 1990a. Giraffe feeding on *Acacia* flowers: predation or pollination? *African Journal of Ecology* 28:63–68.

du Toit, J. T. 1990b. Home range–body mass relations: a field study on African browsing ruminants. *Oecologia* 85:301–303.

du Toit, J. T. 1992. Winning by a neck: some trees succeed in life by offering giraffes a meal of flowers. *Natural History* 8/92:29–33.

du Toit, J. T., J. P. Bryant, and K. Frisby. 1990. Regrowth and palatability of *Acacia* shoots following pruning by African savanna browsers. *Ecology* 71:149–154.

du Toit, J. T., and D. H. M. Cumming. 1999. Functional significance of ungulate diversity in African savannas and the ecological implications of the spread of pastoralism. *Biodiversity and Conservation* 8:1643–1661.

Elton, C. 1930. *Animal ecology and evolution.* Oxford, UK: Oxford University Press.

Galil, J. 1977. Fig biology. *Endeavour* 1:52–56.

Hall, C. A. S., J. A. Stanford, and F. R. Hauer. 1992. The distribution and abundance of organisms as a consequence of energy balances along multiple environmental gradients. *Oikos* 65:377–390.

Hesse, P. R. 1955. A chemical and physical study of the soils of termite mounds in East Africa. *Journal of Ecology* 43:449–461.

Holechek, J. L., M. Vavra, and R. D. Pieper. 1982. Methods for determining the nutritive quality of range ruminant diets: a review. *Journal of Animal Science* 54:363–376.

Huntly, N. 1991. Herbivores and the dynamics of communities and ecosystems. *Annual Review of Ecology and Systematics* 22:477–503.

Huntly, N. 1995. How important are consumer species to ecosystem functioning? Pages 72–83 in C. G. Jones and J. H. Lawton (eds.), *Linking species and ecosystems.* New York: Chapman & Hall.

Inouye, R., N. Huntley, D. Tilman, and J. Tester. 1987. Gophers (*Geomys busarius*), vegetation, and soil nitrogen along a successional sere in east central Minnesota. *Oecologia* 72:178–184.

Jones, C. G., J. H. Lawton, and M. Shachak. 1994. Organisms as ecosystem engineers. *Oikos* 69:373–386.

Kearns, C. A., and D. W. Inouye. 1997. Pollinators, flowering plants, and conservation biology. *BioScience* 47:297–307.

Keet, D. F., N. P. J. Kriek, M. L. Penrith, A. Michel, and H. Huchzermeyer. 1996. Tuberculosis in buffaloes (*Syncerus caffer*) in the Kruger National Park: spread of the disease to other species. *Onderstepoort Journal of Veterinary Research* 69:239–244.

Kemp, A. C. 1995. *The hornbills, Bucerotiformes.* Oxford, UK: Oxford University Press.

Kemp, A. C., J. J. Herholdt, I. J. Whyte, and J. Harrison. 2001. Birds of the two largest national parks in South Africa: a method to generate estimates of population size for all species and assess their conservation ecology. *South African Journal of Science* 97:393–403.

Kitchell, J. F., C. Boggs, X. He, and C. J. Walters. 1999. Keystone predators in the central Pacific. Pages 665–704 in *Proceedings of the Wakefield Symposium on*

Ecosystem Considerations in Fisheries Management. Anchorage: University of Alaska Sea Grant.

Lamprey, H. F. 1967. Notes on the dispersal and germination of some tree seeds through the agency of mammals and birds. *East African Wildlife Journal* 5:179–180.

Laws, R. M. 1970. Elephants as agents of landscape and habitat change in East Africa. *Oikos* 21:1–15.

Magnuson, J. J. 1990. Long-term ecological research and the invisible present. *BioScience* 40:495–501.

McNaughton, S. J. 1979. Grazing as an optimization process: grass-ungulate relationships in the Serengeti. *American Naturalist* 113:691–703.

McNaughton, S. J. 1984. Grazing lawns: animals in herds, plant form, and coevolution. *American Naturalist* 124:863–886.

McNaughton, S. J. 1988. Mineral nutrition and spatial concentrations of African ungulates. *Nature* 334:343–345.

McNaughton, S. J., R. W. Ruess, and S. W. Seagle. 1988. Large mammals and process dynamics in African ecosystems. *BioScience* 38:794–800.

Meyer, V. W. 1997. *Distribution and density of mound-building termites in the northern Kruger National Park.* Unpublished M.Tech. thesis, Technikon Pretoria, Pretoria, South Africa.

Meyer, V. W. 2001. *Intracolonial demography, biomass and food consumption of* Macrotermes natalensis *(Haviland) (Isoptera: Termitidae) colonies in the northern Kruger National Park, South Africa.* Unpublished Ph.D. dissertation, University of Pretoria, Pretoria, South Africa.

Miller, M. F. 1994. The costs and benefits of *Acacia* seed consumption by ungulates. *Oikos* 71:181–187.

Naiman, R. J. 1988. Animal influences on ecosystem dynamics. *BioScience* 38:750–752.

Naiman, R. J. (ed.). 1992. *Watershed management: balancing sustainability and environmental change.* New York: Springer-Verlag.

Naiman, R. J., and K. H. Rogers. 1997. Large animals and the maintenance of system-level characteristics in river corridors. *BioScience* 47:521–529.

National Research Council. 2001. *Grand challenges in environmental science.* Washington, DC: National Academy Press.

Owen-Smith, R. N. 1988. *Megaherbivores: the influence of very large body size on ecology.* Cambridge, UK: Cambridge University Press.

Owen-Smith, R. N., and P. A. Novellie. 1982. What should a clever ungulate eat? *The American Naturalist* 119:151–178.

Pastor, J., B. Dewey, R. J. Naiman, P. F. McInnes, and Y. Cohen. 1993. Moose browsing and soil fertility in boreal forests of the Isle Royale National Park. *Ecology* 74:467–480.

Pastor, J., and R .J. Naiman. 1992. Selective foraging and ecosystem processes in boreal forests. *American Naturalist* 139:690–705.

Pellew, R. A. 1983. The impacts of elephant, giraffe and fire upon the *Acacia tortilis* woodlands of the Serengeti. *African Journal of Ecology* 21:41–74.

Penzhorn, B. L. 1982. Soil-eating by Cape Mountain zebras. *Koedoe* 25:83–88.

Power, M. E., D. Tilman, J. A. Estes, B. A. Menge, W. J. Bond, L. S. Mills, G. Daily, J. C. Castilla, J. Lubchenco, and R. T. Paine. 1996. Challenges in the quest for keystones. *BioScience* 46:609–620.

Prins, H. H. T., and H. P. van der Jeugd. 1993. Herbivore population crashes and woodland structure in East Africa. *Journal of Ecology* 81:305–314.

Rodwell, T. C., I. J. Whyte, and W. M. Boyce. 2001. Evaluation of population effects of bovine tuberculosis in free-ranging African buffalo (*Syncerus caffer*). *Journal of Mammalogy* 82:231–238.

Scholes, R. J., and B. H. Walker 1993. *An African savanna: synthesis of the Nylsvlei study*. Cambridge, UK: Cambridge University Press.

Simmons, R. 1994. Conservation lessons from one of Africa's richest raptor reserves. *Gabar* 9:2–13.

Sinclair, A. R. E., and P. Arcese (eds.). 1995. *Serengeti II: dynamics, management and conservation of an ecosystem*. Chicago: Chicago University Press.

Slater, K., and T. J. du Toit. 2002. Seed dispersal by chacma baboons and syntopic ungulates in southern African savannas. *South African Journal of Wildlife Research* 32:75–79.

Speight, M. R., M. D. Hunter, and A. D. Watt. 1999. *Ecology of insects: concepts and applications*. Oxford, UK: Blackwell Science.

Tarboton, W. R., and D .G. Allan. 1984. The status and conservation of birds of prey in the Transvaal. *Transvaal Museum Monographs* 3:1–115.

Taylor, W. A., P. A. Lindsey, and J. D. Skinner. 2002. The feeding ecology of the aardvark *Orycteropus afer*. *Journal of Arid Environments* 50:135–152.

Thrash, I. 2000. Determinants of the extent of indigenous large herbivore impact on herbaceous vegetation at watering points in the north-eastern lowveld, South Africa. *Journal of Arid Environments* 44:61–72.

Trollope, W. S. W., A. L. F Potgieter, and N. Zambatis. 1989. Assessing veld condition in the Kruger National Park using key grass species. *Koedoe* 32:67–93.

Turner, M. G., R. H. Gardner, and R. V. O'Neill. 1995. Ecological dynamics at broad scales. *BioScience* Supplement 45:S29–S35.

van Oudtshoorn, F. P., W. S. W. Trollope, D. M. Scotney, and P. J. McPhee. 2002. *Gids tot Grasse van Suid-Afrika*. Pretoria: Briza Publikasies.

Venter, F. J. 1990. *Land classification for management planning in the Kruger National Park*. Unpublished Ph.D. dissertation. University of South Africa, Pretoria, South Africa.

Ware, A. B., P. T. Kaye, S. G. Compton, and S. van Noort. 1993. Fig volatiles: their role in attracting pollinators and maintaining pollinator specificity. *Plant Systematics and Evolution* 186:147–156.

Waterhouse, D. F. 1977. The biological control of dung. Pages 314–322 in T. Eisner and E. O. Wilson (eds.), *The insects, Scientific American*. San Francisco: W. H. Freeman.

Wedin, D. A. 1995. Species, nitrogen, and grassland dynamics: the constraints of stuff. Pages 253–262 in C. G. Jones and J. H. Lawton (eds.), *Linking species and ecosystems*. New York: Chapman & Hall.

Western, D. 1975. Water availability and its influence on the structure and dynamics of a savanna large mammal community. *East African Wildlife Journal* 13:265–286.

Yeaton, R. I. 1988. Porcupines, fire and the dynamics of the tree layer of the *Burkea africana* savanna. *Journal of Ecology* 76:1017–1029.

Chapter 11

Vegetation Dynamics in the Kruger Ecosystem

ROBERT J. SCHOLES,
WILLIAM J. BOND, AND HOLGER C. ECKHARDT

Vegetation dynamics are the changes in vegetation over time. The topic has attracted less research attention in Kruger than it deserves, given the heterogeneity of the vegetation in space and time and its importance in management. Outside Kruger, South African ecologists have made significant contributions, in particular in the area of savanna dynamics (Walker et al. 1981; Scholes and Walker 1993; Bond and van Wilgen 1996). The vegetation research carried out in Kruger has been largely descriptive, as documented in Chapter 5.

Why Africa Is Different

Southern African ecologists have a different perspective on the important issues in vegetation ecology, in comparison to their Northern Hemisphere, temperate climate counterparts. Is this purely an accident of history, resulting from the small size of the scientific community, its isolation, and a strong founder effect? Or does it reflect something more fundamental—a real difference in the key drivers of ecosystem functioning in the environments in which they work? We believe the latter (Table 11.1). Fire, the effects of mammalian herbivory, and variability in time are major themes in the vegetation dynamics of semiarid savannas in Kruger.

The Diversity and Prominence of Mammalian Herbivores

Africa has not been glaciated since the Ecca period, 360 million years ago. After the breakup of Gondwanaland, beginning about 160 million years ago, Africa moved only slightly northward (in contrast to, for instance, Australia and the Indian subcontinent, which moved through many degrees of latitude). The consequence of this tectonic stability is a biota that has accumulated over a long

TABLE 11.1

A summary of some of the ways in which tropical savannas, particularly of southern and East Africa, are believed to differ from the Northern Hemisphere temperate forests and grasslands on which most vegetation ecology theory is based.

ISSUE	AFRICAN SAVANNAS	OTHER SAVANNAS	TEMPERATE ECOSYSTEMS
Humans	3-My association with hominids	60 ky in Australia	<20 ky in most
Fire	Frequent, low intensity		Rare, high intensity
Large mammal herbivores	Diverse, numerous, and extant	Less diverse and numerous than Africa, sometimes extant	Sometimes previously abundant, now greatly reduced except domestic stock
Plant diversity	High	Medium to high	Low
Phenology	Controlled by water availability		Temperature driven
Nutrients	Variable, often low	Mostly low	Mostly high
Disturbance	Seldom system resetting		Storms, fires, and glaciation reset succession
Rainfall variability	High to very high		Low
Dynamics	Seen as inherently structurally unstable, with a wide range of possible tree cover and rapid changes		Seen as tending to a stable climax of forest or grassland

period, often occupying ancient land surfaces. The rise of mammalian herbivory, particularly the diversity of antelopes, has occurred over the past 60 million years. The species that contribute to African herbivore diversity have changed over the millennia (Vrba 2000), but the continuous presence of a diverse assemblage has been a constant and unique feature of southern and East Africa. North America had such an assemblage up until about 17,000 BP but lost it, possibly as a consequence of the arrival of *Homo sapiens* on that continent.

The radiation of mammalian herbivores is paralleled by a rise in the dominance and diversity of grasses, although the current dominance by grasses with a C4 photosynthetic pathway may have occurred only in the last few million years (Cerling et al. 1997). There are coevolutionary features in the antelope-grass relationship, such as the development of abrasive silica bodies in the leaves of grasses and the development of hypsodonty (continuously growing, high-crowned teeth) in ungulates, but the chicken-or-egg debate on whether grass biodiversity stimulated grazer diversity or vice versa is unresolved (Stebbins 1981).

Human Influence

The Kruger landscape is not a pristine or natural one, where those terms are defined to exclude human activities. In key respects it is a product of human activities, both currently and in the recent and distant past. Fire in southern Africa has been under human influence possibly for more than a million years (Brain and Sillen 1988). Stone Age hunter-gatherers probably had only a modest impact on herbivores, but the arrival of Iron Age pastoralists and cultivators about 2,000 years ago led to significant increases in hunting pressure on both herbivores and carnivores, competition for grazing by domestic stock, and wood cutting. European colonization in the past 150 years initially decimated the large mammals (assisted by the rinderpest outbreak of 1896) and then led to the establishment of large conservation areas, from which people were removed (Carruthers 1995). The upshot is that there is little reason to believe that the vegetation dynamics in Kruger at the time of its declaration in 1898 were in an equilibrium state, unperturbed by human actions, or that they are now.

Vegetation Dynamics in Relation to Climate

The climate of the Kruger region varies at several time scales. Over the past million years it has alternated many times between moist, warm phases (such as the present) and arid, cool phases associated with global ice ages. Over the past century the rainfall has tended to cycle between moist and dry periods of about ten years each. Within the year, the seasonal variation of rainfall and temperature controls vegetation structure and function, and the day-to-day stochastic variation in rainfall drives the plant and microbial physiology in pulses.

The Savanna Climate

The broad features of the savanna climate regime (alternation of a warm dry season with a hot wet season, each 5–7 months long) probably have been in place in the Kruger area ever since the current configuration of continents (and thus ocean currents) was established, some 14 million years ago (Tyson and Partridge 2000). The main features of this climate in Kruger and its evolution over the past million years are outlined in Chapter 5. From a vegetation dynamics point of view, the key features are as follows:

- The warm dry season provides an annual opportunity for fires to occur (although a given location burns once every 3–10 years on average).
- The large interannual variability of rainfall, partly associated with a cycle linked to the El Niño–Southern Oscillation, leads to multiyear sequences of dry and wet periods (Gertenbach 1980).
- The episodic nature of the convective storms, coupled with the high daily evaporation, creates discrete periods of soil moisture during the growing season, each lasting a few days to a few weeks. This heterogeneity has a degree of spatial and temporal coherence.

Phenology

Variations in plant function associated with the annual cycle of day length, temperature, and rainfall are called phenology. They are predictable from year to year, although their intensity and start and end dates vary. Phenological properties include the germination of seeds, the display of leaves (i.e., the dates of leaf bud break and leaf senescence and the area of leaf exposed), flowering, fruit maturation, and the growth of twigs, stems, and roots.

The broad patterns of plant phenology in Kruger are known but undocumented; the underlying physiological timing mechanisms are not known. The dry season (which coincides with winter but seldom experiences freezing or physiologically limiting temperatures) is a period of physiological inactivity for most plants. In some tree species, such as knobthorn (*Acacia nigrescens*) and weeping boer-bean (*Schotia brachypetala*), flowering precedes the first rains by as much as 2 months. The timing mechanism probably involves either day length or temperature; discriminating between the two using observations of the date of flowering alone is difficult because they are highly correlated. Other trees flower only after the commencement of the rainy season, from about October onward. Fruit maturity in trees occurs about 2 months after flowering. Grasses and forbs begin to flower 30 to 60 days after tiller growth begins, after the spring rains.

Leaf expansion in trees may precede the rains, especially if the previous rainy season was exceptionally wet or prolonged, leaving residual water in the soil. Early leaves in some species (e.g., jackal-berry [*Diospyros mespiliformis*] and red-leaved rock fig [*Ficus ingens*]) are bright red because anthocyanin pigments predominate over chlorophyll (Ludlow 1987). Once leaf expansion begins, maximum leaf area, determined by the tree basal area, is reached within weeks (Figure 11.1). Grasses begin to increase their leaf area only after the spring rains, and the leaf area of grasses per unit ground area continues to increase as long as there is available water in the soil. Grass growth can exhibit several pulses during one growing season, separated by intraseasonal droughts lasting for a few days or several weeks. Thus the peak grass leaf area (and thus the total system leaf area) is in January or February.

Grass curing (leaf senescence) is associated with drying out of the soil and postflowering tiller death and cold in frost-prone areas. It occurs throughout the late wet season and early dry season. Tree leaf senescence, seldom accompanied by autumn colors, occurs later (except in the case of a prolonged midseason drought). The leaves gradually turn brown or straw colored and fall from the branches between May and September. Terminal twig extension in trees and shrubs occurs in the first months of the wet season (Rutherford and Panagos 1982). Radial expansion of the stem (annual increment) begins with a rapid spurt because of rehydration of the trunk, but the laying down of new wood

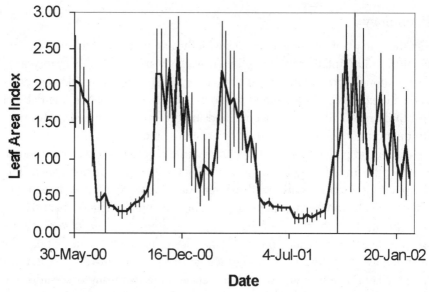

FIGURE 11.1. The pattern of leaf area index at the Skukuza flux site over time. Data are from the MODIS (Moderate Resolution Imagine Spectrometer) sensor on the Terra satellite. The error bars are the standard deviation of 49 pixels, each 500 x 500 m. Note the loss of leaf area in the dry winter months and the decline (mainly in grass green leaf) during mid–wet season drought.

continues throughout the summer. Few species exhibit easily observed and reliable annual growth rings.

The phenological cycle of grasses and, to a lesser extent, trees is associated with changes in their palatability. In particular, the nitrogen content of young tissue is high, and the fiber content is low (Chapter 6, this volume). As the plant grows, the ratio of leaf nitrogen to fiber falls. In plants on soils with a low nitrogen-supplying capacity (such as granitic uplands, especially at rainfalls above 600 mm and altitudes above 500 m, on sandstones, and on the Lebombo rhyolites) and without the benefit of symbiotic biological nitrogen fixation, the nitrogen content drops below the threshold for ruminant digestion (about 0.8–1 percent N) by the late summer. These grasslands are called sourveld (Ellery et al. 1995) and cannot support a high grazer biomass throughout the year. Historically, sourveld was burned by hunters and pastoralists during the dry winter season to promote a flush of palatable new leaf.

The leaf N content of grasses on soils with a high N-supplying capacity (such as the granitic bottomlands, or soils derived from basalt, dolerite, gabbro, or Ecca sediments) do not drop below this threshold and can sustain grazers throughout the year. They are called sweetveld. This nutrient phenology distinction is believed to be the main factor underlying differences in herbivore biomass and composition between the granite and basalt landscapes in Kruger, which in turn has important consequences for the grass fuel load and thus fire intensity. Herbivores within the fertile-infertile mosaic of the granitic landscapes are able to move between the bottomlands and uplands, or between burned and unburned patches, to satisfy their nutritional needs.

Evergreen species are mostly phreatophytes, found in locations where they have perennial access to water, such as along drainages or faults. A small proportion (<5 percent) of the woody plant species in nonriparian areas are evergreen through drought tolerance, with sclerophyllous (and often distasteful or indigestible) leaves; *Euclea* spp. are examples.

Interannual Variations

Aboveground grass net primary production in semiarid savannas is linearly related to the rainfall received in the growing season (Rutherford 1980; Scholes 1994). The slope of the relationship is steeper on clayey soils than on sandy soils, and the *y*-intercept on clayey soils typically is negative, whereas on sandy soils it is positive (Scholes 1987, 1994). A mechanistic explanation is that the slope is related to soil fertility, which is higher on clays than on sands, whereas the intercept is related to the water retention, infiltration, and evaporation characteristics of the soil. At low rainfalls, clayey soils are effectively more arid than sandy soils. The consequence of these differences is that grass production on the clayey soils of the basalt landscapes or granitic bottomlands is more variable between years than production on sandy soils. In years of above-average rainfall, the grass standing crop on clayey soils can exceed 5,000 kg/ha (Trollope and Potgieter 1985), leading to occasional very intense fires.

Large mammal herbivores need to be mobile to exploit the spatial heterogeneity, especially in dry years. Elephants are examples of such high mobility: there are few records of elephant mortality during dry spells (Walker et al. 1987). Wildebeest historically may have moved between the dry plains and the wetter foothills of the escarpment, in an east-west direction. Kruger's wildebeest population is now sedentary and is much smaller than it was before the erection of the western fence. Sedentary species such as impala and warthog show high variability in population numbers.

Woody plant recruitment for some species and in some places in Kruger appears to be episodic (Figure 11.2). Seed production varies between years and sites (Shackleton 1998) but typically is far greater than the number of seedlings that germinate and survive their first year. Most seedlings are believed to be killed by a combination of competition for water and light while in the grass layer, browsing (which may be by grazing species inadvertently eating young

FIGURE 11.2. Kruger has a system of several hundred locations at which photographs of the vegetation are taken every few years. In this sequence of images, a gradual increase (left series, from a site in the mopane shrubveld south of Shingwedzi [23°26.247′ S, 31°30.338′ E] for 1982, 1987, and 1996, respectively) and decrease (right series, from the marula-knobthorn savanna southeast of Satara [24°26.983′ S, 31°52.043′ E] for 1977, 1987, and 2002) in woody plant density and biomass is apparent.

trees among the grass), and fire. Some seedlings survive but are kept in a stunted state, perhaps for years. In an exceptionally wet year immediately after an extended drought, however, small trees have neither strong competition from grasses (because the grass basal area is severely reduced) nor high grazing pressure because the animals have died or moved away. If fires are excluded for several years, the saplings are able to grow above the flame zone and become established as a cohort of young trees (Bond and van Wilgen 1996).

Vegetation Dynamics in Relation to Disturbance

Disturbance is here defined as an event that results in a change in vegetation cover. The ecological effect of a disturbance can persist for a few days to several decades, and the spatial footprint can vary from less than 1 m^2 to many square kilometers. The dominant agents of vegetation change in Kruger are fire and large herbivores. Both interact strongly with rainfall variability, particularly the occurrence of prolonged droughts. The outcomes of fire and herbivory are altered by management practices, such as the burning regimes applied and the provision of water where it was previously unavailable.

Fire

This topic is extensively covered elsewhere (Chapter 7, this volume), so it is only briefly noted here. Fires in Kruger are of low intensity (<3,000 kW/m on average; Trollope and Potgieter 1985) and therefore are seldom stand replacing. Few plants are killed (Trollope et al. 1998), but those entirely in the flame zone may need to regrow from ground level. Above the flame zone, leaves may be scorched and fall off, but the buds survive to produce a new leaf crop.

The majority of grasses in Kruger are perennial and resprout from existing tuft bases after a fire. Annual grasses regenerate from seed. If the soil profile is even slightly moist with water carried over from the previous wet season, seepage from upslope, or recent rain, forage is available within days, and grazing mammals move onto the burned patches immediately. Thus there is a possibility of very intense herbivory if a small fraction of the landscape is burned, particularly if the fire is preceded and followed by a long dry period. Kruger's long-term fire trials, which are unfenced, represent exactly such a case, especially the early and annually burned treatments. Observed shifts in grass species composition on the trial plots may have less to do with fire impacts per se than with the interaction of fire and concentrated herbivory.

The long-term experimental burning plots illustrate the effects that fire frequency and seasonality have had on Kruger's vegetation (Enslin et al. 2000; Brönn et al. 2001; Chapter 7, this volume). In general, it appears that high-intensity fires have resulted in lower numbers of shrubs per hectare in the *Sclerocarya birrea–Acacia nigrescens* veld on basalt soils, but on the *Colophospermum mopane* veld on similar soils the shrub density increased through coppicing. Frequent fires, on the other hand, tend to suppress the woody layer by keeping it at lower heights on all soil types.

Analyses of aerial photographs covering the period 1940–1998 revealed different trends in the percentage woody cover for the granites and basalts. The central parts of Kruger witnessed an increase in woody cover of 12 percent on the granites, whereas on the basalts woody cover decreased by 64 percent (Eckhardt et al. 2000). A similar study was conducted for the mopane veld of the northern half of Kruger (Eckhardt and Biggs 2001), indicating overall increases in woody cover on both granites and basalts. Ground-level photographs, taken horizontally every few years at fixed locations, reveal changes in tree demography. In general there has been a decline in the density of the large tree component and an increase in the shrub layer since the photos began in 1980. These findings are consistent with observations by Coetzee et al. (1979), Engelbrecht (1979), Viljoen (1988), and Trollope et al. (1998).

Physical Damage to Woody Plants by Elephants

About half of the food intake by elephants is derived from woody plants. This fraction rises in drought periods, during the dry season, and in environments with a high tree:grass leaf ratio (e.g., mopane). Given that the metabolic needs of an average adult elephant (3,000 kg body mass, nonlactating) are 450 MJ/day (Meissner 1982), the digestibility in the elephant gut of the typical mixture of twigs and leaves consumed is about 40 percent, and its energy content is 18 MJ/kg, annual consumption is about 23 tons (dry matter) of plant material per elephant. For an elephant population of 10,500 individuals, this translates to 240,000 tons/year for the entire Kruger, or 13 g/m^2. We estimate that the amount of plant material broken off the plant during elephant feeding is twice the amount ingested. Thus the combined consumption and leaf stripping by elephants represents about 10 percent of the average leaf, twig, and fruit production by woody plants in Kruger.

It is commonly observed that elephant feeding is highly concentrated in the landscape, with the herds focusing on a small area within an apparently equally acceptable vegetation type. Elephants also have a preference for certain communities, such as the riparian fringe. The fraction of net primary productivity consumed therefore can rise locally, substantially above the average given earlier. It is speculated that this ultimately leads to stunting and death of the target species, forcing the elephants to abandon the area. The result is a patchwork landscape with stands in different stages of recovery from intensive use by elephants. It has been suggested (Caughley 1976) that this is the mechanism for the regional scale regulation of elephant populations. At the patch scale, elephant density and impact would be highly variable in time, but at the regional scale over long periods of time it may represent a form of equilibrium known as a stable limit cycle. There is no substantive evidence for this hypothesis, nor is there any indication of how large the region might have to be to achieve cyclic stability.

Damage by elephants takes three broad forms: the tearing off of leaves and branches, the stripping of bark from the main stem, and the pushing over of an entire tree. The gouging of baobab trees (*Adansonia digitata*), which may ultimately lead to their collapse and death, is a fourth, species-specific impact.

Branch stripping is easily survived by the plant, although it may make the stem more susceptible to infection by pathogens. Bark stripping typically occurs during the late dry season. It can lead to death of the plant if the tree is debarked around its entire circumference (girdled). It also makes the tree more susceptible to insect and fungal attack and fire.

Tree pushing may be motivated by the desire to reach foliage or fruits in the upper canopy, but sometimes it is done for no apparent reason. On average, each elephant pushes over one to four trees per day, although only a fraction of them are killed and many continue to grow with their trunks semihorizontal (common in red bushwillow [*Combretum apiculatum*]). Others resprout from the base as a new coppice (mopane *Colophospermum mopane*) or as a new stem from near the top of the broken stem (e.g., marula [*Sclerocarya birrea*]), resulting in a characteristic kink in the stem about 1 m above the ground. Resprouting reduces the height profile of the canopy (in extreme cases, changing woodlands into bushlands or even shrublands), altering their suitability as habitat for birds and smaller mammals, their susceptibility to fire, and their visual attractiveness to tourists. Some species (e.g., knobthorn [*Acacia nigrescens*]) apparently lack the ability to resprout if the main stem of a mature plant is severed, and they die if damaged in this way.

There are, on average, 300 mature trees per hectare in Kruger (Scholes et al. 2001). Elephants can be expected to push over about 1 percent in a given year. Given that the time to maturity of the average Kruger tree is about 30 years (or, alternatively, the mean annual mortality is about 3 percent; Shackleton 1998), this should be a supportable level of damage. If the elephant density rises substantially or is concentrated on particular species or portions of the landscape, elephant damage can transform woody species composition and structure.

Analyses of the large tree component in Kruger showed that between 1940 and 1960 no significant changes had occurred on the granitic soils, whereas a moderate decline had been observed in the areas with basaltic soils. However, between 1960 and 1989, there was a decline in the density of large trees in all major landscapes. As in other African studies, it is concluded that the combined effect of elephants and fire, rather than each factor on its own, alters the vegetation structure (Buechner and Dawkins 1961; Laws 1970; Thomson 1975; Dublin et al. 1990). Because trees taller than 5 m normally are not affected by fire, current declining trends in large tree numbers are ascribed primarily to elephant damage and indirectly to fire through the inhibition of recruitment (Eckhardt et al. 2000).

Physical Disturbance of the Soil

Soil disturbance involves the disruption of the surface and the exposure of subsurface material. It can occur at many scales, from the hoof imprint of an antelope to the scar produced by a hillslope slump. It usually involves uprooting of plants and thus the creation of a reduced-competition space in which new individuals can become established. The soil turnover brings dormant seeds to the surface and provides a favorable germination environment. Soil disturbance is a major mechanism of species turnover in the herbaceous layer of savannas (Yeaton 1988) but has not been specifically studied in Kruger other than in relation to waterholes (Thrash et al. 1991, 1993).

Erosion and Deposition

Erosion, the movement of soil under the influence of water and gravity, is an important and desirable disturbance process if soil export at a landscape scale does not greatly exceed the rate of soil formation. Soil formation in the granitic landscapes of Kruger is estimated to be about 1 ton/ha/year (inferred from van Tienhoven et al. 1995); reasoning based on geomorphology and mineralogy suggests that the rate is higher on basalts and lower on rhyolites. There is little evidence of significant accelerated net erosion from Kruger as a whole, although there may be localized examples of soil loss exceeding soil formation (Chappel and Brown 1993). Most soil movement is within the landscape, at a small scale. The small drifts of sediment that form after heavy rain behind obstacles on the soil surface, such as fallen logs or tufts of grass, are important sites for germination of grass and other seeds (Veenendal 1991).

Defoliation by Insects

Broad-leaved, nutrient-poor savannas are characterized by occasional outbreaks of caterpillar larvae (Scholes and Walker 1993). These insects usually are specific to their host species; a well-known example is mopane caterpillar (*Gonimbrasia berlina*) on mopane (*Colophospermum mopane*). The caterpillars almost completely defoliate individual plants and often remove the majority of leaf area for that species from large areas of the landscape. They briefly accelerate the cycling of nutrients through the ecosystem via their frass. The leaves regrow within weeks. There is sometimes a second, much smaller outbreak in the same season, but typically several years elapse before another outbreak in the same location.

Drought

Prolonged drought affects the woody vegetation in a localized and selective way (Viljoen 1995). Considering the woody vegetation of Kruger as a whole, damage is not severe. Scholes (1985) recorded 5 percent mortality of mopane (*Colophospermum mopane*). Shackleton (1998) showed tree mortality increasing from the long-term average of 3 percent per annum to up to 30 percent after extremely dry years.

Prolonged summer drought has a major effect on mortality of perennial plants in the herbaceous layer (Scholes 1985; O'Connor 1991, 1995; Viljoen and Zambatis 1995). Years of below-average annual rainfall before the drought exacerbate the impact of the drought. Perennial grasses are most severely affected, especially in the more arid area north of the Olifants River and on the basalts in general. Drought can cause major changes in grass species composition, especially among decreasers (i.e., species that decrease in abundance under grazing pressure). O'Connor and Pickett (1992) showed that severe drought in combination with high grazing pressure could lead to the local extinction of certain species of desirable, perennial grasses that produce low numbers of seeds (red grass [*Themeda triandra*] and spear grass [*Heteropogon contortus*]).

Floods

Floods are the primary source of disturbance in the riparian ecosystems of Kruger (Junk et al. 1989; Rountree et al. 2000). This topic is dealt with in detail in Chapter 9. It is sufficient to note that in contrast to the situation on dry land, classic successional patterns (albeit of some complexity) are observed after major floods.

Top-Down versus Bottom-Up Control of Ecosystem Processes

We began this chapter by noting that Africa is different primarily because parts of the continent, and Kruger in particular, retain an essentially intact Pleistocene megafauna. Is the observed pattern of vegetation heterogeneity a consequence of the presence of large mammals? Or is the large mammal biomass controlled by the abiotic template of water and nutrients? These alternatives have been called top-down and bottom-up control of ecosystem structure and function, respectively (Power 1992). They have been most studied in aquatic systems, where top-down control has proven to be of great importance in understanding eutrophication of lake systems.

Owen-Smith (1988) has argued that top-down control is also of major importance in terrestrial vegetation. He suggested that megaherbivores (animals heavier than 1,000 kg) were major agents structuring terrestrial vegetation before humans caused their extinction in most parts of the world. Megaherbivores, the argument goes, are too large for their numbers to be controlled by predators and therefore would be limited only by food availability. Owen-Smith suggested that major changes in terrestrial vegetation occurred as a result of the extinction of these creatures in most parts of the world some 10,000–50,000 years ago (Owen-Smith 1989).

A much older idea of top-down control is vegetation change through the agent of fire. Fire is analogous to a generalist herbivore. Changes in fire regime, including suppression of fires, can cause ecosystem shifts in South Africa and elsewhere (Bond and van Wilgen 1996; Bond 1997). Fire has not been incorporated into the conceptual literature on trophic ecology and top-down versus bottom-up control. There is as yet no set of ideas on the spatial and temporal interplay between fire-dominated and herbivore-dominated systems along rainfall or nutrient gradients. Kruger is one of the few areas left on Earth where it is still possible to test whether megaherbivores have the keystone properties attributed to them by Owen-Smith.

There has been very little research dealing explicitly with these questions in Kruger (or other South African savannas), and some relevant work is unpublished. Therefore, the discussion that follows is unavoidably speculative.

Top-Down Control by Carnivores

Lion and other predators could in principle influence local densities of herbivores and, through them, vegetation structure. There is no evidence as yet for large-scale regional effects of predators on herbivores and therefore indirectly on ecosystem properties. Mills et al. (1995) attempted to discriminate between rainfall, a surrogate for primary production, and predation as determinants of population size of several key herbivore species. Using census data, they were able to show that variance in population size was controlled mostly by rainfall variability (closely related to grass productivity, or bottom-up control) rather than predation (top-down). The only exception was buffalo, which showed some evidence for population control by lion predation. The herbivores fell into two groups, one with populations increasing in dry periods (wildebeest, zebra) and the other favoring wet periods (kudu, waterbuck). There has been no analysis of the mechanisms underlying these patterns, but it is interesting to note that dry-period species prefer shortgrass habitats and wet period species prefer tallgrass habitats (Page and Walker 1978). White rhino, a grazer thought to be able to create shortgrass patches (Owen-Smith 1988), was lost from Kruger a century ago and reintroduced in the 1960s. An interesting topic for future research in Kruger is to determine whether such patches reflect soil-determined nutrient cycling or are created by herbivory.

Top-Down Control by Megaherbivores

Megaherbivores in Kruger are elephants, white and black rhinos, hippos, and giraffes. Elephants, as discussed earlier, have the capacity to alter vegetation structure in Kruger. Rhino numbers are still low relative to the size of the park. Hippo populations are confined to areas within foraging distance of permanent water (typically less than 2 km but up to 20 km). Giraffes are locally abundant and have an obvious visual impact on their preferred browse species, trimming them off sharply below 6 m or pruning them into a dumbbell shape. Giraffes have been shown to lead to the local elimination of some *Acacia* species, but not others, 25 years after they were introduced to a savanna in the eastern part of South Africa (Bond and Loffell 2001). In this study the least affected species was umbrella thorn (*A. tortilis*), a common tree in South and East African savannas supporting large giraffe populations. Giraffes have been shown to have a strong impact on tree dynamics in East Africa (Pellew 1983).

Structural changes to woodlands can have cascading effects on biodiversity. For example, species that need intact woodlands disappeared from *miombo* woodlands exposed to heavy elephant damage in Zimbabwe (Cumming et al. 1997). Elephants are also likely to have altered plant species distribution by locally eliminating woody species that cannot coppice after damage and promoting highly resilient coppicing trees. The distinctive tree flora on koppies, rock outcrops too steep to be easily accessible to elephants, may owe more to relative tolerance to elephant damage than any physiological need to grow on rocks.

Top-Down Control by Fire

The extent to which fire influences ecosystem structure, independently from climate, remains controversial (e.g., Frost et al. 1986; Scholes and Archer 1997). Vegetation models make it possible to explore the relative importance of climate and fire as determinants of ecosystem properties. Dynamic global vegetation models (DGVMs) simulate the potential vegetation that can grow under given climate conditions based on a mechanistic understanding of limitations on plant growth. Some DGVMs include fire modules. These allow simulation experiments in which vegetation under a given set of environmental conditions can be constructed with and without fire (Woodward et al. 1995). The results of such a simulation for 1 × 1-degree grid squares of Kruger (Figure 11.3) suggest that fire has little effect on aboveground biomass below about 600 mm but has major effects as rainfall increases above this threshold.

These results are broadly consistent with the widely used classification of savannas into arid (<600 mm) and mesic (>600 mm) forms by pasture scientists and ecologists in South Africa. They are also broadly consistent with fire experiments that have been maintained for nearly 50 years in Kruger. The experiments include a range of fire frequencies, from annual burns to no burning at all. The driest site, in mopane (*Colophospermum mopane*) shrublands at ca 450 mm/year, shows negligible changes in tree biomass (or species composition) despite 50 years of fire exclusion. In contrast, the wettest site, Pretoriuskop

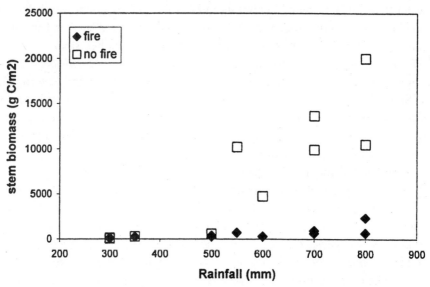

FIGURE 11.3. Simulated aboveground biomass along a rainfall gradient with and without fire in the Kruger region. The simulations were made using the Sheffield dynamic global vegetation model (Woodward et al. 1995) using averaged climate data for 1° squares covering Kruger. Differences in soil texture would influence plant available moisture but were not included in the simulations (Bond et al. 2003).

(~750 mm/year) shows major changes between frequently burned treatments and complete fire exclusion. Not only are there much higher densities of trees in the exclusion treatments, but tree species composition has also changed with invasion of floristically distinct dry forest species. Forest is too dense to support a continuous grass cover, so forest patches burn infrequently, but when they do, they may have a high-intensity fire in the tree crown. They represent an alternative ecosystem state to the open savanna. Similar shifts from fire-prone grassy ecosystems to fire-sensitive and fire-excluding forests and thickets have occurred in many other long-term fire exclusion experiments in South Africa, but only in higher-rainfall areas (Bond 1997; O'Connor and Bredenkamp 1997). Most of Kruger receives less than 550 mm of rain on average per year. Top-down control by fire is likely only in the higher-rainfall areas in the south of the park.

Both the simulation results and the evidence from the burning experiments at Kruger thus support the idea of two distinct savanna forms. Arid savannas could be controlled from the bottom up by scarce water. Mesic savannas can potentially form forest but are kept in the savanna form by top-down fire con-

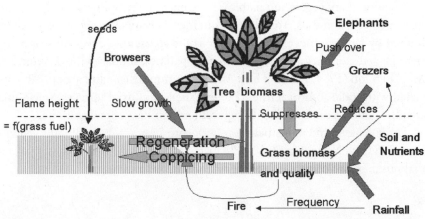

FIGURE 11.4. Key mechanisms and factors controlling the temporal and spatial heterogeneity of savannas in Kruger. Grass production is controlled by rainfall, soil type (nutrients), and tree cover. The amount accumulating as fuel for fires is determined by grazing intensity, which is linked to forage quality (nutrients). The fuel load is the main factor controlling fire intensity, which determines flame height. Young woody plants in the flame zone can escape to become grass-suppressing mature trees only if the return period between fires is sufficiently long relative to their growth rate. Their growth rate is slowed by competition with grasses and by browsing. Once large trees are established, physical damage by megaherbivores (principally elephants) is the main disruption to the system. This system has alternative forms: an open grassy form and a more closed woodland form. The proportions of the forms and the ease with which they interconvert depend on soil and climate but also on herbivore biomass and type.

trol. Both savanna types appear to have distinct suites of plant and animal species, suggesting a long evolutionary history.

Bottom-Up Control

The moist savannas are also associated with ancient land surfaces and low soil fertility (particularly a low nitrogen supply, which may ultimately have its root cause in low available phosphorus), whereas the arid savannas are on younger, more fertile soils (many references; see Scholes 1997 for a summary). In moist, infertile savannas the grass is indigestible in winter and therefore accumulates and burns. It has been speculated that the nitrogen leak that frequent burning imposes is instrumental in maintaining the low-nutrient status. Therefore, the top-down (fire and herbivory) mechanism described earlier may reflect a more fundamental bottom-up control. Analyses of large mammal diversity, based on changing needs for food quality and quantity with different body size, suggest that mammals are responding to resource availability rather than determining their own ecosystem properties (Olff et al. 2002).

The dominant vegetation patterns in Kruger closely follow variation in climate, geology, and soils. Tree densities in mesic savannas in some cases can be limited by factors other than fire, such as the sparse tree cover on seasonally waterlogged seeplines and *vleis*. Lawn grass patches, similar to those studied by McNaughton (1984, 1985) in East Africa, occur in Kruger and are closely associated with sodic soils or fertile locations such as alluvial terraces, suggesting edaphic controls on their distribution rather than heavy grazing alone.

Large-scale correlations between herbivore biomass, rainfall, and geology (East 1984; McNaughton et al. 1989; Fritz and Duncan 1994) suggest that animal numbers are controlled primarily from the bottom up.

Conclusion

In a coupled system, it is illogical to say that control of structure and function is exclusively top down or bottom up. The spatially heterogeneous and time-varying vegetation patterns observed in Kruger ecosystem necessitate a conceptual model of moderate complexity (Figure 11.4). The system illustrated has several simultaneously acting determinants and exhibits multiple equilibria that are a likely consequence of the interplay of the key factors believed to determine vegetation dynamics within the nonriparian parts of Kruger. Future vegetation research must emphasize dynamics, and a combination of models such as this one and those of Pickett et al. (Chapter 2, this volume) will provide fruitful direction.

References

Bond, W. J. 1997. Fire. Pages 421–446 in R. M. Cowling and D. Richardson (eds.), *Vegetation of southern Africa*. Cape Town: Cambridge University Press.

Bond, W. J., and D. Loffell. 2001. Introduction of giraffe changes acacia distribution in a South African savanna. *African Journal of Ecology* 39:286–294.

Bond, W. J., G. F. Midgley, and F. I. Woodward. 2003. What controls South African vegetation—climate or fire? *South African Journal of Botany* 69:1–13.

Bond, W. J., and B. W. van Wilgen. 1996. *Fire and plants*. Population and Community Biology Series 14. London: Chapman & Hall.

Brain, C. K., and A. Sillen. 1988. Evidence from the Swartkrans cave for the earliest use of fire. *Nature* (London) 336:464–466.

Brönn, A. v Z., P. J. L. Lombard, and A. L. F. Potgieter. 2001. *The effect of long term fire regime and herbivory on the* Colophospermum mopane *shrubveld in the Kruger National Park*. Poster presented at 36th Annual Congress of the Grassland Society of Southern Africa, Aldam, South Africa.

Buechner, H. K., and H. C. Dawkins. 1961. Vegetation change induced by elephants and fire in Murchison Falls National Park, Uganda. *Ecology* 42:752–766.

Carruthers, E. J. 1995. *The Kruger National Park: a social and political history*. Pietermaritzburg, South Africa: University of Natal Press.

Caughley, G. 1976. The elephant problem: an alternative hypothesis. *East African Wildlife Journal* 14:265–283.

Cerling, T. E., J. M. Harris, B. J. MacFadden, M. G. Leakey, J. Quade, V. Eisenmann, and J. R. Ehleringer. 1997. Global vegetation change through the Miocene/Pliocene boundary. *Nature* (London) 389:153–158.

Chappel, C., and M. A. Brown. 1993. The use of remote sensing in quantifying rates of soil erosion. *Koedoe* 36:1–14.

Coetzee, B. J., A. H. Engelbrecht, S. C. J. Joubert, and P. F. Retief. 1979. Elephant impact on *Sclerocarya caffra* trees in *Acacia nigrescens* tropical plains thornveld of the Kruger National Park. *Koedoe* 22:39–60.

Cumming, D. H., M. B. Fenton, I. L. Rautenbach, R. D. Taylor, G. S. Cumming, M. S. Cumming, K. M. Dunlop, A. G. Ford, M. D. Hovorka, D. S. Johnston, M. Kalcounis, Z. Mahlangu, and C. V. R. Portfors. 1997. Elephants, woodlands and biodiversity in southern Africa. *South African Journal of Science* 93:231–236.

Dublin, H. T., A. R. E. Sinclair, and J. McGlade. 1990. Elephants and fire as causes of multiple stable states in the Serengeti-Mara woodlands. *Journal of Animal Ecology* 59:1147–1164.

East, R. 1984. Rainfall, soil nutrient status and biomass of large African savanna mammals. *African Journal of Ecology* 22:245–270.

Eckhardt, H. C., and H. C. Biggs. 2001. *Medium and long-term changes in woody vegetation cover in the Kruger National Park, South Africa*. Unpublished report, South African National Parks, Skukuza, South Africa.

Eckhardt, H. C., B. W. van Wilgen, and H. C. Biggs. 2000. Trends in woody vegetation cover in the Kruger National Park, South Africa, between 1940 and 1998. *African Journal of Ecology* 38:108–115.

Ellery, W. N., M. C. Scholes, and R. J. Scholes. 1995. The distribution of sweetveld and sourveld in South Africa's grassland biome in relation to environmental factors. *African Journal of Range and Forage Science* 12:38–45.

Engelbrecht, A. H. 1979. Olifantinvloed op *Acacia nigrescens*-bome in 'n gedeelte van die Punda Milia-sandveld van die Nasionale Krugerwildtuin. *Koedoe* 22:29–37.

Enslin, B. W., A. L. F. Potgieter, H. C. Biggs, and R. Biggs. 2000. Long term effects of fire frequency and season on the woody vegetation dynamics of the *Sclerocarya birrea/Acacia nigrescens* savanna of the Kruger National Park. *Koedoe* 43:27–37.

Fritz, H., and P. Duncan. 1994. On the carrying-capacity for large ungulates of African savanna ecosystems. *Proceedings of the Royal Society, London* Series B 256:77–82.

Frost, P. G. H., J. C. Menaut, B. H. Walker, E. Medina, O. T. Solbrig, and M. Swift. 1986. Responses of savannas to stress and disturbance. *Biology International* Special issue 10.

Gertenbach, W. P. D. 1980. Rainfall patterns in the Kruger National Park. *Koedoe* 23:35–44.

Junk, W. J., P. B. Bayley, and R. E. Sparks. 1989. The flood-pulse concept in river-floodplain systems. *Canadian Special Publications in Fisheries and Aquatic Sciences* 106:110–127.

Laws, R. M. 1970. Elephants as agents of habitat and landscape change in East Africa. *Oikos* 21:1–15.

Ludlow, A. E. 1987. *A developmental study of the anatomy and fine structure of the leaves of* Ochna pulchra *Hook.* Unpublished Ph.D. thesis, University of the Witwatersrand, Johannesburg, South Africa.

McNaughton, S. J. 1984. Grazing lawns: animals in herds, plant form, and coevolution. *American Naturalist* 124:863–886.

McNaughton, S. J. 1985. Ecology of a grazing system: the Serengeti. *Ecological Monographs* 55:259–294.

McNaughton, S. J., M. Oesterheld, D. A. Frank, and K. J. Williams. 1989. Ecosystem-level patterns of primary productivity and herbivory in terrestrial habitats. *Nature* 341:142–144.

Meissner, H. H. 1982. The theory and application of a method to calculate the forage intake of wild South African ungulates for purposes of estimating carrying capacity. *South African Journal of Wildlife Research* 12:41–47.

Mills, M. G. L., H. C. Biggs, and I. J. Whyte. 1995. The relationship between rainfall, lion predation and population trends in African herbivores. *Wildlife Research* 22:75–88.

O'Connor, T. G. 1991. Influence of rainfall and grazing on the compositional change of the herbaceous layer of a sandveld savanna. *Journal of the Grassland Society of Southern Africa* 8:103–109.

O'Connor, T. G. 1995. Transformation of a savanna grassland by drought and grazing. *African Journal of Range and Forage Science* 12:53–60.

O'Connor, T. G., and G. J. Bredenkamp. 1997. Grassland. Pages 215–257 in R. M. Cowling, D. M. Richardson, and S. M. Pierce (eds.), *Vegetation of southern Africa.* Cambridge, UK: Cambridge University Press.

O'Connor, T. G., and G. A. Pickett. 1992. The influence of grazing on seed production and seed banks of some African savanna grasslands. *Journal of Applied Ecology* 29:247–260.

Olff, H., M. E. Ritchie, and H. H. T. Prins. 2002. Global environmental determinants of diversity in large herbivores. *Nature* (London). 415:901–905.

Owen-Smith, R. N. 1988. *Megaherbivores: the influence of very large body size on ecology.* Cambridge, UK: Cambridge University Press.

Owen-Smith, R. N. 1989. Megafaunal extinctions: the conservation message from 11,000 years BP *Conservation Biology* 3:405–412.

Page, B. R., and B. H. Walker. 1978. Feeding niches of four large herbivores in the Hluhluwe Game Reserve, Natal. *Proceedings of the Grassland Society of Southern Africa* 13:113–122.

Pellew, R. A. P. 1983. The impacts of elephant, giraffe and fire upon the *Acacia tortilis* woodlands of the Serengeti. *Journal of Ecology* 21: 41–74.

Power, M. E. 1992. Top-down and bottom-up forces in food webs: do plants have primacy? *Ecology* 73:724–732.

Rountree, M. W., K. H. Rogers, and G. L. Heritage. 2000. Landscape state change in the semi-arid Sabie River, Kruger National Park, in response to flood and drought. *South African Geographical Journal* 82:173–181.

Rutherford, M. C. 1980. Annual plant production-precipitation relations in arid and semi-arid regions. *South African Journal of Science* 76:53–56.

Rutherford, M. C., and M. D. Panagos. 1982. Seasonal woody plant shoot growth in *Burkea African–Ochna pulchra* savanna. *South African Journal of Botany* 1:104–116.

Scholes, R. J. 1985. Drought-related tree, grass and herbivore mortality in a southern African savanna. Pages 350–353 in J. C. Tothill and J. J. Mott (eds.), *Ecology and management of the world's savannas*. Canberra: Australian Academy of Science.

Scholes, R. J. 1987. *Response of three semi-arid savannas on contrasting soils to the removal of the woody component*. Unpublished Ph.D. thesis, University of the Witwatersrand, Johannesburg, South Africa.

Scholes, R. J. 1994. Nutrient cycling in semi-arid grasslands and savannas: its influence on pattern, productivity and stability. Pages 1331–1334 in *Proceedings of the 17th International Grasslands Congress 1993*. New Zealand Grassland Association, Palmerston North.

Scholes, R. J. 1997. Savannas. Pages 258–277 in R. Cowling, D. Richardson, and S. Pierce (eds.), *Vegetation of southern Africa*. Cape Town: Cambridge University Press.

Scholes, R. J., and S. R. Archer 1997. Tree-grass interactions in savannas. *Annual Review of Ecology and Systematics* 28:517–544.

Scholes, R. J., N. Gureja, M. Giannecchinni, D. Dovie, B. Wilson, N. Davidson, K. Piggot, C. McLoughlin, K. van der Velde, A. Freeman, S. Bradley, R. Smart, and S. Ndala. 2001. The environment and vegetation of the flux measurement site near Skukuza, Kruger National Park. *Koedoe* 44:73–83.

Scholes, R. J., and B. H. Walker. 1993. *An African savanna: synthesis of the Nylsvley study*. Cambridge, UK: Cambridge University Press.

Shackleton, C. 1998. *The prediction of woody plant productivity in the savanna biome, South Africa*. Unpublished Ph.D., University of the Witwatersrand, Johannesburg, South Africa.

Stebbins, G. L. 1981. Coevolution of grasses and herbivores. *Annals of the Missouri Botanical Garden* 68:75–86.

Thomson, P. J. 1975. The role of elephants, fire and other agents in the decline of a *Brachystegia boehmii* woodland. *Journal of the Southern African Wildlife Management Association* 5:11–18.

Thrash, I., P. J. Nel, and J. D. P. Bothma. 1991. The impact of the provision of water for game on the basal cover of the herbaceous vegetation around a dam in the Kruger National Park. *Koedoe* 34:121–130.

Thrash I., G. K. Theron, and J. D. P. Bothma. 1993. Impact of water provision on herbaceous community composition in the Kruger National Park, South Africa. *African Journal of Range and Forage Science* 10:31–35.

Trollope, W. S. W., and A. L. F. Potgieter. 1985. Fire behaviour in the Kruger National Park. *Journal of the Grassland Society of Southern Africa* 2:17–23.

Trollope, W. S. W., L. A. Trollope, H. C. Biggs, D. Pienaar, and A. L. F. Potgieter. 1998. Long-term changes in the woody vegetation of the Kruger National Park, with special reference to the effects of elephants and fire. *Koedoe* 41:103–112.

Tyson, P. D., and T. C. Partridge. 2000. Evolution of Cenozoic climates. Pages 371–387 in T. C. Partridge and R. R. Maud (eds.), *The Cenozoic of southern Africa*. Oxford, UK: Oxford University Press.

van Tienhoven, A. M., K. A. Olbrich, R. Skoroszewski, J. Taaljaard, and M. Zunkel. 1995. Application of critical loads approach in South Africa. *Water, Air and Soil Pollution* 85:2577–2582.

Veenendal, E. 1991. *Adaptive strategies of grasses in semi-arid savanna in Botswana*. Unpublished Ph.D. thesis, Amsterdam Vrije University, Amsterdam, The Netherlands.

Viljoen, A. J. 1988. Long-term changes in the tree component of the vegetation in the Kruger National Park. In I. A. W. MacDonald and R. J. M. Crawford (eds.), *Long-term data series relating to southern Africa's renewable natural resources*. South African National Scientific Programmes Report 157. Foundation for Research Development, Council for Scientific and Industrial Research, Pretoria, South Africa.

Viljoen, A. J. 1995. The influence of the 1991/92 drought on the woody vegetation of the Kruger National Park. *Koedoe* 38:85–97.

Viljoen, A. J., and N. Zambatis. 1995. *Short-term effects of the 1991/92 drought on the herbaceous layer of the vegetation in the Kruger National Park*. National Parks Board Internal Report, Skukuza, South Africa.

Vrba, E. S. 2000. Major features of Neogene mammalian evolution in Africa. Pages 277–304 in T. C. Partridge and R. R. Maud (eds.), *The Cenozoic of southern Africa*. Oxford Monographs on Geology and Geophysics 40. Oxford, UK: Oxford University Press.

Walker, B. H., R. H. Emslie, R. N. Owen-Smith, and R. J. Scholes. 1987. To cull or not to cull: lessons from a southern African drought. *Journal of Applied Ecology* 24:381–401.

Walker, B. H., D. Ludwig, C. S. Holling, and R. S. Peterman. 1981. Stability of semi-arid savanna grazing systems. *Journal of Ecology* 69:473–498.

Woodward, F. I., T. M. Smith, and W. R. Emanuel. 1995. A global land primary productivity and phytogeography model. *Global Biogeochemical Cycles* 9:471–490.

Yeaton, R. I. 1988. Porcupines, fires and the dynamics of the tree layer of the *Burkea africana* savanna. *Journal of Ecology* 76:1017–1029.

Chapter 12

Insects and Savanna Heterogeneity

LEO BRAACK AND PER KRYGER

Most of our planet's terrestrial biodiversity is made up of arthropods (Wilson 1992), and of the approximately 1.5 million organisms thus far described, more than half are insects (Gullan and Cranston 1999; Speight et al. 1999). It is not this species richness in itself that is important but rather the consequences of this extraordinary abundance and diversity. To an extent often not appreciated, it is the presence of insects that effectively creates or facilitates ecosystem diversity, maintains harmonious ecosystem processes and heterogeneity, and effectively allows the continued presence and well-being of many of the more conspicuous ecosystem elements. The functions and impacts of insects on savanna ecosystems form the focus of this chapter.

Overview of Insect Studies in Kruger

A detailed survey of current taxonomic knowledge by Scholtz and Chown (1995) listed 43,565 insect species, in 7,753 genera and 569 families, known for southern Africa. These authors also provided sound reasons for estimating that the total of insect species in the subcontinent is likely to be at least twice that number. Based on insect collections compiled for Kruger and personal observations of many taxonomic specialists who contributed to the national knowledge base on insects, it seems probable that the number of insect species occurring in Kruger is somewhere between 40 and 60 percent of the total number of species present in South Africa (Braack, unpublished records). If the estimate is correct, then about 50 percent of the insect diversity of South Africa is being conserved in an area less than 4 percent of the country.

The abundance and species richness of insects in Kruger is narrowly associated with the geographic position of the park, straddling an extensive north-south axis that covers at least 35 clearly differentiated landscape zones, each with its own habitat peculiarities (Gertenbach 1983). Much of the park is in a fairly pristine state and lies in a tropical or subtropical environment.

Intensive and regular surveys of the Kruger insect fauna commenced in the early 1960s, coordinated by the chief biologist, Dr. U. de V. Pienaar (later chief executive officer of South African National Parks). Progress depended on the available expertise and interest, so studies were sporadic and of varying intensity. However, many thorough systematic surveys were conducted in this early era, particularly on ants (Prins 1963), termites (Coaton 1962), and butterflies (Kloppers and van Son 1978).

Ecological studies involving insects commenced in the 1980s, while taxonomic surveys steadily continued. Much attention was focused on insects of medical and veterinary importance, leading to a comprehensive reference collection and, for example, studies on insects associated with decomposing carrion and the role of blowflies in anthrax transmission (Braack 1990) and various aspects of systematics and ecology of *Culicoides* biting midges (Meiswinkel and Braack 1994). In the 1990s a research focus developed on the systematics, ecology, and seasonal fluctuations of aquatic insects, which could be used as indicators of water quality in Kruger's rivers. Extensive work was carried out on insect biocontrol of a range of invasive alien plants in Kruger, and efforts to control malaria led to a broad program of studies on anopheline mosquitoes. A general guidebook to the insects of Kruger was also produced (Braack 1991). In the mid-1990s concern increased regarding the threat to local Kruger honeybees posed by the human-induced spread of the Cape honeybee (*Apis mellifera capensis*) and spread of the alien *Varoa* mite parasitic on honeybees. This led to an ongoing program of study, yielding insights into the genetics and ecology of honeybees in Kruger.

It should be mentioned that the tsetse flies *Glossina morsitans* and *G. pallidipes* occurred endemically in several areas of Kruger in the nineteenth century (Pienaar 1990). However, the rinderpest pandemic that swept through the African continent and reached southern Africa in 1896 reduced populations of buffalo, warthog, and many other hosts to such low levels that tsetse flies soon disappeared from Kruger. Subsequent control measures by the Zimbabwean (then Rhodesian) and Mozambican (then Portuguese East African) governments caused these blood-sucking and disease-transmitting flies to recede even further from the borders of the Kruger. With reduced funding available in those countries from about the mid-1980s, concern was raised about the possible reentry of tsetse flies into Kruger. The State Veterinary Department, based in Skukuza, has used a range of trapping methods during annual monitoring exercises in the Pafuri area (where tsetse are most likely to enter from the Gonarezhou area of Zimbabwe and Mozambique) since the early 1990s, and to date no evidence has been found of renewed presence by these flies.

Role, Function, and Impact of Insects

The ecological significance of insects in African savannas has received disproportionately little attention when compared with that of large vertebrates, for example. This imbalance belies the fact that insects provide ecological services that are essential to the structure and function of savanna ecosystems.

Pollination

It is no exaggeration to state that removing insect pollinators would dramatically change our planet, and the bulk of our flowering plants would disappear (Buchmann and Nabhan 1996). In this section we will briefly review two key groups: honeybees and fig wasps, which use very different strategies but are critical habitat-creating and even landscape-determining pollinators.

Honeybees comprise nine species of the genus *Apis* (Hymenoptera, Apidae). Globally, some 40,000 plant species have been recorded as having importance as food resources for honeybees, which suggests the importance of these bees on reproduction and survival of those plant species (Crane 1990). The distribution map for races of African honeybees, by Hepburn and Radloff (1998), places the Kruger at the line of convergence between *A. mellifera scutellata* and *A. mellifera litorea*. With their genetic status currently being investigated, honeybees are common and widespread throughout Kruger, and they are prolific at flowers of a wide range of trees, shrubs, and smaller flowering plants. Although it is not yet clear to what extent the dominant plant species depend on honeybees for pollination, because beetles and many other insect species also frequent the same flowers (although in lesser numbers), there can be little doubt that honeybees are a key species without which the abundance and distribution of many of Kruger's plants would be dramatically reduced, possibly to extinction. Such a possibility was the cause of concern in the mid-1990s when bee researchers began suspecting potential multiple threats to honeybees in southern Africa and beyond. The first of these threats perceived at the time—that the widespread practice of commercial beekeepers of introducing the Cape honeybee (*A. mellifera capensis*) across geographic barriers into the range of more northerly subspecies would result in interbreeding and negative changes in pollination capacity—has now been allayed. Thelytokous parthenogenesis of *A. mellifera capensis* workers makes it unlikely that introgression of the Kruger population will occur.

An insidious and serious threat is being posed by the exotic bee mite *Varroa destructor*. These mites coexist as parasites of *A. cerana* in Asia but have invaded *A. mellifera* colonies on a worldwide scale since the early 1950s, causing near extinctions of wild honeybees in North America and Europe. The mites entered South Africa in the mid-1990s and are rapidly spreading from the southern

Cape point of entry further northwards. The first mites were recorded on honeybees in Kruger in August 2000, and in 2002 *Varroa* mites were found in all colonies inspected in the Pretoriuskop area (Kryger, pers. obs.). Invasion of these mites into a susceptible hive results in unsustainable losses of parasitized honeybee offspring and eventual death of the colony. If not controlled, *Varroa* mites are expected to have a significant negative impact on the honeybee populations of southern Africa (Martin and Kryger 2002).

In summary, honeybees should be regarded as a key component of African savanna systems, with extensive coevolution between these bees and many plant species, and much of the landscape depends on pollination by bees. The organisms sustained by these trees and other plants, either directly as a food source or indirectly as nesting or other habitat determinants, depend on continued well-being of healthy honeybee populations.

In a contrasting and fascinating strategy, fig wasps of the family Agaonidae (Hymenoptera) have an extraordinarily narrow and interdependent relationship with trees of the genus *Ficus*. It is well known that each of the more than 750 *Ficus* species depends on these diminutive wasps for survival. Coevolution has resulted in only a few but usually only one species of wasp being capable of pollinating the internal flowers of a particular fig species (Galil 1977; Ware et al. 1993). This mutual interdependence often has been cited as the extreme, or ultimate, example of coevolution (Ware and Compton 1992). The minute flowers of all fig trees are located in the hollow central chamber of the fig "fruit," more correctly called the syconium. The entrance to this chamber is blocked by an elaborate and closely interlocked maze of tiny bracts or scales, and the wasps negotiate their way through this obstacle using spadelike heads and other specialized body adaptations to reach the flowers within.

In a complex cycle, a fertilized female fig wasp with pollen stored in specialized receptacles on her body migrates from the fig tree where she originated to a tree of the same species with young fruits at the appropriate stage of development, using chemical cues as guidance. Upon locating a suitable fruit, she undertakes the laborious task of entering through the narrow ostiole, often losing various body parts in the process. Many females become trapped and die before reaching their objective. Once inside, successful females deposit pollen on the stigmas of some flowers and insert their ovipositors into the ovaries of other flowers, inducing subsequent gall-formation around the developing larvae. Male wasps develop more rapidly than females, and when they emerge they are wingless. These males sense which galls contain females and tunnel through to inseminate the developing females. The males then dig additional tunnels through the side of the fruit (syconium) that the females will use to exit. The males then die. Once the females have emerged, they collect pollen from the fig cavity, make their way through the exit holes made by the males, and fly to another tree to repeat the cycle (Galil and Eisikowitch 1968; Galil 1977).

Clearly, highly specific co-adaptation has occurred between these wasps and their hosts, resulting in a mutual interdependence. The nature of the relationship requires that in any given area of occurrence there must be at least one fig tree of the appropriate species at the right fruiting stage and within flying distance of another tree from which tiny, fragile female wasps are emerging. At present, where habitat fragmentation has reduced conservation areas to small islands, chance stochastic events such as floods can easily disrupt the continuity of such delicate interdependencies, with local extinction a potential result.

It is obvious that each of the 19 species of fig trees in Kruger is threatened by the vagaries of modern human actions. These include the fact that rivers or their branches may no longer be connected within the area of the Kruger, and their junctions may lie in heavily de-treed areas outside its boundaries, with serious consequences if no routes exist for fig wasps or seedlings to recolonize after major floods, fires, or similar events.

Decomposition and Nutrient Cycling

Decomposition is a process through which insects have a profound impact on savanna ecosystems. Three groups are worth particular mention here: termites, dung beetles, and wood-boring beetles.

Termites (Isoptera) are globally important because of their role in carbon recycling, and their impact is particularly significant in savanna ecosystems. They are secretive in habit because most species spend much of their time in the soil and forage at night. Most species feed on dead and decaying plant material such as woody debris, leaf litter, and even dry dung. Digesting lignin and cellulose presents a major challenge to most animals, but termites have cellulose-degrading enzymes in some species, and others have symbiotic microorganisms within the gut, making termites among the most efficient plant decomposers (Speight et al. 1999). It has been estimated that 20 percent of all CO_2 production in savanna systems results from the activity of termites (Holt 1987). It has been calculated that on a global scale termites probably are responsible for up to 20 percent of NH_4 production and 2 percent of CO_2 (Bignell et al. 1997). Furthermore, termites are responsible for consuming up to 55 percent of all surface plant litter in some savanna ecosystems (Wood and Sands 1978).

Of the approximately 1.1 million active termite mounds in the northern half of the Kruger, representing an average density of 1.1 mounds per hectare, 62.4 percent were constructed by *Macrotermes*, 4.3 percent by *Amitermes*, 2.1 percent by *Odontotermes*, 1.4 percent by *Trinervitermes* (Meyer et al. 1999), and 29.8 percent by *Cubitermes* (Meyer et al. 2000a). Two species of *Macrotermes* predominated: *M. natalensis* (0.27 active mounds/ha) and *M. ukuzii* (0.25 active mounds/ha). Studies by Meyer et al. (2000b) indicate that small mounds of *M. natalensis* contained approximately 5,000 individual termites, medium-size mounds more than 45,000 individuals, and large mounds more than 200,000 termites. Meyer (2001) found that *M. natalensis* consumed approximately 20.2 kg/ha/year, which suggests that the collective litter and other woody material consumed by all species (in 22 genera) of termites within Kruger must be considerable. The majority of termite species live inconspicuously below ground, where the food is accumulated, which also implies a major role that termites play in nutrient cycling in Kruger, especially for providing nutrients where they can be accessed by plant roots.

The abundance and impact of termites are most obvious during periods of acute drought, when numbers of certain species rapidly increase, most notably the harvester termite (*Hodotermes mossambicus*). During such episodes the protective soil runways (*Macrotermes*) and soil pyramids at nest holes (*Hodotermes*) are very abundant, and substantial areas can be entirely denuded of vegetation (Braack 1995).

Wood-boring beetles, particularly the larval stages that tunnel into and feed on wood in dead or dying trees, are diverse and numerous in Kruger. Although each species usually has a narrow range of preferred host plants, the great diversity of species of different sizes implies that almost all woody plant species in the park are exploited. Most of these beetles belong to the families Anobiidae, Buprestidae, and Cerambycidae. Although they contribute significantly to the breakdown of wood, this function is almost certainly more effectively performed by fire, which is a frequent and characteristic feature of African savanna systems. Some species of wood-boring beetles can also play a destructive role by rapidly colonizing trees injured by elephant feeding, tunneling into and deepening the damage site, thereby weakening the tree, which often dies. This is especially evident in marula trees (*Sclerocarya birria*), where partial debarking by elephants leads to invasion by wood-boring beetles, resulting in the tree eventually breaking off at the weakened site (Braack, pers. obs.).

Dung beetles (Scarabaeidae) have a prominent and particularly important role in savanna ecosystems frequented by large numbers of herbivorous mammals. This role is perhaps best demonstrated by citing the example of a disrupted natural system, such as in Australia. The indigenous dung beetle fauna is adapted to the somewhat dry and small dung pellets of marsupials. When European settlers introduced cattle in the 1700s, the local Australian dung beetles were not able to exploit this new resource, and dung accumulated on pastures. It has been estimated that a single bovine drops an average of 12 dung pads per day, and if dung is not removed one bovine could cover 0.1 ha/year (Waterhouse 1977). With about 30 million cattle in Australia, this means that accumulated dung could compromise pasture production of about 2.5 million ha/year, in addition to creating an enormous resource for flies that breed prolifically in dung. Only by introducing bovine-adapted dung beetles from Africa was this predicament partially addressed (Speight et al. 1999).

Thus far 123 species of dung beetles have been recorded from Kruger, and these beetles perform a critical role in removing the bulk of dung deposited in a very short time, burying most of it underground as food for adult and larval beetles. In the process, nutrient-rich dung moisture and some dung itself enrich the soil for plant uptake.

Other important insects are those that remove carrion. In a study that focused on the potential role of carrion-frequenting insects in transmitting anthrax in northern Kruger, Braack (1981, 1984a, 1986; 1987; Braack and Retief 1986) investigated the species composition, competition, and succession patterns of insect communities exploiting dead animals under varying seasonal and environmental conditions. These studies revealed that more than 10 percent of antelope carcasses remain undetected by hyenas, vultures, and other vertebrate scavengers in Kruger, mainly because of concealment by vegetation in woodland or riparian environments. In such cases it is usually blowflies and other carrion insects that remove the soft tissues and skin. It was found that through insect action alone, mainly the combined actions of blowfly maggots (Diptera: Calliphoridae) and dermestid beetles, all soft tissues would be removed from a medium-sized carcass such as that of an impala within 7 days under warm and wet summer conditions. This process takes 6 weeks or longer in the cool and dry conditions of winter. Under ideal conditions of warm temperature and high humidity, even buffalo and giraffe carcasses would be stripped of soft tissues within a matter of days, although it would take some weeks to remove the drying skin.

Blowflies, particularly the two species *Chrysomyia marginalis* and *C. albiceps*, are clearly the dominant species during the early and middle stages of decomposition of medium to large animals. Because such carcasses are temporary resources that are irregular and inconsistent in their availability, blowflies have adopted highly effective search strategies to rapidly locate and use such carrion. Using radioactively marked flies, it was found that *C. marginalis* could cover an area of at least 63.5-km radius within 14 days of random search and *C. albiceps* could cover a 37.5-km radius (Braack and Retief 1986; Braack 1990). Based on these mark-release-recapture studies, average densities were calculated at 29 flies/ha for *C. marginalis* and 7.5 flies/ha for *C. albiceps*. This indicates that flies must converge from a large area very rapidly to result in the numbers that are typically found at a carcass within a few hours after death of the animal.

There is also clear resource partitioning among these blowflies. *C. marginalis* and *C. albiceps* typically fully exploit and breed in carcasses with a mass of more that about 500 g, whereas smaller carcasses such as those of small mammals and birds are exploited by sarcophagid and other flies (Braack 1987; Monnig and Cilliers 1944).

Food Provisioning

Given the extraordinary diversity and abundance of insects, it should be obvious that insects themselves serve as a major food resource for a wide range of other organisms. Unpublished data of Braack, obtained by fogging whole trees with pyrethrum knockdown insecticide as described by Erwin (1982), suggested that in mixed knobthorn-marula woodland such as that around Skukuza in southern Kruger, about 2,369.8 kg (roughly 2.4 metric tons) of canopy-dwelling insects could be expected per square kilometer in the warm, wet season. This excludes insects present on and below ground, in water, and in wood or other plants. Compare this figure with an estimated average mass of about 2,332 kg/km^2 for large mammalian herbivores (impala and larger) (P. Viljoen, pers. comm.), and it conveys a sense of how abundant insects are. A calculation that factors in termites, grasshoppers, and other noncanopy insects probably would at least triple this figure for insect biomass. However rough this estimate may be, there is sufficient insect biomass to have enabled vertebrate specialization to exploit this rich resource. In the Kruger there are a large number of bird species (e.g., bee-eaters, rollers, shrikes), lizards and snakes, frogs, and mammals (e.g., aardvark, pangolin, and 39 species of insectivorous bats) that feed almost exclusively on insect prey. Insects therefore directly contribute to biodiversity not only through their own species richness, but also by being an abundant resource that allows other organisms to diversify.

Genetic Selection through Parasitism and Insectborne Diseases

Insects are the single most important group of disease vectors in the animal kingdom in terms of both the number of plant and animal diseases they transmit and the rate at which they do so. In well-integrated, coevolved systems, the host, parasite, and vector have usually established a mutually tolerant relationship, but immunologically inferior individual hosts may succumb to the disease or at least be made more susceptible to predation by being debilitated. Insects therefore are important agents of genetic selection for wildlife by being efficient vectors of disease organisms that eliminate weak individuals from host populations.

Although plant disease vectors remain poorly studied in the Kruger, many surveys and investigations have been conducted on insects ectoparasitic on large mammals (Braack 1984b; Braack et al. 1996; Coetzee et al. 1993a, 1993b; Louw et al. 1993, 1995; Segerman and Braack 1988; Quate et al. 1996). We know that some parasitic insects, usually well adapted to their hosts, may severely debilitate or cause direct or indirect mortality in their natural or aberrant hosts. Perhaps best known is *uitpeuloog,* a condition whereby species of *Oestrus* or *Gedoelstia* flies (Diptera: Oestridae)—which in the larval stage normally live in the nasopharyngeal and sinus cavities of the head of perissodactyl and artiodactyl hosts—end up in maladapted hosts such as horses, where they cause protrusion of the eyes and blindness (Zumpt 1965). Similar examples in which wildlife hosts in the Kruger are tolerant and well-adapted to specific pathogens but overt disease results in domestic stock, include African horse sickness (AHS) and bluetongue, both caused by viruses and transmitted by *Culicoides* biting midges (Diptera: Ceratopogonidae). Nine serotypes of AHS occur in Kruger, and almost 100 percent of zebra (*Equus burchelli*) older than 1 year tested positive for AHS virus (Barnard 1993). AHS usually is fatal in young horses and unvaccinated adults. Antibodies to bluetongue virus have been found in a wide range of ruminants in Kruger. This is an unapparent infection in these wildlife hosts but can be fatal and cause a high percentage of abortions in sheep. An outbreak of Rift Valley fever (RVF), transmitted by Aedine mosquitoes, caused six abortions among boma-confined buffalo (*Syncerus caffer*) in Skukuza in 1996, and buffalo seropositive for RVF have been recorded widely throughout Kruger (Bengis 1999). Insects almost certainly contribute to the genetic fitness of host populations, but little is known of the precise dynamics of this function within the Kruger.

Insects in Food Webs

Harrington (1994) calculated that if all causes of natural mortality were withheld, then the offspring of a single pair of aphids (Hemiptera: Aphidae) could cover the entire surface of our planet to a depth of 149 km within 1 year. This principle applies to numerous other insect species that breed prolifically, and although such academic exercises may seem quaint, they illustrate the vast potential of these insect species. What keeps them from expressing this potential? Inclement climatic factors, diseases, and lack of sufficient food plants do, but also predation and parasitism by other insects.

The rapid spread by alien invasive plants in areas where the insects that naturally control them do not occur provides some evidence of the impact of insects. Attempts to redress this imbalance by introducing insect biocontrol agents often result in impressive control, which reflects the contribution of even single insect species in modulating other components of well-integrated ecosystems. No studies have been done in Kruger to explore the role and impact of naturally occurring insects in exercising inhibitory or modulating influence on other organisms, but it is obvious that, given the rich diversity and abundance of insects in this park, insects feeding and parasitizing on each other and other organisms contribute greatly toward maintaining a stably fluctuating natural system.

Diversity feeds itself, and because certain organisms are present it creates opportunities for other organisms to directly subsist on those already present or to use the products made available by them, thereby expanding the resource base even further. The mere presence of some insects consequently allows other insect species to prey on, parasitize, or scavenge on those already present or to take advantage of their physical actions. Examples are blowfly maggots that open a mammal carcass and expose the nutrient-rich fluids that are then used by a wide range of other Diptera such as Milichiidae, honey badgers using honey, and disused termite mounds that provide shelter for mongooses and a variety of reptile species. The high diversity of insect species is amplified further by the different life stages, such as an egg stage, larval stage (e.g., caterpillar), pupal stage, and adult (moth, very different from larva), each of which attracts a different range of opportunistic or specialized range of predators and parasites. An exponential increase in diversity is thereby enabled. Little or no work has been done to elucidate and describe these aspects in Kruger.

Conclusion

No specific studies have focused on insects as drivers of savanna heterogeneity in Kruger, but from this chapter it should be clear that through their role in pollination (determining landscape vegetational composition), their abundance, and their great diversity, insects provide habitat and nutrient resources for many other organisms, thus contributing to heterogeneity. Insects are landscape architects in a far more enriching, pervasive, and enduring manner than the oft-repeated examples of elephants and fire, and they are enablers of diversity.

Insects contribute direct economic and social benefits to humans through the ecological functions they perform. Furthermore, a less appreciated and understood value of the natural assemblage of insects in Kruger is that they represent a natural laboratory and encyclopedia of reference knowledge, a source of future biocontrol organisms, chemicals, and biomedicines, and resources for as yet undiscovered uses. These aspects remain unexplored in Kruger.

Perhaps the easiest and most relevant example of the use of knowledge regarding insects to manage heterogeneity in Kruger has been the use of insect assemblages as indicators of water quality, or river health. Concern over declining quality and quantity of river water flowing into the Kruger resulted in numerous studies over more than a decade on various aspects of rivers in Kruger, including the insect fauna associated with various aquatic habitats and their seasonal fluctuations. This has enabled the development of reliable indices that correlate aquatic insect samples with water quality (Roux et al. 1994; Thirion et al. 1995; Uys et al. 1996). Such indices have been incorporated into various programs aimed at monitoring fluctuations in ecosystem parameters in Kruger, and insects are among the best and most reliable habitat indicators currently available, particularly for aquatic environments.

References

Barnard, B. J. H. 1993. Circulation of African horsesickness virus in zebra (*Equus burchelli*) in the Kruger National Park, South Africa, as measured by the prevalence of type specific antibodies. *Onderstepoort Journal of Veterinary Research* 60:111–117.

Bengis, R. 1999. *Annual report 1999: state veterinarian Kruger National Park*. Unpublished Report, National Department of Agriculture, Pretoria, South Africa.

Bignell, D. E., P. Eggleton, L. Nunes, and K. L. Thomas. 1997. Termites as mediators of carbon fluxes in tropical forests: budgets for carbon dioxide and methane emissions. Pages 109–134 in A. D. Watt, N. E. Stork, and M. D. Hunter (eds.), *Forests and insects*. London: Chapman & Hall.

Braack, L. E. O. 1981. Visitation patterns of principal species of the insect-complex at carcasses in the Kruger National Park. *Koedoe* 24:33–49.

Braack, L. E. O. 1984a. *An ecological investigation of the insects associated with exposed carcasses in the northern Kruger National Park: a study of populations and communities*. Unpublished Ph.D. thesis, University of Natal, Pietermaritzburg, South Africa.

Braack, L. E. O. 1984b. A note on the presence of the elephant louse *Haematomysus elephantis* Piaget (Mallophaga: Rhynchophthirina) in the Kruger National Park. *Koedoe* 27:139–140.

Braack, L. E. O. 1986. Arthropods associated with carcasses in the northern Kruger National Park. *South African Journal of Wildlife Research* 16:91–98.

Braack, L. E. O. 1987. Community dynamics of carrion-attendant arthropods in tropical African woodland. *Oecologia* (Berlin) 72:402–409.

Braack, L. E. O. 1990. Feeding habits and flight range of blow-flies (*Chrysomyia* spp.) in relation to anthrax transmission in the Kruger National Park, South Africa. *Onderstepoort Journal of Veterinary Science* 57:141–142.

Braack, L. E. O. 1991. *A field-guide to the insects of the Kruger National Park*. Cape Town: Struik Publishers.

Braack, L. E. O. 1995. Seasonal activity of savanna termites during and after severe drought. *Koedoe* 38:73–82.

Braack, L .E. O., I. G. Horak, L. C. Jordaan, J. Segerman, and J. P. Louw. 1996. The comparative host status of red veld rats (*Aethomys chrysophilus*) and bushveld gerbils (*Tatera leucogaster*) for epifaunal arthropods in the southern Kruger National Park, South Africa. *Onderstepoort Journal of Veterinary Research* 63:149–158.

Braack, L. E. O., and P. F. Retief. 1986. Dispersal, density and habitat preference of the blow-flies *Chrysomyia albiceps* (Wd) and *Chrysomyia marginalis* (Wd) (Diptera: Calliphoridae). *Onderstepoort Journal of Veterinary Science* 53:13–18.

Buchmann, S. L., and G. P. Nabhan. 1996. *The forgotten pollinators*. Washington, DC: Island Press.

Coaton, W. G. H. 1962. Survey of the termites of the Kruger National Park. *Koedoe* 5:144–156.

Coetzee, M., R. Hunt, and L. E. O. Braack. 1993a. Enzyme variation at the aspartate aminotransferase locus in members of the *Anopheles gambiae* complex (Diptera: Culicidae). *Journal of Medical Entomology* 30:303–308.

Coetzee, M., R. H. Hunt, L. E. O. Braack, and G. Davidson. 1993b. Distribution of mosquitoes belonging to the *Anopheles gambiae* complex, including malaria vectors, south of latitude 15°S. *South African Journal of Science* 89:227–231.

Crane, E. 1990. *Bees and beekeeping*. Oxford, UK: Heinemann.
Erwin, T. L. 1982. Tropical forests: their richness in Coleoptera and other arthropod species. *Coleopterists Bulletin* 36:74–75.
Galil, J. 1977. Fig biology. *Endeavour* 1:52–56.
Galil, J., and D. Eisikowitch. 1968. On the pollination ecology of *Ficus sycomorus* in East Africa. *Ecology* 49:259–269.
Gertenbach, W. P. D. 1983. Landscapes of the Kruger National Park. *Koedoe* 26:9–121.
Gullan, P. J., and P. S. Cranston. 1999. *The insects: an outline of entomology*. Oxford, UK: Blackwell Science.
Harrington, R. 1994. Aphid layer. *Antenna* 18:50–51.
Hepburn, H. R., and S. E. Radloff. 1998. *Honeybees of Africa*. Berlin: Springer-Verlag.
Holt, R. D. 1987. Prey communities in patchy environments. *Oikos* 50:276–290.
Kloppers, J., and G. van Son. 1978. *The butterflies of the Kruger National Park*. Pretoria: National Parks Board of South Africa.
Louw, J. P., I. G. Horak, and L. E. O. Braack. 1993. Fleas and lice on scrub hares (*Lepus saxatilis*). *Onderstepoort Journal of Veterinary Research* 60:95–101.
Louw, J. P., I. G. Horak, M. L. Horak, and L. E. O. Braack. 1995. Fleas, lice and mites on scrub hares (*Lepus saxatilis*) in northern and eastern Transvaal and in KwaZulu-Natal, South Africa. *Onderstepoort Journal of Veterinary Research* 62:133–137.
Martin, S. J., and P. Kryger. 2002. Reproduction of *Varroa destructor* in South African honey bees: does cell space influence *Varroa* male survivorship? *Apidologie* 33:51–61.
Meiswinkel, R., and L. E. O. Braack. 1994. African horsesickness epidemiology: five species of *Culicoides* (Diptera: Ceratopogonidae) collected live behind the ears and at the dung of the African elephant in the Kruger National Park, South Africa. *Onderstepoort Journal of Veterinary Research* 61:155–170.
Meyer, V. W. 2001. *Intracolonial demography, biomass and food consumption of* Macrotermes natalensis *(Haviland) (Isoptera: Termitidae) colonies in the northern Kruger National Park, South Africa*. Unpublished Ph.D. thesis, University of Pretoria, Pretoria, South Africa.
Meyer, V. W., L. E. O. Braack, and H. C. Biggs. 2000a. Distribution and density of *Cubitermes* Wasmann (Isoptera: Termitidae) mounds in the northern Kruger National Park. *Koedoe* 43:57–65.
Meyer, V. W., L. E. O. Braack, H. C. Biggs, and C. Ebersohn. 1999. Distribution and density of termite mounds in the northern Kruger National Park, with specific reference to those constructed by *Macrotermes* Holmgren (Isoptera: Termitidae). *African Entomology* 7:123–130.
Meyer, V. W., R. M. Crewe, L. E. O. Braack, H. T. Groeneveld, and M. J. van der Linde. 2000b. Intracolonial demography of the mound-building termite *Macrotermes natalensis* (Haviland) (Isoptera, Termitidae) in the northern Kruger National Park, South Africa. *Insectes Sociaux* 47:390–397.
Monnig, H. O., and P. A. Cilliers. 1944. Sheep blow-fly research 7: investigations in the Cape winter-rainfall areas. *Onderstepoort Journal of Veterinary Medicine Science and Animal Industry* 19:71–77.
Pienaar, U. de V. 1990. *Neem uit die Verlede*. Pretoria: National Parks Board of South Africa.

Prins, A. J. 1963. A list of the ants collected in the Kruger National Park with notes on their distribution. *Koedoe* 6:91–108.

Quate, L. W., L. E. O. Braack, A. L. Dyce, and H. A. Standfast. 1996. Species composition and periodicity of phlebotomine sand flies (Diptera: Psychodidae) in the Kruger National Park, South Africa. *African Entomology* 4:271–274.

Roux, D. J., C. Thirion, M. Smidt, and M. J. Everett. 1994. *A procedure for assessing biotic integrity in rivers: application to three river systems flowing through the Kruger National Park, South Africa.* Unpublished Report N0000/00/REQ/0894. Institute for Water Quality Studies, Department of Water Affairs and Forestry, Pretoria, South Africa.

Scholtz, C. H., and S. L. Chown. 1995. Insects in southern Africa: how many species are there? *South African Journal of Science* 91:124–126.

Segerman, J., and L. E. O. Braack. 1988. New records of bat fleas (Siphonaptera) from South Africa. *Journal of the Entomological Society of Southern Africa* 51:149–150.

Speight, M. R., M. D. Hunter, and A. D. Watt. 1999. *Ecology of insects: concepts and applications.* Oxford, UK: Blackwell Science.

Thirion, C., A. Mocke, and R. Woest. 1995. *Biological monitoring of streams and rivers using SASS4: a user manual.* Unpublished Report N0000/00/REQ/1195. Institute for Water Quality Studies, Department of Water Affairs and Forestry, Pretoria, South Africa.

Uys, M. C., P.-A. Goetsch, and J. H. O'Keefe. 1996. *National biomonitoring programme for riverine ecosystems: ecological indicators, a review and recommendations.* National Biomonitoring Programme Report Series 4. Institute for Water Quality Studies, Department of Water Affairs and Forestry, Pretoria, South Africa.

Ware, A. B., and S. G. Compton. 1992. Breakdown of pollinator specificity in an African fig tree. *Biotropica* 24:544–549.

Ware, A. B., P. T. Kaye, S. G. Compton, and S. van Noort. 1993. Fig volatiles: their role in attracting pollinators and maintaining pollinator specificity. *Plant Systematics and Evolution* 186:147–156.

Waterhouse, D. F. 1977. The biological control of dung. Pages 314–322 in T. Eisner and E. O. Wilson (eds.), *The Insects.* San Francisco: Scientific American, W. H. Freeman.

Wilson, E. O. 1992. *The diversity of life.* Cambridge, MA: The Belknap Press of Harvard University Press.

Wood, T. G., and W. A. Sands. 1978. The role of termites in ecosystems. Pages 245-292 in M. V. Brian (ed.), *Production ecology of ants and termites.* Cambridge, UK: Cambridge University Press.

Zumpt, F. 1965. *Myiasis in man and animals in the Old World. A textbook for physicians, veterinarians and zoologists.* London: Butterworths.

Chapter 13

Birds: Responders and Contributors to Savanna Heterogeneity

ALAN C. KEMP, W. RICHARD J. DEAN, IAN J. WHYTE, SUZANNE J. MILTON, AND PATRICK C. BENSON

To what extent is the avifauna of 517 species (Kemp et al. 2001) involved in exploiting and generating heterogeneity in Kruger? Flight allows birds to respond at a variety of temporal and spatial scales to patchiness of resources, even though it constrains them to small body size, biomass, and consumption. Only the flightless ostrich (*Struthio camelus*), uncommon in Kruger (Kemp et al. 2001) and with the size and habits of a large herbivore, is an exception.

There are few data on patchiness of resources in Kruger relative to the distribution or abundance of bird species. Sightings of a few species have been correlated with aspects of vegetation structure (Benn et al. 1995; Hurford et al. 1996; Kemp et al. 1989, 1997, 1998) or the Sabie River (Monadjem 1992, 1996). Sightings of all species have been mapped systematically on a 15° latitude × 15° longitude grid (Tarboton et al. 1987; Harrison et al. 1997) or indicated subjectively on parkwide maps of an even smaller scale (Kemp 1974; Newman 1980, 1991; Sinclair and Whyte 1991). These maps record at least 75 common species with limited ranges in the 20,000 km^2 of Kruger, providing evidence of habitat heterogeneity at this coarse scale, and at least 185 species that occur throughout Kruger, their ranges not obviously affected by patchy resources at this scale.

Subjective estimates of population size, biomass, and status for each of the 517 bird species recorded from Kruger (Kemp et al. 2001) are the only quantitative ecological data available. About half of the species (49 percent) were considered resident, and visitors were divided between breeders (23 percent) and nonbreeders (28 percent). Overall, the maximum annual avifauna of Kruger was estimated at 41 million individuals with a biomass of 1.24 million kg, dominated seasonally in numbers (81 percent) and biomass (54 percent) by the redbilled quelea (*Quelea quelea*). The general biology of most species in Kruger is also well known (Kemp 1974; Newman 1980, 1991; Sinclair and Whyte 1991; Maclean 1993; Kemp and Begg 2001).

In this chapter we explore relationships between avifaunal composition and ecological heterogeneity in two ways and at various scales. First, we ask whether differences in the proportions of the avifauna that reside or breed in Kruger are responses to resource heterogeneity. Second, we identify possible contributions by the avifauna to generation of heterogeneity.

Whether or Not to Reside, Breed, and Be Territorial in a Heterogeneous Savanna

Birds are predicted to make efficient use of their mobility to track, use, and defend the resources they need for successful breeding and survival. Bird species in Kruger show a range of spatial and temporal strategies in resource use (Figure 13.1).

Some species make daily movements of less than 5 km to clumped food sources, best illustrated by those that enter rest camps and picnic sites during the dry season to supplement their natural diet but also by those that clump naturally at dung middens or in mixed-species foraging flocks (Vernon 1980). Others are vagrants to Kruger, apparently as nonbreeding winter visitors from the evergreen forests of the Drakensberg escarpment (green twinspot [*Mandingoa nitidula*], bluemantled flycatcher [*Trochoceros cyanomelas*]), or as sporadic breeders (marabou stork [*Leptoptilos crumeniferus*]; Whyte 1993). Others are subcontinental nomads that irrupt sporadically in numbers, timing, and duration (redheaded finch [*Amadina erythrocephala*], larklike bunting [*Emberiza impetuani*]). Such movements are typical for many bird species of arid regions (Herremans 1993, 1995; Herremans and Herremans-Tonnoeyr 1994a, 1994b, 1994c, 1995; Herremans et al. 1993; Dean 1995; Tyler 2001), and subcontinental nomads make up a large proportion of both breeding visitors (61 percent, $n = 117$ species) and nonbreeding visitors (52 percent, $n = 147$ species) to Kruger (Kemp et al. 2001). Finally, many visiting species are regular summer migrants, either intra-African (39.3 percent of breeding visitors, 7.5 percent of nonbreeding visitors) or Palearctic migrants (40.1 percent of nonbreeding visitors). However, excluding red-billed quelea, visitors make up only about 24 percent of the avifauna by numbers (20 percent as breeders) and 23 percent by biomass (17 percent as breeders; Kemp et al. 2001).

The red-billed quelea exemplifies adaptation to a heterogeneous environment (Box 13.1), and its dominance of the avifauna attests to both the success of its strategy and the patchiness of its resources. The patches that it exploits, stands of dense bush for nesting and patches of grass seeds and small insects for feeding chicks, are available seasonally, locally, and erratically. The exact location and area covered by nesting colonies and timing of onset of breeding vary from summer to summer in Kruger, by a considerable but unrecorded

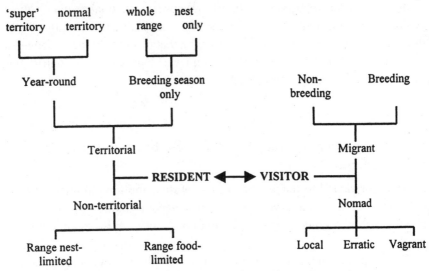

FIGURE 13.1. The main temporal residency and spatial organization options exhibited by the avifauna of Kruger.

amount, and any number of nonbreeders may be present during the dry winter. Only the insectivorous wattled starling (*Creatophora cinerea*) has similar flocking and colonial nesting behavior but occurs in much smaller numbers in Kruger.

Yet despite the obvious advantages of mobility, about half of the bird species that breed in Kruger were classified as year-round residents (Kemp et al. 2001), and many of these are also territorial for all or part of the year. Mobility has its risks, such as predation or starvation while en route, and residency has its advantages, such as prior occupancy when patchy resources develop. However, the choice of residency is independent of any temporal or spatial decisions on territoriality. Some species defend year-round territories, others only during the breeding season, and others not at all, and some form territories at the scale of their entire home range, others only in the immediate vicinity of their nest. Some even establish separate nonbreeding territories, including a number of Palearctic migrants, some of which are annually philopatric (Berthold 1993; Herremans and Herremans-Tonnoeyr 1995) despite annual variation in the quality of their territories. However, many resident species in Kruger hold year-round territories, at scales that range from 10 to 15,000 ha per breeding pair (Tarboton et al. 1987).

> **BOX 13.1**
> *Biology of the Red-Billed Quelea*
>
> This 20-g weaverbird breeds as colonies of millions of birds, the highest numerical concentrations of any bird species in the world (Bruggers and Elliott 1989). For the rest of the year it wanders as large flocks in search of small-grain foods, including agricultural crops, or gathers in huge numbers at water points and communal roosts. Flocks move rapidly over long distances and can breed at short notice (Thompson 1993), signaled by the female's bill turning yellow. Males build nests within a day, clutches of two to four eggs are laid synchronously by all females, the incubation period is the shortest of any bird (10 days), and the nestling period is only 11–13 days, so that within a month the chicks have fledged and the colony is deserted.
>
> Breeding colonies create their own patches. Seeds of annual grasses and small insects are grazed over a radius of about 6 km (range 2–11 km, approximate area 13–380 km^2; Bruggers and Elliott 1989). Various predators are attracted (Pienaar 1969; Biggs 2001; Kemp 2001), some using queleas to raise their own young (small hawks, eagles) or lay down fat for migration (lesser spotted eagles [*Aquila pomarina*]; Kemp 2001).

Each of these territorial residents is expected to have aspects of its biology that ameliorate, at an appropriate scale, any patchiness that exists in its resources. Residency and territoriality depend on details of the biology of each species: neither can be predicted from the ecological success of a species as indicated by its annual contribution to the avian biomass of Kruger (Table 13.1). Most of the estimated biomass (57.7 percent) comes from species (1, 3, 4, and 16) that feed on seeds and arthropods taken near the ground and that, including the red-billed quelea, are seasonal visitors whose biomass is supported ecologically only during their visit. Most of the resident biomass (15.5 percent) comes from large-bodied species (2, 6, 8, 12, 15, 17, 19) that also feed mainly near the ground on seeds, arthropods, and foliage, including the seed-eating doves (5, 10). Two resident hornbills (11, 18; 1.7 percent biomass) feed mainly on arthropods, also on the ground, as does the large-bodied stork (9; 1.2 percent biomass), which is the only nonbreeding visitor. Exceptions are three resident species (7, 14, 20; 2.8 percent biomass) that are arboreal and feed mainly on fruit, and the resident carrion-feeding vulture (13; 0.9 percent biomass). It is notable that most avian biomass survives off the highest concentrations of nutrients in the savanna ecosystem: propagules of plants (especially of grasses and forbs available mainly during summer) and the bodies of the principal primary consumers, arthropods, also mainly available during summer, and large mammals.

TABLE 13.1

Bird species, ranked by their contribution to 80 percent of the estimated annual avian biomass in Kruger (from Kemp et al. 2001) and annotated for their residency, territoriality, and diet.

SPECIES	ESTIMATED BIOMASS (%)	RESIDENT	TERRITORIAL	MAIN BREEDING DIET
1. Red-billed quelea	53.5	No	No	Seeds and arthropods
2. Helmeted guineafowl (*Numida meleagris*)	7.1	Yes	Yes	Seeds and arthropods
3. Kurrichane button quail (*Turnix hottentotta*)	1.9	No	Yes	Seeds and arthropods
4. Harlequin quail (*Coturnix delegorguei*)	1.6	No	Yes	Seeds and arthropods
5. Cape turtledove (*Streptopelia capicola*)	1.6	Yes	No	Seeds
6. Ostrich	1.4	Yes	Yes	Seeds and arthropods
7. Grey lory (*Corythaixoides concolor*)	1.4	Yes	Yes	Fruit
8. Black-bellied korhaan (*Eupodotis melanogaster*)	1.3	Yes	No	Arthropods and seeds
9. White stork (*Ciconia ciconia*)	1.2	No	No	Arthropods
10. Laughing dove (*Streptopelia senegalensis*)	1.1	Yes	No	Seeds
11. Yellow-billed hornbill	1.0	Yes	Yes	Arthropods
12. Red-crested korhaan (*Eupodotis ruficrista*)	0.9	Yes	No	Arthropods and seeds
13. White-backed vulture (*Gyps africanus*)	0.9	Yes	No	Carrion
14. Black-eyed bulbul (*Pycnonotus barbatus*)	0.8	Yes	Yes	Fruit and arthropods
15. Swainson's francolin (*Francolinus swainsoni*)	0.8	Yes	Yes	Seeds and arthropods
16. Wattled starling (*Creatophora cinerea*)	0.7	No	No	Arthropods
17. Egyptian goose (*Alopochen aegyptiacus*)	0.7	Yes	Yes	Foliage and seeds
18. Red-billed hornbill	0.7	Yes	Yes	Arthropods
19. Natal francolin (*Francolinus natalensis*)	0.7	Yes	Yes	Seeds and arthropods
20. Green pigeon (*Treron calva*)	0.6	Yes	No	Fruit
Total (1,239,600 kg)	79.9	No 5, Yes 15	No 9, Yes 11	

What Contributions Do Birds Make to Savanna Heterogeneity?

The avian component of a savanna ecosystem is in itself heterogeneous by virtue of spatial and temporal variation across species in their uses of foods and habitats. But birds also generate and maintain further ecological heterogeneity through the mediation of various ecosystem processes, some of which have been studied in Kruger.

Seed Dispersal

Many woody plants in Kruger produce fruits that are attractive to and eaten by birds (Knight 1984, 1986). These plants may owe their existence and location to avian dispersal and are patchily distributed. The variety and numbers of frugivorous birds in Kruger (Kemp et al. 2001; Table 13.1) also suggest that most plants are not disperser-limited, even though few bird species are obligate frugivores.

Which fruit species are dispersed by which frugivores is unrecorded for Kruger. A number of common indigenous woody plants (genera *Diospyros, Trichilia, Ekebergia, Xanthocercis, Zizyphus, Ficus, Lannea, Commiphora, Euclea, Grewia, Boscia, Antidesma, Ozoroa, Cassine,* and *Securinega*) are widely eaten by birds. Many of these are most abundant along riparian vegetation (or planted around staff and tourist accommodation), where the widest variety of dispersers (including primates, bats, and large herbivores) and germination sites occur. Plants found away from watercourses may have a high probability of being dispersed by birds, especially when found below a prominent perch, such as a hilltop, emergent tree, or termitarium, or where surrounding woodlands provide nest sites for such frugivores as African green pigeon (*Treron calva;* Tarboton and Vernon 1971). Mistletoes in the Loranthaceae (Box 13.2) have a close relationship with birds, depending on them for pollination and for dispersal to suitable establishment sites on host trees.

Ironically, the role of birds in dispersal of alien plants (*Lantana camara, Solanum mauritianum,* and *Melia azedarach*) in Kruger is more obvious (Macdonald 1988; Macdonald and Gertenbach 1988) because they attract large numbers of frugivorous birds as their principal dispersal agents.

Other methods of seed dispersal by birds are less obvious. Some seeds are dispersed inadvertently in the prey of raptorial birds (Dean and Milton 1988), either during consumption or when seeds are disgorged later in a pellet at a nest or roost. Other seeds (plumed, woolly, or attached to flexible peduncles) are incorporated in birds nests (Chippendall 1946; Collias 1964; Skead 1967, 1975, 1995; Dean et al. 1990; Steyn 1997; Tarboton 2001), grasses, (*Galium tomentosum, Galopina circaoides, Lepidium, Leyssera, Clematis,* and *Tarchonanthus*), and cotton (*Gossypium hirsutum,* native to Kruger but possibly contaminated by cultivars grown outside Kruger; van Jaarsveld 1998). Subsequent

> **BOX 13.2**
> *Mistletoes: A Mutualistic Relationship*
>
> Mistletoes (Loranthaceae) are one of the few plant families both pollinated and dispersed by birds and are well represented in Kruger as parasites on other plants (Wiens and Tölken 1979; Polhill and Wiens 1998). Pollination is mainly by sunbirds (Nectariniidae) and white-eyes (Zosteropidae; Gill and Wolf 1975). Mistletoe flowers split explosively when their sides are probed by a bird's bill, causing them to release a cloud of pollen and tap the stamen and stigma on the bird's forehead (Wiens and Tölken 1979). Mistletoe fruits are dispersed primarily by tinkerbirds (Capitonidae) as they wipe the sticky seeds, coated with viscin, against a branch (Godschalk 1983a), but other bird species also eat the fruits (Godschalk 1985).
>
> For two mistletoe species, the mean retention time by the yellow-fronted tinkerbird (*Pogoniulus chrysoconus*) between swallowing the fruit and regurgitating the seed (36 and 58 seconds; Godschalk 1983b) suggests an average dispersal distance from the parent plant of 50 m (Godschalk 1985). Mistletoe plants on large trees are more numerous, attract more birds, and have more fruit eaten and seeds dispersed in the immediate area. This leads to clustering of mistletoes (Godschalk 1983a), which in turn facilitates resource defense and territoriality by the sunbirds, with further influence on gene exchange between patches.
>
> Mistletoes grow best on hosts with large amounts of nitrogen (Dean et al. 1995), are themselves high in water and nitrogen (Lamont 1982; Ehleringer et al. 1985), and leaf and flower independently of host or season. This attracts selective browsing by herbivores (Midgley and Joubert 1991), and those with good access (giraffe [*Giraffa camelopardalis*], African elephant [*Loxodonta africana*], and rock hyrax [*Procavia capensis*]) may strip mistletoes, even removing all from a single tree (Dean, pers. obs.). Where browsing mammals are excluded, as in tourist camps, trees may carry more mistletoes. High parasite loads may reduce vigor in host trees, although evidence from southern Africa is inconclusive (Dean, unpublished data). Damage to trees, including cavity formation and branch or tree mortality, may have cascade effects on other fauna.

decomposition of such nests often places seeds in favorable environments for germination. In semiarid South Africa, 58 seed species were found in 120 nests of 31 bird species; individual nests with more than 1,000 seeds were found, but only three plant species were used by 14 or more bird species (Dean et al. 1990; Dean and Milton 1993). Nest decomposition within 12 months varied from minimal to total breakdown ($n = 34$ nests).

Generally, birds do not damage seeds when they eat fruit, but some may play a role as seed predators of grasses and forbs (by Passeridae, Ploceidae including the red-billed quelea, Estrildidae, and Emberizidae) and some fruits (*Parinari curatellifolia* by the grey-headed parrot *Poicephalus fuscicollis*; Symes 2001). Seed-eating species are more numerous among breeding visitors to Kruger (19 percent, even when red-billed quelea are excluded) than among breeding residents (7 percent), although both classes contribute importantly to annual avian biomass (visitors 26 percent, residents 32 percent). This avian seed-eating community is similar to that of nearby Swaziland (Monadjem 2001), where their trophic impact was equivalent to that of sympatric rodents but insignificant compared with larger mammals.

Dispersal of Nutrients and Diseases from Carrion

The greatest biomass of primary consumers in Kruger is large herbivorous mammals. When they die, their carcasses provide food for diverse predators, scavengers, and decomposers, including vultures (Box 13.3). These secondary consumers convert and redistribute much of the biomass, nutrients, parasites, and diseases in these herbivores. The role of vultures in anthrax dispersal has been especially well studied (Chapter 17, this volume).

The diverse means by which carcasses become available also make carrion one of the most patchy and unpredictable resources of Kruger. Avian scavengers, through their elevation and mobility, are well suited to track this resource, each species with differences in biology that determine when, where, and what sort of carrion each finds. The temporal pattern of herbivore carcass availability, as predicted by seasonal dynamics of their populations, has not been established for Kruger.

Vultures locate carcasses by watching the activities of other mammalian and avian predators or scavengers (Kruuk 1972; Schaller 1972; Houston 1979) or the movements of helicopters involved in culling operations in Kruger (Benson, pers. obs.). This patchy information network, including the location and extent of communal roosts (Ward and Zahavi 1973), is expected to be important in locating patchy carrion. After feeding, the location of bath, roost, and nest sites and the stage in the breeding cycle determine exactly where nutrients are transported and diseases dispersed.

The scale of vulture movements is extensive. A yearling lappet-faced vulture was tracked over the length and width of the northern fifth of Kruger in a single day, despite its return to its nest at midday in search of parental provisioning (Kemp and Begg 2001). Adults in the same area included domestic livestock, obtained outside Kruger, in food that they delivered to nests. Three of 12 yearling lappet-faced vultures marked as nestlings in northern Kruger were observed, one in the Luangwa Valley, Zambia (1,107 km away; Benson 1994–1995; Oatley 1995), one at three different locations in Zimbabwe (furthest 800 km), and a third at Thabazimbi in Limpopo Province, South Africa

BOX 13.3
Avian Scavengers and Their Use of Patchy Resources in Kruger

Four vulture species are common in Kruger and differ in how they find and consume carcasses (Kruuk 1967, 1972; Schaller 1972; Kemp and Kemp 1975; Houston 1979, 1983). The white-backed vulture (*Gyps africanus*), dominant avian scavenger by biomass (Table 13.1), finds carcasses of sedentary large ungulates mainly by watching other scavengers (Houston 1979, 1983), is social, nests colonially in trees, and arrives at carrion in such numbers that it displaces competitors (Schaller 1972). The larger Cape vulture (*G. coprotheres*), uncommon in Kruger, has a similar biology but nests on cliffs and probably locates migratory ungulates more efficiently. A few *Gyps* vultures can demolish an impala (*Aepyceros melampus*) carcass in 2–3 minutes, whereas a dead buffalo (*Syncerus caffer*) or elephant (*Loxodonta africana*) may provide food for hundreds of birds and last for days.

The more solitary lappet-faced vulture (*Torgos tracheliotus*) and white-headed vulture (*Trigonoceps occipitalis*) are territorial and find more carcasses of sedentary small animals. Food remains found at nests of lappet-faced vultures in Kruger include birds, reptiles, hares, felids, and smaller ungulates, including domestic livestock (Benson, pers. obs.), similar to those of the white-headed vulture except that lappet-faced vultures consume fewer large ungulates. The hooded vulture (*Necrosyrtes monachus*), solitary or social, eats mainly small scraps of meat, insects, and even mammal dung. Other avian scavengers in Kruger include the bateleur (*Terathopius ecaudatus*; Watson 1986) and various opportunistic storks, kites, and eagles (Kemp and Kemp 1975).

The tree species used as vulture nest sites in Kruger are also patchy and are used to a different extent by each species (Benson 1994–1995). White-backed vultures nest over a wide range, on top of a variety of tree species, often along rivers, especially when the surrounding vegetation supports few large trees (e.g., the Luvuvhu River), but with many nests away from watercourses. Lappet-faced vultures nest mainly in open bushveld on the basalt plains of eastern Kruger (Kemp et al. 1998), on the flat crown of low thorny trees. Hooded vultures nest below the canopy, often in evergreen riparian trees.

The smaller nests of white-backed vultures are most prone to desertion, when droughts or closures of water points alter nearby movements of their large ungulate prey or floods remove riparian trees (Kemp and Benson, pers. obs.). Their nest spacing also varies between years, from solitary to loosely colonial, probably related to local effects of predators and disease on patchiness of large herbivores. The large nests of lappet-faced vultures are solitary, used repeatedly and often with old nests nearby. This also indicates differences in the nutrients from droppings and food remains that can be expected to accumulate below nests and in the effects of these droppings on plants and secondary scavengers below.

(450 km). Movements of *Gyps* vultures outside Kruger are also known to involve hundreds of kilometers (Benson 1994–1995; Snyman 1999; Scott and Scott 2000). Such mobility suggests major challenges in the scale at which ecology and management of resources, nutrients, and diseases must be addressed (Benson 1998, 2000).

Removal of Ectoparasites from Large Herbivores

Two species of oxpeckers (Sturnidae) interact with especially patchy resources, mammalian hosts from which they glean their food, ectoparasites on the mammals that make up their principal diet, and any diseases organisms they consume or transmit. Both species occur sympatrically, obtain almost all of their food from the surface of their mammalian hosts, nest in natural cavities, and breed in cooperative groups (Stutterheim 1976, 1982a, 1982b; Maclean 1993; Mundy 1997). The more common red-billed oxpecker (*Buphagus erythrorhynchus*) uses at least 15 species of large ungulate as host, especially giraffe, and also domestic livestock adjacent to Kruger. It usually roosts communally in trees, reeds, or aloes, rarely alone or in small groups on a host. The larger, more heavily billed yellow-billed oxpecker (*B. africanus*) associates mainly with large, short-haired mammals such as buffalo, rhino, and cattle (Mooring and Mundy 1996); it usually roosts on its hosts and lives in cooperative groups. The latter species previously went extinct in Kruger but subsequently recolonized from Zimbabwe (Whyte 1987).

Both oxpecker species and most of their hosts are obligate drinkers, so they have a high probability of finding each other at water. Otherwise, oxpeckers must locate mobile hosts whose density and movements are governed by such factors as availability of fodder and pressure from predators, with variable patterns in space and time. Differences in roosting behavior suggest that to locate various hosts in small herds, red-billed oxpeckers travel long distances to safe sites that may also serve as information centers (Ward and Zahavi 1973), whereas yellow-billed oxpeckers remain on their larger hosts, often in large herds, rather than search for them daily. Ectoparasites are also expected to have patchy distributions, but how this is related to patterns of oxpecker, herbivore, and disease distribution is unknown.

Predation Pressure on Small Animals

Habitats adjacent to Kruger support some of the highest densities of avian predators on small vertebrates recorded anywhere in Africa (Tarboton and Allan 1984; Leslie 1993; Simmons 1994; Davison 1998; Roche 2001), with up to 51 pairs/100 km^2 of diurnal raptors and scavengers (Sabie-Sand Reserve; Simmons

TABLE 13.2

Percentage of species in Kruger that use different basic foraging strategies, based on their status as residents or breeders.

FORAGING STRATEGY	BREEDING RESIDENTS	BREEDING VISITORS	NONBREEDING VISITORS	TOTAL
Arboreal gleaning	29.6	17.9	15.6	23.0
Terrestrial gleaning	41.9	35.9	27.2	36.4
Aquatic gleaning	4.3	18.8	25.9	13.7
Perch hunting	16.4	8.5	12.2	13.2
Aerial hunting	7.5	18.8	19.0	13.3
Sample sizes	253	117	147	517

1994). This indicates a diverse and dense prey base under considerable predation and selection pressure, especially if small mammalian carnivores are included. No figures exist for Kruger, but the Serengeti National Park, of similar size and species diversity of birds and small carnivores, is estimated to support 668 mongooses/100 km^2 for only three species (Sinclair and Arcese 1995). Predation by diurnal raptors is also suggested as an important selection pressure for living in cooperative groups (du Plessis et al. 1995) and forming mixed-species bird parties (Vernon 1980), both of which are especially prevalent in wooded savannas.

A large proportion of bird species in Kruger also feed mainly on arthropods (63 percent of breeding residents, 60 percent of breeding visitors, and 61 percent of nonbreeding visitors; Kemp et al. 2001). Species that feed on arthropods also provide the main biomass of nonbreeding visitors (95 percent), compared with less than half the biomass of breeders (residents 46 percent, visitors 45 percent). Species classified as omnivorous are also evenly distributed across these divisions (16, 14, and 12 percent, respectively), feeding small animals to nestlings but eating mainly other foods, such as seeds, nectar, or fruit, when not breeding. All these predators use a wide range of foraging techniques to capture prey (Table 13.2).

Overall, it appears that the structure, composition, mobility, and behavior of the avifauna may be determined to a large extent by the variety, abundance, and availability of small animals they encounter, especially arthropods. Arthropod availability is known to be higher during the warm wet season than the cool dry season (Kemp 1974, 1976; Lack 1980, 1985, 1987; Rautenbach et al. 1988), which may explain many of the patterns of avian residence and abundance, including seasonal formation of mixed-species bird parties (Vernon 1980). Arthropod population dynamics, biomass, predation, and roles in parasite transfer certainly deserve more study than they have attracted.

References

Benn, G. A., A. C. Kemp, and K. S. Begg. 1995. The distribution, size and trends of the saddlebilled stork *Ephippiorhynchus senegalensis* population in South Africa. *South African Journal of Wildlife Research* 25:98–105.

Benson, P. C. 1994–1995. *Vulture foraging range, patterns and behaviour: a study of vulture movements as a basis for management of avian scavengers and implications of wildlife disease.* Progress reports Nos. 6, 7, 9. South African Department of Environment Affairs, Pretoria, South Africa.

Benson, P. C. 1998. Status of vultures in the Northern Province, South Africa. Pages 21–29 in A. F. Boshoff, M. D. Anderson, and W. D. Borello (eds.), *Vultures in the 21st century: proceedings of a workshop on vulture research and conservation in southern Africa.* Johannesburg: Vulture Study Group.

Benson, P. C. 2000. Causes of mortality at the Kransberg Cape vulture colony, a 17 year update. Pages 77–86 in R. D. Chancellor and B.-U. Meyburg (eds.), *Raptors at risk.* Hancock House, UK: World Working Group on Birds of Prey and Owls.

Berthold, P. 1993. *Bird migration: A general survey.* Oxford, UK: Oxford University Press.

Biggs, D. 2001. Eagles feast at quelea colony in Kruger. *Africa Birds and Birding* 6:16–17.

Bruggers, R. L., and C. C. H. Elliott (eds.). 1989. Quelea quelea: *Africa's bird pest.* Oxford, UK: Oxford University Press.

Chippendall, L. K. A. 1946. Contributions to the grass flora of Africa. Part 1. Three new grasses from tropical Africa and *Panicum lanipes* in South and South West Africa. Part 2. A brief survey of some lesser known dispersal mechanisms in African grasses. *Blumea* Supplement 3:25–41.

Collias, N. E. 1964. The evolution of nests and nest-building. *American Zoologist* 4:175–190.

Davison, B. 1998. *Raptor communities in hill habitats in South-East Zimbabwe.* Unpublished M.Sc. thesis, University of Natal, Pietermaritzburg, South Africa.

Dean, W. R. J. 1995. *Where birds are rare or fill the air: the protection of the endemic and the nomadic avifaunas of the Karoo.* Unpublished D.Sc. thesis, University of Cape Town, Cape Town, South Africa.

Dean, W. R. J., J. J. Midgley, and W. D. Stock. 1995. The distribution of mistletoes in South Africa: patterns of species richness and host choice. *Journal of Biogeography* 21:503–510.

Dean, W. R. J., and S. J. Milton. 1988. Dispersal of seeds by raptors. *African Journal of Ecology* 26:173–176.

Dean, W. R. J., and S. J. Milton. 1993. The use of *Galium tomentosum* (Rubiaceae) as nest material by birds in the southern Karoo. *Ostrich* 64:187–189.

Dean, W. R. J., S. J. Milton, and W. R. Siegfried. 1990. Dispersal of seeds as nest material by birds in semiarid Karoo shrubland. *Ecology* 71:1299–1306.

Ehleringer, J. R., E. D. Schulze, H. Ziegler, O. L. Lange, G. D. Farquhar, and I. R. Cowan, 1985. Xylem tapping mistletoes: primarily water or nutrient parasites? *Science* 227:1479–1481.

du Plessis, M. A., W. R. Siegried, and A. J. Armstrong. 1995. Ecological and life-history correlates of cooperative breeding in South African birds. *Oecologia* 102:180–188.

Gill, F. B., and L. L. Wolf. 1975. Foraging strategies and energetics of East African sunbirds at mistletoe flowers. *American Naturalist* 109:491–510.

Godschalk, S. K. B. 1983a. Mistletoe dispersal by birds in South Africa. Pages 117–128 in M. Calder and P. Bernhardt (eds.), *The biology of mistletoes*. Sydney: Academic Press.

Godschalk, S. K. B. 1983b. The reproductive phenology of three mistletoe species in the Loskop Dam Nature Reserve, South Africa. *South African Journal of Botany* 2:9–14.

Godschalk, S. K. B. 1985. Feeding behaviour of avian dispersers of mistletoe fruit in the Loskop Dam Nature Reserve, South Africa. *South African Journal of Zoology* 20:136–146.

Harrison J. A., D. G. Allan, L. G. Underhill, M. Herremans, A. J. Tree, V. Parker, and C. J. Brown (eds.). 1997. *The atlas of southern African birds*. Johannesburg: BirdLife South Africa.

Herremans, M. 1993. Seasonal dynamics in sub-Kalahari bird communities with emphasis on migrants. Pages 555–564 in *Proceedings 8 Pan-African Ornithological Congress*. Johannesburg: South African Ornithological Society.

Herremans, M. 1995. *Population dynamics of passerines in SE-Botswana 1992–1995: seasonal variation in abundance and seasonality of breeding and moulting*. Gaborone, Botswana: Department of Wildlife and National Parks.

Herremans, M., C. J. Brown, W. D. Borello, and D. Herremans-Tonnoeyr. 1993. The abundance of European rollers *Coracias garrulus* in Botswana and Namibia. *Ostrich* 64:93–94.

Herremans, M., and D. Herremans-Tonnoeyr. 1994a. Influx of Dickinson's kestrel *Falco dickinsoni* in northern Botswana in winter 1994. *Babbler* 28:29–30.

Herremans, M., and D. Herremans-Tonnoeyr. 1994b. Regional and temporal variation in the relative abundance of blackshouldered kites *Elanus caeruleus* in Botswana. *Babbler* 28:33–35.

Herremans, M., and D. Herremans-Tonnoeyr. 1994c. Seasonal patterns in abundance of lilacbreasted rollers *Coracias caudata* and purple rollers *Coracias naevia* inferred from roadside counts in eastern and northern Botswana. *Ostrich* 65:66–73.

Herremans, M., and D. Herremans-Tonnoeyr. 1995. Non-breeding site-fidelity of redbacked shrikes *Lanias collurio* in Botswana. *Ostrich* 66:145–147.

Houston, D. C. 1979. The adaptations of scavengers. Pages 263–286 in A. R. E. Sinclair and M. Norton-Griffiths (eds.), *Serengeti: dynamics of an ecosystem*. Chicago: Chicago University Press.

Houston, D. C. 1983. The adaptive radiation of the griffon vultures. Pages 135–152 in S. R. Wilbur and J. A. Jackson (eds.), *Vulture biology and management*. Berkeley: University of California Press.

Hurford, J. L., A. Lombard, A. C. Kemp, and G. A. Benn. 1996. Geographical analysis of six rare bird species in the Kruger National Park, South Africa. *Bird Conservation International* 6:133–153.

Kemp, A. C. 1974. The distribution and status of the birds of the Kruger National Park. *Koedoe* monograph 2:1–352. Pretoria: National Parks Board.

Kemp, A. C. 1976. A study of the ecology, behaviour and systematics of *Tockus* hornbills (Aves: Bucerotidae). *Transvaal Museum Memoir* 20:1–125.

Kemp, A. C. 2001. Concentration of non-breeding lesser spotted eagles *Aquila pomarina* at abundant food: a breeding colony of red-billed queleas *Quelea quelea* in the Kruger National Park, South Africa. *Acta Ornithoecologica*, Jena 4:325–329.

Kemp, A. C., and K. S. Begg. 2001. Comparison of time-activity budgets and population structure for 18 large-bird species in the Kruger National Park, South Africa. *Ostrich* 72:179–184.

Kemp, A. C., K. S. Begg, G. A. Benn, and P. Chadwick. 1997. A visual assessment of vegetation structure for the Kruger National Park. *Koedoe* 40:117–121.

Kemp, A. C., G. A. Benn, and K. S. Begg. 1998. Geographical analysis of vegetation structure and sightings for four large bird species in the Kruger National Park, South Africa. *Bird Conservation International* 8:89–108.

Kemp, A. C., J. J. Herholdt, I. J. Whyte, and J. Harrison. 2001. Birds of the two largest national parks in South Africa: a method to generate estimates of population size for all species and assess their conservation ecology. *South African Journal of Science* 97:393–403.

Kemp, A. C., S. C. J. Joubert, and M. I. Kemp. 1989. Distribution of southern ground hornbills in the Kruger National Park in relation to some environmental factors. *South African Journal of Wildlife Research* 19:93–98.

Kemp, A. C., and M. I. Kemp. 1975. Observations on the white-backed vulture *Gyps africanus* in the Kruger National Park, with notes on other avian scavengers. *Koedoe* 18:51–68.

Knight, R. S. 1984. Patterns of seed dispersal in southern African trees. *Journal of Biogeography* 11:501–514.

Knight, R. S. 1986. Inter-relationships between fruit types in southern African trees and environmental variables. *Journal of Biogeography* 13:99–108.

Kruuk, H. 1967. Competition for food between vultures in east Africa. *Ardea* 55:171–193.

Kruuk, H. 1972. *The spotted hyena.* Chicago: University of Chicago Press.

Lack, P. C. 1980. *The habitats and feeding stations of birds in Tsavo National Park, Kenya.* Unpublished Ph.D. thesis, University of Oxford, Oxford, UK.

Lack, P. C. 1985. The ecology of the land-birds of Tsavo East National Park, Kenya. *Scopus* 9:2–23, 57–96.

Lack, P. C. 1987. The structure and seasonal dynamics of the bird community in Tsavo East National Park, Kenya. *Ostrich* 58:9–23.

Lamont, B. 1982. Host range and germination requirements of some South Africa mistletoes. *South African Journal of Science* 78:41–42.

Leslie, B. J. 1993. Breeding density of Wahlberg's eagle *Aquila wahlbergi* in the Stolznek-Malelane region of the Kruger National Park. *Gabar* 8:14–16.

Macdonald, I. A. W. 1988. The history, impacts and control of introduced species in the Kruger National Park, South Africa. *Transactions of the Royal Society of South Africa* 46:251–273.

Macdonald, I. A. W., and W. P. D. Gertenbach. 1988. A list of alien plants in the Kruger National Park. *Koedoe* 31:137–150.

Maclean, G. L. 1993. *Roberts' birds of southern Africa.* 6th edition. Cape Town: John Voelcker Bird Book Fund.

Midgley, J. J., and D. Joubert. 1991. Mistletoes, their host plants and the effects of browsing by large mammals in Addo Elephant National Park. *Koedoe* 34:149–152.

Monadjem, A. 1992. *The habitat factors determining the distribution of ten species of piscivorous birds on the Sabie River in the Kruger National Park in relation to river characteristics.* Unpublished M.Sc. thesis, University of the Witwatersrand, Johannesburg, South Africa.

Monadjem, A. 1996. Habitat associations of birds along the Sabie River, South Africa. *African Journal of Ecology* 34:75–78.

Monadjem, A. 2001. Community structure and composition of birds in *Acacia* savanna in Swaziland. *Ostrich* Supplement 15:132–135.

Mooring, M. S., and P. J. Mundy. 1996. Factors influencing host selection by yellow-billed oxpecker at Matobo National Park, Zimbabwe. *African Journal of Ecology* 34:177–188.

Mundy, P. 1997. Yellowbilled oxpecker *Buphagus africanus*, redbilled oxpecker *Buphagus erythrorhynchus*. Pages 480–483 in J. A. Harrison, D. G. Allan, L. G. Underhill, M. Herremans, A. J. Tree, V. Parker, and C. J. Brown (eds.), *The atlas of southern African birds*, Vol. 2: *Passerines*. Johannesburg: BirdLife South Africa.

Newman, K. 1980. *Birds of South Africa*, Vol. 1, *Kruger National Park*. Johannesburg: Macmillan South Africa.

Newman, K. 1991. *Birds of the Kruger National Park: 1991 Update*. Halfway House, South Africa: Southern Book Publishers.

Oatley, T. B. 1995. Selected recoveries reported to Safring July 1994–December 1994. *Safring News* 4:27–38.

Pienaar, U. de V. 1969. Observation on the nesting habits and predators of breeding colonies of red-billed queleas *Quelea quelea lathami* (A. Smith) in the Kruger National Park. *Bokmakierie* Supplement 21:xi–xv.

Polhill, R., and D. Wiens (eds.). 1998. *Mistletoes of Africa*. Kew, UK: The Royal Botanic Gardens.

Rautenbach, I. L., A. C. Kemp, and C. H. Scholtz. 1988. Fluctuations in availability of arthropods correlated with microchiropteran and avian predator activity along the Luvuvhu River, South Africa. *Koedoe* 31:77–90.

Roche, C. 2001. Bird of prey nest survey at Ngala. *Africa Ecological Journal* 3:34–36.

Schaller, G. B. 1972. *The Serengeti lion*. Chicago: University of Chicago Press.

Scott, H. A., and R. M. Scott. 2000. Cape vulture monitoring project: Western Cape, South Africa. Report on fieldwork: January 2000. *Vulture News* 43:37–42.

Simmons, R. 1994. Conservation lessons from one of Africa's richest raptor reserves. *Gabar* 9:2–13.

Sinclair, A. R. E., and P. Arcese (eds.). 1995. *Serengeti 2: dynamics, management and conservation of an ecosystem*. Chicago: Chicago University Press.

Sinclair, I., and I. Whyte. 1991. *Field guide to the birds of the Kruger National Park*. Cape Town: Struik Publishers.

Skead, C. J. 1967. *The sunbirds of southern Africa*. Cape Town: A. A. Balkema.

Skead, C. J. 1995. *Life-history notes on East Cape bird species, 1940–1990*, Vol. 1. Port Elizabeth, South Africa: Algoa Regional Services Council.

Skead, D. M. 1975. Ecological studies of four estrildines in the central Transvaal. *Ostrich* Supplement 11:1–55.

Snyman, J. 1999. Rüppell's griffon breeding in the Blouberg. *Vulture News* 41:31–32.

Steyn, P. 1997. *Nesting birds: the breeding habits of southern African birds*. Cape Town: Fernwood Press.

Stutterheim, C. J. 1976. *The biology of the red-billed oxpecker, Buphagus erythrorhynchus (Stanley, 1814) in the Kruger National Park, South Africa*. Unpublished M.Sc. thesis, University of Pretoria, Pretoria, South Africa.

Stutterheim, C. J. 1982a. Breeding biology of the redbilled oxpecker in the Kruger National Park. *Ostrich* 53:79–90.

Stutterheim, C. J. 1982b. Timing of breeding of the redbilled oxpecker (*Buphagus erythrorhynchus*) in the Kruger National Park. *South African Journal of Zoology* 17:126–129.
Symes, C. T. 2001. *Biology of the grey-headed parrot* Poicephalus fuscicollis suahelicus *Reichenow*. Unpublished M.Sc. thesis, University of Natal, Pietermaritzburg, South Africa.
Tarboton, W. 2001. *A guide to the nests and eggs of southern African birds*. Cape Town: Struik Publishers.
Tarboton, W. R., and D. G. Allan. 1984. The status and conservation of the birds of prey in the Transvaal. *Transvaal Museum Monograph* 3:1–115.
Tarboton, W. R., M. I. Kemp, and A. C. Kemp. 1987. *Birds of the Transvaal Museum*. Pretoria: Transvaal Museum.
Tarboton, W. R., and C. J. Vernon. 1971. Notes on the breeding of the green pigeon *Treron australis*. *Ostrich* 42:190–192.
Thompson, T. T. 1993. Opportunistic breeding by the redbilled quelea in eastern Kenya. *Ostrich* 64:32–37.
Tyler, S. 2001. Movements by birds in *Acacia* savanna in southeast Botswana. *Ostrich* Supplement 15:98–103.
van Jaarsveld, M. 1998. Transgenic cotton plants: the answer for the future? *Plant Protection News* 53:8–10.
Vernon, C. J. 1980. Bird parties in central and South Africa. Pages 313–325 in *Proceedings 4 Pan-African Ornithological Congress*. Cape Town: South African Ornithological Society.
Ward, P., and A. Zahavi. 1973. The importance of certain assemblages of birds as "information-centres" for food-finding. *Ibis* 115:517–534.
Watson, R. T. 1986. *The ecology, biology and population dynamics of the bateleur eagle* Terathopius ecaudatus. Unpublished Ph.D. thesis, University of the Witwatersrand, Johannesburg, South Africa.
Whyte, I. J. 1987. The status and distribution of the yellowbilled oxpecker in the Kruger National Park. *Ostrich* 58:88–90.
Whyte, I. J. 1993. Marabou nesting in the Kruger National Park. *Ostrich* 64:186.
Wiens, D., and H. R. Tölken. 1979. Viscaceae and Loranthaceae. *Flora of South Africa* 10:1–59.

Chapter 14

Large Herbivores and Savanna Heterogeneity

Johan T. du Toit

In taxonomic terms, the large herbivores (>5 kg) of Kruger represent some 30 species drawn from three orders (Artiodactyla, Perissodactyla, Proboscidea) and seven families (Bovidae, Elephantidae, Equidae, Giraffidae, Hippopotamidae, Rhinocerotidae, Suidae). In functional terms there are two guilds: grazers (14 species), which feed mostly on monocots, and browsers (11 species), which feed mostly on dicots. An additional few switch back and forth between guilds (five species) depending on the quantity and quality of plant types available to them for food. A distinctive feature of each guild is that syntopic species tend to vary in body size (Owen-Smith 1985; McNaughton and Georgiadis 1986). For example, in any one part of Kruger during the dry season the browsing guild may include the world's heaviest and tallest extant land animals (elephant [*Loxodonta africana*] and giraffe [*Giraffa camelopardalis*], respectively), together with a size-graded variety of others that would include some of the smallest ruminants (steenbok [*Raphicerus campestris*] and grysbok [*R. melanotis*]). Such an assemblage typically would represent a body mass range spanning three orders of magnitude. Although this species richness is impressive in comparison with that of large herbivore assemblages in other biomes and on other continents, it is a typical (and indeed definitive) feature of the African savanna biome (Huntley 1982; Scholes and Walker 1993). Inevitably, questions about why so many large herbivore species evolved and how the extant ones coexist have formed a long-standing theme of inquiry in African savanna ecology (Lamprey 1963; Owen-Smith 1985; McNaughton and Georgiadis 1986; Vrba 1992). Here I extend that theme, drawing on research results from Kruger to explore how ungulate guild structure is influenced by savanna heterogeneity at multiple spatial and temporal scales.

Species Richness among Large Herbivores and Heterogeneity in African Savannas: Is There a Link?

Because the African savanna biome includes more extant ungulate species than are found on any other continent, even after allowing for the late Pleistocene extinctions in Eurasia and the Americas (Owen-Smith and Cumming 1993), it is logical to look for clues to ungulate speciation in the features that characterize the African savanna biome. This biome, which extends over a large and continuous landmass between the Sahel and Karoo semideserts in the north and south, respectively, straddles the equator where the equatorial forest biome partially separates it into northwestern, eastern, and southern distributions. Because it is an essentially tropical biome, it is to be expected that high species richness would result from the latitudinal gradient in species diversity that applies to almost all taxa around the world (reviewed by Rohde 1999). There are additional factors at work, however, because a west-east transect along the equator reveals a sharp and sustained rise in the density of ungulate species at the ecotone between the West African forests and the East African savannas (du Toit and Cumming 1999). Indeed, across the entire African continent, the distribution of ungulate species richness is closely associated with the distribution of the savanna biome (Turpie and Crowe 1994).

Evolutionary explanations for the exceptional ungulate species richness in African savannas (reviewed by Jarman 1974; Vrba 1992; Owen-Smith 1998; du Toit and Cumming 1999) focus on ungulate species being specialized to varying degrees on particular habitats and savannas being characterized by high spatial heterogeneity at the ungulate habitat scale. Additional heterogeneity is added by cyclic paleoclimatic change, causing the savanna biome to expand and contract through evolutionary time so that some habitat boundaries advance and retreat and some habitats appear or disappear, depending on topography. In effect this temporal heterogeneity in climatic variables, driving spatial heterogeneity at the ungulate habitat scale and causing alternating episodes of allopatry and sympatry, has acted as a "species pump" with regard to African savanna ungulates. Indeed, a significant proportion of contemporary sub-Saharan antelope species made their first appearance in the fossil record some 2.5 million years ago, signaling a major speciation pulse that shortly followed a phase of global climatic cooling and expansion of the open savannas in Africa (Vrba 1992). From the biome's evolutionary history it is apparent that African savannas are characterized by heterogeneity occurring at temporal and spatial scales to which the ungulate fauna is particularly sensitive. Consequently, this chapter focuses on the ways in which large herbivore ecology is influenced by heterogeneity in African savannas at feeding patch, habitat, and seasonal scales, using Kruger examples wherever possible.

What Can Kruger Studies Contribute to Large Herbivore Ecology?

The published literature on the feeding ecology of large herbivores in Kruger includes a concentration of papers on the interactions between browsing ruminants and their food resources, based mainly on research in the central region of the park (Tshokwane). This is because a long-term study on the feeding ecology, population dynamics, and life history of kudus (*Tragelaphus strepsiceros*) was initiated there in the 1970s (Owen-Smith 1979, 1990, 1993a, 1993b; Owen-Smith and Novellie 1982), and this stimulated further research on the browsing guild (du Toit and Owen-Smith 1989; du Toit et al. 1990; du Toit 1990a, 1990b, 1990c, 1993, 1995; Woolnough and du Toit 2001), which is ongoing. This line of research on savanna browsing ecology, covering steenbok, impala (*Aepyceros melampus*), kudu, giraffe, and their food plants, may be seen as running parallel to that on grazing ecology in the Serengeti ecosystem of East Africa (Gwynne and Bell 1968; Bell 1971; Sinclair and Norton-Griffiths 1979; McNaughton 1985; Sinclair and Arcese 1995; Murray and Illius 2000). Although grazers have not been completely neglected in Kruger (Harrington et al. 1999), nor have browsers in the Serengeti (Jarman and Sinclair 1979; Pellew 1984; Dublin et al. 1990), this chapter focuses on the browsers in Kruger to provide a different view of how an ungulate guild interacts with the spatially and temporally heterogeneous vegetation patterns that characterize the savanna environment.

Effects of Spatial Heterogeneity on Large Herbivore Ecology

Individuals, populations, and species of large herbivores are influenced by spatial heterogeneity in their environments at scales ranging from the feeding patch to the biome. Here the responses of large herbivores to such heterogeneity are considered for grazing and browsing ungulates in African savannas, with a central theme being body size variation within each guild.

Feeding Patch Scale

Large herbivores select their diets from a foraging environment made up of plants or plant parts that vary in terms of accessibility, dietary (nutrient) value, and antiherbivore defense. These factors are interconnected; for example, the accessibility of high-quality leaf material to a small- or medium-bodied grazing ungulate is influenced by the amount of indigestible stem material in the sward (Bell 1971; Murray and Illius 2000). If the primary antiherbivore defense of a plant is physical (e.g., spinescence), then it will influence the accessibility of leaf material to a browser (Cooper and Owen-Smith 1986), but if it is chemical (e.g., toxicity) then it will influence its dietary value (Cooper et al. 1988). In African savannas it is particularly difficult to tease out the key factors in plants that influence their consumption by large herbivores, given the exceptional heterogeneity in their physical and chemical properties (reviewed by Owen-Smith 1982).

In the grazing guild, variation in leaf:stem ratio across feeding patches (*sensu* Senft et al. 1987) may result in the spatial separation of grazers of different body size when grazer densities are locally high. This outcome is determined by the allometric relationships by which incisor arcade width, intake rate, and specific metabolic needs all scale with body size, with the smallest grazers generally being most efficient at feeding on short swards with high leaf:stem ratios (Illius and Gordon 1987). In the grassland savannas of East Africa this body size effect is demonstrated by the grazing succession phenomenon, with a succession of grazing ungulate species moving across the landscape with the larger species (zebra [*Equus burchelli*] and wildebeest [*Connochaetus taurinus*]) in front and the smaller species (gazelles [*Gazella* spp.]) in the rear (Bell 1971). One hypothesis is that the larger species facilitate improved grazing conditions for the smaller species by reducing the sward height and exposing high-quality plant parts (Gwynne and Bell 1968), so the smaller species are drawn along at the back of the succession. McNaughton (1976) has shown that the feeding actions of wildebeest can induce regrowth (within days) that benefits gazelles. Nevertheless, an analysis of 20 years of Serengeti census data found that zebra, wildebeest, and gazelle population trends contradicted the facilitation hypothesis (Sinclair and Norton-Griffiths 1982). An alternative hypothesis is that when various grazers share the same patch, the smaller species (within the guild) or age-sex classes (within a species) feed more selectively on the high-quality grass parts and thereby reduce the overall quality of the sward, outcompeting the larger grazers and forcing them to move on to a new patch (Clutton-Brock et al. 1987; Illius and Gordon 1987; Murray and Illius 2000). Although this competition hypothesis is gaining empirical and theoretical support, the facilitation-competition debate remains unresolved for the grazing guild (see also Conradt et al. 1999). If a unifying hypothesis is to be found to explain feeding interactions in size-structured ungulate guilds, it should apply to browsers and grazers alike, and it is therefore worth turning to the browsing guild to seek evidence for either hypothesis there.

In terms of spatial distribution at the feeding patch scale, a major distinction between the grass and browse resources used by large herbivores is that grass is distributed in a two-dimensional carpet, whereas woody browse is distributed in three dimensions. Canopy height and architecture are additional components of heterogeneity with which browsers contend. Because the browsing ungulate guild comprises species that are graded in body size (Owen-Smith 1985) and therefore can feed up to different heights in the canopy, there is limited scope for the feeding actions of the larger species to facilitate browsing for the smaller species unless the larger species also feed at the low levels in the canopy to which the smaller species have access. What we find in Kruger, however, is that giraffes feed mainly above the levels used by the other guild members (du Toit 1990a), and there is no clear way in which the feed-

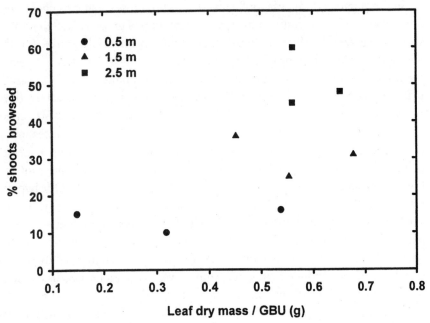

FIGURE 14.1. Large browsers such as giraffes preferentially prune shoots yielding a high leaf biomass per bite, and these occur above the reach of the smaller browsers. Here browsing pressure (percentage of shoots freshly browsed) is plotted against leaf biomass per giraffe browse unit (GBU) at three heights (0.5 m, 1.5 m, and 2.5 m) up the canopy in *Acacia nigrescens* trees in central Kruger (redrawn using data from Woolnough and du Toit 2001).

ing actions of a large browser could improve the feeding efficiency of a smaller browser, even if they fed at the same height levels. Recent research in Kruger (Woolnough and du Toit 2001) has found that giraffes derive a bite-size advantage when feeding on *Acacia nigrescens* trees (staple giraffe browse in central Kruger) at a height (about 2.5 m) above that used by kudus and smaller guild members. Leaf chemistry does not vary across canopy heights, but leaf biomass per giraffe browse unit (GBU) is highest in the upper browsing levels, and shoots with high leaf biomass clearly are preferred by browsers (Figure 14.1). The reason for the lower leaf biomass per GBU in the lower browsing levels could be that the smaller browsers selectively remove individual leaves from shoots within their reach and thereby reduce the overall quality of the browse resource for giraffes, which consequently browse on more leafy shoots at the higher levels. If this is found to be the case (an exclosure experiment is in progress), then it will support the competition hypothesis and give credence to the notion that interspecific competition between different-sized guild members drives the grazing succession from behind and maintains browsing stratification from below. However, such processes are likely to become apparent only when each guild's shared resources are limited during dry periods.

An additional component of heterogeneity with which browsers contend at the feeding patch scale is variation between potential browse plants in the chemical constituents of their tissues (Table 14.1). This is a more significant factor for browsers than for grazers because woody plants and forbs produce defensive compounds of far higher diversity and toxicity than are found in graminoids (Owen-Smith 1982). In a moist dystrophic savanna (Nylsvlei, South Africa) the condensed tannins in plant tissues have a deterrent effect against browsing ruminants (Cooper and Owen-Smith 1985), especially if the relative concentration of leaf protein is low (Cooper et al. 1988). Evidence from central Kruger indicates that the same applies on the eutrophic basalt plains, where the feeding preferences of steenbok, impala, and kudu are negatively associated with condensed tannin concentrations in leaves of a range of common woody plant species (Figure 14.2). Indeed, partial correlation analysis showed that avoidance of condensed tannin has a stronger effect on feeding preference than selection for leaf nitrogen or phosphorus (du Toit 1988). It appears that learning to avoid toxicity through postingestive feedback is the dominant process by which browsers develop their feeding preferences in heterogeneous plant communities (reviewed by Provenza 1995). Furthermore, ruminants can accurately regulate their intake of specific toxins (du Toit et al. 1991) to match their detoxification abilities, which could explain why browsers take many small meals from a wide range of plants and plant parts, as hypothesized by Freeland and Janzen (1974). Steenbok in central Kruger, for example, may feed on the bright yellow and toxic fruits of *Solanum panduraeforme* but will seldom if ever take more than one from a fruit-laden bush (du Toit 1988).

Habitat Scale

Current understanding on the evolutionary history of Africa's species-rich large mammal fauna rests on the premise that terrestrial mammals are habitat specific, and a phase of allopatry is needed for speciation to occur (Vrba 1992). If we view the African savanna biome as a dynamic mosaic of habitats constantly changing (on an evolutionary time scale) in terms of connectivity, shape, size, and presence or absence, with an assemblage of habitat-specific large mammals in each, then in theory the needs for evolutionary radiation are clearly met. However, although it is convenient for evolutionary theory to assume that all species are habitat specific, we need empirical evidence of this. Studies on ungulate–habitat relationships in African savannas, and in Kruger in particular, have provided some significant insights.

TABLE 14.1

Leaf chemistry data for a selection of dicotyledonous plants from the browsing guild's resource base in central Kruger. Values are expressed as the percentage of dry mass and are means (±SE) of 10 replicates in each case (from du Toit 1988).

PLANT SPECIES	GROWTH FORM[a]	CONDENSED TANNIN	TOTAL PHENOLICS	NITROGEN	PHOSPHORUS
Maytenus heterophylla	T	25.6 (0.38)	12.8 (0.20)	1.24 (0.02)	0.10 (0.05)
Combretum hereroense	T	14.0 (0.29)	25.0 (0.46)	1.73 (0.03)	0.11 (0.0)
Maytenus senegalensis	S	11.7 (0.33)	4.07 (0.09)	1.34 (0.03)	0.13 (0.0)
Lonchocarpus capassa	T	6.06 (0.24)	2.46 (0.05)	3.99 (0.05)	0.28 (0.01)
Acacia nigrescens[b]	T	6.45 (0.23)	6.38 (0.14)	2.64 (0.06)	0.13 (0.0)
Acacia nigrescens[c]	T	3.68 (0.33)	5.31 (0.13)	5.05 (0.08)	0.44 (0.0)
Combretum imberbe	T	2.01 (0.08)	15.0 (0.17)	2.22 (0.04)	0.14 (0.0)
Securinega virosa	S	1.44 (0.23)	10.6 (0.45)	3.06 (0.01)	0.23 (0.0)
Acrotome hispida	F	0.99 (0.04)	2.02 (0.08)	3.93 (0.02)	0.31 (0.01)
Heliotropium steudneri	F	0.97 (0.04)	5.79 (0.16)	4.48 (0.23)	0.19 (0.0)
Ipomoea obscura	C	0.93 (0.11)	5.72 (0.06)	2.46 (0.04)	0.34 (0.0)
Hibiscus praeteritus	F	0.92 (0.15)	1.88 (0.01)	2.98 (0.13)	0.76 (0.0)
Justicia flava	F	0.43 (0.16)	0.95 (0.01)	3.43 (0.03)	0.23 (0.0)

[a]T = tree; S = shrub; F = forb; C = creeper.
[b]Mature leaves.
[c]New leaves.

Across the 14 habitat types (distinct vegetation communities) mapped in the Tshokwane area of central Kruger (du Toit 1988) it was found that giraffe, kudu, and steenbok differed in their habitat specificities, with diversity of habitat use increasing with increasing body size (du Toit and Owen-Smith 1989). This finding is consistent with predictions of the Jarman-Bell principle, which states that an increase in herbivore size is associated with an increase in dietary tolerance (Bell 1971; Jarman 1974; Demment and van Soest 1985). In terrestrial ecosystems with high spatial heterogeneity at the ungulate habitat scale, we therefore expect the smallest members of herbivore guilds to be the most constrained by disjunctions in forage quality at habitat edges and thus the most habitat specific. This does indeed appear to be the case, with small ruminants such as klipspringer (*Oreotragus oretragus*) and bushbuck (*Tragelaphus scriptus*) being extreme habitat specialists and large ones such as eland (*Taurotragus oryx*) and giraffe being generalists (Skinner and Smithers 1990).

At the ungulate community level this size-dependent variation in habitat specificity has important implications for the partitioning of energy flow through species populations. Because the larger species are able to feed in a wider range of habitats and therefore are more evenly spread across the landscape, they should have an advantage in being able to consume a larger fraction of the

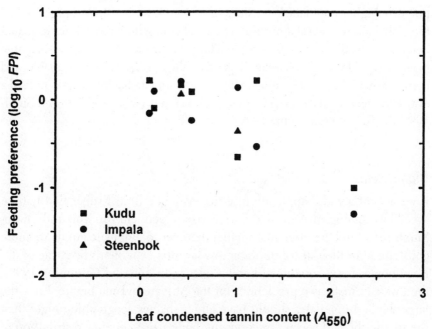

FIGURE 14.2. Feeding preference indices (FPIs) of kudu, impala, and steenbok are inversely related to leaf condensed tannin content in central Kruger ($r = -.81$, $p < .005$; du Toit 1988). Here the absorbance value (at 550 nm) in the proanthocyanidin assay is used as an index of condensed tannin concentration (for methods, see du Toit et al. 1990). The seven woody plant species used in this analysis are indicated as triplets or couplets of data points in the plot and are as follows (from left to right, in order of increasing condensed tannin content): *Securinega virosa, Combretum imberbe, Acacia nigrescens, Lonchocarpus capassa, Maytenus senegalensis, Combretum hereroense,* and *Maytenus heterophylla.*

standing crop of vegetation that is shared locally across all species in the ungulate community. Expressed more specifically, population metabolism (the product of individual metabolism and population density) should be expected to scale positively with body mass across species populations in African savannas, and there is evidence to support this. Using ungulate census data from 11 wildlife areas, du Toit and Owen-Smith (1989) found that population metabolism increases significantly as a function of increasing body mass. This result contrasts with previous analyses that found the amount of energy used by a species population to be generally independent of body mass, presumably because the larger species have lower population densities, and this counterbalances their higher individual intake needs (Damuth 1987). What particular ecosystem feature enables the larger ungulate species to occur at disproportionately higher population densities in African savannas than is predicted by the global allometric relationship? The most compelling answer lies in spatial heterogeneity: the relative benefits that large herbivores derive from being able to feed in a wide range of habitats depend on the range of habitats available, and particularly high spatial heterogeneity at the ungulate habitat scale is indeed a distinctive feature of African savannas (Scholes and Walker 1993; Owen-Smith 1998; du Toit and Cumming 1999).

Biome Scale

If we accept a wider application of the evidence from Kruger that habitat specificity among African ungulates is size dependent (du Toit and Owen-Smith 1989) and therefore that smaller members of ungulate guilds are more constrained in their distributions across savanna ecosystems by virtue of disjunctions in forage quality at habitat edges (du Toit and Cumming 1999), then we can make two predictions for the African savanna biome. First, the larger-bodied ungulates should have larger species geographic ranges than the smaller-bodied species; second, the body size frequency distribution of species in a speciose family such as the Bovidae should be skewed toward a preponderance of small species because of their higher exposure to episodes of allopatry. Both predictions hold for the savanna biome across its full African distribution (du Toit and Cumming 1999). To conclude, there are strong theoretical and empirical grounds to make a close evolutionary link between the distinctly high species richness of ungulates in African savannas and another distinct feature of the biome: high spatial heterogeneity at the ungulate habitat scale.

Effects of Temporal Heterogeneity on Large Herbivore Ecology

By definition, savannas include marked wet and dry seasons, with at least 60 days per year when rainfall is sufficient to allow plant growth and another 60 (contiguous) days when it is insufficient (Scholes and Walker 1993). Being shallower-rooted, grasses and forbs respond sooner than trees to the loss of soil moisture. Consequently, at upper and intermediate positions on the catenary drainage sequence (see Chapter 5, this volume) the onset of the dry season is characterized by a brown herbaceous layer beneath a green tree layer, and the availability of green tree foliage shrinks downslope as the dry season progresses. For large herbivores this means that their food resources vary markedly in abundance, quality, and distribution through each seasonal cycle.

Seasonal Shifts of Large Herbivore Species between Grazing and Browsing Guilds

For large herbivores in African savannas, Lamprey (1963) was the first to study the differential use of graminoids, forbs, and woody plants in the context of niche separation. His results were reanalyzed by McNaughton and Georgiadis (1986), who used the mean annual proportion of woody plant species in the diet as an axis for ordering ungulate species along a browser-to-grazer continuum. However, if we classify browsers as consumers of dicotyledonous plants (including trees, shrubs, forbs, and creepers) and grazers as consumers of monocots (graminoids, including grasses and sedges), and we compare the diets of syntopic species during any one phase of the seasonal cycle, we find that the continuum falls away. In functional terms there are two guilds: browsers and grazers. The few species of "mixed" feeders (e.g., impalas) effectively switch guilds on a seasonal basis, mainly grazing in the wet season and browsing in the dry season. A preference for green grass is apparent among impalas in the Serengeti, where browse is a dietary fallback option for periods in the seasonal cycle when the availability of green grass is limited by rainfall (Jarman and Sinclair 1979). The same applies in Kruger, where a dietary switch from mainly grazing to mainly browsing occurs when the 2-month running mean of rainfall drops below ~30 mm (Figure 14.3). Therefore, for ungulates in African savannas I propose that identifying members of browsing and grazing guilds and recognizing that species can switch between guilds are of greater functional relevance than the "continuum" concept. From a plant-based perspective the loss of foliage to herbivores occurs in discrete bites that are either grazing or browsing bites; from an animal-based perspective the presence or absence of potential competitors depends on whether syntopic animals are in the browsing or grazing guilds.

Seasonal Shifts in the Forb:Woody Plant Ratio in Browser Diets

Except for giraffes (which feed almost entirely on woody plants), the food resources of browsing ungulates in Kruger consist of both forbs and woody plants. Forbs are abundant during the wet season in Kruger (November–April), when they constitute about 50 percent of the kudu diet and about 80 percent of the steenbok diet, as measured in terms of feeding time allocation, but these values fall to <10 percent and <30 percent, respectively, in the late dry season (Figures 14.4 and 14.5). In terms of dry matter intake, forbs were found to make up about 40 percent of the wet season kudu diet in Nylsvley Nature Reserve, also falling to <10 percent in October (Owen-Smith and Cooper 1989). A high dependence on forbs by a very small and selective ruminant such as steenbok

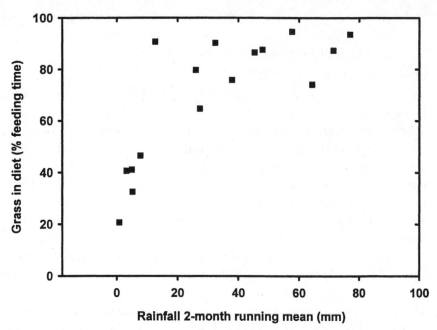

FIGURE 14.3. Representation of grass in the impala diet declines sharply in central Kruger at the onset of the dry season, when the 2-month running mean of rainfall drops below ~30 mm and impalas switch to browsing (du Toit 1988).

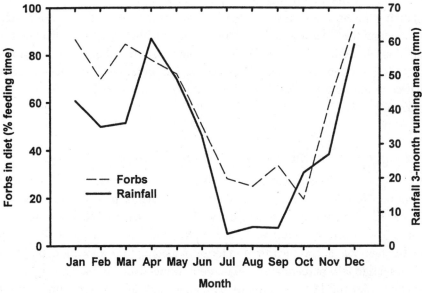

FIGURE 14.4. Forb intake by steenbok closely tracks the 3-month running mean of rainfall in central Kruger, reflecting seasonal variation in the availability of forbs as influenced by soil moisture content. The close association between forb consumption and rainfall ($r = .87$, $p < .005$) indicates reliance on forbs as a high-quality browse resource (redrawn using data from du Toit 1993).

and a close association between the use and the availability of forbs (assuming that the 3-month running mean of rainfall is an appropriate index of soil moisture, which is the prime determinant of forb availability; Figure 14.4) indicate that forbs are a particularly nutritious food class for browsers.

Owen-Smith (1990) found that the annual survival of kudu juveniles, yearlings, and old females in central Kruger was significantly correlated with the preceding wet season's rainfall total and suggested that the mechanism was rainfall-driven variation in the availability of high-quality resources such as forbs. I further suggest that the persistence of forb availability after the end of the wet season determines the onset of each year's lean period for browsing ungulates in Kruger (excluding giraffes), and the length of the lean period may be a stronger determinant of survival than the abundance of food resources during the preceding fat period. Using my data from 1984–1986 (Figure 14.5), it is apparent that a particularly high rainfall total in the 1984–1985 wet season allowed kudus in central Kruger to maintain their forb intake above 20 percent through September in the following dry season, and a lower rainfall total in the 1985–1986 wet season caused forb intake to almost cease by June (despite a temporary increase in August when kudu cow groups increased their use of riparian habitat).

Steenbok in central Kruger depend on dried indehiscent *Acacia tortilis* pods (which they collect off the ground) to offset the reduced availability of forbs in the late dry season (du Toit 1993). Furthermore, they are able to crawl through the recumbent canopies of elephant-felled trees to crop forbs that persist into the dry season by virtue of the humid microclimate and the exclosure effect that prevents other browsing ungulates from gaining access to these forb patches (du Toit 1988).

Seasonal Shifts in Habitat Use across the Catenary Drainage Gradient

Grazing ungulates in African savannas concentrate their feeding in zones that shift up and down the catenary drainage gradient through the seasonal cycle, moving progressively downslope in the dry season as the availability of green grass declines and then switching back to short, nutritious swards on the uplands when the rains commence (Bell 1971). The same cross-catenary movement pattern applies for savanna browsers, as has been found for giraffes, kudus, and impalas in Kruger (Figure 14.6), giraffes in Serengeti (Pellew 1984), and elephants in Chobe (Stokke and du Toit 2002). As a result, there is a concentration of vertebrate herbivory and dense species packing in riparian habitats during the dry (lean) season, when adaptations for resource partitioning should be more strongly selected for than in any other phase of the seasonal cycle. This contradicts previous assertions that resource partitioning among large herbi-

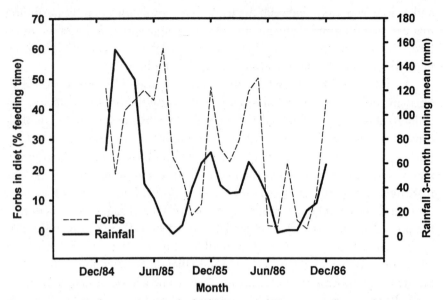

FIGURE 14.5. Forb consumption by kudu in central Kruger is influenced by woody plant phenology and forb availability, which depends on soil moisture content and, ultimately, the 3-month running mean of rainfall (du Toit 1988).

vores in any locally defined African savanna community occurs mainly through variation in species composition across habitats at any given time (Lamprey 1963; McNaughton and Georgiadis 1986). Clearly, there are suites of subtle differences (ecological, behavioral, and physiological) between syntopic species within each large herbivore guild that contribute to resource partitioning. Nevertheless, a robust rule for the research and management of large herbivores in African savannas is that attempts to study competitive interactions between species or quantify the key resources needed to sustain a target species at a given population level should focus on the lower catena in the dry season (see also Illius and O'Connor 2000).

Conclusion

By viewing the results of browsing studies in Kruger together with those from grazing studies in other savanna ecosystems such as the Serengeti, I propose that several advances can be made for a unified understanding of ungulate ecology in African savannas. Because browsers are confronted with an additional component of heterogeneity in the ungulate resource base, with browse being

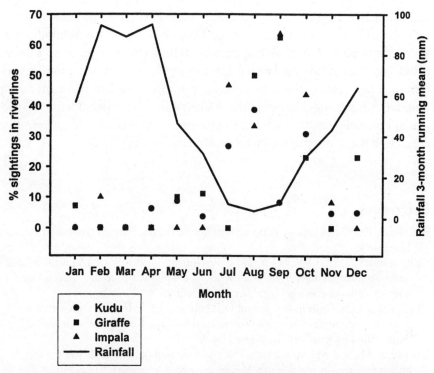

FIGURE 14.6. The browsing guild in central Kruger greatly increases its shared use of low-lying habitats in the late dry season (July–October), when the uplands have dried out. Here the visitation by radiocollared kudus, giraffes, and impalas to riparian habitats (dried watercourses where green woody browse persists through the dry season) is expressed for each species as the percentage of all sightings across all available habitats for each month (data from du Toit 1995).

distributed in three dimensions whereas grass is distributed in a two-dimensional carpet, this provides an additional axis (feeding height) along which to test for mechanisms of resource partitioning. Evidence from the Kruger browsing guild is consistent with the hypothesis that smaller-bodied ungulates may competitively displace larger-bodied guild members away (horizontally for grazers, vertically for browsers) from shared feeding sites. The distinct wet and dry seasons that characterize African savannas impose boom-bust fluctuations in the availability of food resources for ungulates, with the total amount and distribution of rainfall through the wet season determining the persistence of key dietary components (forbs for most browsers, green grass for grazers) into the dry season. Soil catenas interact with rainfall patterns in governing the distribution and abundance of green foliage in the dry season, when the progressive desiccation of soils on the uplands causes most ungulate population distributions to shift downslope. Catenary drainage gradients therefore are significant sources of heterogeneity at the ungulate habitat scale in African savannas. From an evolutionary perspective, the exceptional species richness and size structuring found in African ungulate assemblages suggest that the scales of temporal and spatial heterogeneity to which large herbivore ecology is particularly sensitive are indeed the scales that characterize African savannas.

References

Bell, R. H. V. 1971. A grazing ecosystem in the Serengeti. *Scientific American* 225:86–93.

Clutton-Brock, T. H., G. R. Iason, and F. E. Guinness. 1987. Sexual segregation and density-related changes in habitat use in male and female red deer (*Cervus elaphus*). *Journal of Zoology* (London) 211:275–289.

Conradt, L., T. H. Clutton-Brock, and D. Thomson. 1999. Habitat segregation in ungulates: are males forced into suboptimal foraging habitats through indirect competition by females? *Oecologia* 119:367–377.

Cooper, S. M., and N. Owen-Smith. 1985. Condensed tannins deter feeding by browsing ruminants in a South African savanna. *Oecologia* 67:142–146.

Cooper, S. M., and N. Owen-Smith. 1986. Effects of plant spinescence on large mammalian herbivores. *Oecologia* 68:446–455.

Cooper, S. M., N. Owen-Smith, and J. P. Bryant. 1988. Foliage acceptability to browsing ruminants in relation to seasonal changes in the leaf chemistry of woody plants in a South African savanna. *Oecologia* 75:336–342.

Damuth, J. 1987. Interspecific allometry in population density in mammals and other animals: the independence of body mass and energy-use. *Biological Journal of the Linnaean Society* 31:193–246.

Demment, M. W., and P. J. van Soest. 1985. A nutritional explanation for body-size patterns of ruminant and non-ruminant herbivores. *American Naturalist* 125:641–672.

Dublin, H. T., A R. E. Sinclair, and J. McGlade. 1990. Elephants and fire as causes of multiple stable states in the Serengeti-Mara woodlands. *Journal of Animal Ecology* 59:1147–1164.

du Toit, J. T. 1988. *Patterns of resource use within the browsing ruminant guild in the central Kruger National Park*. Ph.D. thesis, University of the Witwatersrand, Johannesburg, South Africa.

du Toit, J. T. 1990a. Feeding height stratification among African browsing ruminants. *African Journal of Ecology* 28:55–61.

du Toit, J. T. 1990b. Giraffe feeding on *Acacia* flowers: predation or pollination? *African Journal of Ecology* 28:63–68.

du Toit, J. T. 1990c. Home range–body mass relations: a field study on African browsing ruminants. *Oecologia* 85:301–303.

du Toit, J. T. 1993. The feeding ecology of a very small ruminant, the steenbok (*Raphicerus campestris*). *African Journal of Ecology* 31:35–48.

du Toit, J. T. 1995. Sexual segregation in kudu: sex differences in competitive ability, predation risk, or nutritional needs? *South African Journal of Wildlife Research* 25:127–132.

du Toit, J. T., J. P. Bryant, and K. Frisby. 1990. Regrowth and palatability of *Acacia* shoots following pruning by African savanna browsers. *Ecology* 71:149–154.

du Toit, J. T., and D. H. M. Cumming. 1999. Functional significance of ungulate diversity in African savannas and the ecological implications of the spread of pastoralism. *Biodiversity and Conservation* 8:1643–1661.

du Toit, J. T., and N. Owen-Smith. 1989. Body size, population metabolism, and habitat specialization among large African herbivores. *American Naturalist* 133:736–740.

du Toit, J. T., F. D. Provenza, and A. Nastis. 1991. Conditioned taste aversions: how sick must a ruminant get before it learns about toxicity in foods? *Applied Animal Behaviour Science* 30:35–46.

Freeland, W. J., and D. H. Janzen. 1974. Strategies of herbivory in mammals: the role of plant secondary compounds. *American Naturalist* 108:269–289.

Gwynne, M. D., and R. H. V. Bell. 1968. Selection of grazing components by grazing ungulates in the Serengeti National Park. *Nature* 220:390–393.

Harrington, R., N. Owen-Smith, P. C. Viljoen, H. C. Biggs, D. R. Mason, and P. Funston. 1999. Establishing the causes of the roan antelope decline in the Kruger National Park, South Africa. *Biological Conservation* 90:69–78.

Huntley, B. J. 1982. Southern African savannas. Pages 101–119 in B. J. Huntley and B. H. Walker (eds.), *Ecology of tropical savannas*. Berlin: Springer-Verlag.

Illius, A. W., and I. J. Gordon. 1987. The allometry of food intake in grazing ruminants. *Journal of Animal Ecology* 56:989–999.

Illius, A. W., and T. G. O'Connor. 2000. Resource heterogeneity and ungulate population dynamics. *Oikos* 89:283–294.

Jarman, P. J. 1974. The social organization of antelope in relation to their ecology. *Behaviour* 48:215–266.

Jarman, P. J., and A. R. E. Sinclair. 1979. Feeding strategy and the pattern of resource-partitioning in ungulates. Pages 130–163 in A. R. E. Sinclair and M. Norton-Griffiths (eds.), *Serengeti: Dynamics of an ecosystem*. Chicago: University of Chicago Press.

Lamprey, H. F. 1963. Ecological separation of the large mammal species in the Tarangire Game Reserve, Tanganyika. *East African Wildlife Journal* 1:63–92.

McNaughton, S. J. 1976. Serengeti migratory wildebeest: facilitation of energy flow by grazing. *Science* 191:92–94.

McNaughton, S. J. 1985. Ecology of a grazing ecosystem: the Serengeti. *Ecological Monographs* 55:259–294.

McNaughton, S. J., and N. J. Georgiadis. 1986. Ecology of African grazing and browsing mammals. *Annual Review of Ecology and Systematics* 17:39–65.

Murray, M. G., and A. W. Illius. 2000. Vegetation modification and resource competition in grazing ungulates. *Oikos* 89:501–508.

Owen-Smith, N. 1979. Assessing the foraging efficiency of a large herbivore, the kudu. *South African Journal of Wildlife Research* 9:102–110.

Owen-Smith, N. 1982. Factors influencing the consumption of plant products by large herbivores. Pages 359–404 in B. J. Huntley and B. H. Walker (eds.), *Ecology of tropical savannas*. Berlin: Springer-Verlag.

Owen-Smith, N. 1985. Niche separation among African ungulates. Pages 167–171 in E. S. Vrba (ed.), *Species and speciation*. Transvaal Museum Monograph no. 4. Pretoria: Transvaal Museum.

Owen-Smith, N. 1990. Demography of a large herbivore, the greater kudu *Tragelaphus strepsiceros*, in relation to rainfall. *Journal of Animal Ecology* 59:893–913.

Owen-Smith, N. 1993a. Age, size, dominance and reproduction among male kudus: mating enhancement by attrition of rivals. *Behavioral Ecology and Sociobiology* 32:177–184.

Owen-Smith, N. 1993b. Comparative mortality rates of male and female kudus: the costs of sexual size dimorphism. *Journal of Animal Ecology* 62:428–440.

Owen-Smith, N. 1998. Ecological links between African savanna environments, climate change, and early hominid evolution. Pages 138–149 in T. Bromage and F. Schrenk (eds.), *African biogeography, climate change and human evolution*. Oxford, UK: Oxford University Press.

Owen-Smith, N., and S. M. Cooper. 1989. Nutritional ecology of a browsing ruminant, the kudu (*Tragelaphus strepsiceros*), through the seasonal cycle. *Journal of Zoology* (London) 219:29–43.

Owen-Smith, N., and D. H. M. Cumming. 1993. Comparative foraging strategies of grazing ungulates in African savanna grasslands. Pages 691–698 in *Proceedings of the 17th International Grasslands Congress, New Zealand*. New Zealand Grassland Association, Palmerston North, New Zealand.

Owen-Smith, N., and P. Novellie. 1982. What should a clever ungulate eat? *American Naturalist* 119:151–178.

Pellew, R. A. 1984. The feeding ecology of a selective feeder, the giraffe (*Giraffa camelopardalis tippelskirchi*). *Journal of Zoology* (London) 202:57–81.

Provenza, F. D. 1995. Postingestive feedback as an elementary determinant of food preference and intake in ruminants. *Journal of Range Management* 48:2–17.

Rohde, K. 1999. Latitudinal gradients in species diversity and Rapoport's rule revisited: a review of recent work and what can parasites teach us about the causes of the gradients? *Ecography* 22:593–613.

Scholes, R. J., and B. H. Walker. 1993. *An African savanna: synthesis of the Nylsvlei study*. Cambridge, UK: Cambridge University Press.

Senft, R. L., M. B. Coughenour, D. W. Bailey, L. R. Rittenhouse, O. E. Sala, and D. M. Swift. 1987. Large herbivore foraging and ecological hierarchies. *BioScience* 37:789–799.

Sinclair, A. R. E., and P. Arcese (eds.). 1995. *Serengeti 2: Dynamics, management, and conservation of an ecosystem*. Chicago: University of Chicago Press.

Sinclair, A. R. E., and M. Norton-Griffiths (eds.). 1979. *Serengeti: Dynamics of an ecosystem*. Chicago: University of Chicago Press.

Sinclair, A. R. E., and M. Norton-Griffiths. 1982. Does competition or facilitation regulate migrant ungulate populations in the Serengeti? A test of hypotheses. *Oecologia* 53:364–369.

Skinner, J. D., and R. H. N. Smithers. 1990. *The mammals of the southern African subregion*. Pretoria: University of Pretoria.

Stokke, S., and J. T. du Toit. 2002. Sexual segregation in habitat use by elephants in Chobe National Park, Botswana. *African Journal of Ecology.* 40:360–371.

Turpie, J. K., and T. M. Crowe. 1994. Patterns of distribution, diversity and endemism of larger African mammals. *South African Journal of Zoology* 29:19–32.

Vrba, E. S. 1992. Mammals as a key to evolutionary theory. *Journal of Mammalogy* 73: 1–28.

Woolnough, A. P., and J. T. du Toit. 2001. Vertical zonation of browse quality in tree canopies exposed to a size-structured guild of African browsing ungulates. *Oecologia* 129:585–590.

Chapter 15

Rainfall Influences on Ungulate Population Dynamics

NORMAN OWEN-SMITH AND JOSEPH OGUTU

The strong influence that climatic variability can have on the population dynamics of ungulates is widely recognized (Coughenour and Singer 1996; Saether 1997; Forschhammer et al. 1998; Post and Stenseth 1998; Milner et al. 1999), but almost all of the long-term datasets that have been investigated for such effects have represented northern species. Substantial changes over time in the abundance of ungulate populations have been documented for many of Africa's national parks and other protected areas (Campbell and Borner 1995; Runyoro et al. 1995; Gasaway et al. 1996; Ottichillo et al. 2000), but trends have been attributed to local factors such as illegal hunting, changes in animal movements, predator impacts, and unknown causes. Assessments considering climatic influences have been focused on single species, such as wildebeest in the Serengeti region (Sinclair 1979; Mduma et al. 1999) and kudu in Kruger (Owen-Smith 1990). The only comprehensive analysis for a multispecies assemblage was for a section of Kruger, by Mills et al. (1995).

In African savanna regions, the prime climatic feature likely to be influential is rainfall, which largely controls annual vegetation growth and hence food production for large herbivores. Fluctuations in the abundance of ungulate populations in response to variations in rainfall, as well as associated interactions with predators, were recognized early in the history of Kruger. James Stevenson-Hamilton, first warden of the Sabi Game Reserve and later of Kruger, noted in annual reports that between 1916 and 1924, wildebeest "had died during the rains," followed by a "rapid and uninterrupted increase" in wildebeest abundance, and between 1932 and 1939 there was a "visible decline in numbers" (cited by Whyte and Joubert 1988). In summarizing these observations, he wrote, "There are annual curves of alternating abundance and want, which affect herbivora and carnivora conversely" (Stevenson-Hamilton 1947, p. 24). He outlined how this operated through periods of drought and good rains and concluded that when "the fat years for the game persist, the lions decrease, until a time comes when the game had again advanced . . . [to become] more easily accessible . . . [and] the cycle repeats itself" (p. 26).

Joubert (1974, cited by Whyte and Joubert 1988) suggested that "ranker" vegetation associated with high rainfall conditions rendered wildebeest more vulnerable to predation, an interpretation supported by Smuts (1978) with respect to zebra as well. Whyte (1985) proposed that, although the proximate mechanism might have been increased vulnerability to predation, the fundamental cause of the substantial declines in wildebeest and zebra populations in the early 1970s was high rainfall (see also Whyte and Joubert 1988). In support of this interpretation, he noted the concurrent increase in buffalo, waterbuck, kudu, giraffe, warthog, and impala numbers, exposed to the same predator abundance.

The detailed study on kudu demography over 1974–1984 revealed the controlling influence of prior annual rainfall on population dynamics, acting primarily through variability in calf survival (Owen-Smith 1990). Mills et al. (1995) confirmed the strong positive influence of annual rainfall on changes in the abundance of kudu, as well as buffalo and waterbuck, in the central region of Kruger. In contrast, they found no significant response by zebra and wildebeest to rainfall over an annual time frame but rather a negative effect of rainfall totaled over the preceding 3 years. They noted also a negative relationship between predator kills ascribed to lions and rainfall for waterbuck and buffalo, whereas for wildebeest, and to a lesser extent for zebra, predation appeared to be positively related to rainfall. No rainfall-related variation in predation was detected for kudu and giraffe. Accordingly, these authors endorsed the suggestion of Whyte and Joubert (1988) that ungulate population changes were driven largely through prey switching by predators, particularly lion, in response to changing prey vulnerability, dependent in turn on rainfall influences on vegetation.

For the migratory wildebeest population inhabiting the Serengeti region of Tanzania, annual mortality and hence population change appeared positively related to dry season rainfall, through its effect on food availability during this critical period, rather than to the total annual rainfall (Sinclair 1979; Mduma et al. 1999). Moreover, the increase in abundance of wildebeest during the 1970s seemed partly explained by the shift toward more rain during the dry season that occurred during this time.

In this chapter, we examine the influence of rainfall variability on the abundance and dynamics of ungulate populations in Kruger, based on the comprehensive dataset provided by all aerial censuses conducted between 1965 and 1997. After 1997, the census procedure changed to sample counts using a different method, making comparisons difficult. Sufficient data were available to encompass 12 ungulate species, listed in Table 15.1. Elephant are not considered because their population was held below a ceiling of 8,000 animals until

TABLE 15.1

Ungulate species considered.

COMMON NAME	SCIENTIFIC NAME	FEEDING HABITS	MEAN CENSUS TOTAL 1980–1993	RANGE IN CENSUS TOTALS 1980–1993
Impala	*Aepyceros melampus*	Mixed feeder	116,000	91,884–137,055
Buffalo	*Syncerus caffer*	Grazer	27,500	15,250–34,912
Zebra	*Equus burchelli*	Grazer	29,400	21,454–33,164
Wildebeest	*Connochaetes taurinus*	Grazer	12,550	8,568–14,601
Kudu	*Tragelaphus strepsiceros*	Browser	7,700	3,127–10,760
Giraffe	*Giraffa camelopardalis*	Browser	4,900	4,122–5,759
Waterbuck	*Kobus elipsiprimnus*	Grazer	3,300	1,419–5,042
Warthog	*Phacochoerus aethiopicus*	Grazer	2,850	721–4,048
Sable	*Hippotragus niger*	Grazer	1,800	856–2,240
Tsessebe	*Damaliscus lunatus*	Grazer	820	222–1,163
Eland	*Taurotragus oryx*	Mixed feeder	670	349–996
Roan	*Hippotragus equinus*	Grazer	250	43–452

recently. Hippopotamus are omitted because their censusing required a different approach, and population changes depend on river levels as well as rainfall. White rhinoceros are excluded because their population was growing exponentially, with little influence from rainfall variability over the period covered. In particular, we compare and contrast the influence of rainfall variation over different temporal scales: decadal, annual, and seasonal. This report summarizes findings from more detailed analyses of the Kruger dataset presented elsewhere (Ogutu and Owen-Smith 2003 and in preparation).

Methods

Our analyses were based on the results of monitoring programs that Kruger scientists have conducted since the 1960s, and from which we extracted data on rainfall and the size and structure of ungulate populations.

Rainfall Records

Rainfall over the period 1960–1997 was derived from four or five representative stations in each of the four regional subdivisions of Kruger: south, between the Crocodile and Sabie rivers; central, between the Sabie and Olifants rivers; north, between the Olifants and Shingwedzi rivers; and far north, between the Shingwedzi and Luvuvhu rivers. Records before 1960 are from fewer recording stations, and only those for Skukuza in the south extend as far back as 1920 (Gertenbach 1980). To standardize regional comparisons, we normalized the records for each station relative to its average rainfall over 1974–1996. We derived annual totals spanning the rainfall year July–June, seasonal totals for the 6 months of the wet season (October–March) or dry season (April–September), and running averages over a period of 5 years before and including the census year.

Population Census Totals

Between 1965 and 1976, park scientists derived ungulate population estimates from irregular aerial surveys by either helicopter or fixed-wing airplane, supported by differential ground counts. The ground counts, conducted by vehicle in each ranger section, were used to correct for differential species visibility in the aerial censuses, relative to some standard species assumed to be counted accurately from the air, using the ground ratio of sightings. Wildebeest, easily sighted because of their dark color and open plain habitat, usually were chosen as the standard species, but where they were less common zebra or even sable antelope were used instead.

From 1977 to 1997, park staff conducted ecological aerial surveys annually, following a standard procedure. A Cessna 206 was used until 1986, and thereafter a twin-engine Partenavia Observer offering high visibility through its large windscreen. The park area was subdivided into 66 census blocks, each surveyed systematically by flying parallel strips 800 m apart at a height of 65–70 m and flying speed of 166–178 km/hour. The survey team consisted of the pilot, a data recorder, and four observers. Flying was divided into two sessions of 1.5 to 3 hours each during the morning between 7 AM and noon, when visibility was best. The census was conducted during the dry season between May and September, when weather conditions were most favorable. Before 1993, censuses commenced in the north in May, continuing systematically through to the south, then being concluded in the far north in August or early September. From 1993 onward, coverage began in the far north. Flying was suspended when cloud cover exceeded about 25 percent and when conditions were very windy or turbulent. About 250 hours of flying time was needed to cover the whole park. In practice, the procedure was fully consistent and parkwide in coverage (apart from the Punda Maria and Pafuri section of the far north) only between 1980 and 1993. Because of their clumped distribution in large herds, buffalo, together with elephant and hippo, were censused independently using systematic helicopter surveys between mid-August and mid-September each year.

Management culling limited the population totals of some species before 1977, as follows: wildebeest, 3,315 animals removed between 1965 and 1972 (Whyte and Joubert 1988); zebra, 3,697 animals removed between 1965 and 1974 (Smuts 1978); impala, about 7,500 cropped between 1968 and 1975 (Smuts 1978). Buffalo continued to be culled through 1992, with an average of about 10 percent of the population removed annually. Kruger's boundary was fenced after 1959 on the west and south and, after 1975, on the east as well. Sections of the western boundary fence adjoining private wildlife reserves were removed before the 1993 census.

Data Transformations

The censuses were attempted total counts but reflected an inevitable and unknown degree of undercount bias. For impala, a substantial discrepancy was evident between the population totals projected using the differential ground counts and those derived subsequently from the ecological aerial surveys. Accordingly, to link their population trend between the two periods we adjusted the latter counts upward by a factor of 1.5 in the graphs, but not for statistical analysis. Counting efficiency also varied between years, depending on flying and weather conditions. In particular, brown species, such as impala and kudu, were less readily visible in dry years, when the ground was also largely brown (P. Viljoen, pers. comm.). These factors resulted in an exaggeration of population declines in some years, followed by unrealistically high rates of population rebound in the succeeding years.

When the census did not encompass the whole park area, we projected park-wide population totals from the proportional changes shown by the counts for the regions that were censused. To suppress the distortion of estimates of annual population changes resulting from variable sampling efficiency, we transformed the census totals for all species by calculating a weighted running average N'_t, using the formula $N'_t = 0.5N_t + 0.25N_{t-1} + 0.25N_{t+1}$, where N_t = census total for year t. For buffalo, the number of animals culled between censuses was added to the census total to derive the inherent population change between years.

Relating Annual Population Changes to Rainfall

We analyzed the census data by region for the five species that were sufficiently abundant, that is, census totals mostly >500, in each of the four regions (excluding data for the Punda Maria and Pafuri section of the far north, which was not counted consistently). For giraffe and warthog, census totals for the north and far north regions were combined. For waterbuck and sable antelope, census data were combined for the northern (north plus far north) and southern (south plus central) halves of the park. For eland, tsessebe, and roan antelope, the analysis was limited to the northern half of the park, where these species were concentrated.

Annual population change was estimated from the $\log_e$ transform of the ratio N'_t/N'_{t-1} and related to rainfall components as well as relative population abundance by multiple linear regression, using Systat 8.0 for Windows. Abundance was normalized relative to the regional means over 1980–1990. Rainfall was normalized relative to regional means over 1960–1999 and then $\log_e$-transformed. Therefore, the underlying postulate is that changes in population abundance were related to deviations in rainfall from the long-term mean. Statistical significance is one-tailed because our hypotheses are that elevated rainfall promotes increased population abundance and that increased density reduces population growth. Accordingly, relative abundance was omitted from the model when the sign of its slope coefficient appeared spuriously positive. Data points that were extreme outliers, based on studentized residuals exceeding 3.0, were excluded.

Demographic Responses

Between 1983 and 1996, the ecological aerial surveys were supported by comprehensive monitoring of the sex and age structure of ungulate populations throughout the park, conducted annually between August and October (Mason 1990). From these records, the recruitment success contributing to annual population change was derived from the ratio of juveniles (aged >1 year) to adult plus subadult (older than 2 years) females. For wildebeest, impala, and waterbuck, yearling females could not be distinguished reliably from adult females, so the yearling proportion was obtained by doubling the number of yearling males. For tsessebe, yearlings of both sexes could not be distinguished from adults, so yearling and adult females were amalgamated. The annual survival rate of adults, yearlings, and calves from the previous year was then calculated by factoring out the contribution of juvenile recruitment to the annual population change as indicated by the census data: $S_t = N'_t(1 - j_t)/N'_{t-1}$, where $S_t =$ survival rate through to year t, $N'_t =$ smoothed census total for year t, and $j_t =$ proportion of juveniles in the total population sample for sex and age structure in year t. Demographic responses were assessed only for species that both responded to rainfall components and yielded sufficiently large samples of population structure.

Results

Our results demonstrate the pervasive influence of rainfall variability on the dynamics of ungulate populations in Kruger. Nevertheless, some ungulate species responded in different ways to rainfall variation and, in certain cases, in contrasting ways to different temporal scales of rainfall variation.

Rainfall Variability

The cyclic component of rainfall variability across eastern South Africa, with a period of about 18 years (Tyson 1986), is clearly evident for the Kruger region from the 5-year running average (Figure 15.1a). Notably, the wet phase extending through the late 1970s was the wettest wet decade on record, and the subsequent dry phase extending through the 1990s was prolonged well beyond the typical 9-year duration. Extreme drought years occurred in 1963–1964, 1972–1973, 1982–1983, and 1991–1992, associated with the warm phase of the El Niño–Southern Oscillation (ENSO). Although extreme El Niño conditions prevailed worldwide in 1997–1998, this was manifested only weakly in Kruger's rainfall.

Overall, about 80 percent of the annual rainfall falls during the wet season months October–March and only 20 percent during the 6 months of the dry season. Although the dry season component was concurrently low during the dry decade of the late 1960s, it became partially decoupled from the wet season contribution thereafter (Figure 15.1b). Conditions of extremely low dry season rainfall persisted from 1988 through 1994, through the latter part of the extended period when wet season rainfall was also very low.

The coefficient of variation (CV) in rainfall was higher in the drier north (annual mean, 1960–1999, 485 mm, CV 0.38) than in the more mesic south (mean 570 mm, CV 0.30). Rainfall patterns were generally consistent from south to north through the park, with just a few deviations (Figures 15.1c and 15.1d). The 1982–1983 drought was broken in the southern half of the park in 1984 but persisted a year longer in the northern half. The seasonal cycles 1986–1987, 1993–1994, and 1997–1998 were drought years in the north but showed less extreme rainfall deficits in the south.

Population Responses to Cumulative Past Rainfall

Contrasting changes in abundance were evident among the ungulate species censused (Figure 15.2). Zebra and wildebeest increased from low numbers in the mid-1970s to peak in abundance during the low-rainfall phase of the late 1980s. By 1985 zebra were 50 percent more numerous than they had been around the end of the dry period of the late 1960s. Giraffe similarly increased steadily in abundance until about 1985 and maintained their numbers subsequently despite the dry conditions. In contrast, populations of buffalo, kudu, waterbuck, warthog, eland, sable, tsessebe, and roan grew toward peak abundance during the wet period extending through the 1970s and declined steeply in abundance after around 1986. Impala were also most abundant around 1985 but subsequently showed a minor decline in numbers.

Notably, the fall-off in abundance of the latter species commenced not at the onset of prevalently dry conditions in 1982–1983, but 4–5 years later in association with a minor drought. This happened to be around the time that dry season rainfall became particularly low. The precipitous decrease in buffalo after

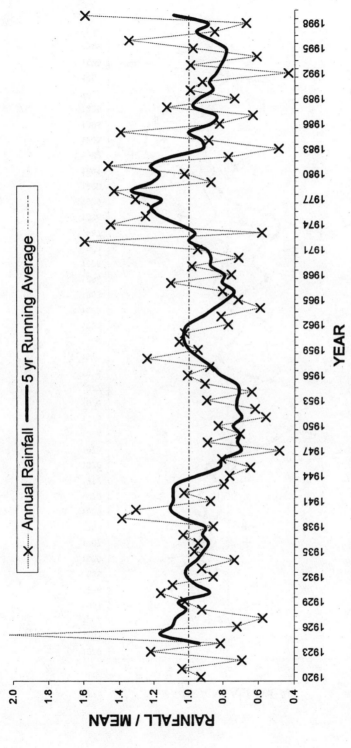

FIGURE 15.1 (A). Rainfall patterns, expressed relative to the long-term mean rainfall. (*a*) Variation in annual rainfall (July–June) and in the 5-year running mean rainfall over the complete period spanned by rainfall records for Kruger. (*b*) Comparative variation in the wet season (October–March) and dry season (April–September) components of rainfall (both annual and 5-year running mean) in Kruger for 1960–1999. (*c*) Comparative variation in the annual rainfall and in the 5-year running mean rainfall among the four regions, from south to north, of Kruger. (*d*) Comparative variation in the dry season component of rainfall, both annual and as the 5-year running mean, among the four regions of Kruger.

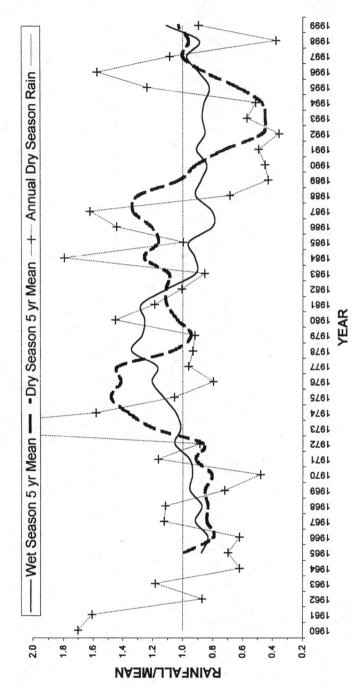

FIGURE 15.1 (B).

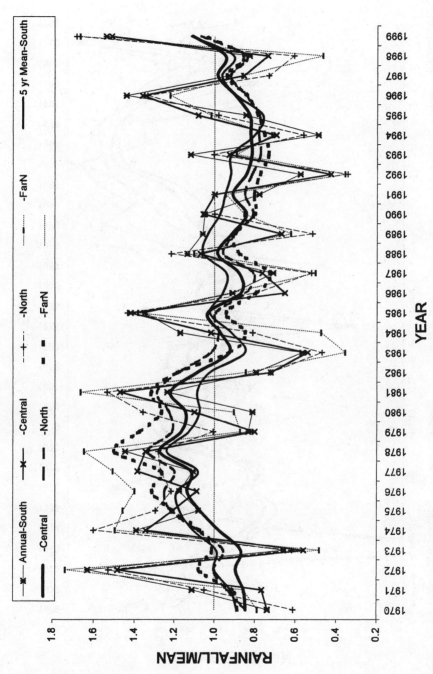

FIGURE 15.1 (c).

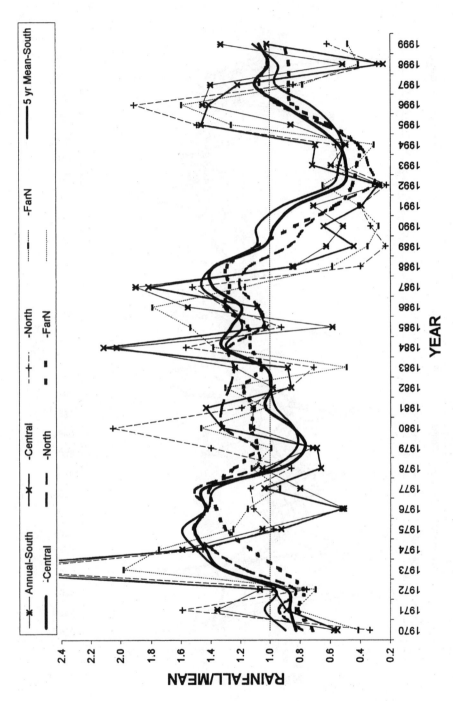

FIGURE 15.1 (D).

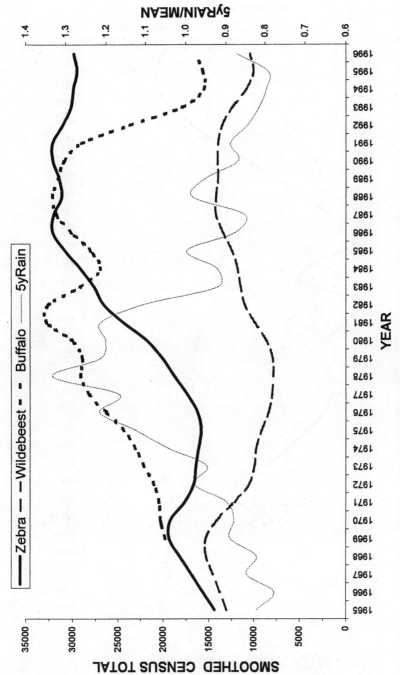

FIGURE 15.2 (A). Changes in censused or projected total populations of the major ungulate species in Kruger since 1965 in relation to the variation in 5-year running mean rainfall. (*a*) Zebra, wildebeest, and buffalo. (*b*) Kudu, giraffe, and impala (census total for the latter divided by 20). (*c*) Waterbuck, warthog, and sable. (*d*) Tsessebe, roan, and eland.

FIGURE 15.2 (B).

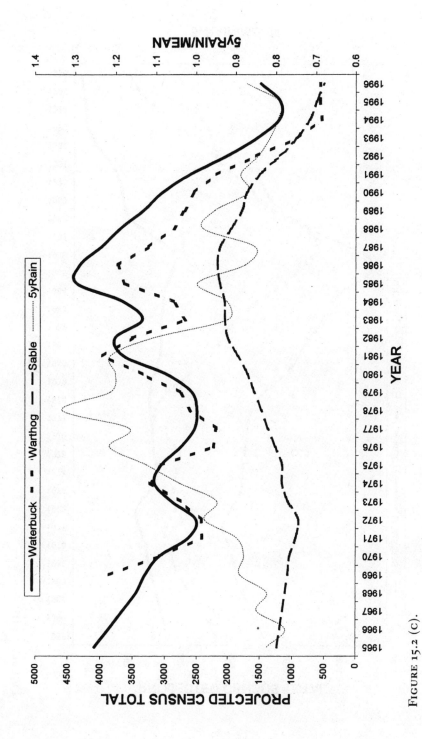

FIGURE 15.2 (c).

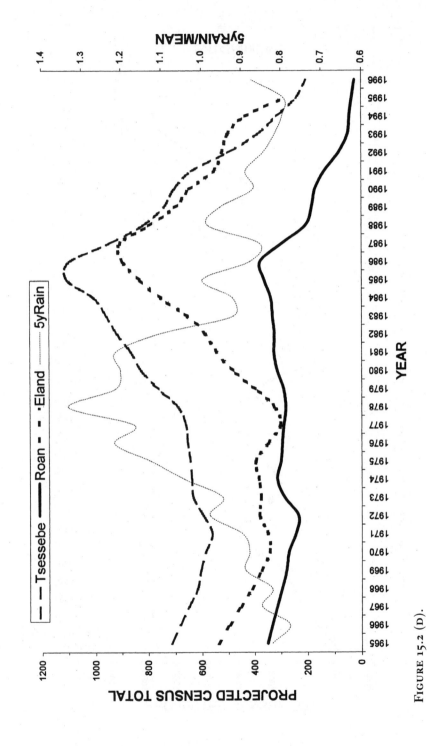

FIGURE 15.2 (D).

TABLE 15.2

Regression relationships between annual population change and annual rainfall total (July–June) for the 12 ungulate species, comparing model with rainfall only and model including both rainfall and relative abundance as factors.

SPECIES	N	ANNUAL RAINFALL ONLY		ANNUAL RAINFALL AND RELATIVE ABUNDANCE	
		REGRESSION COEFFICIENT	R^2	REGRESSION COEFFICIENT	R^2
Declining species					
Buffalo	99	.150***	.106	.141***	.112
Kudu	65	.161***	.242	.165***	.304
Waterbuck	35	.284***	.474	.299***	.510
Warthog	36	.365***	.301	.389***	.347
Sable	34	.120**	.170	.135**	.356
Tsessebe	18	.115+	.157	.123*	.347
Roan	19	.096	.050	.089	.440
Eland	19	.081	.083	.047	.244
Stable species					
Zebra	67	.052*	.075	.027+	.290*
Wildebeest	66	.086*	.055	.077*	.066
Impala	67	.093**	.117	.066*	.217
Giraffe	49	.037*	.063	.020	.279

$^+p < .10$; $^*p < .05$; $^{**}p < .01$; $^{***}p < .001$.

1990 was partly caused by an outbreak of anthrax in the northern half of the park. This disease had a severe impact on the kudu population also but caused only minor mortality among other species.

Population Responses to Annual Rainfall and Seasonal Components

All 12 ungulate species showed a positive relationship between annual population change and annual rainfall, although not always significantly (Table 15.2). The influence of annual rainfall was weakest for the four species that maintained high abundance through the low rainfall period—zebra, wildebeest, impala, and giraffe—and also weak for eland and roan antelope. Density dependence appeared positive rather than negative for warthog, sable, roan, and tsessebe; that is, their populations declined faster when abundance was low, indicating other influences.

The effect of rainfall received during the dry season months (April–September) was more pervasive than that of rainfall during the wet season, being statistically significant or almost significant for all species except giraffe (Table 15.3). Rainfall during the late dry season period (July–September) of the year

TABLE 15.3

Regression relationships between annual population change and seasonal rainfall segments. Relative abundance was omitted as a factor in the model when the sign of its regression coefficient was positive (as indicated by parentheses in the table).

		REGRESSION COEFFICIENTS			
SPECIES	N	WET SEASON	DRY SEASON	RELATIVE ABUNDANCE	OVERALL R^2
Declining species					
Buffalo	99	.105*	.099***	−.072	.212
Kudu	65	.101**	.110***	−.014	.391
Waterbuck	35	.213***	.143***	−.011	.621
Warthog	49	.270***	.188***	(.047)	.430
Sable	35	.066+	.108**	(.161)	.327
Tsessebe	18	.074	.094*	(.130)	.316
Roan	19	.036	.104+	(.322)	.133
Eland	19	−.029	.139***	−.206**	.665
Stable species					
Zebra	67	.017	.022+	−.154***	.310
Wildebeest	65	−.005	.121***	−.008	.303
Impala	67	.036	.059**	−.171**	.302
Giraffe	49	.012	.009	−.214***	.279

+$p < .10$; *$p < .05$; **$p < .01$; ***$p < .001$.

preceding the census appeared most strongly influential, suggesting that mortality losses were concentrated around the end of the dry season. The wet season component had a significant influence only on the dynamics of buffalo, kudu, waterbuck, and warthog. Strong negative density dependence was apparent for giraffe, zebra, impala, and eland. Rainfall plus relative abundance together explained between about a quarter and two-thirds of the annual variance in population abundance for all species except roan antelope and buffalo.

Demographic Responses

Only three species were strongly affected by rainfall variation and also had large enough samples of sex and age structure for a demographic analysis: kudu, waterbuck, and warthog. For these species offspring survival, as assessed by the juvenile:adult female ratio, appeared responsive to both the wet and dry season segments of rainfall (Table 15.4). In contrast, the survival rate of adults and immatures from the previous year depended only on dry season rainfall. Wildebeest and impala showed an overall dependence of population change on dry season rainfall segments but no significant effect of rainfall on either juvenile recruitment or annual survival rate.

TABLE 15.4

Demographic relationships with seasonal rainfall segments, contrasting offspring survival (as indexed by juvenile:adult female ratio) with survival of adult plus immature animals from the previous year.

SPECIES	N	WET SEASON COEFFICIENT	R^2	DRY SEASON COEFFICIENT	R^2
Offspring survival					
Kudu	48	.123	.118*	.103	.224**
Waterbuck	26	.276	.280*	.155	.203*
Warthog	23	.727	.187*	.549	.284**
Impala	43	.044	.008	.117	.115*
Wildebeest	41	.005	.000	.157	.385***
Survival of adults plus immatures					
Kudu	45	.025	.016	.051	.173**
Waterbuck	26	.022	.008	.057	.136+
Warthog	22	−.012	.002	.039	.052
Impala	39	.028	.016	.027	.034
Wildebeest	37	−.008	.003	−.002	.000

$^+p < .10$; $^*p < .05$; $^{**}p < .01$; $^{***}p < .001$.

Discussion

Our analysis shows that wildebeest and zebra reach higher abundance during the dry phase of the decadal rainfall cycle than during extended wet periods, confirming previous findings (Smuts 1978; Whyte and Joubert 1988). Giraffe also conformed to this pattern. Most other ungulate species were favored by the high rainfall conditions extending through the late 1970s and into the early 1980s and declined in abundance to varying degrees during the prolonged dry phase that ensued.

Census records do not extend far enough back in time to reveal the abundance levels maintained by the positively rainfall-dependent species during the high-rainfall phase of the late 1950s. The fact that cumulative rainfall through the wet phase of the late 1970s was substantially greater than previously recorded since 1920 suggests that these species had probably reached record high numbers by 1980, thereby amplifying the relative extent of their population declines during the subsequent dry phase.

However, the three species that maintained high abundance through the sequence of dry years still showed positive responses to annual rainfall variability, albeit much more weakly than the species that showed population declines. This suggests that they were not favored by some negative effect of rainfall on food availability but by some less direct influence. Prey switching by lions away from these species toward other species more susceptible to adverse influences of low rainfall on vegetation, as suggested by Mills et al. (1995), could well be involved. Circumstantial evidence assembled by Harrington et al. (1999) implicated lion predation as the primary cause of the crash of the roan antelope population after 1986, but with immigration by zebra following the addition of water-points contributing to the lion increase. The expansion in the zebra population may have been a response to the ubiquitous provision of artificial waterpoints throughout Kruger, increasing the habitat area that this strongly water-dependent species could occupy (see Chapter 8, this volume).

The substantial population declines exhibited by a suite of species began not at the onset of dry conditions inaugurated by the severe El Niño–related drought of 1982–1983 but 4–5 years later. This could be a sign of the lag in the population response of lions to increased prey vulnerability, particularly for buffalo. The positive density dependence manifested by many of these species could also be an indication of a predation effect, with a higher proportion of the population being taken by a fairly constant predator population as prey population size diminished.

However, the population declines occurred during the period when dry season rainfall remained extremely low. Almost all of the ungulate species, except conspicuously giraffe, seemed more sensitive to lack of rainfall during the dry season than to the major wet season component of the annual rainfall. Dry season rainfall may be directly important by influencing the retention of some green forage during this critical period when malnutrition takes hold. Weakened animals may also become more vulnerable to predation. The dynamics of the Serengeti wildebeest population likewise are driven primarily by rain received during the dry season, although there predation does not seem to be involved (Sinclair 1979; Mduma et al. 1999). Rainfall in the dry season could also have indirect effects by influencing the dependence of animals on waterpoints where lions may lurk.

Furthermore, rainfall conditions can affect the susceptibility of animals to disease outbreaks, particularly of anthrax. The marked population decline by buffalo after 1990, and to some extent also that of kudu, was largely the result of an anthrax outbreak affecting these species, particularly in the northern half of the park. The relationships between anthrax outbreaks and climatic conditions are explained in Chapter 17. The presence of bovine tuberculosis in the southern half of Kruger many also have amplified the susceptibility of the buffalo population to the 1983 drought (Chapter 17, this volume). Competitive interactions may also have influenced some species, especially waterbuck, which inhabit regions close to water likely to be most severely affected by other grazers during dry periods.

Demographically, calf survival to around 0.5 year of age was influenced by both the wet and dry seasonal components of rainfall, among the species that were strongly responsive to rainfall. Therefore, nutritional deficits from inadequate forage quantity or quality strongly affect the juvenile segment, in accordance with the generally greater sensitivity of juvenile recruitment versus adult survival to environmental variability among large mammalian herbivores (Gaillard et al. 1998). Nevertheless, for the species that declined in abundance in Kruger, the survival of older classes was also affected by rainfall variation, specifically in the dry season component. Notably, the older class includes all animals from yearlings upward. Whether this sensitivity is a direct result of nutritional shortfalls during the late dry season or indirectly caused by a relationship between dry season rainfall and predation risk remains unclear.

The long-term dynamics of ungulate populations elsewhere in the world generally have been examined for single species, largely in the absence of predation (Gaillard et al. 2000). Several of these populations show periodic increases to high abundance followed by substantial population dieoffs induced by starvation (Milner et al. 1999; Peterson 1999). Adverse weather conditions may contribute to these population crashes, through cold stress or the effects of snow cover on forage access (Coughenour and Singer 1996; Post and Stenseth 1998). Temperature conditions can also affect forage production (Albon and Clutton-Brock 1988; Post and Stenseth 1999).

Conclusion

For African savanna ungulates, seasonal rainfall components influence both forage production and the food availability through the adverse period of the year. As demonstrated by the Kruger dataset, rainfall variability can result in wide variations in population levels of particular species over time. Whatever the carrying capacity concept represents, there is little evidence of any long-persisting equilibrium between large herbivore populations and their food resource base, and even density-dependent feedbacks may become obscured.

Although the topic of this chapter concerned rainfall influences, it has been necessary to consider also other factors potentially affecting ungulate population dynamics. The effects of temporal heterogeneity in rainfall, over various time frames, on populations of large mammalian herbivores are central and pervasive. Changes in ungulate populations in turn affect the vegetation and ecosystem processes such as nutrient cycling (Chapter 6, this volume). Influences emanating from spatial aspects of heterogeneity cannot adequately be assessed except in the context of this fundamental temporal variability in African savanna ecosystems.

References

Albon, S. D., and T. H. Clutton-Brock. 1988. Climate and the population dynamics of red deer in Scotland. Pages 93–107 in M. B. Usher and D. B. A. Thompson (eds.), *Ecological change in the uplands.* Oxford, UK: Blackwell.

Campbell, K., and M. Borner. 1995. Population trends and distribution of Serengeti herbivores: implications for management. Pages 115–145 in A. R. E. Sinclair and P. Arcese (eds.), *Serengeti 2: dynamics, management, and conservation of an ecosystem*. Chicago: University of Chicago Press.

Coughenour, M. B., and F. J. Singer. 1996. Elk population processes in Yellowstone National Park under the policy of natural regulation. *Ecological Applications* 6:573–593.

Forschhammer, M., N. C. Stenseth, E. Post, and R. Langvatn. 1998. Population dynamics of Norwegian red deer: density dependence and climatic variation. *Proceedings of the Royal Society of London*, Series B 265:341–350.

Gaillard, J.-M., M. Festa-Bianchet, and N. G. Yoccoz. 1998. Population dynamics of large herbivores: variable recruitment with constant adult survival. *Trends in Ecology and Evolution* 13:58–63.

Gaillard, J.-M., M. Festa-Bianchet, N. G. Yoccoz, A. Loison, and C. Toigo. 2000. Temporal variation in fitness components and population dynamics of large herbivores. *Annual Review of Ecology and Systematics* 31:367–393.

Gasaway, W. C., K. T. Gasaway, and H. H. Berry. 1996. Persistent low densities of plains ungulates in Etosha National Park, Namibia: testing the food regulation hypothesis. *Canadian Journal of Zoology* 74:1556–1572.

Gertenbach, W. P. D. 1980. Rainfall patterns in the Kruger National Park. *Koedoe* 23:35–43.

Harrington, R., N. Owen-Smith, P. C. Viljoen, H. C. Biggs, D. R. Mason, and P. Funston. 1999. Establishing the causes of the roan antelope decline in the Kruger National Park, South Africa. *Biological Conservation* 90:69–78.

Mason, D. R. 1990. Monitoring of sex and age ratios in ungulate populations of the Kruger National Park by ground survey. *Koedoe* 33:19–28.

Mduma, S. A. R., A. R. E. Sinclair, and R. Hilborn. 1999. Food regulates the Serengeti wildebeest: a 40-year record. *Journal of Animal Ecology* 68:1101–1122.

Mills, M. G. L., H. C. Biggs, and I. J. Whyte. 1995. The relationship between rainfall, lion predation and population trends in African herbivores. *Wildlife Research* 22:75–88.

Milner, J. M., D. A. Elston, and S. D. Albon. 1999. Estimating the contributions of population density and climatic fluctuations to interannual variation in survival of Soay sheep. *Journal of Animal Ecology* 68:1235–1247.

Ogutu, J. O., and N. Owen-Smith. 2003. ENSO, rainfall and temperature influences on extreme population declines among African savanna ungulates. *Ecology Letters* 6:412–419.

Ottichillo, W. K., J. de Leeuw, A. K. Skidmore, H. H. T. Prins, and M. Y. Said. 2000. Population trends of large, non-migratory herbivores and livestock in the Masai Mara ecosystem, Kenya, between 1977 and 1997. *African Journal of Ecology* 38:202–216.

Owen-Smith, N. 1990. Demography of a large herbivore, the greater kudu, in relation to rainfall. *Journal of Animal Ecology* 59:893–913.

Peterson, R. O. 1999. Wolf-moose interaction on Isle Royale: the end of natural regulation? *Ecological Applications* 9:10–16.

Post, E., and N. C. Stenseth. 1998. Large-scale climatic fluctuation and population dynamics of moose and white-tailed deer. *Journal of Animal Ecology* 67:537–543.

Post, E., and N. C. Stenseth. 1999. Climatic variability, plant phenology and northern ungulates. *Ecology* 80:1322–1339.

Runyoro, V. A., H. Hofer, E. B. Chausi, and P. D. Moehlman. 1995. Long-term trends in the herbivore populations of the Ngorongoro Crater, Tanzania. Pages 146–168 in A. R. E. Sinclair and P. Arcese (eds.), *Serengeti 2: dynamics, management, and conservation of an ecosystem*. Chicago: University of Chicago Press.

Saether, B.-E. 1997. Environmental stochasticity and population dynamics of large herbivores: a search for mechanisms. *Trends in Ecology and Evolution* 12:143–149.

Sinclair, A. R. E. 1979. The eruption of the ruminants. Pages 83–103 in A. R. E. Sinclair and P. Arcese (eds.), *Serengeti 2: dynamics, management, and conservation of an ecosystem*. Chicago: University of Chicago Press.

Smuts, G. L. 1978. Interrelations between predators, prey and their environment. *BioScience* 28:316–320.

Stevenson-Hamilton, J. 1947. *Wild life in South Africa*. London: Cassel.

Tyson, P. D. 1986. *Climatic change and variability in Southern Africa*. Oxford, UK: Oxford University Press.

Whyte, I. J. 1985. *The present ecological status of the blue wildebeest in the Central District of the Kruger National Park*. Unpublished M.Sc. thesis, University of Natal, Pietermaritzburg, South Africa.

Whyte, I. J., and S. C. J. Joubert. 1988. Blue wildebeest population trends in the Kruger National Park and the effects of fencing. *South African Journal of Wildlife Research* 18:78–87.

Chapter 16

Kruger's Elephant Population: Its Size and Consequences for Ecosystem Heterogeneity

IAN J. WHYTE, RUDI J. VAN AARDE, AND STUART L. PIMM

Those charged with managing biodiversity face the immediate questions of what particular goals they must set. American conservationist Aldo Leopold suggested that the first law of intelligent tinkering should be to keep every cog and wheel— a metaphor admonishing managers to not allow species to go extinct. This is a powerful guideline for setting conservation goals because globally extinction rates are now perhaps one thousand times geological background rates (Pimm et al. 1995b). Worldwide, many national parks are losing species found nowhere else; the need to prevent such losses inevitably and sensibly shapes management policics. Kruger is exceptional, suggesting that other goals must shape its policies. Managing for merely rare or globally declining species is an obvious extension of Leopold's dictum because common sense recommends that such species should be protected in as many different places as possible. A second goal might be to manage for a local diversity of species and the heterogeneity of habitats that support them. A third goal might be the restoration of the ecosystem to its original state with a complete set of original species and their habitats. We show that the first two approaches need not be entirely compatible. Moreover, the third approach, which we might sensibly view as a referee when conflicts arise between the first and second, does not give an acceptable answer.

The savanna elephant is locally extinct across much of its former range. Hunting for meat and ivory and the conversion of habitats to crops and other human uses are the major causes. Large-bodied species with valuable body parts and habits that threaten human welfare generally have fared badly. Their protection seems to be a sensible conservation goal. Given these features, elephants are still surprisingly numerous in Africa, with a total population of about 350,000 (Barnes et al. 1999). Where they are protected and confined in fenced reserves, as in Kruger, their numbers can increase rapidly, doubling every decade or so (Whyte 2001; Woodd 1999). Herein lies the problem.

At high densities, elephants reduce the diversity of habitats and hence the diversity of species (Cumming et al. 1997; Herremans 1995; Western and Gichohi 1989). High elephant densities change woodlands to grassland (Bourliére and Hadley 1983; Leuthold 1977). The reverse may occur at low densities (Dublin 1995; Dublin et al. 1990). The confinement and protection afforded to elephant populations in fenced reserves modify plant and animal communities (Cumming et al. 1997; Johnson 1998; Johnson et al. 1999; Lombard et al. 2001; Moolman and Cowling 1994). Structural changes to vegetation caused by elephants in Amboseli National Park (Kenya) have resulted in the extirpation of lesser kudu and bushbuck (Western and Gichohi 1989) and, more recently, gerenuk and giraffe (D. Western, pers. comm.). Elephants living at high densities for extended periods eliminate some plant species (Leuthold 1977; Page 1999). Amboseli has lost more than 50 percent of its plant species (D. Western, pers. comm.; J. Waithaka, pers. comm.) and undoubtedly many other species associated with these plants.

To Kruger's managers, whose goal is ecosystem heterogeneity, the preservation of these other species is no less a desirable goal than the preservation of elephants. Moreover, the homogenization of habitats at which these losses strongly hint suggests that high elephant densities seriously harm many less obvious species. Such considerations suggest that there might be some optimal balance—a "right number" of elephants, as it were. This, in turn, begs our asking what the historical numbers of elephants in Kruger were. Perhaps, armed with historical insights, managers can better maintain the park according to current biodiversity objectives.

We shall show that elephants persisted at low levels for centuries before being shot to extinction, before Kruger's proclamation as a game reserve in 1899. Reasons for these low levels are less certain and may range from climate change to anthropogenic influences, with the latter seeming the more likely. After proclamation, the population grew through births and immigration to about 7,000 in 1967. Managed culls maintained it at that level (with fluctuations between 6,900 and 8,800) until 1994. A moratorium on culling pending a review of elephant management policy saw the population increase to 10,500 in 2002.

So what elephant numbers are desirable? Certainly, "no elephants" seems to be an unnatural condition, a consequence of late-nineteenth-century ivory hunting. The persistence of very few elephants historically seems at odds with the ability of the elephants to increase at their rapid rate in the last century. This begs the question not so much of whether few or many elephants is more natural but of what has changed and why. The present high densities also demand our asking what the consequences to Kruger's plant and animal communities are. Will the problems reported elsewhere be repeated? After reviewing elephant history, we consider the present state of Kruger's plant communities, showing that concerns for them are well founded. We conclude by considering possible management actions.

History of Kruger's Elephant Population

Although earlier authors (Rowland-Jones 1955; Pienaar 1990, 1996) thought that elephants were abundant in the Kruger area, the evidence suggests otherwise. We organize that evidence chronologically.

The Pre-European Era

The San (Bushmen) lived in the Kruger area between 7,000 BC and AD 300 (Cooke 1969; Eloff 1990a, 1990b). Elephants were a popular theme for San paintings signifying the close symbolic relationship between women, rain, and large herbivores (Lewis-Williams 1996; Lewis-Williams and Dowson 1989; Woodhouse 1992, 1996). Interestingly, only three of the 109 shelters with rock art in the Kruger area show elephant (English 1990). This suggests that they were rare, but why? Perhaps climatic conditions made water sources fewer or less reliable than at present. Alternatively, perhaps even low levels of hunting targeted vulnerable individuals (such as young females) that would exert efficient controls on small elephant numbers.

Between AD 200 and 500, Bantu peoples penetrated the area and remained until Kruger became a park (Joubert 1986). From AD 900 to 1220, trade in gold and ivory increased between the indigenous people inhabiting the Limpopo River area and the Islamic empire. The near absence of ivory from archaeological sites combined with the presence of gold in the uppermost levels of the excavation suggests that by the twelfth century, the trade had switched to gold and the trade routes had moved farther north, to Great Zimbabwe (Plug and Voigt 1985). This switch to gold, the lack of ivory in the archaeological sites, and the shift in trade routes may have resulted from the scarcity of elephants.

Certainly, we base our contention that elephant densities were low in the Kruger area since perhaps AD 300 (Whyte 2001) on tenuous and perhaps historically ambiguous evidence. The evidence from baobab trees provides additional support. Low elephant densities would have resulted in increased survival of baobabs in northern Kruger. Kelly (2000) concluded that elephants had not shaped the population structure of baobabs, but Whyte et al. (1996) found no evidence of recent seedling recruitment. In northern Kruger, most small baobabs (diameter >1 m) occur in the Makuleke area (Pafuri), which was added to Kruger in 1969. Before this, elephants were excluded, suggesting that the current population of baobabs germinated at a time when elephant densities were lower than at present. Dates carved in the bark of baobabs in Kruger are still clearly visible after 140 years (Pienaar 1990), and scars caused by elephants should also persist for a long time. Many of the older baobabs in the Makuleke area of Kruger bear no scars from elephant damage, and those that do show recent rather than old damage. The rapid near eradication of baobabs from Tsavo National Park in Kenya (Leuthold 1977) and the fact that they are still abundant in northern Kruger (the southern tip of their range in Africa) indicate that elephants could never have achieved the high densities in Kruger that they did in Tsavo.

In sum, we conclude that elephant densities in Kruger and adjacent areas probably were low during the pre-European era.

After Colonization (Arrival of the Europeans)

In 1725 de Cuiper and his party penetrated the extreme southeastern region of the Kruger area seeking to establish trade. According to his account, the area was well populated by indigenous people (Punt 1990). These people informed them that they had little ivory but that at Phalaborwa there was copper, and Zimbabwe held much gold and ivory (Punt 1990). The correctness of the information on gold and copper supports the notion that few elephants lived in the Kruger region at that time.

Louis Trichardt crossed the Kruger area on his 1838 trek from the Soutpansberg to Lorenzo Marques (now Maputo). He mentions a hunt in the Soutpansberg area and an unsuccessful hunt after their arrival in Lorenzo Marques (Preller 1917). Had opportunity presented itself, he would almost certainly have hunted elephants and recorded that fact.

João Albasini was the first white settler in the Kruger area. In 1845, he established himself in the Phabeni area in the southern Kruger, where he reputedly traded in ivory. After 2 years there, he abandoned the store and finally settled at Schoemansdal in the northern Transvaal (Pienaar 1990) in 1853. The outcome of his ventures hints at elephant numbers too low to sustain trade.

H. T. Glynn (undated) was one of the few lowveld pioneers who wrote of his experiences between 1873 and 1925. He hunted extensively in the region but speaks only of elephant hunting much further north in Mozambique. Regular hunting parties started to exploit the Kruger area only from about 1870. Elephants still apparently lived in the Kruger area around 1880, but by 1896 they were no longer found (Vaughn Kirby 1896).

In his chronicle of the decline of the elephant populations of southern Africa, Bryden (1903) gives specific accounts of many hunters and the areas in which they operated. He makes no mention of the lowveld. Significantly, none of the hunters who left any record (e.g., Selous 1881; Finaughty 1916) hunted elephants in the Kruger area. Those who did hunt there (Vaughn Kirby 1896; Glynn, undated) did not record elephant sightings.

TABLE 16.1

Estimates of the elephant population in Kruger Park from 1903 to 1967.

YEAR	NUMBER	NATURE OF ESTIMATE	SOURCE
1903	0	Estimate	Stevenson-Hamilton (1903a, 1903b)
1905	10	Estimate	Stevenson-Hamilton (1905)
1908	25	Estimate	Stevenson-Hamilton (1909)
1925	100	Estimate	Stevenson-Hamilton (1925)
1931	135	Estimate	c.f. Pienaar et al. (1966)
1932	170	Estimate	Stevenson-Hamilton (1932)
1933	200	Estimate	Stevenson-Hamilton (1933)
1936	250	Estimate	Stevenson-Hamilton (1936)
1937	400	Estimate	Stevenson-Hamilton (1937)
1946	450	Estimate	Sandenbergh (1946)
1947	560	Estimate	c.f. Pienaar et al. (1966)
1954	740	Estimate	Steyn (1958)
1957	1,000	Estimate	Steyn (1958)
1960	1,186	Aerial survey	c.f. Pienaar et al. (1966)
1962	1,750	Fixed-wing survey	Pienaar (1963)
1964	2,374	Partial helicopter count	Pienaar et al. (1966)
1967	6,586	Helicopter count	Pienaar (1967)

In sum, it seems likely that the few elephants inhabiting what is now Kruger were shot to extinction between 1880 and 1896. There is no record of the number of elephants that may have inhabited the Kruger area before the arrival of Europeans or of the amount of ivory that may have come from the area. This, along with the evidence from San paintings and baobabs, suggests that elephant densities were low, even during the Mapungubwe era around AD 900. Those that lived at the time of the arrival of Europeans were extirpated by them before the proclamation of the Sabi Game Reserve at the beginning of the century.

After Proclamation as a Game Reserve in 1898

For the first few years of the reserve's existence, its warden reported no elephants (Stevenson-Hamilton 1903a, 1903b). Stevenson-Hamilton (1905) first reported their presence near the confluence of the Letaba and Olifants rivers. Thereafter numbers increased rapidly until 1967 (Table 16.1). Immigration from Mozambique probably contributed to the dramatic increase between 1960 and 1967.

Figure 16.1 shows the recolonization of Kruger. The dates and localities are from the diaries and reports of the earlier rangers and wardens. Northward colonization took until 1945 (at an average rate of 7.3 km/year). Southward colonization was slightly slower, taking until 1958 (at an average rate of 5.4 km/year).

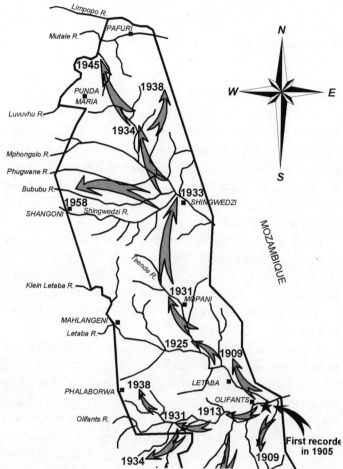

FIGURE 16.1. The recolonization of Kruger by elephants after the arrival of the first warden in 1903.

The Elephant Management Era

As early as the 1940s, Kruger staff discussed the control of elephant numbers (Steyn 1942). In 1965, a symposium on overprotection convened by the National Parks Board of Trustees recommended that the elephant population should be held at its current level. At that time, the estimate for the elephant population was 6,586 (this figure was later rounded to 7,000). This recommendation was approved (National Parks Board of Curators 1966). Culling started in 1967, and although Smuts (1974) reviewed the logistics of the operation, the management policy itself was documented only in 1986 (Joubert

1986). The policy lasted until 1994, when culls ceased pending a review of the policy. This review was finalized in 1998 (Whyte et al. 1999), but at the time of writing the policy still awaits authorization.

Elephant Management between 1967 and 1994

Culls resulted in the killing or removal of 16,201 elephants from Kruger between 1966 and 1994. Between 1994 and 2001, a further 465 individuals were removed from the park (Table 16.2). However, the mean intrinsic rate of increase for the 31 years between 1967 and 2000 was 7.5 percent per year. After the isolation of the park through fencing in 1976, this rate declined to 6.6 percent, suggesting that population growth may have been enhanced by immigration from Mozambique.

Culling as a management option raises ethical, social, and economic problems (see Cumming et al. 1997; Butler 1998; Whyte et al. 1998; Whyte 2001; Whyte and Fayrer-Hosken in press). The culling of elephants from 1967 to 1994 kept the population at between 7,000 and 8,500 individuals, but was this the optimum population range? If so, was the strategy used the best way to achieve that range?

Annual aerial surveys provided the information to determine culling quotas to keep elephant densities within the suggested range (Joubert 1986). From 1967 to 1984, culls removed elephants from areas throughout the Kruger, but from 1985 to 1994, the park was divided into four management regions. The year's quota came from only one region, and it removed 6–32 percent of a region's subpopulation. The fraction of the total numbers removed through culling in a given year ranged from 0.2 percent to 20.9 percent. These culls did not halt elephant population growth, and successive years of decrease in numbers were followed by successive years of increase in numbers.

There is evidence of density dependence. At densities >0.37 elephants/km^2, the numbers tended to decrease in the year after the count (van Aarde et al. 1999). This density corresponds to ~8,000 in the park as a whole. This was within the limits set for the population by park managers (Smuts 1975; Joubert 1986). This density is much lower than the 0.57 elephants/km^2 proposed by Fowler and Smith (1973) and the 1.19 elephants/km^2 suggested by Ambruster and Lande (1993) for semiarid regions. The time that elapsed since the cull also affected the regional population change (van Aarde et al. 1999). Relative to the numbers immediately after the cull, numbers declined sharply in the year of the cull. Mean values a year later were higher than expected, and values during the third and fourth year were approximately the same. It seems that in addition to having been reduced by the cull, elephants initially moved out of regions where others had been culled. However, 1 year later this was followed by an increase in population numbers well beyond the levels expected

TABLE 16.2

Annual elephant census totals and culling quotas in Kruger since the initiation of the census and culling programs in 1966 and numbers removed from the population.

YEAR	CENSUS TOTAL	CULLING QUOTA	TOTAL CULLED	JUVENILES TRANSLOCATED	FAMILY UNITS TRANSLOCATED	ADULT BULLS TRANSLOCATED	TOTAL REMOVED AFTER CENSUS
1966	—	—	—	26	—	—	26
1967	6,586	650	355	—	—	—	355
1968	7,701	1,230	460	—	—	—	460
1969	8,312	1,408	1,160	—	—	—	1,160
1970	8,821	2,093	1,846	—	—	—	1,846
1971	7,916	889	602	—	—	—	602
1972	7,611	618	608	—	—	—	608
1973	7,965	738	732	—	—	—	732
1974	7,702	853	764	—	—	—	764
1975	7,408	601	567	—	—	—	567
1976	7,275	350	285	—	—	—	285
1977	7,715	663	544	26	—	—	570
1978	7,478	392	348	35	—	—	383
1979	—	380	322	48	—	—	370
1980	7,454	395	356	55	—	—	411
1981	7,343	71	16	0	—	—	16
1982	8,051	555	427	46	—	—	473
1983	8,678	2,229	1,290	66	—	—	1,356
1984	8,273	1,890	1,289	88	—	—	1,377
1985	6,887	369	268	101	—	—	369
1986	7,617	495	404	94	—	—	498
1987	6,898	305	245	59	—	—	304
1988	7,344	367	273	83	—	—	356
1989	7,468	367	281	85	—	—	366
1990	7,287	367	232	132	—	—	364
1991	7,470	367	218	140	—	—	358
1992	7,632	350	185	150	—	—	479
1993	7,834	577	308	74	66	—	390
1994	7,806	600	177	31	66	—	356
1995	8,064	0	44	0	169	—	169
1996	8,320	0	18	0	60	—	60
1997	8,371	0	5	0	6	34	40
1998	8,869	0	0	0	14	20	34
1999	9,152	0	0	0	7	17	24
2000	8,356	0	0	0	51	22	73
2001	9,276	0	0	0	47	18	65
2002	10,459	0	0	0	—	—	—
Total	—	20,169	14,629	1,339	486	111	16,666

from reproduction alone, probably as a consequence of further movements within the population as a whole. These densities exceeded those before the cull and could have resulted in an increase in regional elephant impacts and therefore negating any expected benefits derived from the cull. Culling clearly disrupts local dynamics.

Impact of Elephants on Kruger Ecosystem Heterogeneity

The first mention of any significant elephant impact was of the almost complete eradication of the stands of aloes (*Aloe marlothi*) from the Doispane and Sabie River areas in 1959 (Biologiese Afdeling 1959). Concerns over the increasing impacts of elephants on vegetation led to a study in Kruger on this subject by van Wyk and Fairall (1969). They concluded that utilization was generally low to moderate and only in a few small areas did it reach alarming proportions. Furthermore, because the most severe damage occurs in areas utilized by large concentrations of elephants during the dry season, they proposed that the highest number of elephants that could be carried would be $0.29/km^2$ (i.e., 6,000 elephants) if total destruction of the vulnerable areas near water was not to result. This conclusion may have been sensible, but the authors lacked the data to support it.

The damage on marula trees (*Sclerocarya birrea*) was suggested to be a recent phenomenon, and elephant impacts were higher closer to roads than further away (Coetzee et al. 1979). In some stands, 6.5 percent of all trees were felled or ring-barked in a single season, but this was not considered to constitute an immediate threat. Rangers' reports of severe elephant damage to knobthorn trees (*Acacia nigrescens*) in the Punda Maria area led to an investigation by Engelbrecht (1979). Of a sample of 951 trees, 64.3 percent were damaged and 27.7 percent were dead or dying. He concluded that this was also a recent phenomenon and that if the rate of damage continued, elephants would destroy most of the large trees.

Aerial photographs showed that from 1944 to 1974, the number of mature trees (mainly marulas and knobthorns) declined to 6.4 percent of their 1944 densities in the Satara area of Kruger (Viljoen 1988). In both areas most of the decline occurred between 1965 and 1974, when elephant densities increased markedly. He concluded that increasing elephant numbers could have contributed to the decline in tree numbers and that fire prevented their regeneration.

The changes in the density of large trees on four of Kruger's major landscape types from aerial photographs were compared for 1940 and 1960 and for 1960 and 1986–1989 (Trollope et al. 1998). The landscapes (after Gertenbach 1983) selected were

Landscape 5: Mixed *Combretum* and *Terminalia sericea* woodland on granitic soils
Landscape 12: *Colophospermum mopane* and *Acacia nigrescens* savanna on granite
Landscape 17: *Sclerocarya birrea* and *Acacia nigrescens* savanna on basaltic soils
Landscape 23: *Colophospermum mopane* shrubveld on basaltic soils

These four landscapes represent 95 percent of Kruger's area. On granitic soils, there were no significant changes in the density of large trees between 1940 and

1960, whereas a moderate decline occurred on basalts. Conversely, between 1960 and 1986–1989, there was a dramatic decline in density in all four landscapes. These changes in the woody vegetation did not involve a decrease in species diversity but rather a change in structural diversity in which the woody vegetation was converted to short woodland because of a declining density of large trees. Trollope et al. (1998) concluded that elephants had been primarily responsible for the killing of trees >3 m in height in the period after 1960, and the recruitment of large trees was prevented by an increase in the frequency of planned burning. They wrote, "If it is desirable to prevent further structural changes to the woody vegetation, then the current density of elephants should not be allowed to increase, and the frequency of burning should be significantly reduced" (p. 109), and "this conclusion is drawn mindful of the fact that the current practice of limiting the elephant population to approximately 7,000 has not prevented a decline in the structural diversity of the woody vegetation of the Kruger" (p. 109).

The changes in the woody cover and density of shrubs and trees have been assessed from aerial and fixed-point photographs (Eckhardt et al. 2000). Cover increased by 12 percent on granitic soils but declined by 64 percent on basalts because of a 38 percent decline in trees >5 m high between 1984 and 1996. The density of tall trees also decreased on granites but only by 15 percent. They attributed the increase in overall cover on granites to increases in density of small trees and shrubs. Increased mortality (mainly by elephants) and declining recruitment (mainly by fire) appeared to have caused these trends.

The marula was studied in the same four Kruger landscapes, and it was concluded that in Landscape 23, the marula population had become extinct (although these trees did occur in the Nwaxitshumbe rare antelope enclosure, where they are protected from elephants; Jacobs and Biggs 2002a, 2002b). In Landscape 12, there was no recruitment of individuals into the upper canopy. Existing mature trees are predominantly at the end of their life cycle, and extinction is a concern as the seed resource disappears. Landscapes 5 and 17 appear to have a healthy population structure, but more than 60 percent of the trees >8 m high suffered "extreme elephant damage" and 70 percent showed some damage. Therefore, the marula is at risk in all of these landscapes, which is in sharp contrast to the earlier findings of Coetzee et al. (1979). The continued high elephant population densities have had a severe impact on marulas, which is likely to increase as long as the new elephant management plan awaits authorization for implementation.

What Should Be the New Elephant Policy?

The results presented in this chapter pose an obvious dilemma. Historical records all point to there being very few elephants in Kruger. Yet, paradoxically, elephants can rapidly achieve high densities at which they effect marked changes that simplify and homogenize plant and animal communities.

Why were the numbers so low historically? Indigenous Africans were capable of killing elephants (Alpers 1975), but we do not know what impact they had on elephant populations. Elsewhere in the world, human invasions of the Americas, Madagascar, and Australia eliminated almost all of the larger vertebrate species quickly, extensively, and with only Stone Age technologies (Flannery 2001). The Polynesian expansion across the Pacific even eliminated the majority of the region's land birds and almost all its large ones (Pimm et al. 1995a). Globally, the interesting question has always been why Africa's megafauna survived rather than why these other continents lost theirs. The history of the region's elephants suggests that some species did not survive locally. It is not difficult to posit the mechanisms involved. There is no evidence to suggest that the Kruger area may have been more arid during the past few hundred years, but a selective kill targeted at young female elephants would need to remove only 3 to 6 percent of the population to stabilize it (Whyte et al. 1998) and slightly higher percentages to cause its decline. That does not seem an insurmountable task even for a small human population, who probably found young females easier targets than young males or adults.

What are the consequences of these results for the goal of restoring original elephant numbers and the associated ecosystem heterogeneity? The state of Kruger before hominids had the tools to hunt is so far back in time that it lacks any relevance to modern conditions of climate and human settlement. Restoring Kruger to an ecosystem with few San or other indigenous people controlling the numbers of a few elephants is a romantic myth worthy of Rousseau, not a practical recommendation to managers. The option of maintaining very low elephant numbers by culls eventually would entail few animals being slaughtered. However, the transition to those low numbers would entail a large cull and, for that reason alone, undoubtedly generate strong opposition. Put another way, the demands of ecotourists, the need to protect a species that humanity has exterminated across most of its range, and the vigorous opposition to culls of even modest numbers of elephants by some elements of society have required Kruger to maintain historically unprecedented numbers of elephants. Managers should explicitly recognize the consequences of that choice.

After rejecting the one historically defensible option of very few elephants, managers have also rejected a long-held second option: culls to restrain the elephant numbers within some predetermined bounds. We have commented on this elsewhere (Whyte et al. 1998; van Aarde et al. 1999; Pimm and van Aarde 2001). In brief, we first show that killing or sterilizing a small number of prepubertal females could stabilize the population. Although this avoids the large kills of the past, it does not avoid them altogether, nor the considerable operation that this entails.

Second, the culls maintained a high rate of population growth. Rotating the culls across four regions of Kruger showed that regionally high densities often did decline naturally through density-dependent mortality. Holding off culls until a second census found high numbers would sometimes have let such mortality naturally reduce the numbers without human intervention. That said, since culling stopped, the numbers have grown to their highest levels ever (Tables 16.1 and 16.2). Density-dependent mortality probably is an imperfect mechanism on which Kruger's plant and animal communities may not be able to rely; some species may be eradicated before the elephant population stabilizes.

Finally, any suggestion that contraception is a practical alternative does not apply to a large population in a large area such as Kruger. The numbers that would need to be treated far exceed Kruger's logistic and economic resources. Whereas the previous policy (Joubert 1986) aimed at limiting elephant numbers across the whole of Kruger, the new policy (Whyte et al. 1998) has divided Kruger into six management regions. Two of these regions, at the northern and southwestern extremities of the park, will serve as botanical reserves to protect rare, endangered, or otherwise botanically important species and communities. Two other regions in the central part of the park are to be high–elephant impact zones where no management of the elephants will take place. If density-dependent mortality is not effective, the elephant population of these zones might increase at around 7 percent per year. The final two, situated between the high–elephant impact zones and the botanical reserves, will be classified as low–elephant impact zones in which the populations will be systematically reduced at a rate of 7 percent per year. By design, the increasing and declining elephant populations will have important and different impacts. The species that tolerate high elephant densities will be different from those that thrive where densities are low, and it is hoped that the full spectrum of species indigenous to the Kruger will be accommodated. This solution is likely to appease everyone: it reduces the need for culling while protecting some plant and animal communities from the depredations of elephants. And it probably will please no one because some culls will continue and some communities will not be protected. The singular advantage is that it is explicitly experimental. It will allow elephant numbers to increase to whatever limit density dependence sets, allowing an assessment of the damage high numbers may cause while protecting some areas in the event that this damage is unacceptably high. Shifts in the ecosystem's heterogeneity will be monitored through programs developed to detect such change. Limits to change will be set under the concept of thresholds of potential concern (TPCs; Whyte et al. 1999). In many instances, the knowledge or understanding of the system is not adequate to know how much elephant-induced change can be regarded as acceptable. TPCs recognize that such knowledge or understanding is limited, and a limit to change is defined on the best available knowledge. A TPC still allows environmental change up to a predetermined threshold at which managers must officially become concerned, and the situation must then be reviewed. The outcome of such a review could result in the resetting of the TPC (to allow for further change), or management of some form must be implemented to prevent further change. It is envisaged that once TPCs are exceeded, as a result of elephant densities that are either too high or too low, then the management options in the zones should be reversed: the high–elephant impact zones will become low–elephant impact zones and vice versa. Should the policy prove successful, this will allow for long-term fluctuations in elephant population densities and the associated fluctuations in their impacts. To be successful, we suggest the caveats that follow from our discussion so far. First, there is abundant evidence that elephants can simplify habitats in ways that diminish biodiversity. Careful quantitative monitoring of the long-term effects of high densities is essential. Second, experience suggests that any culls will be controversial. Smaller interventions are possible than those in the past if managers select which animals to remove and to allow time for the effects of density dependence.

Conclusion

Throughout its history, the Kruger ecosystem has experienced a wide range of elephant densities, from complete extinction, through low densities for much of its history, up to the current high levels. The desire to protect a species exterminated over large parts of its historical range motivates the choice to maintain historically unprecedented elephant densities. In making this choice, managers run the risk of simplifying Kruger's ecosystems with a concomitant loss of species. Preserving those species and, indeed, heterogeneous ecosystems and the species that depend on them, is a sensible management goal. Furthermore, no species is known to have been lost from this system as a result of elephant impacts. This might be only a matter of time. As elephant numbers have increased since Kruger was established, there have been significant changes. Some of these could ultimately result in a major simplification of habits and the consequent loss of species. Recognizing this, a new elephant management policy allows different directional changes that will contribute to ecosystem heterogeneity and patchiness. Although authorization for the implementation of this policy is pending, it still demonstrates a willingness to conform to the principles of adaptive management, a characteristic we consider crucial to the long-term maintenance of Kruger's ecosystems.

References

Alpers, E. A. 1975. *Ivory and slaves.* Berkeley: University of California Press.
Ambruster, P., and R. Lande. 1993. A population viability analysis for African elephant (*Loxodonta africana*): how big should reserves be? *Conservation Biology* 7:602–610.

Barnes, R. F. W., G. C. Craig, H. T. Dublin, G. Overton, W. Simons, and C. R. Thouless. 1999. *African elephant database.* Occasional Paper of the IUCN Species Survival Commission No. 22. Gland, Switzerland: IUCN.

Biologiese Afdeling. 1959. *Jaarverslag, Nasionale Krugerwildtuin.* Unpublished internal memorandum. South African National Parks, Skukuza, South Africa.

Bourliére, F., and M. Hadley. 1983. Present day savannas: an overview. In F. Bourliére (ed.), Ecosystems of the World Series, Vol. 13: *Tropical savannas.* Amsterdam: Elsevier.

Bryden, H. A. 1903. The decline and fall of the South African elephant. *Fortnightly Review* 79:100–108.

Butler, V. 1998. Elephants: trimming the herd. *BioScience* 48:76–81.

Coetzee, B J., A. H. Engelbrecht, S. C. J. Joubert, and P. F. Retief. 1979. Elephant impact on *Sclerocarya caffra* trees in *Acacia nigrescens* tropical plains thornveld of the Kruger National Park. *Koedoe* 22:39–60.

Cooke, C. K. 1969. *Rock art of southern Africa.* Cape Town: Books of Africa.

Cumming, D. H. M., M. B. Fenton, I. L. Rautenbach, R. D. Taylor, G. S. Cumming, M. S. Cumming, J. M. Dunlop, A. G. Ford, M. D. Hovorka, D. S. Johnston, M. Kalcounis, Z. Mahlangu, and C. V. R. Portfors. 1997. Elephants, woodlands and biodiversity in southern Africa. *South African Journal of Science* 93:231–236.

Dublin, H. T. 1995. Vegetation dynamics in the Serengeti-Mara ecosystem: the role of elephants, fire and other factors. Pages 71–90 in A. R. E. Sinclair and P. Arcese (eds.), *Serengeti 2: dynamics, management and conservation of an ecosystem.* Chicago: University of Chicago Press.

Dublin, H. T., A. R. E. Sinclair, and J. McGlade. 1990. Elephants and fire as causes of multiple stable states in the Serengeti-Mara woodlands. *Journal of Animal Ecology* 59:1147–1164.

Eckhardt, H. C., B. W. van Wilgen, and H. C. Biggs. 2000. Trends in woody vegetation cover in the Kruger National Park, South Africa, between 1940 and 1998. *African Journal of Ecology* 38:108–115.

Eloff, J. H. 1990a. Swart indringing en kolonisasie in die Laeveld. Pages 28–48 in U. de V. Pienaar (ed.), *Neem uit die verlede.* Pretoria: South African National Parks.

Eloff, J. H. 1990b. Toe die Laeveld nog woes en leeg was. Pages 5–17 in U. de V. Pienaar (ed.), *Neem uit die verlede.* Pretoria: South African National Parks.

Engelbrecht, A. H. 1979. Olifantinvloed op *Acacia nigrescens* bome in 'n gedeelte van die Nasionale Krugerwildtuin. *Koedoe* 22:29–38.

English, M. 1990. Die rotskuns van die Boesmans (San) in die Nasionale Krugerwildtuin. Pages 18–24 in U. de V. Pienaar (ed.), *Neem uit die verlede.* Pretoria: South African National Parks.

Finaughty, W. 1916. *Recollections of William Finaughty: elephant hunter.* Bulawayo, Zimbabwe: Books of Rhodesia.

Flannery, T. 2001. *The eternal frontier: an ecological history of North America and its peoples.* New York: Atlantic Monthly Press.

Fowler, C. W., and T. Smith. 1973. Characterising stable populations: an application to African elephant populations. *Journal of Wildlife Management* 37:513–524.

Gertenbach, W. P. D. 1983. Landscapes of the Kruger National Park. *Koedoe* 26:9–122.

Glynn, H. T. Undated. *Game and gold.* London: Dolman Printing.

Herremans, M. 1995. Effects of woodland modification by African elephant (*Loxodonta africana*) on bird diversity in northern Botswana. *Ecography* 18:440–454.

Jacobs, O. S., and R. Biggs. 2002a. The impact of the African elephant on marula trees in the Kruger National Park. *South African Journal of Wildlife Research* 32(1):13–22.

Jacobs, O. S., and R. Biggs. 2002b. The status and population structure of the marula in the Kruger National Park. *South African Journal of Wildlife Research* 32:1–12.

Johnson, C. F. 1998. *Vulnerability, irreplaceability and reserve selection of the elephant-impacted flora of the Addo Elephant National Park, Eastern Cape, South Africa.* Unpublished M.Sc. thesis, Rhodes University, Grahamstown, South Africa.

Johnson, C. F., R. M. Cowling, and P. B. Phillipson. 1999. The flora of the Addo Elephant National Park, South Africa: are threatened species vulnerable to elephant damage? *Biodiversity and Conservation* 8:1447–1456.

Joubert, S. C. J. 1986. *Masterplan for the management of the Kruger National Park.* Unpublished internal memorandum. South African National Parks, Skukuza, South Africa.

Kelly, H. L. P. 2000. *The effect of elephant utilisation on the* Sterculia rogersii *and* Adansonia digitata *populations of the Kruger National Park.* Unpublished M.Sc. thesis, University of Pretoria, Pretoria, South Africa.

Leuthold, W. 1977. Changes in tree populations of Tsavo East National Park, Kenya. *East African Wildlife Journal* 15:61–69.

Lewis-Williams, J. D. 1996. *Discovering southern African rock art.* Cape Town: David Philip.

Lewis-Williams, J. D., and T. A. Dowson. 1989. *Images of power: understanding Bushman rock art.* Johannesburg: Southern Book Publishers.

Lombard, A. T., C. F. Johnson, R. M. Cowling, and R. L. Pressey. 2001. Protecting plants from elephants: Botanical reserve scenarios within the Addo Elephant National Park, South Africa. *Biological Conservation* 102:191–203.

Moolman, H. J., and R. M. Cowling. 1994. The impact of elephant and goat grazing on the endemic flora of South African succulent thicket. *Biological Conservation* 68:53–61.

National Parks Board of Curators. 1966. *Minutes of the meeting of 22 March 1966.* Typescript document. National Parks Board, Pretoria, South Africa.

Page, B. R. 1999. *Detecting extirpation and changes in abundance in woody species: a case study from the N.E. Tuli Block, Botswana.* Poster presentation in Workshop on Long Term Ecological Monitoring in southern Africa, Skukuza, South Africa.

Pienaar, U. de V. 1963. Large mammals of the Kruger National Park: their distribution and present-day status. *Koedoe* 6:1–137.

Pienaar, U. de V. 1967. *'n Lugsensus van olifante en ander grootwild in die hele Krugerwildtuin gedurende September 1967.* Unpublished internal memorandum. South African National Parks, Skukuza, South Africa.

Pienaar, U. de V. 1990. Baanbrekers en jagters in die Laeveld. Pages 172–201 in U. de V. Pienaar (ed.), *Neem uit die verlede.* Pretoria: South African National Parks.

Pienaar, U. de V. 1996. *Neem uit die verlede.* Pretoria: South African National Parks.

Pienaar, U. de V., P. van Wyk, and N. Fairall. 1966. An aerial census of elephant and buffalo in the Kruger National Park and the implications thereof on intended management schemes. *Koedoe* 9:40–107.

Pimm, S. L., M. P. Moulton, and J. Justice. 1995a. Bird extinctions in the central Pacific. Pages 75–88 in J. H. Lawton and R. M. May (eds.), *Extinction rates.* Oxford: Oxford University Press.

Pimm, S. L., G. J. Russell, J. L. Gittleman, and T. M. Brooks. 1995b. The future of biodiversity. *Science* 269:347–350.

Pimm, S. L., and R. J. van Aarde. 2001. Population control: African elephants and contraception. *Nature* 411:766.

Plug, I., and E. A. Voigt. 1985. Archaeozoological studies of Iron Age communities in southern Africa. *Advances in World Archaeology* 4:189–238.

Preller, G. S. 1917. *Dagboek van Louis Trichardt (1836–1838)*. Bloemfontein, South Africa: Het Volksblad-drukkerij.

Punt, W. H. J. 1990. Die eerste blankes besoek die Laeveld. Pages 67–76 in U. de V. Pienaar (ed.), *Neem uit die verlede*. Pretoria, South Africa: South African National Parks.

Rowland-Jones, M. 1955. *Agenda of the Meeting of the South African National Parks: 29–30 June 1955. Item 37: Memorandum oor die verspreiding van Olifante in die Nasionale Krugerwildtuin*. Unpublished internal memorandum. South African National Parks, Skukuza, South Africa.

Sandenbergh, J. A. B. 1946. *Kruger National Park, Warden's Annual Report 1946*. Unpublished internal memorandum. South African National Parks, Skukuza, South Africa.

Selous, F. C. 1881. *A hunter's wanderings in Africa*. London: Richard Bentley and Sons.

Smuts, G. L. 1974. Game movements in the Kruger National Park and their relationship to the segregation of sub-populations and the allocation of culling compartments. *Journal of the Southern African Wildlife Management Association* 4:51–58.

Smuts, G. L. 1975. Reproduction and population characteristics of elephants in the Kruger National Park. *Journal of the Southern African Wildlife Management Association* 5:1–10.

Stevenson-Hamilton, J. 1903a. *Game preservation: Transvaal administration reports for 1903*. Unpublished internal memorandum. South African National Parks, Skukuza, South Africa.

Stevenson-Hamilton, J. 1903b. *Report on Singwitsi Game Reserve. Transvaal administration reports for 1903*. Unpublished internal memorandum, Sabie Bridge. South African National Parks, Skukuza, South Africa.

Stevenson-Hamilton, J. 1905–1925. *Annual reports on the government game reserves*. Unpublished internal memoranda. South African National Parks, Skukuza, South Africa.

Stevenson-Hamilton, J. 1932–1937. *Kruger National Park, warden's annual reports*. Unpublished internal memoranda. South African National Parks, Skukuza, South Africa.

Steyn, L. B. 1942. *Extract from diary, 21 May, 1942*. South African National Parks Archives, South African National Parks, Skukuza, South Africa.

Steyn, L. B. 1958. *Jaarverslag van die Opsiener: Nasionale Krugerwildtuin vir die tydperk 1 April 1957 tot 31 Maart 1958*. Unpublished internal memorandum. South African National Parks, Skukuza, South Africa.

Trollope, W. S. W., L. A. Trollope, H. C. Biggs, D. Pienaar, and A. L. F. Potgieter. 1998. Long-term changes in the woody vegetation of the Kruger National Park with special reference to the effects of elephants and fire. *Koedoe* 41:103–112.

van Aarde, R .J., I. J. Whyte, and S. Pimm. 1999. Culling and the dynamics of the Kruger National Park elephant population. *Animal Conservation* 2:287–294.

van Wyk, P., and N. Fairall. 1969. The influence of the African elephant on the vegetation of the Kruger National Park. *Koedoe* 12:66–75.

Vaughn Kirby, F. 1896. *In haunts of wild game*. Edinburgh: Blackwood and Sons.

Viljoen, A. J. 1988. Long term changes in the tree component of the vegetation in the Kruger National Park. Pages 310–315 in I. A. W. Macdonald and R. J. M. Crawford (eds.), *Long-term data series relating to southern Africa's renewable natural resources*. South African National Scientific Programmes Report No. 157. Council for Scientific and Industrial Research, Pretoria, South Africa.

Western, D., and H. Gichohi. 1989. Segregation effects and the impoverishment of savanna parks: the case for ecosystem viability analysis. *African Journal of Ecology* 31:269–281.

Whyte, I. J. 2001. *The conservation management of the Kruger National Park elephant population*. Unpublished Ph.D. thesis. University of Pretoria, Pretoria, South Africa.

Whyte, I. J. 2001. Headaches and heartaches: the elephant management dilemma. Pages 293–305 in D. Schmidtz and E. Willot (eds.). *Environmental Ethics: What really matters, what really works*. New York: Oxford University Press.

Whyte, I. J., and R. Fayrer-Hosken (in press). Playing elephant god: ethics of managing wild elephant populations. In K. Christen and C. Wemmer (eds.) *Never forgetting: Elephants and Ethics*. Washington: Smithsonian Press.

Whyte, I. J., H. C. Biggs, A. Gaylard, and L. E. O. Braack. 1999. A new policy for the management of the Kruger National Park's elephant population. *Koedoe* 42:111–132.

Whyte, I. J., P. J. Nel, T. M. Steyn, and N. G. Whyte. 1996. *Baobabs and elephants in the Kruger National Park: preliminary report*. Scientific Report 5/96. South African National Parks, Skukuza, South Africa.

Whyte, I. J., R. van Aarde, and S. L. Pimm. 1998. Managing the elephants of Kruger National Park. *Animal Conservation* 1:77–83.

Woodd, A. M. 1999. A demographic model to predict future growth of the Addo elephant population. *Koedoe* 42:97–100.

Woodhouse, H. C. 1992. *The rain and its creatures*. Rivonia, South Africa: William Waterman.

Woodhouse, H. C. 1996. Elephants in rock art. *The Rhino and Elephant Journal* 10:24–27.

Chapter 17
Wildlife Diseases and Veterinary Controls: A Savanna Ecosystem Perspective

ROY G. BENGIS, RINA GRANT, AND VALERIUS DE VOS

Accounts of diseases and veterinary controls in wildlife management regimes are seldom written from the ecosystem perspective. Here we strive to present such a perspective for Kruger and explore the role of wildlife diseases as responders to and drivers of heterogeneity at different scales. Recent generalized models of adaptive cycles in social, economic, and natural systems (Holling and Gunderson 2002) show long, slow predictive buildups followed by a sudden release and a less predictable reorganization phase. This chapter suggests that disease can act as such a releasing mechanism, allowing ecosystem components such as vegetation communities and nutrients to start new trajectories (Chapter 10, this volume). The importance of regional disease status in the proclamation and maintenance of Kruger is discussed. This is followed by a general account of how disease spreads and interacts with the ecosystem, spatially and temporally. The life histories and behaviors of selected diseases are then discussed in terms of this framework. Veterinary interventions are discussed as being influential drivers in their own right, and newer techniques that minimize these effects are described. For reference throughout the chapter, Table 17.1 summarizes mechanisms of spread for each disease and their spatial and temporal patterns. A disease grouping is proposed that facilitates understanding of ecological links.

History

Historically, many conservation areas in Africa are thought to have acquired their status as a result of the previous or persistent presence of regionally endemic animal or human diseases or vectors, which made these areas unsuitable for animal husbandry–based agriculture or human habitation. The distribution of tsetse flies (*Glossina* spp.), the biological vectors of trypanosomiasis (nagana, sleeping sickness), profoundly limited the distribution of livestock in

TABLE 17.1

Summary of diseases

TYPE OF DISEASE	DISEASE	RATE OF REPRODUCTION	EFFICIENCY OF TRANSMISSION	METHOD OF TRANSMISSION	EPIDEMIC SIZE OF POPULATION NEEDED
Rapidly spreading, contagious infections	Rinderpest	High	High	Aerosol and contact	High (to replace lost hosts)
	Foot-and-mouth disease	High	High	Aerosol	Medium to high
Fatal intermediate spreading, infections	Rabies	Moderate	Moderate	Bite or mucous membrane contamination	Medium to high
	Anthrax	High	Moderate	Ingestion or inoculation	Moderate to high
Vectorborne and other intermediate spreading infections	African horse sickness	High	Moderate	*Culicoides* vector	Medium to high
	African swine fever	High	Moderate	Tampan vector	Medium to high
	Rift Valley fever	High	Moderate	Mosquito vector	Medium to high
	Bovine brucellosis	Moderate	Moderate—long latent infection	Contact	Not critical
	Malignant catarrhal fever	Moderate	Moderate	Contact and congenital	Not critical
Slowly spreading, contagious	Bovine tuberculosis	Moderate	Moderate	Aerosol or ingestion	Medium to high
	Sarcoptic mange	Low	Moderate	Contact	Not critical
Always present	Helminth infections	Many eggs are laid	High	Not critical Ingestion or percutaneous penetration	
Ubiquitous	Other Ectoparasites (ticks)	High	High	Cutaneous contamination	Not critical

discussed in the text.

POTENTIAL CARRIERS INVOLVED	MORBIDITY AND MORTALITY IN WILDLIFE HOST	TEMPORAL SCALE		SPATIAL SCALE AFFECTED (WILDLIFE ONLY)	PRESENT IN KRUGER
		PERSISTENCE IN INDIVIDUAL	ENVIRONMENTAL STABILITY		
None	Acute and often fatal in certain species	Short	Poor	Global, regional, and management unit	Historically
African Buffalo	Clinical signs in noncarrier species	Long in carrier species only	Poor at high temperatures, low humidity, and high ultraviolet radiation	Global, regional, and management unit	Yes
Unknown	Long incubation, acute death	Until death	Poor	Regional to local	No
No	Peracute to acute fatal infection	Spores persist in carcass remains	Extremely stable in environment	Regional to local	Yes
Zebra	No clinical signs in zebra	Intermediate	Poor	Regional	Yes
Tampans and wild porcines	No clinical signs in wild porcines	Intermediate	Intermediate	Regional and management unit	Yes
No	Some abortions recorded	Short	Poor	Regional and management unit	Yes
Persistent infection in some wild ruminants	Some abortions recorded	Long	Poor	Regional and management unit	Yes
Wildebeest	No	Long	Poor	Regional and management unit	Yes
Buffalo, kudu, lechwe	Chronic wasting, coughing, swollen glands	Long	Intermediate	Regional and management unit	Yes
Yes, many	Eliminates weak animals	Long	Intermediate	Local may be regional	Yes
Yes, many	May not cause clinical signs but with heavy burdens will cause morbidity	Long	Stable	Regional and management unit	Yes
Many during short phases of life cycle	Heavy burdens may cause morbidity. May transmit infectious diseases				

Africa and severely inhibited agricultural development and expansion in many regions of the continent (Connor 1994). Many erstwhile or current-day tsetse endemic areas became conservation areas (e.g., Mkuze, South Africa; Moremi, Botswana), resulting in the maintenance of biotic diversity in a heterogeneous patchwork of protected areas.

In part, Kruger also owes its origin to the presence of tsetse fly–transmitted trypanosomiasis and anopheline mosquito–transmitted malaria in the lowveld region of South Africa in the latter half of the nineteenth century (Pienaar 1980). Malaria and Nagana took a heavy toll on colonial settlers and their livestock moving into this area. Hunting in this area and pastoral transhumance from higher elevations were limited to the winter months, when disease vector activity was low. A combination of these vectorborne diseases and persistent tribal conflict in southeastern Africa resulted in the lowveld region being sparsely populated by humans in the 1800s, and the influence of humans and domestic livestock on the habitat was minimal. Kruger also partially owes its continuing existence to the endemic presence of several important wildlife-associated diseases of livestock, including foot-and-mouth disease (FMD), theileriosis, malignant catarrhal fever (MCF), anthrax, African swine fever (ASF), and African horse sickness (AHS), which still make this area unsuitable for livestock farming.

The patchy distribution of protected sub-Saharan savanna ecosystems appears to have increased in the twentieth century as animal health officials attempted to contain and control these wildlife-associated diseases and vectors. In many areas, they succeeded in pushing back the limits of tsetse fly distribution, thus opening up new areas for commercial and subsistence agriculture with resultant shrinkage of wildlife habitats on a regional scale.

Diseases as Drivers or Responders in Savanna Ecosystems

In ecological terms, indigenous diseases can be equated to micropredators, having a similar role to the traditional predators of an ecosystem. Both disease organisms and megapredators may function as population regulators (Prins and Weyerhaeuser 1987) by decreasing animal densities at climatically opportune times or periods of relative overabundance, acting as ecosystem responders (Chapter 2, this volume). On the other hand, certain alien diseases may act as independent, indirect ecosystem drivers (controllers), having major population impacts on herbivores and through them on vegetation (Prins and Weyerhaeuser 1987) and on nutrients (Chapter 10, this volume).

Diseases may affect animal populations at a local, regional, continental, or even global scale. The result may be reduction in absolute numbers or density of animals or species or a skewing of population structure caused by age or sex predilection. Significant disease-related morbidity or mortality of herbivores in a certain geographic area may in turn benefit local predator and scavenger populations in the short to medium term. Both the spatiotemporal distribution of hosts and parasites and their genetic diversity determine how much disease will influence such populations (Figure 17.1).

Heterogeneity in Host Resistance

Endemic stability (the constant presence of a disease in a population, *sensu* Thrushfield 1995) is associated with the genetic selection of resistant animals. In most natural ecosystems, a dynamic relationship is maintained between parasitic organisms, their indigenous hosts, and the environment. High biotic diversity and low population densities generally result in a low incidence of infectious disease and low to moderate macro-parasite burdens.

Not all indigenous or endemic diseases follow the endemic stability pattern. Some exceptions are certain multispecies diseases such as anthrax and rabies, for which completion of life cycles is inherently fatal in most hosts. Clinical outbreaks tend to be temporally and spatially clustered and often are driven by population density. These diseases also are historically and ecologically part of a variety of African ecosystems (de Vos 1994; Swanepoel 1994) and appear to contribute to animal population patch dynamics.

Parasitic Diversity

Sub-Saharan Africa has a large number and diversity of mammals that have coevolved with a much larger diversity of macroparasites and microparasites and their arthropod vectors. Some parasites, such as the anthrax bacillus, are genetically homogenous and stable, with minimal mutation over hundreds of years (Smith et al. 2000). Other microparasites (e.g., FMD viruses and trypanosomes) are genetically heterogeneous and unstable, having a high mutation index, with new subtypes evolving over short time scales (Vosloo et al. 1996). Occasionally, these new strains are more pathogenic than the parent strain, which results in selection pressure on traditional hosts.

Vertebrate hosts, including carnivore-herbivore links, often play an essential role in the life cycle, survival, biodiversity, and heterogeneity of macroparasites because many of these parasites are host- or taxon-specific. In natural ecosystems with a wide variety of species sharing pastures, all may not be susceptible to the same parasite, and the diversity of grazers may lead to a reduction in available parasitic ova and larvae via ingestion by nonpatent hosts.

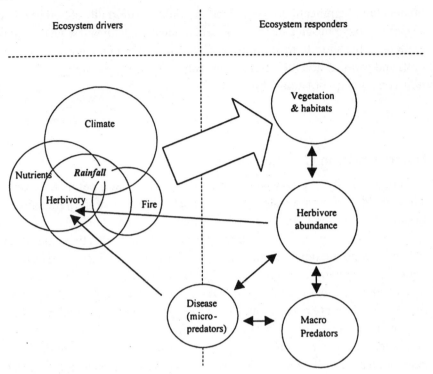

FIGURE 17.1. Diseases can be viewed as the functional equivalents of micropredators in the context of savanna ecosystem dynamics.

When Do Diseases Act as Drivers or Responders?

The state of relative equilibrium between endemic disease and host may be destabilized by climatic factors, by introduction of an alien parasite, by certain host factors, or by environmental disturbance.

Climatic Extremes

Disease patterns often are linked to spatial and temporal characteristics of climatic variation. Seasonal and longer-term climatic cycles and fluctuations affect animal numbers, densities, and distribution, vegetation, habitat and nutrition, fire frequency and intensity, and parasite and disease vector abundance. During dry seasons and cycles many individuals of different species are forced to congregate in habitats with residual nutritious and palatable vegetation and around permanent water-points, creating ideal conditions for disease transmission. Herbivores under some degree of nutritional stress are even more susceptible to infection, and aggregation of animals facilitates close contact disease transmission via body contact, respiratory aerosols, or ingestion of contaminated feed or water. Epidemic FMD and anthrax typically are associated with dry periods in the Kruger savanna system (Bengis et al. 1994; de Vos and Bryden 1996).

On the other hand, during wet cycles an increase in the abundance of winged (arthropod) and flightless (tick and rodent) disease vectors results in the efficient short-, intermediate-, and long-range transmission of infectious agents. Examples of wet season cycle diseases include Rift Valley fever, AHS, theileriosis and other tickborne diseases, and encephalomyocarditis.

The close relationship between ticks and climatic cycles in Kruger is described by Horak in Box 17.1.

Introduction of an Alien Disease

Certain highly contagious alien diseases such as rinderpest, although initially density-fueled, may continue to cause morbidity and mortality in partially depleted populations. Rinderpest is characterized by continuous spatial and temporal spread in associated wildlife as a result of efficient transmission mechanisms in immunologically naive populations.

Another example is bovine tuberculosis (BTB), which has recently entered several African conservation areas, including Kruger. The potential long-term impact of this disease on wildlife populations at sustained high prevalence rates is unknown.

Canine distemper (alien) and rabies (indigenous) may also threaten low-density species or relict threatened populations, as has been reported in wild dogs (Alexander and Appel 1994; van de Bildt et al. 2002; Gascoyne et al. 1993). These small and endangered populations cannot themselves maintain and perpetuate most microparasites and therefore are likely to be threatened only by microparasites persisting in larger reservoir populations such as domestic dogs.

Host Factors

The entry of a foreign host, such as domestic livestock, may introduce alien diseases or parasites into the ecosystem. Conversely, indigenous infections silently cycling in wildlife may be unmasked and cause mortality in domesticated livestock entering the system. This may prompt certain control measures, such as erection of fences or depopulation campaigns, which negatively affect indigenous wildlife.

Another host factor is animal density. When animal densities pass a certain upper threshold, nutrition and social stress are major cofactors that may trigger or sustain a disease outbreak. There is strong evidence from Kruger that anthrax outbreaks are related to kudu densities (de Vos and Bryden 1996). The 1990 outbreak terminated with a 20 percent reduction in the kudu population.

> **BOX 17.1**
> *Ticks, Their Hosts, and Environmental Variation*
>
> IVAN G. HORAK
>
> During the period 1988–2002, 50.8 percent of tick larvae collected from vegetation in the south of Kruger were of the three-host tick *Amblyomma hebraeum*, followed by the one-host tick *Boophilus decoloratus* (21.8 percent) and various other species (27.4 percent altogether). *B. decoloratus* is more specific in its host preference than *A. hebraeum*. Of the tick larvae collected from zebras (*Equus burchelli*), wildebeest (*Connochaetes taurinus*), impalas (*Aepyceros melampus*), bushbuck (*Tragelaphus scriptus*), and kudus (*Tragelaphus strepsiceros*) in the south of Kruger, 56.8 percent were identified as *B. decoloratus* compared to 22.2 percent *A. hebraeum* and 20.9 percent all other species combined. Over a similar period *B. decoloratus* larvae represented only 0.27 percent of tick larvae found on various large and small carnivores, warthogs (*Phacochoerus aethiopicus*), scrub hares (*Lepus saxatalis*), and crowned guineafowls (*Numida meleagris*), while 78.6 percent were *A. hebraeum*. No *A. hebraeum* or *B. decoloratus* larvae were collected from red veld rats (*Aethomys chrysophilus*) or bushveld gerbils (*Tatera leucogaster*) in the same area.
>
> Tick populations in Kruger undergo wide fluctuations in lag-response to climatic variations, with rainfall governing firstly the abundance of hosts and secondly the availability of suitable questing habitat. The severe 1991/92 drought resulted in a 75 percent reduction in impala numbers along the Nwaswitshaka River and in 1992/93 a 65 percent reduction was recorded in the density of all questing ticks collected from the vegetation in that area (Horak et al. 1995). After the same drought, followed by several years of below-average rainfall, the questing *Rhipicephalus appendiculatus* population in the Lower Sabie region had declined to <20 percent of its pre-drought density in 1993/94 and to <1 percent by 1996/97 (Horak et al. 2000). A rodent population eruption, confirmed by an increase in trapping success from 2.9 percent in 1990 to 56 percent in 1994, was followed by a 5-fold increase in adults of the three-host tick *R. simus* (and a 3-fold increase in *R. turanicus*; both species use rodents as hosts for their immature stages), on the vegetation during 1995, compared to their long-term mean densities (Horak et al. 2000). A third way in which climatic variation can influence tick population dynamics is through their life cycles. An unusually warm winter in 1998 (1–2°C warmer than the long-term average) allowed *B. decoloratus* larvae to peak twice in southern Kruger, once in June–July (unusual) and again November–December (usual), implying that the ticks had undergone an extra life cycle that year (Horak et al. 2000).

Outbreaks of sarcoptic mange in many species are linked to density related nutritional or social stress (Pence and Ueckermann 2002), and in Kruger numerous mortalities of entire lion litters were recorded in the 1970s, when lion populations peaked. Sporadic outbreaks of sarcoptic mange, in response to similar environmental stressors, have also been documented in black-backed jackal, impala, giraffe, buffalo, and wildebeest in Kruger. Therefore, although both anthrax and sarcoptic mange can be fatal, they tend to act as ecosystem responders because their clinical expression is dependent and related to other factors such as climate and host density. In other regions of Africa, outbreaks of rabies in jackals, bat-eared foxes, mongooses, and kudu appear to be related to high population densities (Swanepoel 1994; Barnard and Hassel 1981; Box 17.2). Occasionally, under high population densities or other conditions, an organism may expand its host range and be highly pathogenic in a naive host. An example is the recently recognized rodent-associated encephalomyocarditis virus, which crossed over into elephants in Kruger (Grobler et al. 1995).

Environmental Degradation

Practices such as fencing, excessive animal populations, reduction in flow of rivers caused by excessive upstream extraction, poorly sited artificial water provision, excessive or inadequate burning, and any excessive exploitation of natural resources can result in disease effects. A common example, seen on small, fenced game farms that are overstocked, is a density-related increase in ectoparasites and endoparasites with resultant verminosis, tick toxicoses, and infectious disease outbreaks as environmental degradation progresses. The term *sick habitat syndrome* has been coined to describe this unnatural multifactorial environmental health problem (Bengis et al., in press).

Dynamics of Disease Spread

The basic reproductive rate (R_0) at which any parasite multiplies and spreads in a host population determines its epidemiology. In the case of microparasites (e.g., viruses, bacteria, protozoa), R_0 is the average number of secondary infections produced when one infected individual is introduced into a host population in which every individual is susceptible. For macroparasites (helminths and some arthropods), R_0 is the average number of female offspring produced throughout the lifetime of a mature female parasite (Anderson and May 1979). For any parasite to be maintained in a host population, R_0 must equal at least 1. The greater the magnitude of R_0, the more likely the disease is to spread in a population. Diseases with high R_0 values tend to be episodic or epidemic in nature, with frequent, marked amplitude fluctuations; when R_0 values are close to unity, infections exhibit little variation in incidence amplitude, and when R_0 is below 1 the infection dies out.

The host population size necessary for a disease to become established and persist has a direct effect on the likelihood of parasite transmission and therefore on the magnitude of R_0. Every parasite needs a minimum number or den-

BOX 17.2
Rabies

ROY G. BENGIS

Rabies is yet another multispecies disease that is endemic in many areas of sub-Saharan Africa, with sporadic epidemic cycles. It has been diagnosed in 33 carnivorous species and 23 herbivorous species, with a regional variation of dominant epidemiological role players (Swanepoel 1994). Rabies epidemics appear to be linked to periods of overabundance of potential hosts, and it is thought that this disease has not managed to establish itself in the Kruger ecosystem because of the balanced and low-density mix of small predators. In other African conservation areas, particularly where an interface with infected domestic or feral dogs is present, canid rabies has managed to establish itself in wild carnivores, particularly communal burrow-dwelling species such as bat-eared foxes and jackals. Rabies can have a major impact when infection spills over into fragmented populations of endangered social carnivores such as the wild dog and the Ethiopian wolf. Rabies and canine distemper are thought to be partially responsible for the demise of the wild dog population in the Serengeti ecosystem (Gascoyne et al. 1993).

Two rabies outbreaks were also responsible for major mortalities in a recently introduced population of wild dogs in Madikwe National Park in South Africa. After the first outbreak, additional introductions were protected by immunization, but after successful breeding, the recruitment of pups was hampered by a second outbreak of this disease, which appeared to be jackal associated (Hofmeyr et al. 2000).

A rabies outbreak was also documented in kudu in Namibia in 1981 (Barnard and Hassel 1981). A unique nonbite transmission of disease was described, as a result of salivary contamination of browse. This epidemic caused the deaths of 10,000–15,000 kudu over a 3-year period, and abnormally high kudu population densities at the time were cited as being a major contributing factor.

---

sity of hosts to become established (called N_T, and at this level, $R_0 = 1$). Generally, the magnitude of N_T varies inversely with the efficiency with which a parasite is transmitted (Wobeser 2002). More efficiently transmitted parasites need smaller populations for persistence. However, with some efficiently transmitted microparasitic diseases, after infection rate has peaked, disease incidence progressively decreases as infected animals die or become immune, and once R_0 falls below 1, the disease dies out. Large populations with sufficient recruitment of susceptible individuals are needed for these diseases to persist. Other microparasitic diseases with long incubation periods (bovine brucellosis), environmental stability (anthrax), true carrier hosts (FMD), alternate overwintering or amplifier hosts (Rift Valley fever), biological vector transmission (arthropod-borne infections), persistent clinical infection (BTB), and latent infections such as bovine malignant catarrh are able to persist successfully in much smaller populations. Helminths, with a long life span in the individual host and persistent transmission stages in the environment or intermediate host, can persist in even small host populations.

Epidemic Potential

A disease epidemic represents a spatial and temporal clustering of cases (Thrushfield 1995). Diseases with a high reproductive rate (R_0), which spread via aerosols, direct contact, or efficient winged vectors, cause propagating epidemics with high morbidity or mortality in contiguous populations. The course of the disease in the individual usually is short (e.g., rinderpest). Diseases with intermediate epidemic potential often are not directly contagious, and epidemics often assume a focal, clustered, and irregular spatial pattern. The course of the disease in the individual usually is brief (e.g., anthrax). However, others, such as tuberculosis, although highly contagious, are slow-spreading diseases, having a prolonged temporal course in the population, with each infected individual being a potential long-term source of infection. Macroparasitic infections in open systems tend to have low epidemic potential.

Diseases as Drivers of Habitat Change

Herbivores respond to spatial and temporal variation in the environment, from fine to broad scales (McNaughton et al. 1988). The reciprocal interaction between herbivores and their habitats is a process-pattern relationship that is manifest in many landscapes. In Kruger, there is historic evidence that the landscape has changed dramatically in the past 60 years. There appears to have been a progressive increase in woody vegetation cover and a reduction in the areas of open savanna (Chapter 11, this volume). In contrast to the current situation, historical records indicate that sable antelope and other selective feeders were common in the 1930s–1950s, and impala and giraffe were relatively rare. We hypothesize that this can be related to the rinderpest pandemic, which decimated the artiodactyl populations in the lowveld around 1900. This epidemic, coupled with overexploitation by hunters, reduced ungulate populations to the extent that the tsetse flies died out because of the lack of preferred hosts in this region. The increased grass biomass and accumulated fuel load as a result of the reduction in high-density grazers may have increased the intensity and frequency of fires, giving rise to the open savannas recorded in the first few decades of the twentieth century. This habitat was more suitable for bulk (buffalo), shortgrass (wildebeest), and selective (sable antelope) grazers than for mixed feeders (impala) and browsers (giraffe and kudu). Conversely, in East Africa, Prins and van der Jeugd (1993) relate the establishment of *Acacia* seedlings to punctuated epidemics of anthrax. These epidemics reduced numbers of impala, which normally limited seedling establishment.

Disease and System Heterogeneity—The Kruger Experience

Within the conceptual framework discussed above, the Kruger experience presents the following robust examples and illustrations.

Rapidly Spreading Contagious Diseases

Rinderpest, a highly contagious viral disease, is believed to have been introduced into Eritrea by infected cattle from India brought by the Italian armed forces in the late 1800s. This caused a major pandemic, which spread mainly via the movement of infected draught cattle on trade routes in sub-Saharan Africa (1888–1905). It moved westward and southward through contiguous populations and reached southern Africa in 1896 (Mack 1970). In South Africa this disease peaked in 1897–1898. Buffalo, tragelaphids, and wild suids were most severely hit, although alcelaphines and hippotragids were also affected, leaving only relict populations in some areas (Rossiter 1994). In the Kruger lowveld ecosystem this outbreak followed decades of overhunting that had severely depleted the wild ruminants and suids (Stevenson-Hamilton 1957). The timing and other circumstantial evidence link the rinderpest pandemic with the disappearance of the tsetse fly from this region. On a fine scale, rinderpest reduced biodiversity and heterogeneity of wildlife because it affected a wide range of species and may even have caused local extinctions of certain ungulates. On a broader scale, rinderpest resulted in the patchy distribution of isolated subpopulations, thus increasing heterogeneity.

Infectious Diseases with Intermediate Spread Rates

The northern sand and panveld areas of Kruger, in the vicinity of the Levubu and Limpopo drainages, are persistently infected with anthrax organisms. After a susceptible host ingests spores, the bacilli enter the bloodstream and undergo exponential replication, resulting in rapid fatal septicemia in herbivores. Outside the endemic areas, localized to extensive epidemics occur in the Kruger ecosystem, in cycles of 6–20 years (de Vos and Bryden 1996). Extensive anthrax epidemics were documented in 1959–1960, 1970, and 1990–1991 in the north (Figure 17.2) and in 1993 and 1999 in the central district of the park. These epidemics occurred during dry cycles and in seasons when a relative overabundance of herbivores was congregated around stagnant surface water.

Kudu and blowflies play a pivotal role in these epidemics. Blowflies and their larvae (maggots) feed on an infected carcass. Infected discard droplets produced by the blowflies contaminate leaves, which are browsed by kudu that are then infected and die. The adult blowflies that emerge from pupae in the vicinity of the carcass after a few weeks, feed on the new kudu carcasses, reestablishing the infection cycle. Vultures, and to a lesser extent hyenas, are also important disseminators of anthrax because they contaminate water-points after feeding on infected carcasses (de Vos 1994). The creation of artificial surface water appears to facilitate this transmission mode and spread anthrax more homogenously during epidemics (de Vos and Bryden 1996).

Most mammalian species in Kruger are susceptible to anthrax, but kudu and buffalo are particularly vulnerable to infection, due to feeding and drinking behavior. North of the Olifants River, kudu made up 42 percent and buffalo 38 percent of all anthrax-positive carcasses found in 1990. These species constituted 3.7 percent and 15.9 percent, respectively, of the large herbivores counted during the aerial census of 1990–1991. Kudu alone made up 51.7 percent of the total of 1,550 positive carcasses found in 1991.

In Kruger, only locally rare and highly susceptible species such as roan antelope have ever been threatened with local extinction by this disease. Carnivores generally are more resistant to anthrax and tend to develop the localized buccal form of the disease, affecting the soft tissues of the head. Anthrax-naive lions may die acutely from septicemia or develop a protracted (7- to 10-day), fatal infection with typically swollen heads and oral pathology. Lions that have survived initial exposure appear to develop immunity to infection. Anthrax usually spreads throughout an area in an irregular wavelike front, with many spatially clustered cases at the leading edge of the front. The first spring rains result in a dramatic decrease in or total cessation of anthrax cases through physical rinsing and runoff of spores and through herbivore dispersal.

Slow-Spreading Contagious Diseases

The ecosystem effects of diseases such as BTB are not clear. BTB is an alien bacterial disease that entered Kruger about 40 years ago (Bengis et al. 1996). This disease, initially transmitted to buffalo by infected cattle on the southern boundary, slowly spread north at about 5 km/year. The incubation period of BTB is about 2 months. The disease is progressive in the individual, in the herd, and

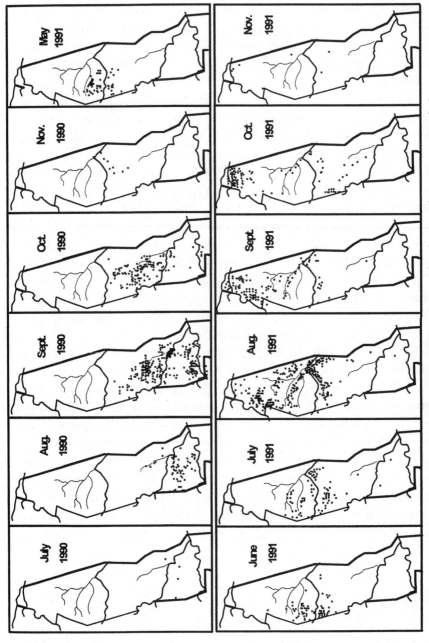

FIGURE 17.2. Spatiotemporal dynamics of the 1990–1991 anthrax outbreak in northern Kruger. Each dot represents a case or a focal cluster of cases. The epidemic was interrupted between November 1990 and May 1991 by the summer rainy season.

between herds and may take years to kill its host. There is no sexual predilection, but the disease prevalence increases with age (de Vos et al. 2001). In the heavily infected (>60 percent) herds, infected calves and yearlings may not survive to reproductive age. BTB infection has spilled over into kudu, baboons, lions, leopards, cheetah (Keet et al. 1996, 2000, 2001), warthogs, genet, and honey badger (Bengis and Keet, unpublished data, 2002). However, only buffalo, kudu, and possibly warthog appear capable of maintaining the infection in the system (Bengis et al. 1996; de Vos et al. 2001; Keet et al. 1996, 2001).

The long-term effects of BTB on animal populations are difficult to predict. There are indications in buffalo that sustained high prevalence of BTB may change population structure by removing the mature and older animals as well as some of the prepubertal cohort. In social and territorial animals with a dominance hierarchy such as lions, the loss of coalition males and dominant pride females could be socially disruptive (Keet, pers. comm., 2002). Sarcoptic mange also falls into this category, and a similar disease caused by the mite *Notoedres cati* also occurs in cheetahs in Kruger.

Much remains to be learned about whether, when, and how to manage diseases and their interactions with other factors (such as surface water availability) in such a way as to maintain desirable ecosystem heterogeneity. Parallel management cues may exist between an alien disease such as BTB and certain alien invasive plants or fish.

Veterinary Controls

Since the emergence of government veterinary services in Africa some ten decades ago, epidemic diseases received the most attention because of the early bias toward livestock farming among both indigenous and colonial stockmen. Many of the silent African endemic infections cycling in wildlife caused major morbidity and mortality in these foreign hosts. The presence of serious or "trade sensitive" livestock diseases that are maintained by wildlife remains a significant barrier to agricultural development (R. A. Kock et al. 2002). In retrospect, many of the disease control measures used to protect livestock at the time would by today's standards be considered immoral and excessive. These included placing disease control fences (M. D. Kock et al. 2002), shooting wildlife disease carriers, bush clearing, and aerial spraying with long-residue pesticides (du Toit 1954). These techniques were successful in that they homogenized the system to favor domestic livestock. Today, improved epidemiological knowledge and skills, better diagnostic tests, and more environmentally friendly drugs allow for more pragmatic control of African endemic diseases.

The first state veterinarian was appointed in Kruger in 1961 to control and contain diseases such as FMD, anthrax, and theileriosis and to research AHS,

ASF, brucellosis, and MCF. Additional research included study of other wildlife infections and parasites (Boomker 1985; Boomker 1990), disease pathology, evaluation of new drugs, and collection of baseline physiological data (Bryden and de Vos 1994).

Fences

Fences usually are erected at the interface between wildlife and domestic livestock populations, separating two land-use practices, to prevent the bidirectional transmission of diseases (Bengis et al. 2002). The Kruger boundaries were fenced between 1960 and 1976. The western and southern boundary fences were erected to prevent the spread of FMD and theileriosis from buffalo to neighboring cattle populations. In pastoral systems in East Africa it has been shown that animal diseases become difficult to control or contain where wildlife and cattle share range (R. A. Kock et al. 2002). This lack of control limits the access of small farmers to international trade markets (Thambi et al., in press). Responsible fencing of remaining conservation areas will allow wildlife-based ecotourism enterprises to exist alongside neighboring communities where livestock disease control is practiced. The Kruger fences have reduced the number of FMD disease outbreaks in neighboring cattle, with no outbreaks recorded between 1983 and 2000.

In the past, fences that had been erected across important wildlife migration routes had major effects on both the migrating animals and plant communities (M. D. Kock et al. 2002). The Kruger western boundary fence, erected in 1961 to control FMD, cut off a large part of the summer grazing areas of the wildebeest and the traditional winter migration routes from the east to the southwest. The dams built in the reduced Kruger summer grazing area, coupled with the large buildup of wildebeest numbers, led to severe trampling of the vegetation. This in turn led to reduction campaigns initiated in 1965, cropping 3,155 wildebeest over 7 years (Whyte and Joubert 1988). This fence was later moved to the outer boundary of the adjoining lowveld private nature reserves in 1994, reopening a traditional movement route.

Wildlife Translocation

Wildlife translocation is used for species introductions, reintroductions, and population augmentation. Translocation is an important conservation tool for reestablishing species distribution ranges and ecosystem biodiversity. To limit the risk of spreading animal diseases, postcapture quarantining of wild animals at place of origin, disease screening, and treatment for ectoparasites and endoparasites are essential requirements for ensuring responsible translocations. All living animals are "biological packages" consisting of the host animal with its attendant passenger macroparasites and microparasites. These parasites may or may not be pathogenic to the host or to other sympatric species at destination, and their translocation should be prevented at all costs. The release of a new infectious agent into a natural ecosystem must be considered almost irreversible.

Since the late 1950s, veterinary controls have prevented the translocation of cloven-hoofed ungulates out of Kruger because of its endemic FMD, ASF, and theileriosis status. Only since the early 1980s, after pachyderms such as elephant, rhinoceros, and hippopotamus were shown to play no role in the epidemiology of FMD (Bengis et al. 1984), was translocation of these species out of the lowveld region permitted.

Since the discovery of BTB in Kruger, only non–cloven-hoofed species in which BTB has never been detected (e.g., elephant, rhino, hippo, and zebra) may be translocated from the Park. Carnivores and primates in which the disease has been detected may be translocated to other areas of South Africa only after appropriate quarantine and disease testing.

The success of wildlife translocation exercises also relies on identifying potential parasites present in the ecosystem at destination. Many mortalities have been recorded as a result of excessive tick burdens, tick toxicoses, and tickborne diseases, when animals from climatically or ecologically unrelated regions have been introduced into the lowveld. Of the 15 species of mammals that have been reintroduced into Kruger, only 3 species have thrived to the point where the exercise can be considered a success. These species (white and black rhino and nyala) were all sourced from Kwazulu-Natal, an area with similar habitats and parasites.

Other Techniques Reducing Conflict between Wildlife and Livestock

Several alternative strategies have been developed that increase the likelihood of wildlife and livestock coexisting with less overt disease transmission risks:

- Accelerated genetic selection of livestock (such as cattle breeds resistant to trypanosomes).
- Vaccination of livestock (e.g., against rinderpest; Kock et al. 1999); vaccination of wildlife usually requires parenteral application techniques such as darts or biobullets, which are technically difficult and expensive.
- Environmentally friendly vector control includes pyrethroid cattle dips or sprays for tick and tsetse control (which do not affect tick-eating oxpeckers); where no cattle are present (e.g., in fenced conservation areas) odor-baited targets impregnated with synthetic pyrethroids are used to attract and control tsetse flies (Vale et al. 1988).
- Sterile insect techniques, as used successfully to curb the spread of the alien New World screwworm (*Cochliomyia hominovoarax*) in north Africa, are now being tested for tsetse flies.

- Breeding of specific disease-free wildlife, such as buffalo that are free of FMD, theileriosis, BTB, and bovine brucellosis (Bengis and Grobler 2000) allows expansion of population distribution ranges without the attendant disease risks to domesticated livestock.

Conclusion

Disease can modulate herbivore populations to release (*sensu* Holling and Gunderson 2002) the potential of other ecosystem components such as vegetation communities and nutrients. This effect of disease depends on ecosystem drivers such as climate and herbivores, so disease can be seen as a responder. We have illustrated a disease classification with, at one extreme, a fast-spreading alien disease (rinderpest) producing mainly homogenizing effects at all but the broadest scales and, at the other extreme, a slow-spreading indigenous disease (sarcoptic mange) with generally more patchy effects. A wide range of possibilities exists between these extremes, depending on the particular disease profile and the specific interacting factors discussed. Different levels of exploitation and reorganization result in different levels of patchiness in vertebrate populations and thus in different levels of released potential.

Most endemic parasites rarely cause fatal disease in their natural hosts but often are highly pathogenic in domestic livestock. This often leads to veterinary controls that can induce important ecosystem and heterogeneity effects through land-use changes, fencing, and other interventions. New control strategies are slowly minimizing negative side effects.

When compared with abiotic and other biotic components that determine savanna ecosystem heterogeneity, the animal disease driver or response usually operates in a local or regional spatial context and at short- to medium-term temporal scales. At these scales disease appears to contribute significantly to the shifting spatiotemporal mosaic in large conservation areas, and value judgments concerning its desirability should be examined with this in mind.

References

Alexander, K. A., and M. J. G. Appel. 1994. African wild dogs (*Lycaon pictus*) endangered by canine distemper disease epidemic amongst domestic dogs near Masai Mara National Reserve, Kenya. *Journal of Wildlife Diseases* 30:481–485.

Anderson, R. M., and R. M. May. 1979. Population biology of infectious disease. Part 1. *Nature* (London) 280:361–367.

Barnard, B. J. H., and R. H. Hassel. 1981. Rabies in kudus (*Tragelaphus strepsiceros*) in South West Africa/Namibia. *Journal of the South African Veterinary Association* 52:309–314.

Bengis, R. G., and D. G. Grobler. 2000. Research into the breeding of "disease free" buffalo. Pages 1032–1033 in *Proceedings of the North American Veterinary Conference, 2000*, Orlando, FL.

Bengis, R. G., R. S. Hedger, V. de Vos, and L. R. Hurter. 1984. The role of the African elephant (*Loxodonta africana*) in the epidemiology of foot-and-mouth disease in the Kruger National Park. Pages 39–40 in *Proceedings of the 13th World Congress on Diseases of Cattle, 1984*, Durban, South Africa.

Bengis, R. G., R. A. Kock, and J. Fischer. 2002. Infectious animal diseases: the wildlife/livestock interface. Pages 53–65 in *Infectious diseases of wildlife: detection, diagnosis and management*. O.I.E. Scientific and Technical Review 21.

Bengis, R. G., N. P. J. Kriek, D. F. Keet, J. P. Raath, V. de Vos, and H. F. A. K. Huchzermeyer. 1996. An outbreak of bovine tuberculosis in free-living buffalo (*Syncerus caffer*–Sparrman) population in the Kruger National Park, a preliminary report. *Onderstepoort Journal of Veterinary Research* 63:15–18.

Bengis, R. G., G. R. Thomson, and D. F. Keet. 1994. Foot-and-mouth disease in impala (*Aepyceros melampus*). Pages 8–9 in *Proceedings of the O.I.E. Conference on Foot-and-Mouth Disease, African Horse Sickness, and Contagious Bovine Pleuropneumonia, 1994*, Gaborone, Botswana.

Bengis, R. G., R. A. Kock, G. R. Thomson, and R. D. Bigalke, In press. Infectious Diseases of animals in sub-Saharan Africa: the wildlife/livestock interface. In J. C. Coetzer and R. C. Tustin. *Infectious Diseases of Livestock in Subsaharan Africa*. Cape Town: Oxford University Press.

Boomker, J. D. F. 1985. Helminths in wild ruminants—co-existence or competition? *South African Journal of Science* 81:11, p. 704.

Boomker, J. D. F. 1990. A comparative study of the helminth fauna of browsing antelope in South Africa. DSc. Thesis, Medical University of South Africa, p. 297.

Bryden, H. B., and V. de Vos. 1994. A scientific bibliography on the national parks of S. A. *Koedoe* supplement 1994:1–133.

Connor, R. J. 1994. African animal trypanosomiasis. Pages 167–205 in J. A. W. Coetzer, G. R. Thomson, and R. C. Tustin (eds.), *Infectious diseases of livestock, with special reference to southern Africa*. Cape Town: Oxford University Press.

de Vos, V. 1994. Anthrax. Pages 1262–1289 in J. A. W. Coetzer, G. R. Thomson, and R. C. Tustin (eds.), *Infectious diseases of livestock, with special reference to southern Africa*. Cape Town: Oxford University Press.

de Vos, V., R. G. Bengis, N. P. J. Kriek, A. Michel, D. F. Keet, J. P. Raath, and H. F. Huchzermeyer. 2001. The epidemiology of tuberculosis in free-ranging African buffalo (*Syncerus caffer*) in the Kruger National Park, South Africa. *Onderstepoort Journal of Veterinary Research* 68:119–130.

de Vos, V., and H. B. Bryden. 1996. Anthrax in the Kruger National Park: temporal and spatial patterns of disease occurrence. *Salisbury Medical Bulletin*, special supplement 87:26–31.

du Toit, R. 1954. Trypanosomiasis in Zululand and the control of tsetse flies by chemical means. *Onderstepoort Journal of Veterinary Research* 26:317–385.

Gascoyne, S., M. K. Laurenson, S. Lelo, and M. Borner. 1993. Rabies in African wild dogs (*Lycaon pictus*) in the Serengeti region. *Journal of Wildlife Diseases* 29:396–402.

Grobler, D. G., J. P. Raath, L. E. Braack, D. F. Keet, G. H. Gerdes, B. J. Barnard, N. P. Kriek, J. Jardine, and R. Swanepoel. 1995. An outbreak of encephalomyocarditis-virus infection in free-ranging African elephants in the Kruger National Park. *Onderstepoort Journal of Veterinary Research* 62:97–108.

Hofmeyr, M., J. Bingham, E. P. Lane, A. Ide, and L. Nel. 2000. Rabies in wild dogs (*Lycaon pictus*) in the Madikwe Game Reserve, South Africa. *The Veterinary Record* 146:50–52.

Holling, C. S., and L. H. Gunderson. 2002. Resilience and adaptive cycles. Pages 25–62 in L. H. Gunderson and C. S. Holling (eds.), *Panarchy: understanding transformations in human and natural systems*. Washington, DC: Island Press.

Horak, I. G., V. de Vos, and Leo Braack. 1995. Arthropod burdens of impalas in the Skukuza region during two droughts in the Kruger National Park. *Koedoe* 38:65–71.

Horak, I. G., A. M. Spickett, and Leo Braack. 2000. Fluctuations in the abundance of *Boophilus decoloratus* and three *Rhipicephalus* species on vegetation during eleven consecutive years. Pages 247–251 in M. Kazimirova, M. Labuda, and P. A. Nuttall (eds.), *Proceedings of the Third International Conference on Ticks and Tick-borne Pathogens: into the 21st Century*. Bratislava: Institute of Zoology, Slovak Academy of Sciences.

Keet, D. F., N. P. J. Kriek, R. G. Bengis, D. G. Grobler, and A. Michel. 2000. The rise and fall of tuberculosis in a free-ranging chacma baboon troop in the Kruger National Park. *Onderstepoort Journal of Veterinary Research* 67:115–122.

Keet, D. F., N. P. J. Kriek, R. G. Bengis, and A. Michel. 2001. Tuberculosis in kudus (*Tragelaphus strepsiceros*) in the Kruger National Park. *Onderstepoort Journal of Veterinary Research* 68:225–230.

Keet, D. F., N. P. J. Kriek, M.-L. Penrith, A. Michel, and H. F. A. Z. Huchzermeyer. 1996. Tuberculosis in buffaloes (*Syncerus caffer*) in the Kruger National Park: spread of the disease to other species. *Onderstepoort Journal of Veterinary Research* 63:239–244.

Kock, M. D., G. H. Mullins, and J. S. Perkins. 2002. Veterinary disease control in Botswana: impacts on wildlife health, ecosystems and rural livelihoods. Pages 265–275 in A. A. Aguirre, R. S. Ostfeld, G. M. Tabor, C. House, and M. C. Pearl (eds.), *Conservation medicine: ecological health in practice*. New York: Oxford University Press.

Kock, R. A., B. Kebkiba, R. Heinonan, and B. Bedane. 2002. Wildlife and pastoral society: shifting paradigms in disease control. *Annals of the New York Academy of Sciences* 969:24–33.

Kock, R. A., J. M. Wambua, J. Mwanzia, H. Wamwayi, E. K. Ndungu, T. Barret, M. D. Kock, and P. B. Rossiter. 1999. Rinderpest epidemic in wild ruminants in Kenya, 1993–1997. *The Veterinary Record* 145:275–283.

Mack, R. 1970. The great African cattle plague epidemic of the 1890s. *Tropical Animal Health and Production* 2:210–219.

McNaughton, S. J., R. W. Reuss, and S. W. Seagle. 1988. Large mammals and process dynamics in African ecosystems. *BioScience* 38:794–800.

Pence, D. B., and E. Ueckermann. 2002. Sarcoptic mange in wildlife. Pages 385–398 in *Infectious diseases of wildlife: detection diagnosis and management*. O.I.E. Scientific and Technical Review 21.

Pienaar, U. de V. 1980. Die rol van die tsetsevlieg, malaria en die runderpes-epidemie van 1896–1897, in die ontwikkelingsgeskiedenis van die Laeveld. Pages 279–288 in *Neem uit die Verlede*. Pretoria, South Africa: National Parks Board Publishers.

Prins, H. H. T., and H. P. van der Jeugd. 1993. Herbivore population crashes and woodland structure in East Africa. *Journal of Ecology* 81:305–314.

Prins, H. H. T., and F. J. Weyerhaeuser. 1987. Epidemics in populations of wild ruminants: anthrax and impala, rinderpest and buffalo in Lake Manyara National Park. *Oikos* 49:28–38.

Rossiter, P. B. 1994. Rinderpest. Pages 735–757 in J. A. W. Coetzer, G. R. Thomson, and R. C. Tustin (eds.), *Infectious diseases of livestock, with special reference to southern Africa*. Cape Town: Oxford University Press.

Smith, K. L., V. de Vos, H. Bryden, M. E. High-Jones, and P. Keim. 2000. Bacillus anthracis diversity in Kruger National Park. *Journal of Clinical Microbiology* 38:3780–3784.

Stevenson-Hamilton, J. 1957. *Wildlife in South Africa*. London: Hamilton and Company.

Swanepoel, R. 1994. Rabies. Pages 493–552 in J. A. W. Coetzer, G. R. Thomson, and R. C. Tustin (eds.), *Infectious diseases of livestock, with special reference to southern Africa*. Cape Town: Oxford University Press.

Thambi, E. N., O. W. Maina, and R. Bessin. In press. *Animal and animal products trade in Africa: new development perspectives in international trade for Africa*. Nairobi, Kenya: OAU-IBAR.

Thrushfield, M. 1995. Some general epidemiological concepts and principles, pages 26–56; the transmission and maintenance of infection, pages 81–111 in *Veterinary epidemiology*. 2nd edition. Cambridge, UK: Cambridge University Press.

Vale, G. A., D. F. Lovemore, S. Flint, and G. F. Cockbill. 1988. Odour-baited targets to control tsetse flies, *Glossina* spp. (Diptera: Glossinidae), in Zimbabwe. *Bulletin of Entomology Research* 78:31–49.

van de Bildt, M. W. G., T. Kuiken, A. M. Visee, S. Lema, T. R. Fitzjohn, and A. D. M. E. Osterhaus. 2002. Distemper outbreak and its effect on African wild dog conservation. *Emerging Infectious Diseases Journal* 8:1–5.

Vosloo, W., A. D. Bastos, E. Kirkbribe, J. J. Esterhuysen, D. Janse van Rensburg, R. G. Bengis, and G. R. Thomson. 1996. Persistent infection of African buffalo (*Syncerus caffer*) with SAT–type foot-and-mouth disease viruses: rate of fixation of mutations, antigenic changes and interspecies transmission. *Journal of General Virology* 77:1457–1467.

Whyte, I. J., and S. C. J. Joubert. 1988. Blue wildebeest population trends in the Kruger National Park and the effects of fencing. *South African Journal of Wildlife Research* 18:78–87.

Wobeser, G. 2002. Disease management strategies for wildlife. Pages 159–178 in *Infectious diseases of wildlife: detection, diagnosis and management*. O.I.E. Scientific and Technical Review 21.

Chapter 18

Large Carnivores and Savanna Heterogeneity

MICHAEL G. L. MILLS AND PAUL J. FUNSTON

Because of its large size and heterogeneity, Kruger supports a rich diversity of large (>20 kg) carnivores, including five extant and one locally extinct species (Figure 18.1). The brown hyena (*Hyaena brunnea*) went extinct as a breeding species about 50 years ago. At the top of the food chain large carnivores may exert an influence on other species, influencing spatial heterogeneity through their impact on prey populations and mediating the role herbivores have on patch dynamics by regulating or limiting their populations (Sinclair et al. 1985; Fryxell et al. 1988; Mills and Shenk 1992; Caughley and Sinclair 1994). Through competition they also influence each other's populations (Macdonald 1983; Mills 1990; Cooper 1991). Equally at various hierarchical levels patch dynamics influences carnivore behavior in terms of hunting success, interspecific competition, and social dynamics (Smuts 1978; van Orsdol et al. 1985; Funston et al. 2001; Donkin 2000).

Large carnivores may also respond to temporal heterogeneity (Scheel and Packer 1995), particularly to the vagaries of rainfall so important to the Kruger ecosystem (Tyson and Dyer 1975; Gertenbach 1980). They exhibit population fluctuations (Mills 1995) and influence populations of other species (Mills and Shenk 1992; Mills et al. 1995). Therefore, they are an important component in ecosystem dynamics and of the flux of nature paradigm (Chapter 3, this volume). They are also an important consideration in strategic planning of conservation managers and have often been at the center of management decisions in Kruger (Smuts 1978; Chapter 19, this volume).

Much of what is known about large African carnivores comes from the Serengeti Plains in Tanzania, atypical of most of the African savannas in that it is a vast open plain with long-distance migrations of large herbivores (Sinclair 1979). It posses a different scenario of ecosystem heterogeneity at spatial and temporal scales from the more wooded Kruger ecosystem, with a largely resident prey base, which is more typical of the savanna regions of Africa. There-

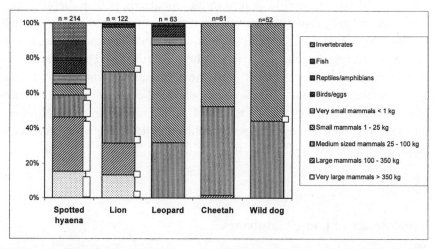

FIGURE 18.1. Percentage occurrence of different food categories in the diets of large carnivores in Kruger (n = number of food items eaten). The narrow histogram on the side of a species' histogram shows the proportion of items that were scavenged.

fore, the behavior and ecological relationships of large carnivores on the Serengeti Plains might not be typical for most of Africa. From an evolutionary and a wildlife management viewpoint, it is important to understand these relationships under a range of conditions.

In this chapter we discuss the effect of ecological processes on lion spatial demography, coexistence of the large carnivores, the role of lion predation, and intraguild competition. We give examples of how these interactions contribute to the role of carnivores as responders to and determinants of savanna heterogeneity and evaluate the management consequences. We also compare aspects of their ecology and behavior with their conspecifics on the Serengeti Plains.

Lion Spatial Demography

Donkin (2000) investigated the effect of habitat structure, prey availability, and rainfall on lion (*Panthera leo*) group dynamics from a long-term dataset of lion sightings across Kruger. The study highlighted the importance of a number of variables that affect lion spatial socioecology.

Four land types based on geological and soil formations were identified: thickets, woodlands, mountain areas, and open tree savanna. Lion were found in larger groups of adults and more often than expected in the open savanna, the area favored by buffalo (*Syncerus caffer*) and other important prey species. However, subadult and cub group sizes peaked in the woodlands, perhaps because they need cover for protection from infanticidal males. Although group dynamics did not differ significantly across averaged mean annual rainfall regions or across seasons, it did between regions of varying variability in rainfall and between extreme years of rainfall. Lion were found in larger groups in more variable environments. This was taken as evidence of risk-prone behavior because it is more advantageous for lion to form larger groups to fulfill their minimum daily needs. However, this is possible only because of the availability of buffalo, which are particularly vulnerable to predation during droughts (Mills et al. 1995).

Coexistence of Large Carnivores

The manner in which the large carnivores respond to the diverse food items and habitats of Kruger is an important factor in their coexistence.

Food Habits

Mammals are the basis of each species' diet (Figure 18.1), with the spotted hyena (*Crocuta crocuta*) having the most varied diet, including a range of other vertebrates and even some invertebrates (Mills and Biggs 1993). It also scavenges nearly 50 percent of its food in Kruger (Henschel and Skinner 1990).

The diets of the three cats and the wild dog (*Lycaon pictus*) overlap strikingly with respect to medium-sized and small mammals (Mills 1992), particularly impala (*Aepyceros melampus*). However, because of the diversity of mammals, particularly herbivores, there is also separation. Common duiker (*Sylvicapra grimmia*) is important for leopard (*Panthera pardus*) and cheetah (*Acinonyx jubatus*), steenbok (*Raphicerus campestris*) for spotted hyena and cheetah, kudu (*Tragelaphus strepsiceros*) for spotted hyena and wild dog, and warthog (*Phacochoerus aethiopicus*) for leopard and lion. Lion take blue wildebeest (*Connochaetes taurinus*) and zebra (*Equus burchelli*) almost exclusively, as does leopard small carnivores and primates (Table 18.1).

Within a species there may be further segregation in diet. Individuals sometimes specialize on a prey species; for example, one lion pride was responsible for 14 of 15 porcupine (*Hystrix africaeaustralis*) kills observed. Additionally, there may be segregation between the sexes. Although impala are the most frequently killed prey species for cheetah, adult males also hunt the juveniles of larger herbivores such as kudu, wildebeest, and zebra, whereas females hunt smaller species such as steenbok and duiker (Mills et al. in press). With lion, Funston et

TABLE 18.1

Percentage occurrence of prey types in carnivore diets in Kruger (prey types contributing <10 percent of kills are not shown).

	PERCENTAGE OCCURRENCE IN DIET				
PREY	SPOTTED HYENA ($N = 27$)	LION ($N = 111$)	LEOPARD ($N = 63$)	CHEETAH ($N = 61$)	WILD DOG ($N = 52$)
Zebra		16			
Adult		5			
Foal		11			
Wildebeest		14			
Adult		12			
Calf		2			
Kudu	15				12
Impala	15	29	28	44	54
Reedbuck					12
Warthog		13	15		
Duiker			14	13	
Steenbok	22			13	
Porcupine		13			
Small carnivores			11		
Primates			10		

al. (1998) showed that male Kruger lions are significant hunters, concentrating on buffalo, whereas females select medium-sized ungulates, particularly wildebeest and zebra.

Habitat Use at the Landscape Level

Observations of radiocollared carnivores in three large-scale habitat types in southeastern Kruger revealed differences in the manner in which they respond to different habitats (Mills and Biggs 1993; Figure 18.2). The habitats based on Gertenbach's (1983) landscape types were *Acacia* and *Combretum* woodland, marula and knobthorn tree savanna, and the Lebombo hills. A simple habitat preference ratio for each species was measured (Mills and Biggs 1993). A ratio >1 indicates a preference for the habitat and <1 an avoidance.

Woodland was selected by all carnivore species except cheetah but most strongly by wild dog and spotted hyena. Tree savanna was preferred by cheetah and also selected by lion and hyena but was strongly avoided by wild dog. Hills were preferred by wild dog and leopard but avoided by the other species.

More detailed analyses of habitat use in cheetah and lions revealed differences between the responses of males and females (Broomhall et al. 2003; Funston, et al. 2001). Although both cheetah and lion showed a preference for tree savanna, female cheetah and male lion selected thicker bush, including drainage lines in tree savanna, more so than members of the opposite sex. In the

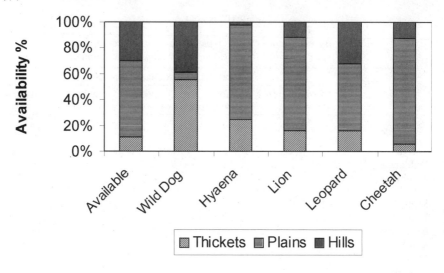

FIGURE 18.2. Percentage use of three main habitats by large carnivores and the availability of these habitats in southeastern Kruger.

case of cheetah this is the preferred habitat of their preferred prey, impala. Drainage lines are particularly good hunting habitat for cheetahs because they can stalk the prey through the thick bush on the sides and then reach high speeds during the chase on the more open areas in the drainage line. Females therefore appear to maximize encounter rates with impala by using denser habitat more often than males. Tree savanna is the preferred habitat of lion prides and spotted hyena (Mills and Gorman 1997), so female cheetah accompanied by cubs may avoid contact with dangerous predators and kleptoparasites (Schaller 1972; Laurenson 1994) by selecting habitats that provide greater concealment. However, male lion also favored drainage lines for patrolling and scent marking (Funston 1999), probably because they often lead to focus areas such as water-points and have a higher density of suitable (dense broad-leaved) shrubs and low trees for spraying urine on. A coalition of three territorial cheetah males used more open areas than females and concentrated their activities along roads where they preferred to scent mark. Roads and tracks are landscape features likely to be traversed by other male cheetah. Territorial male cheetah also typically spray-urinate at the base of large, conspicuous trees or leave feces on fallen trees in open areas. Additionally, male cheetah use a wider prey base than females, hunting the young of wildebeest, zebra, and waterbuck, which prefer more open areas. They may also be able to stand up to kleptoparasites more efficiently than females.

Hunting Behavior and Habitat Use at the Small Patch Scale

Carnivores must capture fleet-footed and wary prey, and the manner in which they respond to habitat structure at the small scale is important for hunting success. That leopard need good cover for stalking prey is illustrated by the fact that 90 percent of kills were in dense vegetation and only 10 percent in open areas (Bailey 1993).

Generalized linear models revealed that seven variables, mainly associated with habitat structure, have significant independent influences on lion hunting success (Funston et al. 2001). Specific variables associated with the environment and habitats were moon brightness and grass height, respectively, showing that concealment from prey is also important to lion. The only significant second-order interaction between the ten variables tested was the interaction between shrub cover and grass height. In open, moderate, and dense shrub cover lion hunting success increased with increasing grass height, most significantly in dense shrub, where the probability of success increased markedly from short to long grass. The third-order interaction between lion sex, prey species, and shrub cover also resulted in significant effects. For male lion only the probability of catching small (<65 kg, other than impala) prey species increased from open to moderate shrub cover, whereas that of killing impala increased sharply from moderate to dense bush areas. The probability of success decreased slightly for females hunting medium-sized (65–250 kg) prey from open to moderate or dense bush, whereas it remained low for males in all shrub densities. Medium-sized prey were also encountered most often in more open terrain. For buffalo hunts, however, males had consistently high success rates in all classes of shrub cover, whereas females had higher success in open areas only.

Predator-Prey Relationships and the Influence of Ecological Conditions

The role of large carnivores as determinants of heterogeneity through their influence on the prey is a contentious issue. African ecosystems involve complex multispecies predator-prey systems (Schaller 1972; Mills 1990; Mills and Biggs 1993). In southern Africa these are further complicated by dramatic fluctuations in environmental conditions, especially rainfall (Tyson and Dyer 1975). Several studies have shown that predators can limit resident low-density ungulate populations, whereas resident or migratory high-density populations are regulated by lean season food abundance, not predation (Sinclair et al. 1985; Fryxell et al. 1988).

The Dynamics of Lion Predation

Smuts (1978) showed that wildebeest and zebra populations declined during years of exceptionally high rainfall. He speculated that predation by lion and spotted hyena was the major cause as vulnerability of these species to predation increased because of increased hunting cover and fragmentation of herds. As a result, the largest systematic culling operation in Kruger's history took place: 445 lion and 375 spotted hyenas were removed in a 5-year period from an area of about 1,500 km^2. The operation was terminated in 1980 when it was found that their reduction had no detectable influence on the population densities of lions or on the population trends of the wildebeest and zebra (Whyte 1985).

Subsequently, Mills and Shenk (1992) used simulation models to measure the impact of lion predation on wildebeest and zebra populations in the southeast of Kruger during a period of high rainfall. The results suggested that lion predation regulated the resident, low-density wildebeest population but had little effect on the semimigratory zebra population. This was because lion selected adult wildebeest, whereas they selected juvenile zebra (Table 18.2) and because of the sedentary behavior of the wildebeest. However, because of lowered predation rates, lions did not affect the wildebeest population as much in a drier period, and there was a slight increase in the population from the wet to the dry period (Funston 1999).

Long-term datasets from the central and southeastern regions of Kruger showed that buffalo population fluctuations were positively correlated with rainfall, being most heavily influenced by predation during population declines associated with drought conditions (Mills et al. 1995). A model developed by Starfield et al. (1992) accurately predicted buffalo population trends, but only after the predation parameters (as determined by environmental conditions) had been adjusted by using actual predation data to calculate mortality. Initially the extent of predation under below-average rainfall and particularly drought conditions was underestimated. Mills et al. (1995) experienced similar difficulty, showing how strongly the effects of environmental conditions on buffalo are mediated by lion predation.

Lion prey more frequently and have a bigger impact on the buffalo population in drought conditions; the converse is true for wildebeest in above-average rainfall periods (Figure 18.3). Thus it seems that the wildebeest and buffalo populations are regulated through a dynamic relationship between periods of high and low rainfall and the manner in which lions respond to these conditions. This suggests an intriguing relationship between spatial and temporal heterogeneity manifest through suitable grazing biomass and hunting conditions. Although lion prey more heavily on zebra in higher rainfall periods, the predation rates even in wet periods do not seem to have a strong limiting effect.

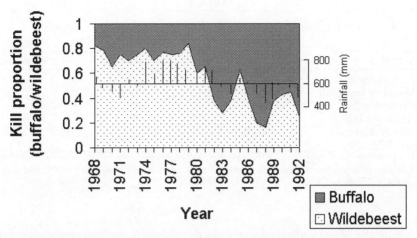

FIGURE 18.3. Lion kill proportions for buffalo and wildebeest in wet and dry cycles. Rainfall is expressed as the 3-year running mean (the mean of the current and 2 previous years; after Mills et al. 1995).

It is still unclear what limits zebra numbers in Kruger. The role of lion in limiting prey and hence as determinants of heterogeneity therefore is strongly influenced by environmental conditions. The consequences of these findings for management suggest that control of both predator and prey populations should not be taken lightly and that fluctuations in prey numbers should be seen as part of natural ecosystem functioning.

The Possible Role of Predation in the Decline of Roan Antelope: A Management Conundrum

Between 1986 and 1993 the roan antelope population in Kruger dropped from a calculated 452 to 42 (Harrington et al. 1999). Competition from zebra and wildebeest, which moved into the roan's range after the introduction of water-points, exacerbated by habitat deterioration caused by an extended drought, is believed to be an important factor for this decline (Harrington et al. 1999; Grant et al. 2002). However, Harrington et al. (1999) noted that the population crash was associated with an increase in adult mortality but little apparent change in calf survival, suggesting that nutritional factors were not the prime cause. They suggested that increased predation on adult roan caused by a buildup of lion numbers following the wildebeest and zebra influx was the proximate cause of the decline, although the ultimate cause was the provision of numerous artificial water-points. We suggest that this was further exacerbated by the habitat changes that made the roan more vulnerable to predation.

TABLE 18.2

Ranked habitat preferences for wild dog, lion, and spotted hyena
(1 = most preferred, 6 = least preferred).

HABITAT	WILD DOG	LION	SPOTTED HYENA
Sour bushveld	2	6	5
Mountain bushveld	3	4	6
Combretum bushveld	4	2	2
Acacia thickets	5	3	1
Marula savanna	6	1	3
Lebombo hills	1	5	4

Spearman rank correlation coefficients (two-tailed): wild dog and lion, $r_s = -.89$, $p < .05$; wild dog and hyena, $r_s = -.6$, $p > .05$; lion and hyena, $r_s = .6$; $p > .05$.

Closure of some artificial water-points resulted in the movement away from the prime roan habitat of many of the zebra and lions (Grant and van der Walt 2000). However, the expected positive response by the roan population did not materialize despite good rains (Grant and van der Walt 2000). This is probably because the roan numbers had dropped so low that even mild predation became a major block to population increase. Caughley and Sinclair (1994) predicted that populations at critically low numbers exposed to even low rates of predation would remain so or go extinct. However, culling of predators to help the roan was not considered to be a viable option because even halving the number of lion may not have stopped the few predation events per year needed to keep the population suppressed. In addition, other predators might also be involved, and it was unacceptable to cull all species.

Interguild Relationships

Carnivore populations usually are limited by the food supply (Macdonald 1983). However, kleptoparasitism and predation have been mentioned as important forms of interference competition among large African carnivores (Laurenson 1995; Carbone et al. 1997). Therefore, interguild relationships can see carnivores as both responders to and determinants of heterogeneity.

Factors Affecting the Distribution and Density of Wild Dogs

The wild dog density in Kruger is low compared with that of most of the other large carnivores (Table 18.3). Impala are the dominant prey of wild dog, but there is a negative relationship between the number of impala in a pack's ter-

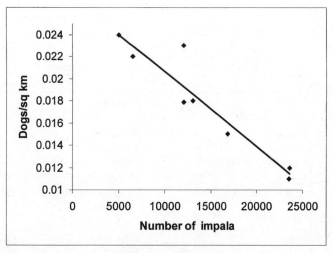

FIGURE 18.4. The relationship between Kruger wild dog densities in different territories and the number of impala in those territories ($r^2 = .883$; $p > .01$).

ritory and the density of dogs in that territory (Figure 18.4). The density of wild dogs in Kruger therefore does not appear to be influenced by the dispersion pattern of their prey at the territorial scale (Mills and Gorman 1997).

Analyses showed a strong negative correlation between the habitat preferences of wild dog and lion but this was not significant for wild dog and spotted hyena (Table 18.2). Lion favor the habitats selected by their most important prey, wildebeest, buffalo, and zebra, but wild dog avoid those habitats favored by lion irrespective of prey availability. Lion predation is an important cause of wild dog mortality (van Heerden et al. 1995). This appears to be the major factor causing wild dog to avoid certain high-density impala habitats. Similar findings have been found elsewhere (Creel and Creel 1996). Spotted hyena were not recorded as agents of mortality for wild dog in Kruger (van Heerden et al. 1995) and rarely deprived the dogs of food (Mills and Biggs 1993).

The Brown Hyena: A Victim of Previous Management Strategies?

The brown hyena used to be regarded as common in Kruger, especially on the western boundary north of the Letaba River (Pienaar 1963), and Stevenson-Hamilton (1947) recorded the species breeding along the Sabie River in the 1940s. Over the last 40 years there have been very few records of the brown

TABLE 18.3

Estimated densities (animals/100 km²) of large carnivores in various landscapes in Kruger and Serengeti.

	KRUGER		SERENGETI	
	AREA	DENSITY	AREA	DENSITY
Cheetah	Whole park	0.9[a]	Ecosystem	2.3[b]
	Marula tree savanna	2.2[c]	Plains	2.2[d]
Leopard	Whole park	3.5[e]	Ecosystem	3.2–4.0[d,f]
	Sabie River	30.3[e]	Woodlands	4.5[f]
	Other riparian zones	10.5[e]	Plains	2.2
Lion	Whole park	7.5–12.5[g,h]	Plains	8.7–10.0[f,i]
	Marula tree savanna	14.9[h]	Corridor and northern woodlands	30.0[f]
	Central region	12.7[g]		
Spotted hyena	Whole park	13.9[j]	Ecosystem	12.0–20.6[b,k]
			Plains	84[l]
Wild dog	Whole park	0.8–2.0[m]	Plains	0–1.2[n]

[a]Bowland (1994), [b]Borner et al. (1987), [c]Mills, unpublished data, [d]Caro (1994), [e]Bailey (1993), [f]Schaller (1972), [g]Smuts (1978), [h]Funston (1999), [i]Hanby et al. (1995), [j]Mills et al. (2001), [k]Kruuk (1972), [l]Hofer and East (1995), [m]Davies (2001), [n]Burrows (1995).

hyena from Kruger, where it is believed to be extinct as a breeding species. Yet in areas adjacent to Kruger, such as the Messina area, it is reported to be common (Mills and Hofer 1998).

The spotted hyena is dominant to the smaller brown hyena. In the southern Kalahari, where the brown hyena outnumbers the spotted hyena by 2:1, it was found that in areas well frequented by spotted hyena, such as around their dens, brown hyena were less common (Mills 1990). Large carnivores, including the spotted hyena, were persecuted in the Kruger area in the early 1900s (Joubert 1986), which may have favored the less conspicuous brown hyena. However, with the implementation of efficient antipoaching operations and the proliferation of artificial water-points sunk in Kruger from the 1930s through the 1970s, herbivore, lion, and spotted hyena numbers increased and in many areas probably exceeded the densities that they exhibited in earlier times (Smuts 1978). Therefore, it is possible that this increase, particularly in spotted hyena, led to the extinction of the brown hyena.

Comparisons with Serengeti Plains

The Serengeti Plains is a vast, open short grass to tall grass area of 5,200 km² in the southeast of the Serengeti ecosystem, with the majority of the herbivore biomass made up of migratory wildebeest, zebra, and Thomson's gazelle (*Gazella thomsonii*). The woodlands to the north and west are predominantly *Acacia-Commiphora* woodland savannas (Caro 1994). Most carnivore studies in this region have been conducted on the plains. Although the plains are an integral part of the entire Serengeti ecosystem, they can be considered a separate system, and the carnivores in this habitat are subpopulations with their own unique behavior and independent density. In contrast, Kruger is a heterogeneous woodland savanna area of 20,000 km^2 made up of a mosaic of 35 landscapes, none larger than 2,000 km^2, with mainly resident herbivores (Gertenbach 1983).

Because the rainfall is higher and the soils more nutrient rich, the biomass of herbivores in the Serengeti is higher than in the Kruger ecosystem. There are about eight times more prey animals available to predators in the Serengeti ecosystem (76 animals/km^2; Campbell and Borner 1995) than in Kruger (9 animals/km^2; Viljoen et al. 1994). Because of migration, the major prey species are abundant on the Serengeti Plains only during the wet season. The average wet season biomass is more than 20,000 kg/km^2, dropping to less than 1,000 kg/km^2 during the dry season (Maddock 1979; Campbell and Borner 1995). The average biomass of herbivores available to large carnivores in the Kruger ecosystem is about 1,750 kg/km^2 throughout the year.

Carnivore Densities

In a number of large carnivores it has been found that those with a migratory prey base have larger home ranges than those with a sedentary prey base (Hanby et al. 1995; Broomhall et al. 2003). In addition, the density of lions generally is correlated with the lean season biomass of suitable prey (van Orsdol et al. 1985), and it is reasonable to assume that this might hold for other species that are food limited as well. Therefore, one would expect the interaction between these relationships to result in higher large-carnivore densities in areas with sedentary high-density prey than those with migratory high-density prey. Additionally, woodland savannas, through the enhanced cover they provide, might support higher densities of carnivores that are not food limited because they will have a wider selection of habitats to choose from to escape competing carnivores. Therefore, Kruger could support higher densities of these less aggressive large carnivores than the homogeneous Serengeti Plains despite the overall higher prey densities on the plains.

Although the average cheetah density in the Serengeti ecosystem is higher than in Kruger (Table 18.3), cheetah density is similar in their preferred but limited marula and knobthorn savanna habitat in Kruger and on the Serengeti Plains. VORTEX modeling (Lacy 1993) suggested that cheetah population viability was greater in a woodland savanna than a grassland savanna (Broomhall 2001). With exceptionally high cub mortality on the Serengeti Plains (Laurenson 1995), this population may be a sink, and the woodland population, with lower cub mortality and predicted high dispersal rates, may be a source.

In both ecosystems there is substantial variation in leopard numbers in various habitats (Table 18.3). In Kruger about 50 percent of the leopard population is estimated to inhabit perennial river riparian zones, which constitute only 16 percent of the park's total surface area (Bailey 1993). The riparian zone along the Sabie River is particularly rich because it has exceptionally high impala densities, averaging around 95/km^2. Riparian zones along other rivers support a lower leopard density. The biomass of leopard prey in other regions of Kruger is only 5 to 25 percent of the prey biomass in the riparian thickets. By these criteria Bailey (1993) conservatively estimated a leopard population of 670 for Kruger. Similarly, leopards occur at higher densities in the adjacent *Acacia* woodlands in the Serengeti ecosystem than on the plains. Thus, although the average densities of leopards in the Kruger and Serengeti ecosystems are similar, the average density in Kruger is higher than that of the Serengeti Plains.

In Serengeti, lion predation changes between seasons, across habitats, and from year to year. Most of this variation can be attributed to the annual migration of wildebeest, zebra, and gazelle. Predation on buffalo is consistently greater in the woodlands than on the plains. Thus, the woodland lions may be buffered against seasonal changes in prey during seasons in which migrant species are locally scarce (Scheel and Packer 1995). Lion home ranges on the Serengeti Plains average around 200 km^2 and shift each year with the arrival of the migration, with the result that there is overlap between prides in their annual ranges (Schaller 1972; Hanby et al. 1995). Pride ranges in the central and southern regions of Kruger are smaller (100–150 km^2) and overlap less than those of Serengeti Plains prides (Whyte 1985; Funston 1999). Because of the generally more sedentary prey in Kruger, there is little difference in the relative proportions of wildebeest and zebra killed by lion prides in wet and dry seasons (Mills and Shenk 1992), although there is a difference during wet and dry cycles when buffalo take over in dry cycles (Mills et al. 1995). The cumulative effect of these responses is that lions on the Serengeti Plains occur at lower densities than their average density in the central and southern districts of Kruger. The larger prey base created by higher rainfall and richer soils ensures a higher lion population density for the Serengeti ecosystem as a whole (Table 18.3).

The spotted hyena lives in clans on the Serengeti Plains, occupying fairly small territories of about 50 km^2. When the migration passes through a clan's territory, there is an abundance of food, but when it moves on the food supply in the territory decreases drastically. Then hyenas leave the territory to follow distant migratory herds (Hofer and East 1993a). In this way they are able to maintain an extremely high density, although the overall density for the Serengeti ecosystem is much lower (Table 18.3). In Kruger densities vary in different broad habitats from 2.6 to 21.1 hyenas/100 km^2, with the overall density 13.9 hyenas/100 km^2 (Mills et al. 2001).

Wild dog live at low population densities (Fuller et al. 1992) and are given to wide fluctuations (Burrows 1995; Davies 2001). Their numbers on the Serengeti Plains declined in the late 1980s before they disappeared from the area by 1992, apparently as the result of disease (Burrows 1995). The openness of the plains makes the dogs vulnerable to kleptoparasitism by lion and particularly spotted hyena. An increase in number of these competitors in the 1960s and 1970s (Hanby and Bygott 1979) may have started the decline because wild dogs appear to be particularly sensitive to kleptoparasitism (Fanshawe and FitzGibbon 1993; Carbone et al. 1997; Gorman et al. 1998). In addition, demographic stochasticity, food shortages caused by the migration, and emigration contributed to the initial low population density (Schaller 1972; Fanshawe and FitzGibbon 1993; Burrows 1995). In Kruger the wild dog density has fluctuated over the last decade (Table 18.3), but the large size of the area and the diversity of landscapes have ensured its survival (Mills and Gorman 1997).

With the exception of the spotted hyena, large carnivores appear to do better in the heterogeneous wooded savannas of Kruger than on the higher prey biomass grassland plains of the Serengeti. The sedentary as opposed to migratory nature of the prey in Kruger and the diverse habitats combine in different ways for each species and appear to provide more ecological niches or refuges for the various carnivores to escape competition or to provide hunting opportunities.

Comparative Behavior

A number of studies have illustrated flexibility in carnivore behavior and social organization. This is often influenced by ecological conditions, particularly the dispersion pattern of food (Kruuk 1972; Macdonald 1983; Mills 1990); structural features of the habitat may also be important.

Adult female cheetah are solitary with their latest litter of cubs, and adult males are solitary or in coalitions of two to four. Female cheetah home ranges on the Serengeti Plains (mean = 833 km^2; Caro 1994) are significantly larger than was found in southern Kruger (mean = 186 km^2; Broomhall et al. 2003). In the Serengeti, up to 20 females might collect in an area as small as 50 km^2 for a short time (Caro 1994), whereas in Kruger no more than three females have been found to have overlapping home ranges. Female cheetahs on the Serengeti Plains follow the migratory Thomson's gazelle from the short grass plains in the wet season northwest to the long grass areas in the dry season. In Kruger the impala are sedentary.

Most male cheetahs on the Serengeti Plains do not follow the migration. They set up very small (mean = 37.4 km^2) territories in hotspot areas on the plains, contingent on adequate cover and a reasonable number of gazelle, that is, in areas that might be expected to attract females (Caro 1994). Kruger males inhabit territories of similar size to and overlapping those of females (mean = 173 km^2; Broomhall et al. 2003). In each area, therefore, males position their territories and adjust the size for optimum female acquisition.

Lion prides live in territories in which their movements are limited by dependent young. Therefore, breeding adults on the Serengeti Plains must stay in their territories and defend their young during lean periods. Consequently, cub survival rates are low, with 30 percent surviving to 1 year of age (Hanby et al. 1995). In comparison, 84 percent of cubs survived the first year in a study in Kruger (Funston et al. 2003). Subadult male lions disperse from their natal prides. By 48 months 69 percent had left a 2,000-km^2 study area on the Serengeti Plains (Pusey and Packer 1987). Access to greater hunting and scavenging opportunities in the neighboring woodlands probably is the main reason for this. In contrast, all subadult males born in prides in southeastern Kruger were encountered in or next to their natal territories at 48 months of age (Funston et al. 2003). Lower cub production rates due to higher cub survival appear to result in male cubs being tolerated for longer in the natal pride. Furthermore, the ability of subadult males to kill resident buffalo (Funston et al. 1998) makes it possible for them to remain close to their natal pride territory and even to become pride males close by (Funston et al. 2003).

The spotted hyena is the dominant predator on the Serengeti Plains, outnumbering the lion by about 9:1. Its ability to commute from territories to the migration concentrations, even in the case of lactating females (Hofer and East 1993b), is responsible for its success in this ecosystem. In Kruger, lion and spotted hyena are more equally represented. Its ability to hunt and scavenge equally enables the spotted hyena to take advantage of the large supply of carrion left from lion kills there (Henschel and Skinner 1990; Mills and Biggs 1993).

The cooperative hunting and killing behavior of the wild dog on the Serengeti Plains has been well documented (Khume 1965; Estes and Goddard 1967; Malcolm and van Lawick 1975). In contrast, in the wooded savannas of Kruger the degree of cooperation is much less. When a pack encounters impala, each dog tends to individually pursue the prey so that one or two dogs make the kill. The killers then lead the others to the carcass (Mills, pers. obs.).

Conclusion

Large carnivores respond to heterogeneity by using landscape features and the diverse prey species they produce differently. Their role as predators and competitors contributes to the determination of heterogeneity, although this is mediated through changing ecological conditions, both natural and human induced. They have played an important role in management strategies, but their behavioral flexibility makes both spatial and temporal extrapolation difficult and necessitates a conservative management approach.

References

Bailey, T. N. 1993. *The African leopard: ecology and behaviour of a solitary felid.* New York: Columbia University Press.
Borner, M., C. D. FitzGibbon, M. Borner, T. M. Caro, W. K. Lindsay, D. A. Collins, and M. E. Holt. 1987. The decline of the Serengeti Thomson's gazelle population. *Oecologia* 73:32–40.
Bowland, A. E. 1994. *The 1990/1991 cheetah photographic survey.* Scientific report 6/94. Skukuza, South Africa: National Parks Board.
Broomhall, L. S. 2001. *Cheetah* Acinonyx jubatus *ecology in the Kruger National Park: a comparison with other studies across the grassland-woodland gradient in African savannas.* Unpublished M.Sc. thesis, University of Pretoria, Pretoria, South Africa.
Broomhall, L. S., M. G. L. Mills, and J. T. du Toit. 2003. Home range and habitat use by cheetahs in the Kruger National Park. *Journal of Zoology* (in press).
Burrows, R. 1995. Demographic changes and social consequences in wild dogs. Pages 400–420 in A. R. E. Sinclair and P. Arcese (eds.), *Serengeti 2: dynamics, management, and conservation of an ecosystem.* Chicago: University of Chicago Press.
Campbell, K., and M. Borner. 1995. Population trends and distribution of Serengeti herbivores: implications for management. Pages 117–145 in A. R. E. Sinclair and P. Arcese (eds.), *Serengeti 2: dynamics, management, and conservation of an ecosystem.* Chicago: University of Chicago Press.
Carbone, C., J. T. du Toit, and I. J. Gordon. 1997. Feeding success in African wild dogs: does kleptoparasitism by spotted hyaena influence hunting group size? *Journal of Animal Ecology* 66:318–326.
Caro, T. M. 1994. *Cheetahs of the Serengeti Plains.* Chicago: University of Chicago Press.
Caughley, G., and A. R. E. Sinclair. 1994. *Wildlife ecology and management.* New York: Blackwell Science.
Cooper, S. M. 1991. Optimal hunting group size: the need for lions to defend kills against loss to spotted hyaena. *African Journal of Ecology* 29:130–136.
Creel, S. R., and N. M. Creel. 1996. Limitation of African wild dogs by competition with larger carnivores. *Conservation Biology* 10:1–15.
Davies, H. T. 2001. *The 1999/2000 Wild Dog Photographic Survey.* Scientific Report 3/2000. Skukuza, South Africa: National Parks Board.
Donkin, D. A. 2000. *Lion spatial socio-ecology: the effect of habitat on lion group dynamics.* Unpublished M.Sc. thesis, University of Natal, Durban, South Africa.
Estes, R. D., and J. Goddard. 1967. Prey selection and hunting behaviour of the African wild dog. *Journal of Wildlife Management* 31:52–70.
Fanshawe, J. H., and C. D. FitzGibbon. 1993. Factors influencing the hunting success of an African wild dog pack. *Animal Behaviour* 45:479–490.
Fryxell, J. M., J. Grever, and A. R. E. Sinclair. 1988. Why are migratory ungulates so abundant? *American Naturalist* 131:781–798.
Fuller, T. K., P. W. Kat, J. B. Bulger, A. H. Maddock, J. R. Ginsberg, R. Burrows, J. W. McNutt, and M. G. L. Mills. 1992. Population dynamics in African wild dogs.

Pages 1125–1139 in D. R. McCullough and H. Barrett (eds.), *Wildlife 2001: populations*. London: Elsevier Applied Science.

Funston, P. J. 1999. *Predator-prey relationships between lions and large ungulates in the Kruger National Park*. Unpublished Ph.D. thesis, University of Pretoria, Pretoria, South Africa.

Funston, P. J., M. G. L. Mills, and H. C. Biggs. 2001. Factors affecting the hunting success of male and female lions in the Kruger National Park. *Journal of Zoology* (London) 253:419–431.

Funston, P. J., M. G. L. Mills, H. C. Biggs, and P. R. K. Richardson. 1998. Hunting by male lions: ecological influences and sociological implications. *Animal Behaviour* 56:1333–1345.

Funston, P. J., M. G. L. Mills, P. R. K. Richardson, and A. S. van Jaarsveld. 2003. Reduced dispersal and opportunistic territory acquisition in male lions. *Journal of Zoology*. 259:131–142.

Gertenbach, W. P. D. 1980. Rainfall patterns in the Kruger National Park. *Koedoe* 23:35–43.

Gertenbach, W. P. D. 1983. Landscapes of the Kruger National Park. *Koedoe* 26:9–121.

Gorman, M. L., M. G. L. Mills, J. P. Raath, and J. R. Speakman. 1998. High hunting costs make African wild dogs vulnerable to kleptoparasitism by hyaenas. *Nature* (London) 391:479–481.

Grant, C. C., T. Davidson, P. J. Funston, and D. Pienaar. 2002. Challenges faced in the conservation of rare antelope: a case study on the northern basalt plains of the Kruger National Park. *Koedoe* 45:45–66.

Hanby, J. P., and J. D. Bygott. 1979. Population changes in lions and other predators. Pages 249–262 in A. R. E. Sinclair and P. Arcese (eds.), *Serengeti 2: dynamics, management, and conservation of an ecosystem*. Chicago: University of Chicago Press.

Hanby, J. P., J. D. Bygott, and C. Packer. 1995. Ecology, demography and behaviour of lions in two contrasting habitats: Ngorongoro Crater and Serengeti Plains. Pages 315–331 in A. R. E. Sinclair and P. Arcese (eds.), *Serengeti 2: dynamics, management, and conservation of an ecosystem*. Chicago: University of Chicago Press.

Harrington, R., N. Owen-Smith, P. C. Viljoen, H. C. Biggs, D. R. Mason, and P. J. Funston. 1999. Establishing the causes of the roan antelope decline in the Kruger National Park, South Africa. *Biological Conservation* 90:69–78.

Henschel, J. R., and J. D. Skinner. 1990. The diet of the spotted hyaena *Crocuta crocuta* in Kruger National Park. *African Journal of Ecology* 28:69–82.

Hofer, H., and M. L. East. 1993a. The commuting system of Serengeti spotted hyaenas: how a predator copes with migratory prey. 1. Social organization. *Animal Behaviour* 46:547–557.

Hofer, H., and M. L. East. 1993b. The commuting system of Serengeti spotted hyaenas: how a predator copes with migratory prey. 3. Attendance and maternal care. *Animal Behaviour* 46:575–589.

Hofer, H., and M. L. East. 1995. Population dynamics, population size and the commuting system of Serengeti spotted hyaenas. Pages 332–363 in A. R. E. Sinclair and P. Arcese (eds.), *Serengeti 2: dynamics, management, and conservation of an ecosystem*. Chicago: University of Chicago Press.

Joubert, S. C. J. 1986. *Masterplan for the management of the Kruger National Park*. Skukuza, South Africa: National Parks Board.

Khume, W. D. 1965. Communal food distribution and division of labour in African hunting dogs. *Nature* (London) 205:443–444.

Kruuk, H. 1972. *The spotted hyena: a study of predation and social behavior.* Chicago: University of Chicago Press.

Lacy, R. C. 1993. VORTEX: A computer simulation model for population viability analysis. *Wildlife Research* 20:45–65.

Laurenson, M. K. 1994. High juvenile mortality in cheetahs (*Acinonyx jubatus*) and its consequences for maternal care. *Journal of Zoology* (London) 234:387–408.

Laurenson, M. K. 1995. Implications for high offspring mortality for cheetah population dynamics. Pages 385–399 in A. R. E. Sinclair and P. Arcese (eds.), *Serengeti 2: dynamics, management, and conservation of an ecosystem.* Chicago: University of Chicago Press.

Macdonald, D. W. 1983. The ecology of carnivore social behaviour. *Nature* (London) 301:379–384.

Maddock, L. 1979. The "migration" and grazing succession. Pages 104–129 in A. R. E. Sinclair and P. Arcese (eds.), *Serengeti 2: dynamics, management, and conservation of an ecosystem.* Chicago: University of Chicago Press.

Malcolm, J. R., and H. van Lawick. 1975. Notes on wild dogs (*Lycaon pictus*) hunting zebras. *Mammalia* 39:231–240.

Mills, M. G. L. 1990. *Kalahari hyaenas: the comparative behavioural ecology of two species.* London: Unwin Hyman.

Mills, M. G. L. 1992. A comparison of methods used to study food habits of large African carnivores. Pages 1112–1124 in D. R. McCullough and R. H. Barrett (eds.), *Wildlife 2001: populations.* London: Elsevier Applied Science.

Mills, M. G. L. 1995. Notes on wild dog *Lycaon pictus* and lion *Panthera leo* population trends during a drought in the Kruger National Park. *Koedoe* 38:95–99.

Mills, M. G. L., and H. C. Biggs. 1993. Prey apportionment and related ecological relationships between large carnivores in Kruger National Park. Pages 253–268 in N. Dunstone and M. L. Gorman (eds.), *Mammals as predators: symposium of the Zoological Society of London* 65. Oxford, Clarendon Press.

Mills, M. G. L., H. C. Biggs, and I. J. Whyte. 1995. The relationship between lion predation, population trends in African herbivores and rainfall. *Wildlife Research* 22:75–88.

Mills, M. G. L., and M. L. Gorman. 1997. Factors affecting the density and distribution of wild dogs in the Kruger National Park. *Conservation Biology* 11:1397–1406.

Mills, M. G. L., and H. Hofer. 1998. *Hyaenas: status survey and conservation action plan.* Gland, Switzerland: IUCN.

Mills, M. G. L., J. M. Juritz, and W. Zucchini. 2001. Estimating the size of spotted hyaena *Crocuta crocuta* populations through playback recordings allowing for non-response. *Animal Conservation* 4:335–344.

Mills, M. G. L., and T. M. Shenk. 1992. Predator-prey relationships: the impact of lion predation on wildebeest and zebra populations. *Journal of Animal Ecology* 61:693–702.

Mills, M. G. L., L. S. Broomhall, and J. T. du Toit (in press). Cheetah ecology across African savanna habitats: do cheetahs prefer hunting on open plains? *Wildlife Biology.*

Pienaar, U. de V. 1963. The large mammals of the Kruger National Park: their distribution and present-day status. *Koedoe* 6:1–37.

Pusey, A. E., and C. Packer. 1987. The evolution of sex-biased dispersal in lions. *Behavior* 180:262–296.

Schaller, G. B. 1972. *The Serengeti lion: a study of predator-prey relations*. Chicago: University of Chicago Press.

Scheel, D., and C. Packer. 1995. Variation in predation by lions: tracking a moveable feast. Pages 299–314 in A. R. E. Sinclair and P. Arcese (eds.), *Serengeti 2: dynamics, management, and conservation of an ecosystem*. Chicago: University of Chicago Press.

Sinclair, A. R. E. 1979. Dynamics of the Serengeti ecosystem. Pages 1–30 in A. R. E. Sinclair and M. Nornton-Griffiths (eds.), *Serengeti: dynamics of an ecosystem*. Chicago: University of Chicago Press.

Sinclair, A. R. E., H. T. Dublin, and M. Borner. 1985. Population regulation of Serengeti wildebeest: a test of the food hypothesis. *Oecologia* 65:226–268.

Smuts, G. L. 1978. Interrelations between predators, prey and their environment. *BioScience* 28:316–320.

Starfield, A., M. Qualding, and J. Venter. 1992. *A management-orientated buffalo population dynamics model for the Kruger National Park*. Unpublished report, National Parks Board, Skukuza, South Africa.

Stevenson-Hamilton, J. 1947. *Wildlife in South Africa*. London: Cassell.

Tyson, P. D, and T. G. J. Dyer. 1975. Mean annual fluctuations of precipitation in the summer rainfall regions of South Africa. *South African Geographical Journal* 57:104–110.

van Heerden, J., M. G. L. Mills, M. J. van Vuuren, P. J. Kelly, and M. J. Dreyer. 1995. An investigation into the health status and diseases of wild dogs (*Lycaon pictus*) in the Kruger National Park. *Journal of the South African Veterinary Association* 66:18–27.

van Orsdol, K. G., J. P. Bygott, and J. D. Bygott. 1985. Ecological correlates of lion social organisation. *Journal of Zoology* (London) 206:97–112.

Viljoen, P. C., M. A. Rochat, and C. A. Wood. 1994. *Ecological aerial survey in the Kruger National Park: 1993*. Scientific Report 1/94. National Parks Board, Skukuza, South Africa.

Whyte, I. J. 1985. *The present ecological status of the blue wildebeest* (Connochaetes taurinus taurinus, *Burchell, 1823*) *in the central district of the Kruger National Park*. Unpublished M.Sc. thesis, University of Natal, Pietermaritzburg, South Africa.

PART IV
Humans and Savannas

Humans are integral to African savannas. Since several million years ago, successive forms of rapidly evolving hominids have hunted and gathered in landscapes such as those visited by tourists in Kruger today. The impalas, marula fruits, and other resources on which early humans depended are unlikely to have behaved, tasted, or looked very different from those occurring there now. But we humans have moved on to become the most powerful agents of ecological change on the planet. In the process some of us have learned that ecosystems respond to human influences in ways that are seldom gradual, linear, or reversible and that ecosystems cannot be managed in isolation from human effects. We have also learned that large nature reserves such as Kruger increase the resilience of regional social-ecological systems, affording people and enterprises a greater capacity to adapt and self-restore when challenged by major disturbances. This part of the book therefore adopts an outward-looking perspective to consider Kruger in the context of a natural treasure embedded in a human-dominated matrix.

The Kruger experience exemplifies the human propensity to trigger unpredictable and nonlinear ecosystem responses. For example, toward the end of the nineteenth century the South African lowveld was ravaged by two unprecedented, widespread, and effectively simultaneous anthropogenic disturbances, of which one was completely unexpected. An episode of excessive hunting for meat and ivory (Whyte et al., Chapter 16) was immediately followed by the rinderpest pandemic, caused by an exotic cattleborne virus that decimated artiodactyl populations across sub-Saharan Africa (Bengis et al., Chapter 17). Kruger's existence as a national park is rooted in this calamitous period and its aftermath, so the actions of early management were understandably focused on internal restoration. What followed was recovery of wildlife within, intensification of agriculture and human settlement without, delineation of boundaries between, and the inevitable self-reinforcing set of inside/outside, conserved/degraded, them/us, privileged/impoverished dichotomies that typify the dilemma of protected area management in all parts of the world.

There are no "silver bullet" solutions in this part of the book, although, as with all vexing human issues, it helps to discuss them. We explain that systems are in place to integrate the science being conducted in and around Kruger to improve interdisciplinarity and promote the synthesis of results. There is a reorientation underway to manage the park from the inside out, for example, by establishing partnerships far beyond Kruger's boundaries to work toward a cooperative stewardship of the lowveld's main river systems. Finally, the park's management hierarchy recognizes that Kruger is more than a big game reserve. It is a self-renewing source of material, economic, and aesthetic benefits that can be marketed regionally, nationally, and globally for improved livelihoods in the land-use matrix in which Kruger is embedded.

Chapter 19

Anthropogenic Influences at the Ecosystem Level

STEFANIE FREITAG-RONALDSON AND
LLEWELLYN C. FOXCROFT

Anthropogenic changes are many and varied, exerting different levels of influence at different scales. In the global environment these changes cause complex responses strongly linked to ecosystem processes and are leading to unprecedented changes in global biodiversity. Sala et al. (2000) identified the main agents of change in savanna ecosystems as land use change, elevated CO_2, increased nitrogen deposition, climate change, and alien biota introductions. At the same time, aquatic systems are expected to be hugely affected by land-use change, alien invasions, and climate change. These drivers increasingly affect Kruger, which is also influenced by regional and local anthropogenic impacts that act at finer scales and over differing temporal time frames.

Kruger, the crown jewel of South African National Parks (SANParks), provides considerable financial revenues. Since its inception and sporadic development, management has been driven by a desire to minimize human influences and maintain "pristine" characteristics, no doubt shaped by the romanticized European view of the natural landscape before twentieth-century modernization (Carruthers 1995). However, even without overarching global influences this noble intention is contradictory because the Kruger ecosystem has been and still is affected by human presence, direct and indirect use, and management policies and actions. The myriad positive and negative human-induced impacts have all played some part in shaping the ecological and spiritual landscape of the Kruger we have today. These are overlaid on the geological template, geomorphological history (Chapter 5, this volume), prevailing climate, and ongoing spatial and temporal redefinition through the forces of nature, which in themselves are being shaped indirectly by humans. This chapter explores the varying human influences not covered by other chapters and examines their influences on heterogeneity.

Impact of Early Humans on the Kruger System

Humans affect the environment through physical presence and in an intangible social manner through decision making, induced conflict, religion, and other factors. Low population densities and low-intensity resource use by Stone Age hunter-gatherers probably would have constituted a low-impact period in the Kruger's prehistory, and it is accepted that early humans did not shape the environment in a permanent way (Table 19.1). Although savanna burning regimes would have had some influence on the Kruger area, little is known (Chapter 7, this volume). A medium-impact period followed during the Iron Age, dominated by metalworking skills and a more residential lifestyle based on hunting and pastoralism (Plug 1987; Carruthers 1995; Table 19.1). Population numbers in what is today Kruger are thought to have peaked around 15,000 in this period, resulting in localized homogenization of the ecosystem through agricultural practices. Nevertheless, Kruger is considered to have been a marginal or transitional area in terms of cultural-historical occupation and farming, with population fluctuations driven by climatic factors and disease (Meyer 1986).

From Colonialism to Conservation

The first non-Africans influencing the Kruger area were Arabian traders up to the eighteenth century, followed by the Portuguese control of the gold and slavery trade through East African ports and Dutch and Voortrekker pioneers a century later (Pienaar 1990; Carruthers 1995). The tsetse fly, carrying blood trypanosomes, no doubt slowed down exploration, exploitation, and settlement of the lowveld by Europeans in the nineteenth century. Although the tsetse fly was severely decimated during the rinderpest epizootic of 1896–1898 and ensuing drought (Smuts 1982), malaria and horse sickness in the wet summer months also posed a significant deterrent (Chapter 17, this volume).

The development of Kruger had its beginnings in the recognition that the impacts of humans on lowveld game populations and hunting prospects in the late 1800s and early 1900s were unsustainable and that the game needed protection (Chapter 20, this volume). Impacts of professional hunting in the erstwhile Transvaal were enormous (Carruthers 1995), and the decimation of elephant populations must have influenced structural heterogeneity in the region, but to what degree is difficult to assess.

Establishment of the Sabi Game Reserve in 1898 saw the first separation of the human component from the landscape when scattered villages were resettled west of the boundary (Pienaar 1990; Chapter 20, this volume). Over time Kruger became renowned through its wildlife research and management programs, mirrored by staffing structures, but the basic philosophy of protection-

TABLE 19.1

Generalized representation of the probable impact by early humans on the Kruger ecosystem.

PERIOD	LIFESTYLE AND MAIN RESOURCES USED	PREFERRED HABITAT	KRUGER OCCUPATION AND ESTIMATED NUMBERS
Stone Age	Hunting and gathering, using wild fauna and flora, firewood, and probably fire to drive game.	Wooded, hilly, game-rich areas close to water.	Probably the whole park, but southern Kruger concentrations estimated at 1 group per 7,000 ha (i.e., approx. 1 person/700 ha). Minimum of 150 sites (approx. 1,500 persons) and maximum of 300 sites (approx. 3,000 persons).
Iron Age	Hunting, farming, metalworking, and trading. Would have used natural fauna and flora, also for medicinal uses. Started domesticating plants and animals and used firewood, natural building material (wood, thatch, stone, soil), minerals (iron, copper, gold).	Good grazing areas with suitable soils, water, and climate for agriculture such as riverbanks. Suitable places for villages, with a preference for hills and rivers. Availability of wood crucial for domestic and metalworking purposes. Preferred disease-free areas for both humans and animals.	Almost the whole park, with settlement patterns centered on cattle and staple grain production. Settlements consisted of villages, cattle posts, and sometimes capitals. ~350 recorded sites spanning ~1,800 years. Gradual increase in human numbers to ~15,000 people. Occupation commenced by migration along the main rivers. During the earlier period, five high-density centers were located mainly in valleys of the Limpopo, Letaba, and Sabie rivers (Meyer 1986). Later, this era was characterized by conflict and forced migrations, with resettlement in the 1800s in the western parts, including Makahane, Masorini, Pretoriuskop, and Nsikazi and Crocodile rivers. The mid-1800s–1900 were an unstable period, with tribal wars, droughts, and raids resulting in only a few remaining temporary villages. Low human densities with impacts of tsetse flies and rinderpest.

ism prevailed. Nature conservators set the standards and norms based on biotic and abiotic associations, generally excluding the human component from the historical or management landscape. In contrast, it was soon noted that for conservation objectives to be met, the public had to be allowed access to Kruger. This resulted in the development of tourism facilities and infrastructure, in a manner aimed at maintaining the natural qualities of Kruger as far as possible. This essential paradox continues to this day, often with tension between activities (such as road construction) and the intended philosophy (minimum interference) behind them.

Proclamation of Kruger and its forerunners rested on two main drivers, politics and concern about the overexploitation of the region's wildlife (Carruthers 1995), although populations of species such as eland and buffalo were decimated by the rinderpest panzootic, not hunting pressures (Chapter 17, this volume). This resulted in boundary definition, fencing and protection, and road and firebreak development to facilitate much-desired control over the area. Since that time the Kruger ecosystem has been shaped by evolving game management practices, including water provision (Chapter 8, this volume), fire management (Chapter 7, this volume), elephant management (Chapter 16, this volume), and disease (Chapter 17, this volume). Changing management philosophies have affected how these impacts were dealt with and the intensity of their outcomes in time and space (Chapter 4, this volume). Management paradigms evolved with increased understanding of the ecosystem, scientific advances, and theoretical developments. This includes the notion of stable states, the balance of nature versus fluctuations, and the implicit goal of managing for ecosystem health and heterogeneity rather than homogeneity, as well as a perceived notion of untouched primeval naturalness and wildness (Chapter 1, this volume).

Boundaries and Fencing

The Sabi and Shingwedzi reserves were characterized by low species densities, local extinctions, sparse settlement, and seasonal use by humans. In 1926, proclamation of the Kruger National Park resulted in western boundary definition through negotiations with landowners, without considering ecological boundaries. Main drivers of boundary fencing were the protectionist segregation philosophy to conservation (Carruthers 1995; Chapter 20, this volume), veterinary control requirements (Chapter 17, this volume), and political boundary definition with Mozambique and Zimbabwe. By 1976 Kruger was entirely fenced, although segments were again removed from the mid-1990s onward as neighboring conservation-oriented holdings went into agreements with Kruger. Although some boundary adjustments and land swaps have had political implications only, others have resulted in biodiversity losses. Of significance were the 1960–1961 and 1967–1968 southwestern boundary changes that excised areas west of the Nsikazi River, including Numbi Hill, to facilitate provincial road construction. This resulted in loss of unique habitats from Kruger, including the last permanent refuge of red duiker, the only suitable habitat for oribi, and some of the best grey rhebuck and favored roan and sable habitat in Kruger (Pienaar 1990). Fencing not only affected these locally rare species but also affected east-west migratory patterns of wildebeest and zebra populations (Whyte and Joubert 1988).

Compounding impacts of fencing include highly local effects such as injury, maiming, and death of individuals, subpopulation extinctions (Joubert 1986), and problem animal control. Although these effects pose significant risks to the individual, they are not big when considered at the population dynamics scale. In contrast, positive spinoffs include facilitating boundary patrols through associated road construction, inhibition of poaching activities, prevention of straying or luring of rare or human-conflict species, and barriers to spreading of disease (Chapter 17, this volume). Nevertheless, effectiveness depends on adequate fence maintenance, which often is not the case. Similarly, the exact positioning of Kruger's legal boundary is ill defined along certain rivers and remains a controversial point of friction between Kruger and riparian neighbors.

Fences have separated land management practices, resulting in sharp fence-line contrasts. These are generally unquantified in the Kruger context, and visual interpretation of changes in heterogeneity may appear harsh, although plant diversity (all species weighted equally) may be higher in adjacent communal lands (Shackleton 2000). With or without fences, the effective size of Kruger is gradually shrinking in certain areas through land-use change and encroaching development, whereas the "Greater Kruger Park" expands in other areas as private and provincial conservation areas are incorporated.

Animal Population Management and Poaching

This has historically been a central focus of conservation and research activities in Kruger. Large herbivore responses to protection measures defined management actions, and by 1912 it was believed that most populations responded well. Stevenson-Hamilton's game laws prohibited illegal killing of wildlife, rangers were appointed to enforce legislation, and by 1925 herbivore populations had recovered to pre-exploitation levels of around 1880. Herbivore population fluctuations were monitored and ascribed to environmental conditions or management actions, with severe population declines or increases resulting in population management activities (Joubert 1986).

Population Augmentations

The Game Reserves Commission made provision for reintroductions and artificial regulation of game numbers in 1918, with no formal policy until 1949. This policy vacuum resulted in consideration of exotic species introductions, although these proposals were not implemented because funding was insufficient. Later, numerous introductions, reintroductions, and translocations, particularly of antelope species, took place (Table 19.2). Most occurred after the advances in wild animal capture and care techniques of the 1960s, which paved the way for acceptable and safe means of effecting translocations. Rhino conservation has been at the forefront of these efforts. The last naturally occurring

TABLE 19.2

Wildlife reintroductions and other augmentation efforts in Kruger.

SPECIES	DATE	REINTRODUCTION	SUCCESS
White and black rhino	Numerous	Numerous	Highly successful.
Crested guinea fowl	1930	20 sent to Kruger	Apparently successful.
Lichtenstein's hartebeest	1985	6 from Malawi	Successful but with much effort: 65 released from Nwatshitsumbe camp, 50 from Hlangwine camp. Still breeding in the camps.
	1986	15	As above.
Roan antelope	Unrecorded	12 from Zimbabwe to Hlangwine camp	Unsuccessful; did not adapt, primarily because of tick burdens. Population depleted by anthrax and drought in 1960s and 1970s.
	1984	To Rietpan area	Unsuccessful.
Mountain reedbuck	1974–1976	370 from Mountain Zebra National Park to Stolznek	Population healthy, often seen.
Grey rhebok	1978	20 from Golden Gate Highlands National Park to Khandizwe plateau	Limited success. Suitable habitat is limited in Kruger; this is probably the only suited area (west of Malelane).
Oribi	1962	29 from Badplaas to Fayi camp (Pretoriuskop)	Unsuccessful. Unsuitable habitat (Novellie and Knight 1994).
	1972–1973	98 from Badplaas released between Stolznek and Pretoriuskop	Unsuccessful. Did not adapt because of limited suitable highveld grassland habitat and fire regimes.
Aardwolf	1962	Badplaas (with the oribi)	Unknown.
	1979	3 released at Punda Maria from Natal	Unknown.

(continued)

white and black rhinos were seen in the region in 1896 and 1936, respectively, and reintroductions of white rhino began in 1960 and were followed by black rhino reintroduction in 1972 (Pienaar 1970; Pienaar 1994). Kruger's healthy, expanding white rhino population, estimated at around 5,000 animals, is now used for translocation to other areas as part of the white rhino conservation program, with 40 to 100 white rhino moved out of Kruger annually. Most other Kruger reintroduction efforts have been unsuccessful (Table 19.2). Although this can be attributed to exclusion of suitable habitat by boundary changes over time in some instances, limited data are available to adequately assess reasons for success or failure. Success rates are highly dependent on habitat suitability and size and composition of the founder group, but documentation of dates, numbers of animals, and success rates remains poor in conservation circles as a whole (Novellie and Knight 1994).

TABLE 19.2 *(continued)*

SPECIES	DATE	REINTRODUCTION	SUCCESS
Suni	1982	4 released at Punda Maria from Natal	Unknown.
	1989	20 from Natal, 9 from de Wildt to Skukuza pens	Did not do well in captivity.
Red duiker	1972	27 from Mariepskop to Shabinkop, Newukop, and Sabie River	Unknown.
	1981	21 released between Sabie and Sand rivers	Unknown.
Nyala	1980	21 brought from Natal to Sabie River	Successful. Regularly seen along Skukuza–Lower Sabie tarred road.
Eland	1971	27 from Addo Elephant National Park to Hlangwine camp	Vulnerable to high tick counts and few calves raised; released from the camp and supplemented with bulls from northern Kruger.
	1972	6 from Mountain Zebra National Park to Hlangwine camp	As above.
Sable antelope	1970	19 from Crocodile River Estates at Hoedspruit to northern Kruger	11 released into Nwashitsumbe camp bred well, and 64 animals were released in 1976.
Samango monkeys	1982–1988	95 from Entabeni near Louis Trichardt released along the Luvuvhu River	Sometimes seen in the Pafuri area.
Cheetah	1968–1969	34 from Namibia to Tshokwane, Malelane, and Crocodile Bridge areas	Very limited success.
Tsessebe	1972	Unrecorded numbers taken from the north of Kruger and Hoedspruit to Hlangwine enclosure	Heavy mortalities, but 4 survived.
	1978	2 bulls from Percy Fyfe Reserve to Hlangwine camp	Unrecorded.

Population Reductions

With an early approach to management described as pragmatic intervention, population reductions of various species were a regular feature. Initially this targeted predators (considered overabundant vermin, threatening stock, human lives, and conservation efforts), with control peaking in 1911–1920 and 1951–1960 in response to concerns about small or declining ungulate populations (Figure 19.1). Aims and objectives of culling changed under the influence of different wardens (Smuts 1982), officially ceasing in 1975. Impacts of these actions on lion and other predator populations in Kruger are speculative. Although they would have led to reduced populations at specific times, high reproductive rates and turnover may have buffered the population from extreme impacts, and knock-on effects probably were not significant in changing herbivore densities and structures. This contention is based on the outcomes of intensive lion and spotted hyena culling between 1975 and 1980, which resulted in negligible population impacts on lions and their prey populations (Smuts 1978; Whyte 1985; Chapter 18, this volume). Today, carnivore populations on Kruger's boundaries result in antagonistic encounters with neighboring communities. Costs of repatriation of animals are high, and often "problem animals" are killed or hunted on or along the boundaries. This edge effect probably is symptomatic of the Kruger source area not having a sink to which excess animals can migrate and naturally be removed from the system.

The 1960s and 1980s were characterized by large herbivore population control (Figure 19.1). In 1965 the board authorized artificial control of elephant (Chapter 16, this volume), buffalo, hippo, giraffe, wildebeest, zebra, and impala, based on the principles that large herbivores are limited by water and grazing during droughts and that high-density species could affect habitat and species diversity and compete with rarer species. Similarly, culling during high-rainfall periods was motivated by the constant fear of drought. From 1985 onward, an alternative to culling, providing financial return, became standard practice through the capture, sale, or donation of species. Numerous animals have been translocated from Kruger, increasing the budget available to SANParks for land acquisitions. The most significantly affected species is the white rhino, with an annual sale and translocation of 1–2 percent of the population. Black rhino are removed on a limited and ad hoc basis, usually to boost populations in other national parks or for political reasons. Although elephant translocations have been limited in the past because availability of alternative ranges is limited, the recent transfrontier agreement has resulted in the initiation of large-scale translocation of this species to Mozambique. Impacts of such actions on the overall Kruger ecosystem are unknown but are surely limited in the current framework of operations.

Poaching and Illegal Exploitation

Poaching has received attention since Sabi Game Reserve proclamation, with Stevenson-Hamilton establishing the first antipoaching unit in 1903. Common practices of snaring and fishing became illegal with the protectionist approach at the end of the eighteenth century and resulted in conflict around access to natural resources (Carruthers 1995; Chapter 20, this volume). These practices

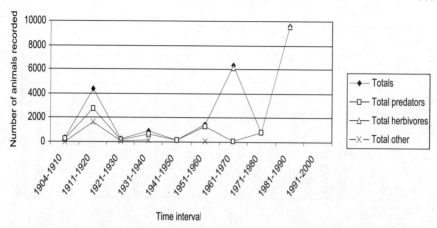

FIGURE 19.1. Total recorded numbers of animals culled in Kruger (excluding elephant). Data from Joubert (1986).

would have had an insignificant impact on game populations, but this changed with the advent of organized rhino and elephant poaching in the late 1970s and early 1980s. Poaching incidences have fluctuated in recent times, linked to increasing poor populations on Kruger's borders and a high demand for rhino horn and ivory (Figure 19.2). Today's antipoaching operations are highly organized and logistically and financially well supported and play a deflective and proactive role outside Kruger's boundaries. Success has been achieved through informer networks, general education, and counterincentives, with poaching operations well under control at present (relative to other conservation areas in Africa) and considered to have limited population effects (Figure 19.2).

Even rough estimates of illegal exploitation of plant resources in Kruger are difficult (Botha 1998). Although Botha et al. (2001) allude to occasional noncommercial harvesting, it is speculated that substantial amounts of medicinal plant materials leave Kruger. The annual South African trade in medicinal plants has been estimated at around $500 million, with an estimated 494–741 tons from 176 species used in Mpumalanga per year (Botha 1998; Botha et al. 2001). Some resources are harvested in Kruger because they have become scarce elsewhere, and others are collected by resident staff with easy access (e.g., bark in Skukuza staff village). Harvesters generally select for size to maximize their returns (Botha et al. 2001), which may affect population recruitment in rarer or slow-growing and maturing species. Increasing demand for such products will exert increasing pressures on wild populations if long-term resource management programs are not established. The exact impact of these practices is unknown but has the potential to lead to local species extinction, as has happened in areas outside Kruger (Botha 1998).

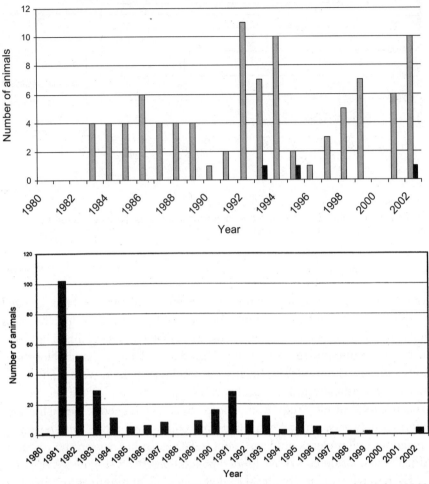

FIGURE 19.2. (a) Kruger rhino poaching incidents between 1980 and October 2002. White and black rhino numbers are indicated by gray and black bars, respectively. (b) Elephant poaching incidents in Kruger between 1980 and October 2002. The Convention on International Trade in Endangered Species ban came into effect in 1989–1990.

Tourism

Kruger's appeal as a tourist destination is enormous, and it is a major driver of economic development in the region. This, together with its annual net income, has led to Kruger being perceived as the goose that lays the golden egg (Ferreira and Harmse 1999). Economic and political pressures to increase revenue generation, attract more visitors, provide benefits to neighboring communities, and become more accessible often are juxtaposed to maintenance of attraction integrity and long-term sustainability of economic and environmental parameters. Tourism development is commonly regarded as an economic stimulus but carries with it impacts and environmental costs that constrain activities and may result in degradation of the resource base on which tourism is centered. Sustainable operations must do more than merely stem and mitigate negative impacts; critical links between tourist numbers, thresholds of use, and environmental impacts must be considered and understood.

Tourism and its role as revenue generator were entrenched in Kruger in 1926. The opening of Skukuza, Satara, and Pretoriuskop camps in 1928 followed the winter opening of the Pretoriuskop section for day trips in 1927. Hot water provision was considered in 1933 only after long debates over whether this constituted an unnecessary luxury, and pit latrines were replaced by waterborne sewerage systems in 1961. Since 1926, tourism has increased dramatically, with peak figures reaching almost 1 million visitors per annum in 1997–1998 (Figure 19.3). Facilities and infrastructure have expanded, and services and experiences have diversified in accordance with the National Parks Act (No. 57 of 1976, as amended), which provides for sustained use of national parks "for the benefit and enjoyment of visitors," a provision that has been debated over time (Braack and Marais 1997). Throughout Kruger's history park managers have expressed concern that tourist pressures were approaching capacity levels (Braack and Marais 1997; Ferreira and Harmse 1999), including issues of traffic congestion, overuse of facilities, and impacts on visitor enjoyment. Peripheral development and social carrying capacity principles were first considered in the 1940s, aimed at siting new developments close to boundaries and avoiding enlargement of centrally located camps. Tourist number control strategies were amended over time and included limiting camp size, developing advance booking systems, limiting overnight and day visitor numbers, implementing vehicle:road length ratios, and zoning Kruger for development and use (Ferreira and Harmse 1999).

Balancing Conservation with Tourism

The importance of balancing conservation achievement, scientific value, economic viability, and cultural value was recognized in 1962. It was argued that conservation success entailed economic returns and that nature conservation was for the benefit of humans rather than for its own protection. The importance of recreational value in its widest sense included spiritual, intellectual, and physical renewal. Therefore, in the mid-1970s interpretive services' role included disseminating knowledge and increasing visitor sensitivity for conservation issues, arguing that the ill effects of tourism could be reduced by reducing visitor ignorance (Joubert 1986).

More recently, the needs of tourism and conservation have been addressed through zoning techniques. Nevertheless, zoning has been a source of debate,

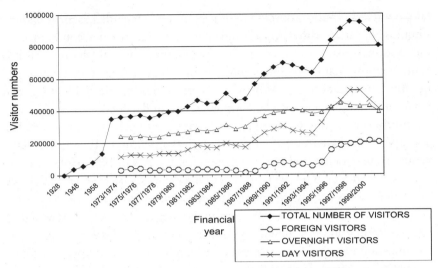

FIGURE 19.3. Numbers of annual visitors to Kruger since 1928. The total includes all foreign, local, overnight, and day visitors, but foreign visitor numbers are included in both the overnight and day categories.

with contention revolving around the basis, definition, and size of zones, their desired qualities, use, and management (Freitag-Ronaldson et al. in press). Perceptions, expectations, and interpretation of zoning no doubt have resulted in much of the debate. Wilderness management and zoning have become a cornerstone objective of the Kruger management framework (Braack 1997) and resulted in the Recreational Opportunities Zoning (ROZ) policy, which guides infrastructural and ecotourism development. This provides a heterogeneous template for directing levels of management and tourism activities. It is based on the premise that grades of wilderness experience and opportunities should be made available to visitors without impeding biodiversity management goals (Venter et al. 1997). This policy is under review because of ongoing challenges and a perception by some that it restricts tourism expansion.

Dealing with Tourism

An annotated list of visitor-influenced wildlife impacts in Kruger points toward some effects on the structural, functional, and compositional aspects of biodiversity (Freitag-Ronaldson et al. in press). Although <3 percent of Kruger is directly disturbed by infrastructure, the overall impact of tourism activities on all aspects of biodiversity is poorly understood and often is not recognized explicitly. Although direct resource use (water, wood, gravel, waste generation) is measurable, it is often not taken into serious consideration when tourism activities or infrastructure projects are planned. For example, the practice of providing firewood from the field for visitors reached such enormous proportions that it was considered unsustainable and was replaced by liquid petroleum gas, commercial charcoal, and wood from exotic species. An indication of the scale of such use is reflected in the 2000–2001 liquid petroleum gas sales, which reached more than $27,000 at picnic spots alone. Other unquantified side effects of tourism in Kruger include road network development, water use, waste generation, and electricity consumption and may even include vehicular emissions. Sustainability of development is being investigated holistically through evaluation of the overall human ecological footprint (resources consumed and waste generated) in all parastatal conservation areas.

Developmental and revenue generation forces increasingly threaten the greater sense of place in Kruger. Added pressures around the borders impinge on this conservation area, effectively reducing the nonimpacted core area and the distribution of wilderness attributes sought by visitors. A zoning system alone will not provide a holistic approach to the protection of tourism qualities, an integrated environmental management approach is needed. Kruger must consolidate a balanced plan to guide further development, combining societal values, biodiversity conservation, precautionary principles, and sustainable development.

Roads

Road ecology (Foreman and Alexander 1998) has received little attention, even though the conspicuous presence of roads fragments the Kruger landscape. Road impacts operate on two primary levels: the individual, species, and population level and the ecosystem process and landscape level. The former impacts influence abundance, distribution, mortality, and colonization rates (Tshiguvho 2000) and includes road kills (with spates of mortality, probably in seasons when vulnerable species forage more frequently on roads or night traffic rates are higher), road avoidance by animals because of noise, and barrier effects with possible demographic and genetic consequences of population subdivision (Forman and Alexander 1998). At the process level, hydrological, erosion, sedimentation, and chemical effects, nutrient cycling, and alien invasion impacts are operating. From observations in Kruger, this seems to be limited mainly to the road edge zone, although this is worthy of investigation. Similarly, little is known about the wider scale effects of habitat and landscape fragmentation, which may affect a variety of species, communities, and ecosystem functions. Certainly the impacts of roads will be species, ecosystem, or landscape specific, varying with road type and associated characteristics (Tshiguvho 2000).

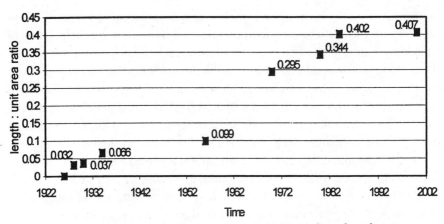

FIGURE 19.4. Kruger road development over time, measured as a length:unit area (km/km^2) ratio. Data taken from Joubert (1986) and information supplied by Kruger Technical Services Division.

There were no vehicles in Kruger in 1926, and the Selati railway, pack donkeys, and horses were used for transportation. The need for roads accompanied increasing management requirements and the advent of tourism, with the first car tracks opened in 1928 and the first road tarred in 1961. Since then, the road network (including firebreaks, management roads, and airstrips) has expanded to 7,926 kilometers, dramatically increasing between 1956 and 1970 (Figures 19.4 and 19.5). Here we conservatively estimate the immediate road verge effect by buffering roads by twice their width (an arbitrarily selected buffer width, with gravel road width = 6.5 m and tar road width = 12 m, compared with maximum road verge transect widths of 20 m in the Tshiguvho et al. [1999] study). This results in a corresponding immediate road-affected area of 29,751 ha, or 1.5 percent of Kruger's surface area. Generally these effects operate at the local and individual scale, with negative effects mainly on invertebrates, small mammals, and timid game species (Pienaar 1968) and positive effects on species such as scavenging ants (Tshiguvho 2000) and plants (e.g., *Colophospermum mopane*, *Dichrostachys cinerea*), which have encroached road verges.

Of greater environmental consequence are fragmentation effects, poorly drained or badly planned road networks, and erosion problems as well as the impacts of maintenance and construction activities such as gravel use. Although no quantitative data exist, a high proportion of unnatural erosion is associated with road-building activities in Kruger, most severe in duplex soil types. Gravel pits reflect the environmental costs of road building in Kruger, with more than 12.5 million m^3 of gravel excavated over the years (Table 19.3). Annual maintenance estimates of 713,000 m^3 of gravel are influenced by rainfall and flood

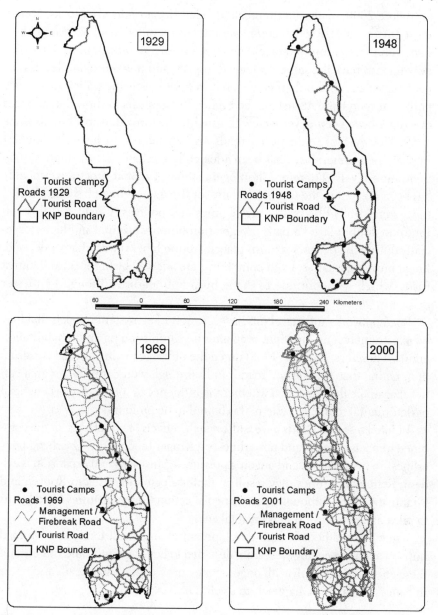

FIGURE 19.5. Tourist and management road networks of Kruger at different times.

events; almost 500,000 m³ gravel was used after the February 2000 floods. Approximately 1,000 gravel pits of differing sizes, depending on gravel content, construction needs, and locality, are largely unrehabilitated. They are unsightly and have other detrimental effects such as withholding water and acting as unnatural water sources throughout the year, sometimes in areas that animals would have otherwise vacated in the dry season (Chapter 8, this volume).

Before 2000, roads and firebreaks divided Kruger into 456 blocks, averaging 4,164 hectares. Road decommissioning has resulted in an increased block size, with three or four old blocks now constituting new burn units. However, not all roads have the same ecological impact, and fragmentation levels may become meaningful only if considered in light of specific policies (e.g., fire policy) or road type. At first glance Figure 19.5 appears to show that all areas of Kruger have been developed with roughly the same density of roads since 1969. However, if main paved roads are considered to have the highest impact, the southern area has been affected longest and has the highest fragmentation levels. Therefore, homogenization of spatial processes probably has been greatest in the south. Road use for fire management has also affected the system. Burn blocks and fire regimes and policies (Chapter 7, this volume) no doubt played a part in vegetation homogenization in blocks (especially during the perimeter burns placed around blocks) with sharply defined, linear burn boundaries. Even point burns are affected by firebreak and tourist roads, which will continue to shape burn boundaries, although to a lesser degree.

Road networks and gravel pits have a variety of ecosystem impacts that may be of extremely long duration, even after decommissioning and rehabilitation efforts. Therefore, Pienaar (1968) urged the utmost care during road construction, noting that "wherever a road is built through wild country, this area can never again be the same, and whether the influence of such a road is beneficial or detrimental to the ecosystem, it is bound to be profound" (p. 174).

Limited research efforts have addressed the effects of powerlines, or the combined impacts of roads and powerlines, on Kruger biodiversity. Overhead powerlines are unattractive, and ongoing pressure against any such expansion is evident. Reticulation lines may result in raptor electrocutions (van Rooyen and Grantham 1998), although some remedial action (design features or additions to prevent electrocution) has been taken.

Kruger would benefit from development of an overall framework on road impacts and road ecology research, designed to better quantify and understand these effects, particularly with ongoing pressure to build more roads and tracks in concession areas (Chapter 1, this volume).

Invasive Alien Species

Invasive alien species pose the second greatest threat to global biodiversity (IUCN 1997). Impacts include replacement of diverse systems with single (or mixed) species stands, alteration of soil chemistry, geomorphological processes, fire regimes, hydrology, extinction of compositional diversity, and threats to indigenous fauna (Cronk and Fuller 1995), displacement by competition,

TABLE 19.3

Conservative estimates of gravel quantities used for construction and maintenance of the Kruger road network (as supplied by the Kruger Roads Department).

ROAD TYPE	DISTANCE (M)	AVERAGE WIDTH (M)	AVERAGE CUMULATIVE GRAVEL DEPTH USED OVER TIME (M)	GRAVEL USED (M^3)
Gravel tourist roads	1,768,000	6.5	0.25	2,873,000
Tar roads	885,000	12	0.8	8,496,000
Graveled management roads (firebreaks)	737,000	6.5	0.25	1,196,000
Airfields	12,800	30	0.5	192,000
Total				12,757,000

reduced structural diversity, increased biomass production, and disruption of prevailing vegetation dynamics (van Wilgen and van Wyk 1999). Furthermore, Richardson et al. (1997) suggest that in southern Africa the destruction of riparian habitats is a key impact, and alien plant invasions have directly resulted in the extinction of species.

Kruger has been invaded by numerous taxa, particularly plants. Undesirable effects of alien plant species were first recognized in 1937 by Stevenson-Hamilton, who stated that the introduction of exotic types of fauna and flora "should be religiously avoided" (Stevenson-Hamilton 1937, p. 260). Nevertheless, policy proposals over time (prohibiting cultivation of exotic species and supporting their removal from Kruger) did not significantly affect resident staff attitudes toward their gardening habits or the growing threat of alien plant invasions. The first alien plant list of six species was compiled by Obermeijer in 1937, and the current Kruger estimate stands at 367 species (Foxcroft and Richardson 2001; Figure 19.6). Most habitat types have been invaded, with disturbed habitats most affected (rivers, rest camps, staff villages, and road verges). In 1997 invasive alien species were highlighted as posing a great threat to biodiversity conservation in Kruger, highlighting the general lack of institutional learning and recognition over time of the long-term threats posed by alien plants (Foxcroft and Richardson 2001).

The overemphasis on compositional diversity has been accompanied by an inadequate awareness of structural and functional diversity and the role of invasive species in driving systems into more heterogeneous or homogenous states. Similarly, Kruger research surveys are not designed to specifically explore direct biodiversity impacts but implicitly assume that such impacts are negative and undesirable. One measured biodiversity response came through an experiment on efficacy of control methods for *Lantana camara*, which showed that there was a significant increase in biodiversity after its chemical removal (Erasmus et

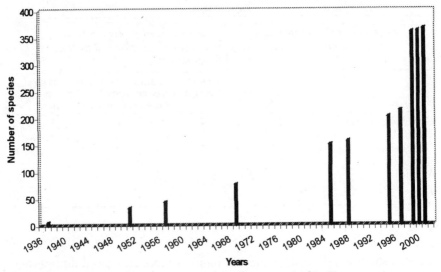

FIGURE 19.6. The number of invasive alien plants reported in Kruger surveys between 1937 and 2001 (data from Obermeijer 1937; Codd 1951; van der Schijff 1957, 1969; Gertenbach 1985; MacDonald and Gertenbach 1988; Anon. 1995, 1996; Foxcroft 1999, 2000, 2001).

al. 1993). At this point *Lantana camara* is considered the most serious riparian invader, covering an area of 30,028 ha, with some patches having attained 100 percent cover in the past. Ongoing control efforts and frequent flooding limit its proliferation, although propagule pressure from surrounding areas facilitates reestablishment. The impact of terrestrial invasives such as the Cactaceae (specifically *Opuntia stricta* in Kruger) is pronounced. It has been shown that *Opuntia stricta* affects vegetation structure in densely infested areas of Kruger (Lotter and Hoffmann 1998). Experiential evidence indicates that in a fire-dominated savanna such species disrupt natural ecosystem cycles where they form dense infestations, excluding other plants such as grasses, affecting fuel loads, fire frequency, and intensity. What we don't yet know is whether species with allelopathic properties inhibit establishment of indigenous species in their close proximity, increasing the disturbance effect and providing a niche for other invasive species or for the expansion of their own range.

Control efforts initially were aimed at eradication, beginning with *Melia azedarach* in 1956 (Joubert 1986). Eradication attempts generally have been unsuccessful. Control efforts now recognize the more realistic target of maintaining infestations at acceptably low levels (the definition of *acceptable* is not rigorously defined but varies by species and refers to the logistical resource needs for controlling alien plant populations). Management operations involved mechanical and chemical means and were improved by the use of registered herbicides, with initial research and funding addressing herbicide efficacy and effects. Contributions of the National Working for Water Program (a large, nationally deployed initiative aimed at poverty relief through employment of people to remove alien plants) have injected more than $4.2 million (at 1998 exchange rates) into Kruger since 1997 and cleared more than 8 million alien plants by 1999 and 92,700 infested ha since then (this change in measurement of effectiveness follows the changed nomenclature used in the program). Most alien species occurring in the field and along river courses (i.e., excluding those found in rest camps and staff gardens) have reached close to maintenance levels, necessitating annual follow-up treatments. Introduction of biological agents by the Plant Protection Research Institute began in 1985 to control *Pistia stratiotes*. Since then an additional 16 biocontrol agents have been introduced, with varying success rates, and have provided an additional tool now used widely in integrated control programs in Kruger (Lotter 1996, 1997; Lotter and Hoffmann 1998). Nevertheless, biocontrol has not always provided the short-term effects demanded by management (Box 19.1), with fluctuating cycles of plant diebacks and recovery as biocontrol agents track their host population demographics.

Management by Objectives

Until recently, charismatic aspects of wildlife management received disproportionate attention and budget allocation, with alien control operations receiving lower priority (Foxcroft and Richardson 2001). The 1997 management-by-objectives approach in Kruger sparked a strategic adaptive management response to invasive alien research and control, integrating science and management more tightly than before. A monitoring and goal maintenance system (Bestbier et al. 2000) provides a structured approach enabling multitiered threshold of potential concern (TPC) assessment (Figure 19.8; Chapter 4, this volume). The first-order TPC states that all invasions into Kruger are undesirable and must be prevented where possible. Therefore, TPC notification follows the first record of an invasive alien plant approaching or entering Kruger. As these plants disperse, second-order prevention-of-spread TPCs are activated, aimed at reducing the dispersion of the species. These are followed up at the third level, which strives for suppression of plant densities (i.e., TPCs come into effect when densities increase by a predetermined amount). Consideration is given to priority taxa most likely to become troublesome, maintaining them at minimum levels, whereas other species are not controlled and fluctuate without intentional intervention. Scale issues are important in successful control program implementation because the definition of priorities changes at different scales. Species significant at the Kruger level therefore may not be as important to control at the regional or national level, and species reported at the level of a ranger section may not carry much weight when viewed across the entire park and weighed against

BOX 19.1
Biological Control in Managing Alien Plants in Kruger
LLEWELLYN C. FOXCROFT AND JOHN H. HOFFMANN

Biological control has been used to manage several alien invasive species in Kruger, including *Opuntia stricta, Lantana camara, Pistia stratiotes, Salvinia molesta, Eichhornia crassipes,* and *Azolla filiculoides*. *O. stricta* is the most problematic invasive species in Kruger, with more than 30,000 ha in the Skukuza region invaded to various extents (Lotter and Hoffmann 1998). Historically, the invasion displayed a typical lag phase, with plants remaining scarce and limited in geographic distribution. However, in recent years the dispersal and densification of the species have been rapid despite intensive control efforts (Figure 19.7). Initial control efforts relied on chemical herbicides, but they failed to produce satisfactory results, largely because of the extent of the infestation and shortage of labor.

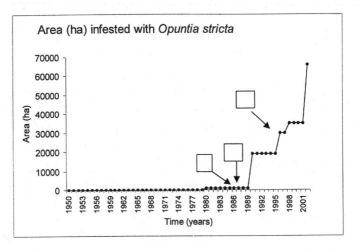

To improve control and limit expansion, biological control was initiated in 1988 with the introduction of *Cactoblastis cactorum* (Lepidoptera: Phycitidae). The release was successful, and the moth readily became established in Kruger although, for unknown reasons *C. cactorum* has not reached effective levels of abundance for *O. stricta* control (Hoffmann et al. 1998a, 1998b). Nevertheless, *C. cactorum* plays a major role in curtailing regrowth of *O. stricta* in infestations treated with herbicides. This strategy provides a rare example of the successful integration of biological and chemical control methods and has achieved substantial savings for Kruger.

A particular biotype of cochineal, *Dactylopius opuntiae* (Hemiptera: Dactylopiidae), was obtained from Australia and released in May 1997 into Kruger, where it readily established in the vicinity of Skukuza. Large, dense clumps of cactus were completely destroyed within 18 months, although cochineal populations declined dramatically during the extreme rainfall experienced in 2000–2001. Cochineal populations have recovered slowly, but periodic heavy rainfall and limited dispersal abilities of the insects have slowed progress (Foxcroft and Hoffmann 2000).

To date the combined impact of *C. cactorum* and *D. opuntia* has not been able to prevent the spread of *O. stricta* (Figure 19.7). However, the insects have curbed the rate of long-range dispersal and densification of *O. stricta* (Hoffmann et al. 1998a, 1998b). Precedents elsewhere indicate that cochineal insects effectively destroy target weeds during periods of below-average rainfall (Moran et al. 1987; Moran and Hoffmann 1987), and it is anticipated that the same will happen in Kruger in the next dry period. Currently *D. opuntiae* is being mass reared under hothouse conditions and released throughout *O. stricta* infestations in and around Kruger.

In general, too little effort is expended on research to evaluate the effectiveness of control strategies for alien invasive plants and other pests. This often results in inappropriate and ineffective control methods, with an enormous waste of time, effort, and resources. The development and implementation of a workable integrated control program against *O. stricta* in Kruger have been possible through research and management integrated in a targeted manner.

the overall priorities. Therefore, managers must match scales to achieve appropriate prioritization and resource allocation.

Implicit in invasive species control are mitigation measures focusing on vectors of dispersal and, where possible, limitation of propagule distribution. It has long been recognized that rivers are the main vectors of invasion into Kruger (MacDonald and Gertenbach 1988; Foxcroft and Richardson 2001), with heavily invaded catchments pressurized by increasing development, commercial farming, and forestry. Until infestations in upper catchments are better controlled, control efforts in Kruger will, at best, remain an attempt to maintain populations at the lowest possible levels. However, internal sources are also important if long-term effective control is to be achieved. Propagule movement by animals and birds implies that before dispersal of an exotic plant species can be successfully limited, all flowering and fruiting populations must be controlled and followed up before the next season of setting seed, as is done in the management of *Opuntia stricta* (Lotter and Hoffmann 1998; Hoffmann et al. 1998a, 1998b).

Human-assisted invasion in Kruger is discussed by Foxcroft (2001), highlighting the role played by staff in introducing species into restcamps and gardens for ornamental and other uses. In 1935 ranger Hoare objected to the planting of flam-

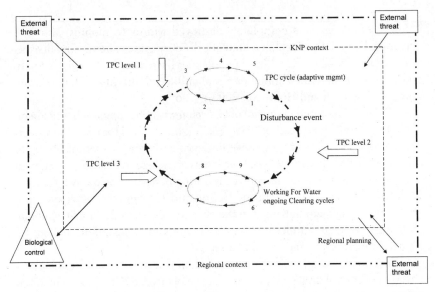

FIGURE 19.8. The flow of thresholds of potential concern (TPCs) indicating the various TPC levels for invasive species management: 1, analyze; 2, notify TPC reached; 3, take management action; 4, provide feedback; 5, measure; 6, contract; 7, clear; 8, monitor; 9, follow up. TPC levels: 1, prevention; 2, measuring dispersal in Kruger; 3, measuring density of species.

boyant *Delonix regia* at Pretoriuskop; this is one of the first efforts to actively prevent an intentional introduction of an alien species (Joubert 1986). MacDonald and Gertenbach (1988) state that 13 of the 156 alien species recorded were in cultivation in gardens and camps, with 2 of those species (*Lantana camara, Melia azedarach*) considered to have high impact on the Kruger system. A survey of the Skukuza camp and staff gardens in 1999 listed approximately 240 exotic species and culminated in a policy that prioritizes species for removal in a phased approach.

Spatial Structure, Lag Periods, and Uncertain Trajectories

Invasive species spread depends on point of origin, transport, and ability to establish and spread into new habitats. Riparian systems are especially sensitive to invasions because they combine high disturbance rates with high fertility and propagule influx (Prieur-Richard and Lavorel 2000). The February 2000 floods altered the structure of the Sabie and other rivers in Kruger, affecting vegetation structure and alien vegetation composition. The initial riparian aliens were quickly replaced by stands of annual herbaceous *Ageratum conyzoides* and *A. haustonianum*, followed and replaced by others. In addition to rapid colonization, the floods provided a vehicle for new alien species to enter Kruger in great numbers, including *Mimosa pigra* and *Chromolaena odorata*. *Chromolaena* formerly was found only sporadically in small numbers in Kruger and has now become widespread along all major rivers. Interestingly, Kruger scientists had already been warned of the threats posed by *C. odorata* because its habitat is comparable to that of *L. camara* (considered the greatest threat to Kruger riparian systems after the 1984 floods). The adaptive management framework adopted in Kruger, and associated TPCs, provide a structured and active approach to ensuring that such warnings are taken seriously in future and proactively addressed.

Humans have contributed to biogeographical migration of species (Hodkinson and Thompson 1997; Vitousek et al. 1997). Alien species began seriously invading the lowveld only in the last 100 years, which is late compared with the invasion of coastal areas (Henderson and Wells 1986) and is linked to the inhospitability to settlement at the time. Temporal issues have important implications and consequences in long-term invasive species impacts and management. An apparently benign lag phase precedes rapid dispersal and invasion (Mack 1985; see Box 19.1), after which invasive species rapidly occupy all available niches at the greatest density allowed by physical and environmental constraints. This sequential trend may be broken in certain cases but is the exception rather than the rule. Hypothetically, when all available space is occupied and a dense monoculture of invasive species exists, the system should be regarded as transformed (space 4, Figure 19.9a), with little chance of reclaiming it to former composition and function. With Kruger's history of alien plant control, geographic distributions and densities of invasive species have been limited, but whether this will hold in the long term remains to be seen. If left unchecked, invasive species exhibit abilities to move ecological systems from natural heterogeneity to homogeneity over a large scale and time frame. However, serious alien invasives (i.e., transformers and others) that are controlled effectively should be suppressed to maintenance levels with tolerable impact, effectively staying within or reverting to the lag phase condition (Figures 19.9b, 19.9c). Species not controlled will fluctuate as environmental conditions dictate, switching between scattered and locally dense infestations.

Variations in temporal invasion are not always fully understood, particularly for seasonal species. Distinctive cycles of recurrence have long been noted, with extensive stands in some years followed by a complete absence in others (e.g., *Abutilon angulatum*, *Zinnia peruviana*, and *Nicotiana glauca*). Are such species responding to environmental parameters that allow them to flourish at some stages but not others, and, if so, how long are these cycles? The extent of species displacement and rearrangement of floristic patterns during invasive cycles are unknown but would help management assessments maximize use of limited resources. The real question therefore is whether fluctuating infestations of some species such as *N. glauca* on sandy stretches

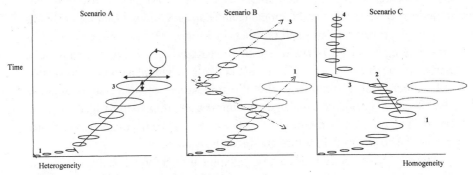

FIGURE 19.9. Scenario A: Trend of plant invasion over time and space, with fluctuations between heterogeneity and homogeneity: 1, lag phase; 2, trend line (tending toward 4); 3, invasion covers a wide area as a homogenous stand but can fluctuate within certain parameters on a local or heterogeneity scale; 4, transformed system with homogeneous stand of invasive plants over a wide area, with little fluctuation in homogeneity (fluctuation would have little impact because of propagule pressure and the seed bank in the entire area). Scenario B: Trend of plant invasions over time with mechanical and chemical control options: 1, trend line heading for transformed state; 2, implementation of mechanical or chemical control operations; 3, trend toward altered state when operations stop for any reason. Scenario C: Trend of plant invasion over time with the release of effective biological control agents: 1, introduction of biocontrol agent; 2, establishment of biocontrol phase; 3, rapid collapse of host plants; 4, tending to central point of fluctuations through seasonal variations of low impact and densities of host plant. Circles represent spatial scale or extent.

of the Olifants and Letaba rivers warrant attention other than for aesthetic reasons. The cyclic nature of these invasions and our lack of understanding of the main drivers make effective control of these species difficult. Control efforts may be better applied by focusing on plants established above the annual flood line, with the rest of the infestation following cycles dictated by environmental influences.

The overall conclusion is that disturbance (both natural and human-induced) drives alien plant invasions in Kruger. Important for management of this problem is an understanding of the temporal lag periods that are both species dependent and environmentally driven. Research and experience in other regions assist management in distinguishing between the long-fuse–big-bang species and those that are unimportant, with limited ecological impacts.

Ranking Anthropogenic Agents of Change

Influences on Kruger have been many and varied, and the majority of them have an anthropogenic origin. Teasing out the exact influences of each agent on landscape heterogeneity, whether this is a direct management or an unintentional side effect, is difficult, but a nonquantitative evaluation is provided in Table 19.4. Landscape-level heterogeneity has been significantly and directly affected by management policies dealing with fire management and water provision. However, similar levels of impact have come from altered river and catchment conditions arising outside Kruger. Medium-level impact activities include invasive alien biota, which, although considered to have had a mild influence to date because of intensive ongoing management and removal programs, has the overall potential to cause significant changes in savanna heterogeneity. Also, many anthropogenic agents of change have had cumulative and synergistic effects, interacting with each other and with natural phenomena such as droughts and floods.

What has been learned over time is that management of this ecosystem must go beyond the fence and incorporate new paradigms of boundary softening such as buffer zone comanagement (e.g., Makuleke and Mdluli Contractual Parks, Marietta Buffer Zone). Nevertheless, Kruger cannot be expected to bear the costs of external impacts and resource use, and techniques must be developed to measure and predict levels of impacts and rates of resource flow. Quantitative estimates of the economic value of the conservation land-use option must also be made and value assigned to the human benefits flowing out of Kruger's continued existence (Chapter 23, this volume).

The Imperative of Responding to Global Environmental Change

Although not specifically discussed elsewhere in this chapter and not ranked in Table 19.4, Kruger recognizes the potential paramount importance of global environmental change. Unprecedented changes in global climate and biogeochemical cycles have resulted in collaborative initiatives at both Kruger (Chapter 22, this volume) and national levels. At a wider level, our participation in the Assessments of Impacts and Adaptations to Climate Change in Multiple Regions and Sectors (AIACC) initiative will lead to improved understanding of the impacts of climate change on the biodiversity sector of southern Africa. The project aims to enhance scientific and technical capacity and provide scenario-planning tools to project climate change vulnerabilities and test and facilitate adaptation strategies. Linked to these are the wide-scale land use and land cover changes and biodiversity reduction. Here Kruger is involved in initiatives aimed at addressing desertification in South Africa and in a carbon emission offset initiative for natural resource rehabilitation in rural areas. Although this is a new response to an objective of Kruger (see Table 4.1, this volume), it is starting to attract the requisite attention.

TABLE 19.4

Ranking anthropogenic agents of change in Kruger.

AGENT OR IMPACT	MAIN EFFECTS AND OUTCOMES	HETEROGENEITY INCREASED OR DECREASED	RELATIVE MAGNITUDE
Fire management	Rigid triennial burn policies between 1957 and 1992 led to homogenization of vegetation; shift from open to closed woodland in higher-rainfall areas through earlier fire suppression; encroachment of some species; structural changes to mopane regions; decline in large trees.	Homogenization at landscape scale	High, especially in conjunction with effects of herbivory, notably by elephants
Water provision policies	Resulted in changed species composition and structure in immediate area of artificial water and impacts of elephant in particular on the woody vegetation; increased patchiness at local level but overall homogenization of Kruger system.	Increased heterogeneity at fine scale; homogenization at Kruger scale	High
River flows and catchment management	Upstream abstractions reduce Kruger surface water, with some perennial rivers becoming seasonal in some years, affecting riparian vegetation and recruitment processes; effects of eutrophication, sedimentation, and decreasing water qualities affect downstream (Kruger) biotic responses.	Homogenization through biodiversity loss and structural and habitat reduction	High
Alien biota	Has high homogenization potential, but this has been severely limited by control efforts, and infestations have still been predominantly within the latent lag phase.	Increased species diversity (although undesirable); local and habitat homogenization (especially riparian systems)	Medium to date (but very high potential)
Tourism activities	Varied effects through associated infrastructure, although biggest concerns are environmental costs not necessarily associated with heterogeneity impacts (through direct and indirect resource use).	Some local homogenization (e.g., wood collection), some local heterogeneity (e.g., effects of artificial wetlands)	Low to medium

(continued)

TABLE 19.4 *(continued)*

AGENT OR IMPACT	MAIN EFFECTS AND OUTCOMES	HETEROGENEITY INCREASED OR DECREASED	RELATIVE MAGNITUDE
Population management and culling	Synergistic effect with fire management; stabilized elephant numbers had an overall homogenization effect at the broadest level; population reductions of herbivores and carnivores had uncertain effects.	Local-scale variability increased, larger-scale homogenization through population stabilization	Medium to high (in conjunction with fire policies)
Disease and disease management	Natural disease assists in maintenance of patchiness by its cyclic nature, operating at focal, local, or regional levels; alien disease generally has homogenizing local effects but may contribute to large-scale heterogeneity.	Ecosystem effects linked to other drivers and may then act synergistically in increasing patchiness or homogenizing at various scales	Medium to low
Roads	Fragmented Kruger into blocks, homogenizing within blocks because of fire effects. Small and localized barrier effects, but large implications for gravel use and effects of resource use.	Localized homogenization, although larger-scale synergistic effect with fire management	Medium to low
Fencing and boundaries	Mainly affects large mammal movements and cuts off migratory routes, resulting in unnatural pressure on vegetation.	Homogenizing effects	Small to medium
Poaching	Large effect at turn of century by extirpating dominant herbivores, but more recently has more localized effects with limited impacts on heterogeneity.	Limited effects	Small (but possibly medium before Kruger proclamation)
Population augmentation and reintroductions	Resulted in increased species diversity, but often attempts were unsuccessful.	Negligible	Small
Early humans	Heterogeneity effects through burning and local-scale enhancement of patchiness.	Increased heterogeneity	Assumed to be small

Conclusion

In a recent study assessing effectiveness in protecting biodiversity, Brunner at al. (2001) noted that parks were generally successful at mitigating the anthropogenic impacts of land clearing, hunting, fire, and grazing. Park effectiveness was most strongly correlated with density of guards and inversely with levels of illegal activities in parks. These findings appear to hold for Kruger. They will also lead to increased motivations for more field staff to patrol and combat activities such as poaching along boundaries. However, this narrow focus may result in short-term protection but longer-term biodiversity loss. Far greater threats that tend to be overlooked are already exerting their invisible forces on the ecosystem. Many of them are linked to global change in its broadest sense and are changing the composition and functioning of ecosystems. Of immediate and direct consequence are the escalating biological invasions. Threats from global environmental agents of change probably will dominate in future and attract increased attention and focus. Human-induced changes in atmospheric deposition, greenhouse gases, and nitrogen deposition must be afforded greater concern and consideration to facilitate a pragmatic and adaptive approach to attaining conservation goals.

References

Anon. 1995. *Alien Biota Section records*. Scientific Services Department, Skukuza. Unpublished records, South African National Parks.

Anon. 1996. *Alien Biota Section records*. Scientific Services Department, Skukuza. Unpublished records, South African National Parks.

Bestbier, R. X., D. L. Jacoby, and K. H. Rogers. 2000. *A goal maintenance system for the management of the Kruger National Park's riverine alien vegetation: developing a protocol and a prototype*. Water Research Commission Report 813/2/00. Water Research Commission, Pretoria, South Africa.

Botha, J. 1998. Developing an understanding of problems being experienced by traditional healers living on the western border of the Kruger National Park: foundations for an integrated conservation and development programme. *Development Southern Africa* 15:621–634.

Botha, J., E. T. F. Witkowski, and C. M. Shackleton. 2001. An inventory of medicinal plants traded on the western boundary of the Kruger National Park, South Africa. *Koedoe* 44:7–46.

Braack, L. E. O. 1997. Vol. 8. *Policy proposals regarding issues relating to biodiversity maintenance, maintenance of wilderness qualities, and provision of human benefits. A revision of parts of the management plan for the Kruger National Park*. Online: http://www.parks-sa.co.za.

Braack, L. E. O., and C. Marais. 1997. Contribution towards a policy on tourism in the KNP. In Vol. 8. *Policy proposals regarding issues relating to biodiversity maintenance, maintenance of wilderness qualities, and provision of human benefits. A revision of parts of the management plan for the Kruger National Park. Online: http://www.parks-sa.co.za.

Brunner, A. G., R. E. Gullison, R. E. Rice, and G. A. B. da Fonseca. 2001. Effectiveness of parks in protecting tropical biodiversity. *Science* 291:125–128.

Carruthers, J. 1995. *The Kruger National Park: a social and political history.* Pietermaritzburg, South Africa: University of Natal Press.

Codd, L. E. W. 1951. *Trees and shrubs of the Kruger National Park.* Botanical Survey Memoir No. 26. Department of Agriculture. Union of South Africa.

Cronk, Q. C. B., and J. L. Fuller. 1995. *Plant invaders: the threat to natural ecosystems.* London: Chapman & Hall.

Erasmus, D. J., K. A. R. Maggs, H. C. Biggs, D. A. Zeller, and R. S. Bell. 1993. Control of *Lantana camara* in the Kruger National Park, South Africa, and subsequent vegetation dynamics. In *Proceedings of Brighton Crop Protection Conference, Weeds. Farnham: BCPC,* 1:399–403.

Ferreira, S. L. A., and A. C. Harmse. 1999. The social carrying capacity of the Kruger National Park, South Africa: policy and practice. *Tourism Geographies* 1:325–342.

Foreman, R. T. T., and L. E. Alexander. 1998. Roads and their major ecological effects. *Annual Review of Ecological Systems* 29:207–231.

Foxcroft, L. C. 1999. *Alien Biota Section records.* Scientific Services Department, Skukuza. Unpublished records, South African National Parks.

Foxcroft, L. C. 2000. A case study of the human dimensions in invasion and control of alien plants in the personnel villages of Kruger National Park. Pages 127–134 in J. A. McNeely (ed.), *The great reshuffling: human dimensions of invasive alien species.* Gland, Switzerland: IUCN.

Foxcroft, L. C. 2001. *Invasive alien plant research programme proposal: assessing the biology and ecology of invading alien plants for optimising control strategies in the KNP.* Scientific Report 03/01. South African National Parks, Skukuza, South Africa.

Foxcroft, L. C., and J. H. Hoffmann. 2000. Dispersal of *Dactylopius opuntiae* Cockerell (Homoptera: Dactylopiidae), a biological control agent of *Opuntia stricta* (Haworth) (Cactaceae) in the Kruger National Park. *Koedoe* 43:1–5.

Foxcroft, L. C., and D. M. Richardson. 2001. Alien plant invasions and management systems in the Kruger National Park, South Africa. In *Proceedings of the 6th International Conference on the Ecology and Management of Alien Plant invasions,* Loughborough University, England.

Freitag-Ronaldson, S., R. H. Kalwa, J. C. Badenhorst, J. du P. Erasmus, F. J. Venter, and F. J. Nel. In press. Wilderness, wilderness quality management and recreational opportunities zoning within the Kruger National Park, South Africa. A. Watson and J. Sproul (eds.), *Science and stewardship to protect and sustain wilderness values.* Proceedings of the 7th World Wilderness Congress Symposium 2001, Port Elizabeth, South Africa.

Gertenbach, W. P. D. 1985. *Alien Plant Section records.* Scientific Services Department, Skukuza. Unpublished records, South African National Parks.

Henderson, L., and M. J. Wells. 1986. Alien plant invasions in the grassland and savanna biomes. Pages 109–117 in I. A. W. MacDonald, F. J. Kruger, and A. A. Ferrar (eds.), *The ecology and management of biological invasions in southern Africa.* Cape Town: Oxford University Press.

Hodkinson, D. J., and K. Thompson. 1997. Plant dispersal: the role of man. *Journal of Applied Ecology* 34:1484–1496.

Hoffmann, J. H., V. C. Moran, and D. A. Zeller. 1998a. Evaluation of *Cactoblastis cactorum* (Lepidoptera: Phycitidae) as a biological control agent of *Opuntia stricta* (Cactaceae) in the Kruger National Park, South Africa. *Biological Control* 12:20–24.

Hoffmann, J. H., V. C. Moran, and D. A. Zeller. 1998b. Long-term population studies and the development of an integrated management programme for control of *Opuntia stricta* in Kruger National Park, South Africa. *Journal of Applied Ecology* 35:156–160.

IUCN (International Union for the Conservation of Nature). 1997. Conserving vitality and diversity. In C. D. A. Rubec and G. O. Lee (eds.), *Proceedings of the World Conservation Congress workshop on alien invasive species*, Canadian Wildlife Service, Ottawa, Canada.

Joubert, S. C. J. 1986. *Master plan for the management of the Kruger National Park.* Skukuza, South Africa: National Parks Board. Online: http://www.parks-sa.co.za.

Lotter, W. 1996. *Strategic management plan for the control of* Opuntia stricta *in the Kruger National Park.* Scientific report 17/96. National Parks Board, Skukuza, South Africa.

Lotter, W. D. 1997. *Management proposals for the alien aquatic biological invasions of the Kruger National Park.* Scientific Report 10/97. National Parks Board, Skukuza, South Africa.

Lotter, W. D., and J. H. Hoffmann. 1998. An integrated management plan for the control of *Opuntia stricta* (Cactaceae) in the Kruger National Park, South Africa. *Koedoe* 41:63–68.

MacDonald, I. A. W., and W. P. D. Gertenbach. 1988. A list of alien plants in the Kruger National Park. *Koedoe* 31:137–150.

Mack, R. N. 1985. Invading plants: their potential contribution to population biology. Pages 127–142 in J. White (ed.), *Studies on plant demography: a festschrift for John L. Harper.* London: Academic Press.

Meyer, A. 1986. *'n Kultuurhistoriese interpretasie van die ystertydperk in die Nasionale Krugerwildtuin.* Unpublished Ph.D. thesis, University of Pretoria, Pretoria, South Africa.

Moran, V. C., and J. H. Hoffmann. 1987. The effects of simulated and natural rainfall on cochineal insects (Homoptera: Dactylopiidae): colony distribution and survival on cactus cladodes. *Ecological Entomology* 12:61–68.

Moran, V. C., J. H. Hoffmann, and N. C. J. Basson. 1987. The effects of simulated rainfall on cochineal insects (Homoptera: Dactylopiidae): colony composition and survival on cactus cladodes. *Ecological Entomology* 12:51–60.

Novellie, P. A., and M. Knight. 1994. Repatriation and translocation of ungulates into South African national parks: an assessment of past attempts. *Koedoe* 37:115–119.

Obermeijer, A. A. 1937. A preliminary list of the plants found in the Kruger National Park. *Annals of the Transvaal Museum* 17:185–227.

Pienaar, D. J. 1994. Kruger's diversity enriched. *Custos* May:22–25.

Pienaar, U. de V. 1968. The ecological significance of roads in a national park. *Koedoe* 11:169–174.

Pienaar, U. de V. 1970. The recolonisation history of the square-lipped rhinoceros in the Kruger National Park. *Koedoe* 13:157–169.

Pienaar, U. de V. 1990. *Neem uit die verlede.* Pretoria, South Africa: National Parks Board.

Plug, I. 1987. Iron Age subsistence strategies in the Kruger National Park, South Africa. *Archaeozoologia* 1:117–125.

Prieur-Richard, A. H., and S. Lavorel. 2000. Invasions: the perspective of diverse plant communities. *Australian Ecology* 25:1–7.

Richardson, D. M., I. A. W. MacDonald, J. H. Hoffmann, and L. Henderson. 1997. Alien plant invasions. Pages 535–570 in R. M. Cowling, D. M. Richardson, and S. M. Pierce (eds.), *The vegetation of southern Africa*. Cambridge, UK: Cambridge University Press.

Sala, O. E., S. Chapin, J. J. Armesto, E. Berlow, J. Bloomfield, R. Dirzo, E. Huber-Sanwald, L. F. Huenneke, R. B. Jackson, A. Kinzig, R. Leemans, D. M. Lodge, H. A. Mooney, M. Oesterheld, N. L. Poff, M. T. Sykes, B. H. Walker, M. Walker, and D. H. Wall. 2000. Global biodiversity scenarios for the year 2100. *Science* 287:1770–1774.

Shackleton, C. M. 2000. Comparison of plant diversity in protected and communal lands in the Bushbuckridge lowveld savanna, South Africa. *Biological Conservation* 94:273–285.

Smuts, G. L. 1978. Interrelations between predators, prey and their environment. *BioScience* 28:316–320.

Smuts, G. L. 1982. *Lion*. Johannesburg: Macmillan.

Stevenson-Hamilton, J. 1937. *South African Eden: from Sabi Game Reserve to Kruger National Park*. London: Cassel and Company.

Tshiguvho, T. E. 2000. *The effects of roads on invertebrate communities in the Pafuri–Punda Maria area, Kruger National Park: ants as indicators*. Unpublished M.Sc. dissertation, University of Venda, South Africa.

Tshiguvho, T. E., W. R. J. Dean, and H. G. Robertson. 1999. Conservation value of road verges in the semi-arid Karoo, South Africa: ants (Hymenoptera: Formicidae) as bio-indicators. *Biodiversity and Conservation* 8:1683–1695.

van der Schijff, H. P. 1957. *Ekologiese studie van die flora van die Nasionale Krugerwildtuin*. D.Sc. thesis, Potchefstroom University for Christian Higher Education, Potchefstroom.

van der Schijff, H. P. 1969. *A check list of plants of the Kruger National Park*. Publikasies van die Universitiet van Pretoria, Nuwe reeks 53:1–100.

van Rooyen, C., and P. Grantham. 1998. *An assessment of the risk that powerlines pose for wildlife in the Kruger National Park with specific emphasis on bird moralities*. Project Report. Endangered Wildlife Trust/Eskom, Johannesburg, South Africa.

van Wilgen, B. W., and E. van Wyk. 1999. *Invading alien plants in South Africa: Impacts and Solutions*. In: People and Rangelands: Building the Future. Proceedings of the 6th International Rangeland Congress, Townsville, Australia (edited by David Eldridge and David Freudenberger), Vol. 2, pages 566–571.

Venter, F., L. Braack, F. Nel, W. Jordaan, F. Gerber, and H. Biggs. 1997. Recreational opportunity zoning within the Kruger National Park. Pages 184–198 in Vol. 8, *Policy proposals regarding issues relating to biodiversity maintenance, maintenance of wilderness qualities, and provision of human benefits. A revision of parts of the management plan for the Kruger National Park*. Online: http://www.parks-sa.co.za.

Vitousek, P. M., H. A. Mooney, J. Lubchenco, and J. Mellilo. 1997. Human domination of earth's ecosystems. *Science* 277:494–499.

Whyte, I. J. 1985. *The present ecological status of the blue wildebeest* (Connochaetes taurinus taurinus, *Burchell, 1923*) *in the central district of the Kruger National Park*. Unpublished M.Sc. thesis, University of Natal, Pietermaritzburg, South Africa.

Whyte, I. J., and S. C. J. Joubert. 1988. Blue wildebeest population trends in the Kruger National Park and the effects of fencing. *South African Journal of Wildlife Research* 18:78–87.

Chapter 20

Beyond the Fence: People and the Lowveld Landscape

SHARON POLLARD, CHARLIE SHACKLETON,
AND JANE CARRUTHERS

Conservation areas do not exist in a vacuum but are nested in a heterogeneous social, economic, environmental, and political matrix that influences their origin and development. Yet for much of the twentieth century, most protected areas worldwide have been managed as units distinct from their surrounding landscapes. However, the preconditions for and discourse on the sustainability of conservation areas is being rewritten as the practice of fortress conservation is increasingly challenged (Pigram and Sundell 1997; Adams and Hulme 2001; Fabricius et al. 2001a). Importantly, much of the earth's biodiversity is found outside protected areas, so conservation advocates are challenged to move the principles and practices beyond the fences through support for conservation by the people (Murphree 1996).

Kruger is no exception to these trends. This global reorientation has dovetailed with a period of political change in South Africa that calls for the reintegration of Kruger into the broader landscape. Indeed, Kruger's survival in its present form may be threatened by circumstances beyond its boundaries and outside its control. But at the same time, Kruger sits on a threshold of opportunity. In particular, building partnerships with neighboring communities that can be empowered by Kruger's existence can secure and enhance its future. To negotiate a future for the park, Kruger's managers need to understand the broader landscape patterns and how and why they have changed. Collaborative opportunities may emerge in the process (Walker et al. 2002). A core purpose of this chapter is to describe changes in the lowveld landscape over the period of Kruger's existence and to present the key drivers of this change.

We start by introducing a framework used for integrating the complexities of multiple influences and outcomes. We then provide a brief overview of the South African lowveld landscape today and examine the theme of a changing landscape heterogeneity, using the central lowveld as a case study.[1] Not only does the history of this area capture many of the contrasts that have so distinguished the southern African lowveld (exemplary biodiversity, harsh and variable climates, and a unique political history), but it also boasts a reasonable research base from the last two decades. These accounts are used to integrate discussions around two environmental themes: water and land. We conclude by examining the emerging issues and trends as a framework for looking forward.

An Integrative Framework

Our overall approach is to examine landscape patterns based on the key imperatives that prevail today, specifically those relating to sustainability. The goal of sustainable development is to create and maintain prosperous social, economic, and ecological systems (Folke et al. 2002). Over the last decade, understanding the state of the landscape and the interrelationships with sociocultural, environmental, economic, and political (SEEP) factors has become the foundation for an integrated, analytical approach (Figure 20.1; Campbell and Olson 1991; Gunderson and Holling 2001). For example, pejorative attitudes and profligate hunting at the turn of the twentieth century precipitated the demise of wildlife in the lowveld. In turn, this situation prompted a shift in protectionist ethics and policies that, coupled with economic factors, supported the increase in conservation areas.

The need to couple SEEP factors into an integrative framework for understanding complex systems (Kinzig et al. 2000) has provided some promising results, notably that of adaptive cycles and resilience (Gunderson and Holling 2001). This principle holds that sociopolitical systems are linked, coevolving complex systems that exhibit periods of stability and marked thresholds of change. Importantly, the supply of ecosystem goods and services depends on the state of the system. However, human activity may compromise the ability to provide these services by reducing the resilience of the system, with consequences for livelihoods and vulnerability. In complex systems, a handful of factors can be identified that increase the vulnerability of socioecological systems to cope with shocks (Holling 2001). Understanding change through a historical lens reveals a great deal about the current system dynamics and how systems may respond to future shocks (Walker et al. 2002). This chapter attempts to unearth some of the key SEEP factors that have triggered change or promoted stability in the lowveld landscape, of which Kruger is part.

The Lowveld Landscape Today

Our use of the term *lowveld* coincides with the descriptions by Stevenson-Hamilton (1929), the first warden of Kruger, who used an altitude of 650 m to delineate

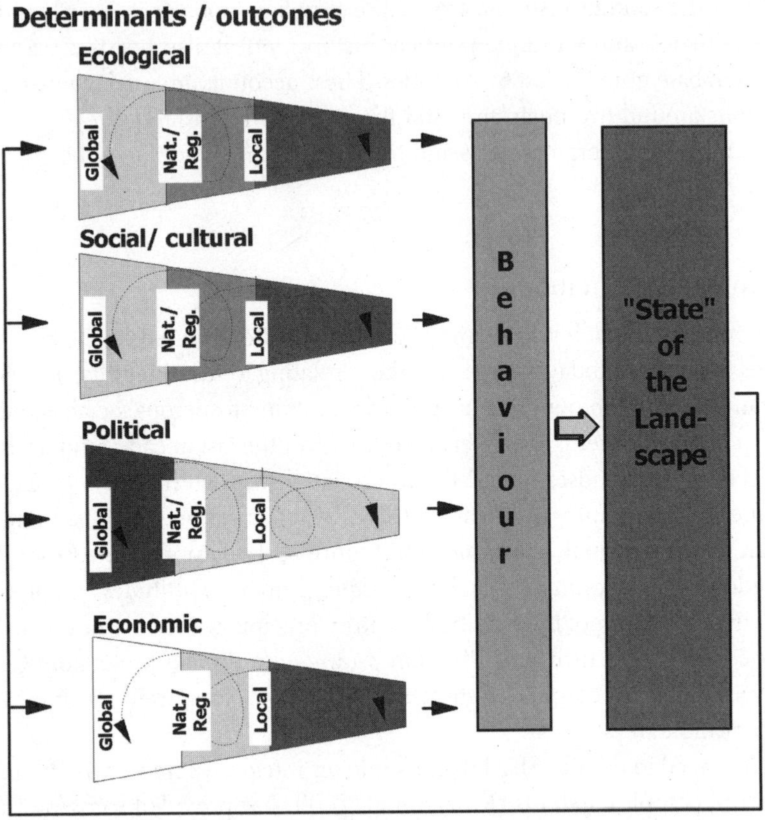

FIGURE 20.1. Schematic of the relationship between the state of the landscape and sociocultural, environmental, economic, and political (SEEP) factors or determinants (adapted from Campbell and Olson 1991; Breen et al. 1997). The state of the landscape at any point in time reflects the consequences of human behavior, which is influenced, in turn, by a range of SEEP factors. These factors operate at different temporal and spatial scales and are themselves determined by the state of the landscape in a feedback loop. The recognition of this interdependency forms the basis for concepts that link social and ecological systems.

its western border. It approximates a 50-km radius around Kruger, bounded in the west by the Escarpment Mountains, in the north by the foothills of the Soutpansberg in South Africa and by Gonarezhou Park in Zimbabwe, and in the south by Swaziland (Figure 20.2a). The landscape embraces a rich variety of landforms, climate, and vegetation, as well as cultures and land uses (Figure 20.2).

The climate is tropical to subtropical, and drought is endemic (Tyson 1986). The Drakensberg escarpment (over 1,800 m) descends rapidly eastward to the lowveld of Kruger and Mozambique, with altitudes of 600 and 400 m (Figure 20.2a). Similarly, rainfall decreases sharply from more than 1,200 mm/annum along the escarpment, and other mountainous areas, such as Venda, to less than 500 mm along the western border of Kruger (Figure 20.2b).

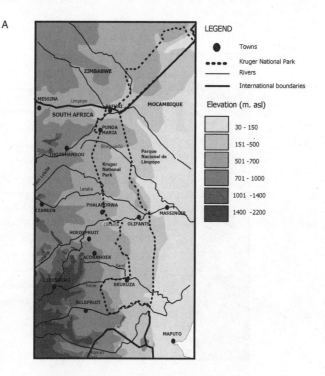

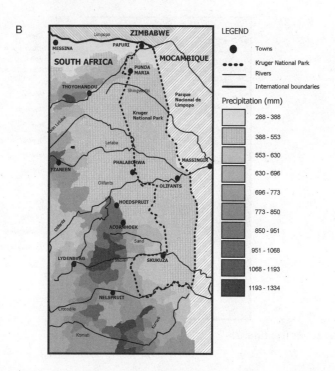

FIGURE 20.2. An overview of the biophysical and land-use patterns of lowveld landscape: *(a)* topography, *(b)* rainfall, *(c)* vegetation, *(d)* geology, *(e)* land use and land cover, and *(f)* conservation areas.

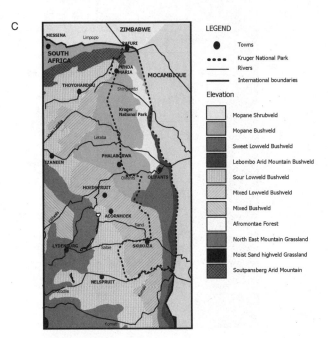
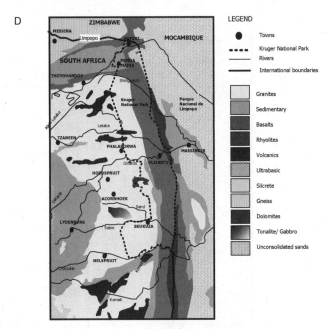

FIGURE 20.02 (continued)

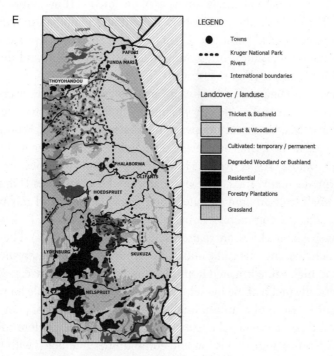

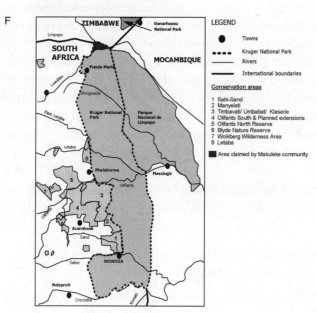

FIGURE 20.02 *(continued)*

The variety of natural vegetation types of the lowveld (Low and Rebelo 1996; Figure 20.2c) reflects the gradients in rainfall and temperature regimes, and underlying geological substrata (Figure 20.2d). Six major rivers transect the lowveld and Kruger, most originating near the escarpment and flowing east toward Mozambique (Figure 20.2a).

These diverse biophysical landscapes are paralleled by equally diverse cultural heterogeneity. At least 2 million people reside within 50 km of the western boundary of Kruger. Major cultural groups are the Vhavhenda, Tsonga, Pedi, and Swazi, concentrated in the former apartheid homelands of Venda, Gazankulu, Lebowa, and KaNgwane, respectively. Human population densities exceed 300 people/km^2 in many areas. The dominant land-use activities (Figure 20.2e) include small-scale cropping, limited commercial farming, and grazing. In contrast, the adjacent commercial farms, currently owned mainly by English- and Afrikaans-speaking whites, are sparsely settled (5–20 people/km^2). The dominant farming activities include cattle and game farming, private conservation (Figure 20.2f), and high-value irrigated tropical crops along the main rivers.

Historically much of the lowveld was regarded as inhospitable for permanent settlement because of erratic rainfall and high temperatures, poor soils, and endemic and sometimes fatal livestock and human diseases. Before colonization in the mid-1800s, indigenous lowveld communities interacted with their landscape, unimpeded by fences, but with colonial intrusion and the introduction of firearms, the socioeconomic matrix was transformed in favor of whites. Events in this period were seminal in establishing wildlife reserves, in particular the Sabi and Singwitsi game reserves, which, in the form of Kruger after 1926, came to occupy most of the eastern part of the lowveld (Carruthers 1995). As with national parks in the United States, these game reserves received public support from those in power because they were regarded as worthless lands, situated in remote parts of the country (Nash 1982). This perception of land value changed in the 1900s when, with the reduction in malaria and demise of the tsetse fly (caused by rinderpest in 1896 and drought between 1897 and 1913), the lowveld opened up for denser settlement.

Land-Use Patterns and Macrohistorical Trends in the Central Lowveld: 1900 to the Present

In the central lowveld, an area contained between the Sabie and Olifants rivers, six major land uses prevail, but two key features dominate the landscape: the large area under wildlife conservation (state and private), juxtaposed with the densely populated, economically impoverished communities of the former homelands of Gazankulu and Lebowa (Table 20.1, Figure 20.2).

TABLE 20.1

Land use in the central lowveld, including central Kruger (Shackleton 1996).

LAND USE	AREA (HA)
State conservation	561,240
Private conservation	287,690
Villages and communal grazing lands	191,200
Private commercial cattle farms	54,540
Plantation forestry (below 650 m)	~36,000
Commercial crop farms	Limited to irrigable belts along the Sabie, Blyde, and Klaserie rivers

Lands under Communal Tenure

Today, the communal lands of the central lowveld are characterized by high population densities (Figure 20.2e), with more than 300 people/km^2 at the base of the escarpment, where the soils are rich and the climate more congenial, to approximately half that density in the drier east. Until the mid-1930s, agriculture was the mainstay, but political and economic policies acted in concert to produce a rural economy that, by the 1940s, depended on migrant remittances and state pensions for cash (Bundy 1988; May 1995; Tollman et al. 1999). Given that only 6 percent of the local cash economy is generated by agriculture (Harries 1989), the direct use values of home consumption from livestock, agriculture, and natural resource harvesting are high, accounting for more than 50 percent of total livelihood streams (Shackleton et al. 2001). The cultivation of maize, intercropped with fruit trees and vegetables, is undertaken at the homestead and, for some, in demarcated arable fields adjacent to villages (High and Shackleton 2000; Dovie 2001). Uncultivated land is used for natural resource harvesting and grazing, where stocking rates are at agricultural carrying capacity (Parsons et al. 1997).

This chapter cannot present a detailed history of racial division in South Africa, but the country's apartheid policies have shaped both Kruger and the other side of the fence. Because the human settlement patterns were strongly controlled by dominant political ideologies, a brief mention of some key events is warranted. From the beginning of white settlement in the seventeenth century, South African society was segregated, and after the Union of South Africa in 1910 a number of laws ensured that whites remained politically and economically dominant. By 1945, those that directly affected the central lowveld were the Native Land Act (1913) and the Natives Trust and Land Act (1936), which stipulated that Africans (the majority) had legal tenure only in designated regions, totaling some 13 percent of South Africa (de Wet 1995). Additionally, various laws excluded Africans from the political process and allowed the government to intervene in African agricultural production (Davenport and Saunders 2000). So-called betterment schemes that entailed the concentration of the population into villages and placed restrictions on livestock and agricultural production (purportedly to rationalize agriculture) ensured that African traditional conservation and farming practices were abandoned in what became a struggle for survival.[2]

In 1948, the apartheid policies of the National Party government introduced a more effectively enforced racist ideology. These policies exalted Afrikaner nationalism and entrenched ethnic segregation by establishing homelands (Bantustans). These ethnically defined areas were facilitated by the Promotion of Bantu Self-Government Act of 1959 and a plethora of other apartheid laws that forced Africans to live in the homelands. Large-scale forced removals occurred, mainly from elsewhere in the Transvaal province, creating overcrowded and impoverished areas in which investment and development were negligible (Fischer 1988; de Wet 1995). The homelands became dumping grounds for what the state regarded as surplus Africans who were not engaged in active service to the white-controlled economy as migrant labor. For example, between 1972 and 1994, the population of Gazankulu (700,000 ha) doubled to more than 500,000 people. As more government interference and social engineering occurred, Africans in the central lowveld increasingly lost their power to make livelihood decisions relating to their own sociocultural, economic, environmental, and political matrix.

In particular, Tsonga-speaking people (Shangaans) faced increasing hardship as they were relocated to the driest eastern districts, which were traditionally used only for seasonal grazing and hunting because of their inhospitable summer climates (Harries 1989; Spenceley 2001). In 1972 the central lowveld was divided into two "self-governing states." Gazankulu was established for the Tsonga people, and by 1985 such were the forced removals and boundary gerrymandering that it contained 43.4 percent of Shangaans (Christopher 1994). Lebowa, adjoining Gazankulu on the western side, was reserved for the Pedi people. Similar scenarios played out to the north and south of the central lowveld, with the creation of the Venda and Swazi homelands, respectively, which also bordered Kruger.

During this period of colonial expansion, Stevenson-Hamilton removed some 2,500 Africans from the Sabi Game Reserve, although this practice was later reversed in the Singwitsi Reserve, where residents were encouraged to seek employment as rangers and laborers. In 1939, many Africans left Kruger when veterinary authorities, against the protests of Stevenson-Hamilton, slaughtered their livestock to prevent the spread of disease (Carruthers 1995). In the 1980s, 400 families were relocated to make way for game reserves in Gazankulu (Fischer 1988).

The high human population on the borders of Kruger has had a number of consequences. First, the limited and marginal land was farmed and harvested more intensively, with a concomitant effect on the natural environment. Second, in response to the attendant unemployment that accompanied the burgeoning population, the homeland governments embarked on a number of infeasible job creation schemes involving agriculture, conservation, and forestry (Fischer 1988). Third, the deleterious environmental and economic situation fostered conflict over land and resources (Stadler 1994), equally evident in the sociocultural and political landscape (Ritchken 1995; Niehaus 2001). Indeed, political instability could best be interpreted as a deliberate result of the apartheid government's homeland policy. The division of this area into two provinces after 1994 and the ongoing political acrimony between them is a consequence of this history (Ramutsindela and Simon 1999).

Subtle contestations of Kruger's fence have existed since its inception. This has taken the form of poaching, widespread harvesting of medicinal plants within Kruger's boundaries, and demands for compensation for crops destroyed by wildlife. Furthermore, people often traversed Kruger (and still do) from Mozambique to escape the crippling civil war there and its subsequent economic difficulties. Since 1994 a number of land claims in Kruger have been gazetted, and the well-known Makuleke land claim was successful (Box 20.1). Several more, such as that of the ba-Phalaborwa for the area between the Olifants and Letaba rivers, are outstanding. However, democratic changes have also prompted moves by communities to enter conservation partnerships with Kruger. The Mdluli land settlement in southwestern Kruger has resulted in a joint tourism venture, and in the area known as the Mariyeta corridor that adjoins the northeastern Kruger boundary, eight communities have explored the option of adding communal land to Kruger for use as an ecotourism opportunity. Although the latter initiative has been thwarted to date by extra-Kruger politics, all these cases point to possibilities for partnerships between Kruger and its neighbors.

The fenceline of Kruger dramatically embodies the historical segregationist political and economic policies. On one side are densely populated, underdeveloped, impoverished landscapes and populations; on the other are largely unaffected systems, generating economic profits. The experience of forced removals, in some cases from conservation areas, and the decreasing autonomy over their own futures has shaped the attitudes of the rural communities of the central lowveld toward Kruger. These sentiments, mirrored in people-park relations in many protected areas throughout Africa (Hulme and Murphree 2001), have shaped expectations under the changed SEEP factors in the newly democratic South Africa.

Land under Individual Tenure

Kruger is not surrounded solely by communal lands. Approximately one-third of the central lowveld is under individual tenure (real or corporate), mostly

> **BOX 20.1**
> ## The Makuleke
>
> The claim of the Makuleke community concerned 23,700 ha of land in Limpopo Province that they had occupied for some 200 years (Figure 20.2f). In 1969, after resisting removal from this area for 30 years, the Makuleke were expelled from the Pafuri Game Reserve, which was then incorporated into Kruger. In 1995 the community applied for repossession of the land under the Restitution of Land Rights Act (1994) and the Communal Property Associations Act (1996). When making the negotiated settlement between interested parties, the presiding judge took into account a number of environmental considerations and the fact that the land in dispute was a national park and an internationally recognized Ramsar wetland. In terms of the agreement and the reversion of land title to the Makuleke, the community agreed not to resettle on the land but to retain certain rights of use and access. In effect, the community manages the area in collaboration with South African National Parks (SANParks).
>
> The Makuleke case attracted publicity for a number of reasons. The claim was fiercely resisted by Kruger management as being a threat to the integrity of Kruger and to its exclusive authority over the park. They were concerned about the precedent it might set for other land claims and possible deproclamation of large sections of many of South Africa's parks. However, it was one of the first land claims to be successfully resolved because the contesting parties eventually shifted their positions to reach a compromise, regarded by some as a win-win situation. On one hand, SANParks underwent policy changes that favored comanagement initiatives and led to protecting biodiversity rather than protecting power structures. On the other hand, although the Makuleke leaders never wavered from their demand for return of title to the land, they became willing to use their restored land for conservation and ecotourism as a contractual park. Although the eventual outcomes have not yet been evaluated, it is regarded as significant that the power relations between the park and local people shifted as a new contract between them was defined (Carruthers 1995; de Villiers 1999; Steenkamp 2000; Glazewski 2000; Ramutsindela 2002).

devoted to ecotourism and conservation since the conversion from farming in the last four decades (Table 20.1, Figure 20.2f).

After the consolidation of the South African Republic north of the Vaal River in the 1860s, large farms (>1,500 ha) were sold or allocated to whites by the Boer government as military or civic rewards. Because of malaria, horse sickness, and tsetse fly, much of the private or state land was used only in the safe winter months for grazing or hunting. Speculation in land by whites, particularly after the discovery of minerals elsewhere in the country, was common practice (Fischer 1988; Harries 1989). Stevenson-Hamilton expanded the Sabi Game Reserve (which consisted only of a small area between the Sabie and Crocodile rivers; see Chapter 1, this volume) into the central lowveld between 1903 and 1926 by leasing a band of private farms between the Sabie and Olifants rivers. Some of the private farms were incorporated into Kruger by expropriation, exchange, or sale, and others were excised from it.

After tsetse fly had disappeared with the outbreak of the rinderpest epizootic in 1896, the twentieth-century lowveld became safe for domestic stock throughout the year. Initially the aridity of the area and the absence of markets limited cultivation on the farms bordering Kruger, and endemic malaria deterred permanent human settlement. Therefore, the farms on the boundary of Kruger remained marginal: some supported small cattle-ranching enterprises, and others were used as private hunting domains in the winter. Although most farmers tolerated game species on their lands for hunting and aesthetic reasons, many species disappeared during this period, including sable, roan, tsessebe, eland, nyala, lion, and rhinoceros (Porter 1970). The reduction of large herbivores and of fire on cattle farms resulted in an increase in the woody vegetation, now called bush encroachment, and today many conservation areas engage in bush clearing (Walker et al. 1987).

By the mid-1950s, the success of Kruger as a tourist venture, improved malarial controls, growing conservation awareness, and marginal profitability of commercial farming in the region catalyzed the establishment of private game reserves. The first private game reserve consortium was the Sabi-Sand Wildtuin, created from a number of the excised farms (Figure 20.2f). The conservation area later expanded with the addition of private reserves, such as Timbavati and Klaserie. This trend has accelerated in the last two decades, which have seen the transformation of most cattle farms in the central lowveld to game and conservation enterprises, offering a variety of ecotourism operations. In many cases, this has been driven mainly by economic imperatives, including declining subsidies for cattle farming and the growing national and international tourism markets, in which South Africa is a favored destination. To maximize economies of scale, both economically and ecologically, and particularly to accommodate the "big five" species as a tourist drawcard, many of these ventures are increasingly turning to share block or cooperative schemes (e.g., encouraging traversing rights on each others' properties). Although this merits further attention, Kruger appears to have exerted a strong nursery effect by attracting similar land use along parts of its borders.

Increasing collaboration to extend the boundaries outward, at least with other established or potential conservation ventures, is seen in a number of initiatives. The fence between Kruger and some of its neighbors was dropped in 1994. More recently, this has taken an international dimension with the proclamation of a Transfrontier Conservation Area, linking Kruger with protected areas in Mozambique and Zimbabwe. If these are successful—and the Greater Limpopo National Park has already been formally approved (Chapter 1, this volume)—Kruger will have additional rural neighbors in Zimbabwe and Mozambique. Furthermore, the Kruger to Canyons Biosphere Reserve, which straddles the central lowveld, is another example of this wider vision. It is intended to facilitate sustainable development that secures the well-being of the environment and people of the area (Newenham 2001).

This transforming social footprint in the lowveld has had a range of impacts on natural resources of the region. The contrasting histories of woodlands and freshwater resources, in particular, provide examples of emerging future trends.

The Transforming Natural Resource Base: Woodland Resources

Households in communal lands of the central lowveld (and in the lowveld generally) are highly reliant on a range of natural resources including fuelwood, fruits, construction materials, and medicinal plants (Shackleton and Shackleton 2000). Most households trade in such resources, mainly for supplementary income, but many do so as their main source of livelihood (Shackleton et al. 1998). Despite recent electrification in some villages, fuelwood is still the primary energy source (Banks et al. 1996).

The impacts of land use on biodiversity are highly variable, with the greater effects evident closest to the village (Sonnenberg 1993; Shackleton et al. 1994). Land transformations together with trapping have eradicated most large animal species, and the bird species richness is reduced (Lewis 1997). Hoffman et al. (1999) consider the degradation status of this area intermediate, and there is a general perception of widespread land degradation in the central lowveld, although Shackleton (1998) has questioned this.

Regardless of its degradation status, there has been a marked decline in woodland resources, which seems to be accelerating (Shackleton et al. 1994; Shackleton 1998; Pollard et al. 1998). Niehaus (2001) suggests that as agriculture assumed a less prominent role in peoples' livelihoods with apartheid planning, cultural beliefs in the powers of nature also declined. Coupled with the forced influx of people and the legislative relocation of power to tribal authorities, much of the traditional management of natural resources was replaced by an increasingly mechanistic approach, effected through chiefs who instituted fines against transgressors. In 1994 this tribal control, which had afforded some degree of protection, albeit autocratic, collapsed as part of the popular challenge to apartheid (Shackleton et al. 1995). The institutional vacuum that has persisted since 1994 has resulted in an open access system in that people view the resource as a public asset that can be commandeered for personal gain, often through threat.

Although the new National Forests Act (1998) provides for community-based institutional controls, no examples of them exist in the central lowveld. Indeed, the issues of ownership and of credible, meaningful management regimes and structures for these woodlands need urgent attention, in which Kruger could participate. It is clear that the current extraction levels for many resources in the communal areas are unsustainable because harvesting (either for local consumption or to supply a growing commercial demand) has increased and local controls have disintegrated (Shackleton et al. 1995). The growing commercialization of resource use is little understood but seems to provide much-needed cash for local households. However, this accelerates the decline in resources, thereby undermining livelihood security in the long run.

Water

The policy reforms that accompanied the democratic transitions in South Africa in 1994 have had direct effects on the lowveld landscape. The story of rivers, as an integrator of land- and water-use practices, is particularly insightful. As the downstream recipient of the waters of six major river systems flowing east into Mozambique, Kruger feels the changes in their flow regimes resulting from landscape transformations.

Some of the major rivers of the lowveld, recorded as perennial in the 1920s by Stevenson-Hamilton (1929), such as the Letaba, Luvuvhu, and Sand, were transformed to intermittent systems by anthropogenic disturbance. Others, such as the Olifants, have suffered a severe reduction in flow (O'Keeffe and Davies 1991). Many systems have deteriorated as a result of organic enrichment, salinization and acid pollution, sedimentation, and the introduction of exotic and invasive plants and animals. These effects have resulted in the loss of four and six fish species in the Olifants and Crocodile rivers, respectively. The Sabie River remains the least perturbed of these rivers, with the exception of the Sand tributary (Russell and Rogers 1989; Pollard 1996; Weeks et al. 1996). In the central lowveld, agriculture, forestry, and the mining industry along the Olifants River have been implicated in many of the large-scale modifications (Chapter 21, this volume).

Partly through the efforts of Kruger staff, government attention was drawn to upstream impacts on these rivers as early as the 1970s, but to little avail. In 1992, the worst drought in recorded history revealed the vulnerabilities of these systems. For instance, despite the fact that water had to be trucked to communities of the Sand River to meet basic survival needs, uneconomical irrigation in the middle reaches continued unchecked (Pollard et al. 1998). This situation highlighted the urgent need for integrated management approaches, which were given full effect through the new National Water Act of 1998. Not only were catchments constituted as the units for water resource management, but also the statutory protection of the right to water was afforded to the environment and to people through the concept of the reserve. These changes necessitated a rethinking of the integration of land and water, with the support of all stakeholders. Unlike in the case of terrestrial resources, Kruger has actively participated in this process.

Redefining the Boundaries: Integrating Kruger into the Lowveld Landscape

The dynamic historical forces described in this chapter have shaped the landscape heterogeneity that we observe today in the central lowveld and Kruger's place therein. Likewise, the future is dynamic, and the patterns of land use will continue to change as internal and external determinants vary. As Kruger grapples to gain wider acceptance by its neighbors, these historical developments and emerging future trends are key to planning its reintegration into the broader SEEP landscape. Indeed, this new path, embraced in the SANParks mission — "biodiversity protection, public use and benefit, and constituency-building" — recognizes the links between Kruger and its surrounds and resonates with contemporary philosophies of holistic ecosystem management.

The Lowveld as a Vulnerable Landscape?

Although untested, the previous account suggests that the socioecological systems in communal lands appear to be increasingly vulnerable (*sensu* Holling 2001) as they become less capable of coping with shocks and stresses. This reflects the persistent effects of political and economic drivers, manifest in chronic poverty, institutional demise, natural resource depletion, and perverse incentives for overexploitation of resources. These vulnerabilities are likely to have ramifications in the private and state-owned conservation areas.

Despite increased engagement with neighboring communities in the last decade, perceptions from both sides of the fence remain guarded, making cooperation between them complex and urgent. As far as Kruger is concerned, the communal areas lack sociocultural and political homogeneity and strong institutional structures. Economically they are extremely poor, a factor that is driving environmental decline. Some managers in Kruger have been apprehensive about land claims under the national land restitution process. On the other hand, many people in the communal areas resent Kruger (although not necessarily conservation per se) because of its authority over valuable natural resources from which they were alienated and because its managers actively collaborated with the apartheid state (Koch et al. 1990; Carruthers 1995). For them, conservation areas maintain an aura of hegemony where, even today, many South Africans are alienated from conservation philosophies. However, several of Kruger's immediate neighbors acknowledge that there are benefits to be gained from its presence as a potential employer or as a regional economic driver.

Thus, a major portion of Kruger's neighbors, who should benefit from and participate in biodiversity conservation according to SANParks' mission, are largely unable or unmotivated to do so. In the context of existing vulnerability and additional socioeconomic threats (such as HIV and AIDS, violence, climate change, economic restructuring, and consequent loss of employment), it is not inconceivable that there will be increasing demands for restitution or benefit sharing. This situation renders conservation areas such as Kruger increasingly vulnerable, and it would be irresponsible not to build a constructive win-win lowveld landscape. How can Kruger, together with its neighbors, reduce this vulnerability?

At the Juncture of Change: Recoupling Socioecological Systems

The new mission of Kruger outlines a commitment to its integration with the broader socioecological landscape, and a number of key ingredients already exist to enhance this process.

First, with the acceleration of policies and programs in the 1990s, many lessons have emerged regarding the relationships between protected areas and their neighbors. Second, this period dovetailed with the political transformation in South Africa. The recent policy changes have provided the legislative framework for an integrated approach to conservation in South Africa, although tenure arrangements in communal lands remain a constraint. Third, both sides of the fence possess different strengths that, if coupled, can contribute to a collaborative vision on conservation and sustainable development. On one hand, people depend heavily on natural resources and are aware of their decline. These societies represent a rich and diverse social matrix with political leverage, traditional knowledge, and skills and, rather surprisingly in the face of historical injustice, have demonstrated a willingness to work with Kruger. On the other hand, conservation areas including Kruger have management and scientific expertise, institutional and economic leverage, and intact natural resources that could be harvested sustainably for local benefit.

Early efforts by Kruger, mainly to address wildlife problems in communal lands, were dealt with through the Community Relations Programme. Additionally, the People and Parks project, initiated and coordinated by a non-government organization, supported the establishment of a number of Kruger-neighbors forums as a channel for dialogue (see also Venter et al. 1994). With the restructuring of SANParks in 1994, the commitment to engage neighboring communities was conceptualized through the establishment of the Social Ecology Department and supported by a number of international agencies. Social Ecology's five core functions (SANParks 2000) are shown in Figure 20.3. In keeping with global thinking that seeks to redress past imbalances (Murphree 1996), most of the initiatives are oriented toward poor rural communities, among which perhaps the greatest enmity exists. Currently there are no examples of community-based natural resource management (CBNRM), and most of the effort is directed at outreach, purportedly without a significant impact on SANParks mission (McKinsey and Company 2002), the main elements of which are listed in Figure 20.3.

Although an evaluation of the social ecology program is beyond the scope of this chapter, certain comments are pertinent. There have been a number of interesting initiatives to extend conservation efforts by adding communal lands to Kruger. This has involved collaborative arrangements with neighboring communities, such as the aforementioned Makuleke contractual park (Box 20.1), the Mdluli land settlement, and the Mariyeta project. In the case of the Mhinga community, lodge development is under way and concrete steps are in place to remove the fence. These examples, and the commercialization program (Chapter 1, this volume), which includes shareholding targets for the previously disadvantaged, are all seen as ways to heal the divide and give communities real equity (H. Magome, pers. comm., 2002).

On the other hand, the functions of the Social Ecology Department are very broad (Figure 20.3), and only some of them are being tackled. For instance, the absence of research and monitoring is problematic not only because it offers the basis for adaptive management, as in Zimbabwe's CAMPFIRE (Communal Areas Management Programme for Indigenous Resources) project (Fabricius et al. 2001a; Hulme and Murphree 2001), but also because the lack of documentation has compromised recent outreach attempts (Tapela and Omara-Ojungu 1999). Aside from the obvious aspect of direct employment, economic empowerment is limited mainly to the marketing of artisanal products. Although entrepreneurship training courses have recently been started, many other initiatives, such as much-needed access for black communities to markets for supplying fresh produce in Kruger, have not materialized. The focus on economic empowerment also overlooks benefits requested by neighbors, whose expectations include learning more about conservation (J. Botha, pers. comm., 2002). There are few links between interpretive cultural and educational tourism inside and developmental education outside Kruger, and these learning processes must be integrated.

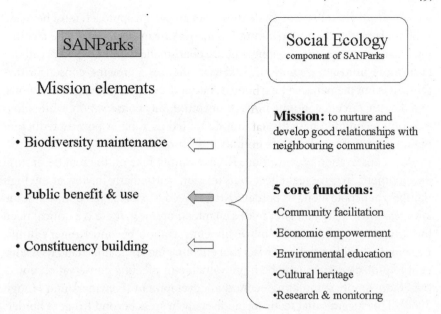

FIGURE 20.3. Overview of the SANParks mission and the mission and five core functions of the Social Ecology Department (SANParks 2000).

Not surprisingly, with such long borders and so diverse a neighbor profile, together with the wide-ranging functions of the Social Ecology Department, the challenges are complex. Nonetheless, a number of key issues emerge from the cumulative experiences of the last decade. They include the need for a clearly articulated vision and conceptual framework that would facilitate the development of strategic and integrated initiatives and the failure to truly integrate constituency building into the wider philosophy and functions of Kruger. For instance, the relationship between the five core functions, and particularly how they can contribute to recoupling social and ecological issues, is unclear. Currently, however, a policy for the Social Ecology Department is being drafted that outlines the basic principles according to which it will function (Fabricius, pers. comm., 2002). Given the negative views on the part of some Kruger staff toward social ecology, a clear policy framework will facilitate the obvious need to address its integration into wider Kruger activities. Constituency building cannot fall victim to a silo mentality (compartmentalized thinking), but in the end success will entail shared vision and commitment and practicable arrangements.

The extensive international literature outlines many examples of pitfalls and successes of outreach and community conservation (Leach et al. 1999; Wily 2000; Fabricius et al. 2001b; Hulme and Murphree 2001). How Kruger, in its local context, plans to incorporate these into an overall approach must be made explicit. Importantly, and in contrast to many African case studies, the fact that lowveld communal lands no longer house charismatic mammal species calls for even more innovative solutions. Likewise, there is a growing consensus that tourism is not the panacea for all conservation-development integration efforts, and that the direct economic impacts of community conservation, while offering some economic benefits, cannot be touted as a major poverty reduction option (Fabricius et al. 2001b; Hulme and Murphree 2001).

As a conservation organization, Kruger cannot take on the mantle of rural development. To give real effect to its mission, particularly in view of the high numbers of people along its borders, Kruger needs strategic direction, and some key areas are touched on later in this chapter. Foremost, there is a critical need for Kruger to work toward biodiversity conservation beyond Kruger's limits. Lessons worldwide point to the fact that encouraging compatible land uses adjacent to protected areas may be more valuable in meeting conservation objectives than simply stimulating economic development (Newmark and Hough 2000). This suggests that securing biodiversity in areas beyond Kruger's borders would give its mission greater effect. Although Kruger has demonstrated a proactive commitment to ecosystem approaches in terms of water, the integrity of woodland systems has been neglected. Ecosystems are contiguous, regardless of fences or tenure. The lowveld woodlands, for example, are not confined to Kruger, nor is such a scenario desirable given their socioeconomic and ecological importance. Yet without a broader ecosystem view by all stakeholders, the degeneration of a common property management system to one of open access poses a real threat to the biodiversity along Kruger's border and to local peoples' livelihoods. Despite warnings (Pollard et al. 1998; Turner 2000), neither government nor Kruger's neighbors are engaging these issues coherently. Even with supportive legislation, few examples of CBNRM institutions exist in the communal lands of the lowveld. Therefore, the need to reinstate rights and responsibilities, so that the locus of power rests with local user groups, is paramount. Once appropriate institutional arrangements exist, conservation agencies can meaningfully engage in numerous initiatives that reflect the core functions of constituency building such as beneficiation and support for sustainable harvesting, training, monitoring, and information needs.

Somewhat paradoxically, these rural communities live adjacent to some of the largest and best-endowed protected areas in the country. There are real opportunities for sustainable use of resources from protected areas as a means of local economic empowerment and as living models for modern conservation policy and practice (Shackleton 1996), but Kruger and most of the private reserves have yet to take advantage of them. Currently, legislative constraints prohibit this in Kruger, yet compelling achievements of conservation-development partnership are available (see Western et al. 1994; Fabricius et al. 2001a).

Kruger cannot tackle its mission alone. Partnerships with community-based organizations and other neighbors and stakeholders, such as non-government organizations, government, business, and the private sector are fundamental to its success (DANCED [Danish Co-operation for Environment and Development] 2000; Hulme and Murphree 2001). Already a number of examples of potential partnerships are in place, including the Trans-Frontier Conservation Areas, various integrated catchment management programs, and the Kruger to Canyons Biosphere Reserve. Lessons from the oft-cited Makuleke example (Box 20.1), which reflects the foundations of successful comanagement (Turner 2000), and from the Mdluli land settlement should prove invaluable in exploring potential collaborative conservation initiatives with neighboring communities. Finally, links with local levels of government, particularly those with an environmental agenda, should be incorporated into partnership efforts.

With a task as new, daunting, and broad as Kruger's mission, the long-term importance of constituency building must not be underestimated, no matter how difficult the challenges. As Hulme and Murphree (2001) argue, this shift is more than just a moral or public relations exercise; it is embraced because it is believed to be the best policy.

Conclusion

As Kruger moves into a new era and seeks reintegration into the broader socio-economic, ecological, and political landscape, one of the biggest challenges it faces, like many African parks, is the social issue. Among other factors, the persistent influences of politics, economics, and conservation ethics in the last 100 years have acted in concert to produce a heterogeneous and dynamic lowveld landscape. Across the fence, the disparities in access to and responsibilities for land and water, coupled with the alienation of lowveld communities from Kruger and its conservation philosophies, are stark. At the same time, South Africa boasts a network of national parks and tourism is generating important revenue, yet biodiversity is at risk. This makes the integration of Kruger and the areas beyond the fence all the more compelling.

If we seek to ensure the integrity of these systems, should we not better equip all South Africans to manage resources outside protected areas and to support conservation in protected areas? Like other parks, Kruger sits on this threshold of opportunity: the democratic changes brought by South Africa's new political identity are supported by the global reorientation toward sustainability. Kruger is not excluded from this, and there is no reason why the historical divide, and at times enmity, between Kruger and neighboring communities should not be bridged and developed into a shared and mutually beneficial future.

Notes

1. Although the South African lowveld landscape is contiguous with that of Mozambique and Zimbabwe, which also border Kruger, information and space limitations preclude discussion of these countries here.
2. Just before the proclamation of Kruger in 1926, the land shortage for black Africans was so acute that a large portion of the Sabi Game Reserve in the central lowveld near Acornhoek was excised as a "native area" (Carruthers 1995).

References

Adams, W., and D. Hulme. 2001. Changing narratives, policies and practices in African conservation. Pages 9–23 in D. Hulme and M. W. Murphree (eds.), *African wildlife and livelihoods: the promise and performance of community conservation.* Oxford, UK: James Curry Ltd.

Banks, D. I., N. Griffin, C. M. Shackleton, S. E. Shackleton, and J. Mavrandonis. 1996. Wood supply and demand around two rural settlements in a semi-arid savanna, South Africa. *Biomass and Bioenergy* 11:319–331.

Breen, C. M., R. Bestbier, D. Kamundi, D. le Maitre, G. Marneweck, and F. Venter. 1997. Integrating socio-economic and governance systems with ecological knowledge of structure and function of riparian systems. Pages 60–66 in K. H. Rogers and R. J. Naiman (eds.), *The ecology and management of riparian corridors in southern Africa.* Proceedings of the International Workshop, Kruger National Park, Skukuza, South Africa.

Bundy, C. 1988. *The rise and fall of the South African peasantry.* 2nd edition. Cape Town: David Philip.

Campbell, D. J., and J. M. Olson. 1991. Framework for environment and development: the Kite. *Occasional Paper of Michigan State University* 10:30.

Carruthers, J. 1995. *The Kruger National Park: a social and political history.* Pietermaritzburg, South Africa: University of Natal Press.

Christopher, A. J. 1994. *The atlas of apartheid.* Johannesburg: Witwatersrand University Press.

DANCED. 2000. *Implementing People and Parks projects: perspectives and constraints.* Report to SANParks, Skukuza, South Africa.

Davenport, R., and C. Saunders. 2000. *South Africa: a modern history.* 5th edition. Basingstoke, UK: Macmillan.

de Villiers, B. 1999. *Land claims and national parks: the Makuleke experience.* Pretoria: Human Sciences Research Council.

de Wet, C. 1995. *Moving together, drifting apart: betterment planning and villagisation in a South African homeland.* Johannesburg: Witwatersrand University Press.

Dovie, B. D. K. 2001. *Woodland resource utilization, valuation and rural livelihoods in the lowveld South Africa.* Unpublished M.Sc. dissertation, University of the Witwatersrand, Johannesburg, South Africa.

Fabricius, C., E. Koch, and H. Magome. 2001a. *Community wildlife management in South Africa: challenging the assumptions of Eden.* Evaluating Eden Series 6. London: International Institute for Environment and Development (IIED).

Fabricius, C., E. Koch, and H. Magome. 2001b. Towards strengthening collaborative ecosystem management: lessons from environmental conflict and political change in southern Africa. *Journal of the Royal Society of New Zealand* 7:831–844.

Fischer, A. 1988. Whose development? The politics of development and the development of politics in South Africa. Pages 22–35 in E. Boonzaaier and J. Sharp (eds.), *South African key words: the uses and abuses of political concepts*. Cape Town: David Phillip.

Folke, C., S. Carpenter, T. Elmqvist, L. Gunderson, C. S. Holling, and B. Walker. 2002. Resilience and sustainable development: building adaptive capacity in a world of transformations. *Ambio* 31:437–440.

Glazewski, J. 2000. *Environmental law in South Africa*. Durban, South Africa: Butterworth.

Gunderson, L., and C. S. Holling (eds.). 2001. *Panarchy: understanding transformations in human and natural systems*. Washington, DC: Island Press.

Harries, P. 1989. Exclusion, classification and internal colonialism: the emergence of ethnicity among the Tsonga-speakers of South Africa. Pages 87–117 in L. Vail (ed.), *The creation of tribalism in southern Africa*. Berkeley: University of California Press.

High, C., and C. M. Shackleton. 2000. The comparative value of wild and domestic plants in home gardens of a South African rural village. *Agroforestry Systems* 48:141–156.

Hoffman, T. S. Todd, Z. Ntshona, and S. Turner. 1999. *Land degradation in South Africa*. National Botanical Institute, Cape Town.

Holling, C. S. 2001. Understanding the complexity of economic, ecological and social systems. *Ecosystems* 4:390–405.

Hulme, D., and M. W. Murphree (eds.). 2001. *African wildlife and livelihoods: the promise and performance of community conservation*. Oxford, UK: James Curry.

Kinzig, A. P., S. Carpenter, M. Dove, G. Heal, S. Levin, J. Lubechenko, S. H. Schneider, and D. Starrett. 2000. *Nature and society: an imperative for integrated environmental research*. Executive summary at http://lsweb.la.asu.edu/akinzig/report.htm.

Koch, E., D. Cooper, and H. Coetzee. 1990. *Water, waste and wildlife: the politics of ecology in South Africa*. Harmondsworth, UK: Penguin.

Leach, M., R. Mearns, and I. Scoones. 1999. Environmental entitlements: dynamics and institutions in community-based natural resource management. *World Development* 27:225–247.

Lewis, S. N. 1997. *Birds as indicators of biodiversity: a study of the avifaunal diversity and composition in two contrasting land-use types in the eastern Transvaal lowveld*. Unpublished M.Sc. thesis, University of East Anglia, Norwich, UK.

Low, A. B., and A. G. Rebelo. 1996. *Vegetation of South Africa, Lesotho and Swaziland*. Pretoria: Department of Environmental Affairs and Tourism.

May, J. 1995. *The composition of poverty in rural South Africa: an entitlement approach*. Report commissioned by the World Bank/Land and Agricultural Policy Centre.

McKinsey and Company. 2002. *SANParks and McKinsey final meeting report, following strategic review*. Report to SANParks, Skukuza, South Africa.

Murphree, M. W. 1996. *Approaches to community participation. African policy consultation, final report*. London: ODA.

Nash, R. 1982. *Wilderness and the American mind*. 3rd edition. New Haven, CT: Yale University Press.

Newenham, J. 2001. *Kruger to Canyons Biosphere Initiative. Application to UNESCO*. Unpublished report.

Newmark, W. D., and J. L. Hough. 2000. Conserving wildlife in Africa: integrated conservation and development projects and beyond. *BioScience* 50:585–592.

Niehaus, I. 2001. *Witchcraft, power and politics: exploring the occult in the South African lowveld.* London: Pluto Press.

O'Keeffe, J. H., and B. R. Davies. 1991. Conservation and management of the rivers of the Kruger National Park: suggested methods for calculating instream flow needs. *Aquatic Conservation in Marine and Freshwater Ecosystems* 1:1–17.

Parsons, D. A., C. M. Shackleton, and R. J. Scholes. 1997. Changes in herbaceous layer condition under contrasting land use systems in the semi-arid lowveld, South Africa. *Journal of Arid Environments* 37:319–329.

Pigram, J. J., and R. C. Sundell. 1997. *National parks and protected areas: selection, delimitation, and management.* Armidale, New South Wales: Centre for Water Policy Research.

Pollard, S. R. 1996. *Social assessment of riverine resource use as input into the instream flow assessment for the Sabie River.* Pretoria: Department of Water Affairs and Forestry.

Pollard, S. R., J. C. Perez de Mendiguren, A. Joubert, C. M. Shackleton, P. Walker, T. Poulter, and M. White. 1998. *Save the Sand. Phase 1 feasibility study: the development of a proposal for a catchment plan for the Sand River Catchment.* Pretoria: Department of Water Affairs and Forestry.

Porter, R. N. 1970. *An ecological reconnaissance of the Timbavati Private Nature Reserve.* Unpublished report, Klaserie, South Africa.

Ramutsindela, M. F. 2002. The perfect way to end a painful past? Makuleke land deal in South Africa. *Geoforum* 33:15–24.

Ramutsindela, M. F., and D. Simon. 1999. The politics of territory and place in post-apartheid South Africa: the disputed area of Bushbuckridge. *Journal of Southern African Studies* 25:479–496.

Ritchken, E. 1995. *Leadership and conflict in Bushbuckridge: struggles to define moral economies within the context of rapidly transforming political economies (1978–1990).* Unpublished Ph.D. thesis, University of the Witwatersrand, Johannesburg, South Africa.

Russell, I. A., and K. H. Rogers. 1989. The distribution and composition of fish communities in the major rivers of the Kruger National Park. Pages 281–288 in S. Kienzle and H. Maaren (eds.), *Proceedings 4th South African National Hydrological Symposium.* Pretoria: University of Pretoria.

SANParks. 2000. *Visions of change: social ecology and South African national parks.* Johannesburg: DCC and SANParks.

Shackleton, C. M. 1996. Potential stimulation of rural economies by harvesting secondary products: a case study from the central eastern Transvaal lowveld, South Africa. *Ambio* 25:33–38.

Shackleton, C. M. 1998. Examining the basis for perceptions of communal land degradation: lessons from the central lowveld. Pages 196–210 in T. D. de Bruyn and P. F. Scogings (eds.), *Communal rangelands in southern Africa: a synthesis of knowledge.* Symposium Proceedings, Fort Hare University, Alice, South Africa.

Shackleton, C. M., N. Griffin, D. I. Banks, J. Mavrandonis, and S. E. Shackleton. 1994. Community structure and species composition along a disturbance gradient in a communally managed South African savanna. *Vegetation* 115:157–168.

Shackleton, C. M., and S. E. Shackleton. 2000. Direct use values of savanna resources harvested from communal savannas in the Bushbuckridge lowveld, South Africa. *Journal of Tropical Forest Products* 6:21–40.

Shackleton, C. M., S. E. Shackleton, and B. Cousins. 2001. The role of land-based strategies in rural livelihoods: the contribution of arable production, animal husbandry and natural resource harvesting from communal areas in South Africa. *Development Southern Africa* 18:581–604.

Shackleton, S. E., C. M. Shackleton, C. M. Dzerefos, and F. R. Mathabela. 1998. Use and trading of wild edible herbs in the central lowveld savanna region, South Africa. *Economic Botany* 52:251–259.

Shackleton, S. E., J. J. Stadler, K. A. Jeenes, S. R. Pollard, and J. S. S. Gear. 1995. *Adaptive strategies of the poor in arid and semi-arid lands: in search of sustainable livelihoods: a case study of the Bushbuckridge district, eastern Transvaal, South Africa.* World Resources Foundation internal report.

Sonnenberg, D. 1993. *Alpha, beta and gamma diversity within the communal lands of the Mhala district of the eastern Transvaal lowveld.* Unpublished honors dissertation, University of Witwatersrand, Johannesburg, South Africa.

Spenceley, A. 2001. Sociocultural impacts of Ngala Private Game Reserve on the Welverdiend community. *Institute for Natural Resources Occasional Paper* 191:1–54.

Stadler, J. 1994. *Generational relationships in a lowveld village: questions of age, household and tradition.* Unpublished M.Sc. thesis, University of Witwatersrand, Johannesburg, South Africa.

Steenkamp, C. 2000. *The Makuleke land claim: power relations and CBNRM.* Evaluating Eden Discussion Paper 18. London: International Institute for Environment and Development.

Stevenson-Hamilton, J. 1929. *The lowveld: its wildlife and its people.* London: Cassell and Company.

Tapela, B. N., and P. H. Omara-Ojungu. 1999. Towards bridging the gap between wildlife conservation and rural development in post-apartheid South Africa: the case of the Makuleke community and Kruger National Park. *South African Geographical Journal* 81:148–155.

Tollman, S. M., K. Herbst, M. Garenne, J. S. S. Gear, and K. Kahn. 1999. The Agincourt demographic and health study: site description, baseline findings and implications. *South African Medical Journal* 89:858–864.

Turner, S. 2000. Department of Environmental Affairs and Tourism with GTZ: the TRANSFORM Project—monitoring and evaluation, 1999. Final summary report. Pages 21–37 in *Towards best practice: communities and conservation.* Proceedings of the SANParks Conference, Berg en Dal, South Africa.

Tyson, P. D. 1986. *Climatic change and variability in southern Africa.* Cape Town: Oxford University Press.

Venter, A. K., C. B. Breen, and C. M. Marais. 1994. *The participative forum approach: integrating the goals of conservation and the development of local indigenous peoples.* Unpublished report, SANParks, Skukuza, South Africa.

Walker, B., S. Carpenter, J. M. Anderies, N. Abel, G. Cumming, M. Janssen, L. Lebel, J. Norberg, G. Peterson, and R. Pritchard. 2002. Resilience management in social-ecological systems: a working hypothesis for a participatory approach. *Conservation Ecology* 6. Online: http://www.consecol.org/vol6/.

Walker, B. H., R. H. Emslie, R. N. Owen-Smith, and R. J. Scholes. 1987. To cull or not to cull: lessons from a southern African drought. *Journal of Applied Ecology* 24:381–401.

Weeks, D. C., J. H. O'Keeffe, A. Fourie, and B. R. Davies. 1996. A *pre-impoundment study of the Sabie–Sand River System, Mpumalanga with special reference to predicted impacts on the Kruger National Park*, Vol. 2. Water Research Commission Report 294/1/96.

Western, D., R. M. Wright, and S. Strum (eds.). 1994. *Natural connections: perspectives in community-based conservation*. Washington, DC: Island Press.

Wily, L. A. 2000. Forest laws in east and southern Africa: moving towards a community based future? *Unasylva* 203:19–26.

Chapter 21

Heterogeneity and Management of the Lowveld Rivers

JAY O'KEEFFE AND KEVIN H. ROGERS

Of all ecosystems, rivers, especially in semiarid regions, probably are the most variable and least predictable (Davies et al. 1994). This may be why river scientists have produced many theoretical frameworks of river structure and functioning (Hildrew et al. 1994), but these frameworks often fail to provide the practical insights necessary for river management. Leopold et al.'s (1964) explanation of fluvial processes in geomorphology, Hynes's (1975) "The Stream and Its Valley," and the river continuum concept of Vannote et al. (1980) all provide deterministic explanations of river processes, components, and relationships. However, they tend to simplify our understanding too much to be useful in managing these complex ecosystems.

The major agent of complexity in river systems is undoubtedly the disturbance regime generated by variability in the flow of water (Townsend 1989), a medium much more dense than the atmosphere of terrestrial systems. The frequency, intensity, and severity of flow-related disturbances determine when, if ever, a community will reach equilibrium and also have major influences on productivity, nutrient cycling and spiraling, and decomposition. In fact, flow-generated disturbance is the dominant organizing agent in stream ecology (Resh et al. 1988).

Disturbances provide a variety of niches and refugia (Townsend 1989) that make the likelihood of catastrophic mortality through the whole system extremely unlikely, even though the disturbance may be propagated throughout the system. The maintenance of such habitat heterogeneity is akin to spreading the risk at the community and population level (Hildrew and Giller 1994).

The dominant effect of inherently variable physical processes causes many stream communities to be in a state of perpetual recovery from frequent disturbances (Reice et al. 1990). This perspective suggests a "clinging to the wreckage" model of community organization in which species are entirely noninteractive or the recurrence interval of disturbances is too short to allow interactions to eliminate species (Hildrew and Giller 1994).

In the past, the themes of variability, change, and stochasticity implicit in disturbance theory have been uncomfortable for managers to deal with. River management traditionally was seen as the taming of these unpredictable fluctuations, the smoothing out of variability, to ensure stability of supply in the river or for off-channel use. Storing water behind dams and regulating flow for irrigation or hydropower needs are classic examples of this desire for predictable and secure supply. The realization that the healthy functioning of river ecosystems depends as much on periodic droughts and floods as it does on normal base flows presents managers and scientists with conceptual and operational challenges that they are only beginning to address.

Policymakers, scientists, and managers in South Africa have made far-reaching strides in this field of environmental management. The 1998 South African Water Act contains the visionary principle of the ecological reserve as a right: "the quantity and quality of water required for the protection of ecological components and processes on which people depend, shall be reserved before any uses (other than the provision for basic human uses such as drinking and washing) are licensed." This in itself is a clear recognition that if the resource is to be protected, the best way to do it is by managing for healthy river ecosystems.

Methods for assessing "environmental flows" for rivers began with the concept of providing a minimum flow but now concentrate on maintaining as much natural variability as possible. Richter et al. (1997) have developed a hydrology-based method for setting streamflow-based river ecosystem management targets that they call the range of variability approach. The natural flow paradigm (Richter et al. 1997) maintains that the full range of interannual and intra-annual variation of hydrological regimes and associated characteristics of timing, duration, frequency, and rate of change are critical elements of flow regime aimed at managing for native biodiversity and integrity of aquatic ecosystems. Another recent development, the flow stressor-response method (O'Keeffe et al. 2002), is based on ensuring that the managed flow regime retains as much of the natural range of ecological stresses as possible.

This chapter describes the natural heterogeneity of the main rivers that flow across the lowveld and through Kruger, and the anthropogenic changes they have experienced. We discuss the implications for the future management, use, and protection of the rivers under new South African environmental and water legislation.

The Geographic Setting

The Kruger is a longitudinal north-south area, with five main river systems flowing from west to east across it. Implications of this geographic arrangement are that the management of the rivers in the park and the adjacent terrestrial systems are fundamentally affected by the management of the catchment upstream of the park and are largely out of its direct control. To understand the implications, imagine the park turned sideways to encompass the entire Sabie catchment, with the watersheds forming the boundaries. No agriculture, no overgrazing, no mining, and no water abstraction would complicate the park's mission to conserve the natural diversity of processes, functions, and components in the landscape.

Rivers globally are recognized as biodiversity hotspots (Naiman et al. 1992), and this diversity is enhanced by the interactions with upland ecosystems. Kruger is generally semiarid (rainfall 6–400 mm per year), but the perennial rivers arise in the escarpment to the west of the park, which receives much higher rainfall (more than 2,000 mm per year). Kruger includes a total of 31,548 km of river (marked on 1:50,000 map sheets and therefore mostly second- or higher-order streams), giving an average density of 1.5 km/km^2 of river flowing across a wide range of geological substratum and rainfall (Chapter 5, this volume). Many of the smaller drainages originate in and are contained within the park, but all of them are in low-rainfall areas and flow only seasonally or intermittently. These extensive drainage systems distribute water throughout the park, acting as foci and hotspots for animal and plant diversity and productivity (Chapter 10, this volume). These figures demonstrate the scale and potential importance of water ways in ecosystem processes. Unfortunately, the focus of almost all river research has been on the 600-km length of perennial river, and the role of more than 30,000 km of ephemeral streams is essentially unknown (Chapter 8, this volume).

The Limpopo River forms the northern boundary of Kruger, but the river itself is outside the park boundary. Four naturally perennial rivers, flowing from west to east, cross the 60-km × 350-km park before flowing into Mozambique and the Indian Ocean. From north to south, these are the Luvuvhu, the Letaba (a tributary of the Olifants), the Olifants, and the Sabie. The Crocodile River forms the southern boundary and is included in the park. The Luvuvhu and the Olifants are part of the Limpopo system, and the Sabie and the Crocodile flow into the Inkomati in Mozambique. This chapter concentrates on these five rivers. The chapter also concentrates on heterogeneity presented by the main rivers, both in the catchments outside Kruger and inside the park. Chapter 9 (this volume) focuses on within-river heterogeneity, mainly inside the park. Figure 21.1 shows the catchments of the main rivers of the park, with some indication of the major land uses, emphasizing the extent of the influences from outside Kruger. The main specific anthropogenic influences that have modified each of the catchments are detailed in this chapter. Table 21.1 provides catchment scale and flow statistics for these rivers, outside and inside Kruger.

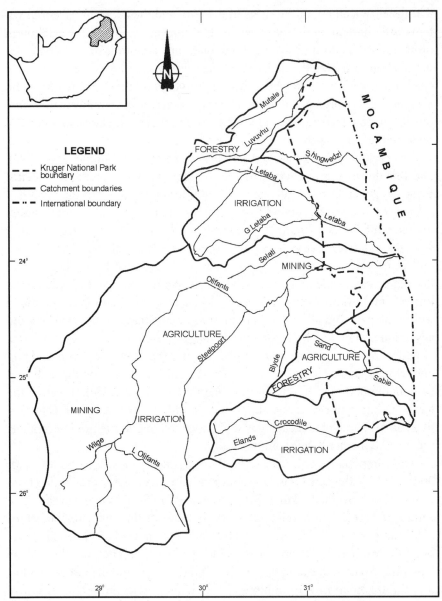

FIGURE 21.1. The catchments of the main rivers that run through Kruger, with some indication of the major land uses.

Physical Heterogeneity

A number of factors interact to generate a wide range of differences in the physical characteristics of the rivers that flow through Kruger. The rivers have a range of catchment sizes and flow across a wide range of geological formations

TABLE 21.1

Some catchment scale variables that describe Kruger's rivers, their catchments, and their flow characteristics (modified from Carter and Rogers 1995). Flow statistics describe the monthly flow data.

VARIABLE	CROCODILE	SABIE	OLIFANTS	LETABA	LUVUVHU
River length^f (km)	320	189	840	573	247
River length$^/$ (km)	115	110	100	100	85
River length$^°$ (km)	204	78	740	475	162
Gradient$^/$ (m/km^{-1})	1.759	2.514	1.540	1.643	2.624
Gradient$^°$ (m/km^{-1})	9.343	20.949	2.001	2.838	3.019
Catchment area^f (km^2)	10,420	6,385	54,815	13,865	5,330
Catchment area$^/$ (km^2)	2,390	4,665	4,445	3,300	3,255
Catchment area$^°$ (km^2)	8,030	1,720	50,370	10,565	2,075
Stream order$^/$	5	5	6	6	5
MAR^f	1,444	666	2,257	819	440
MAR$^/$	73	85	89	26	222
MAR$^°$	1,371	581	2,168	793	218
Flow mean (m^3 × 10^6)$^/$	93.5	49.9	113.9	20.8	10.4
Flow CV	116.1	132.5	150.3	292.2	95.8

Superscripts indicate the catchment or river sections to which variables apply: $^°$ = section upstream of Kruger; $^/$ = Kruger river section; $f = / + °$.
MAR = mean annual runoff.

and through a range of rainfall regimes. When the rivers enter Kruger they are at different distances from the retreating escarpment (Chapter 5, this volume) and therefore have different slopes. All these factors combine to generate between-river heterogeneity.

Geology, Geomorphology, and Flow

Chapter 5 (this volume) and Carter and Rogers (1995) summarize the geological formations and their effect on the main rivers of the park. The geological formations of Kruger occur mainly in bands with a north-south orientation, and the relative positions of differentially erodible materials down the length of the park generate complex longitudinal profiles. The southern four rivers traverse mainly granite in the west of the park and strips of Karoo sediments (mostly sandstone), basalt, and rhyolite in the east. They also cross intrusive dikes of dolerite, sections of gabbro, and minor faults, which add heterogeneity to the physical template. Generally, these rivers display increasing river valley width and sedimentation, decreasing stream gradient and bedrock exposure, and increasing channel braiding and migration downriver. However, the actual expression of this pattern at any given location depends on local geology, and all rivers except the Luvuvhu and Limpopo flow through narrow gorges as they pass through the Lebombo Mountains at the eastern border.

The Luvuvhu River is further from the escarpment than the other four rivers and crosses a different set of geological formations. In the west it flows through deep narrow gorges incised into quartzite and sandstone. It becomes broader and less steep to the east, where it traverses more erodible sandstone and basalt, finally meandering across the alluvial floodplain sediments of the lower Luvuvhu and Limpopo rivers (Carter and Rogers 1995).

The hierarchical pattern of spatial heterogeneity that has emerged (Chapter 9, this volume) emphasizes that, at the largest scale, each river has distinct catchment characteristics resolvable at the scale of hundreds of square kilometers. A north-south rainfall gradient (450–650 mm/year) means that runoff in the southern catchments is greater per unit area than in the north, and the Olifants River has the largest catchment because it has cut through the escarpment and captured streams on the inland plateaus (Figure 21.1). All this complexity leads to much between-river heterogeneity on the lowveld. Within rivers, a hierarchical classification distinguishes a nested sequence of zones, macroreaches, reaches, channel types, morphological units, and microsites. Each of these scales has particular structural and response correlates in hydrological, geomorphological, and ecological spheres (Heritage et al. 1997; Moon et al. 1997; van Coller et al. 2000; Chapter 9, this volume).

All the rivers have a channel-in-channel morphology, with a steep macrochannel bank (10–15 m high) and a wide (200–600 m) macrochannel floor. The macrochannel bank is colonized by riparian forest species differentially distributed along the elevation-inundation gradient. All flows were thought to be contained within these banks, but the 2000 floods overtopped them in the Sabie by several meters. Geomorphologically, the macrochannel bank is stable, but the macrochannel floor is highly dynamic, with frequent alterations of state even at the whole channel type scale (Moon et al. 1997; Rountree et al. 2002). State change is a consequence of fluctuations in flow-mediated sediment storage and transport on a complex bedrock template.

Structurally, the Luvuvhu, Sabie, and Crocodile rivers are the most diverse in their reaches within the park. The Luvuvhu flows through a major gorge and debouches onto a wide floodplain before its junction with the Limpopo. The hydrology of the floodplain is complex, with backflooding when the Limpopo is in flood and flushing from its own flows. The Sabie exhibits wide local anastomosing reaches, single-thread channels, and sandy braided reaches downstream of the confluence with its major tributary, the Sand River. The Crocodile is similar to the Sabie upstream of the Sand confluence.

The Olifants and the Letaba rivers generally are set in much wider macrochannels than the other three, although both are constricted in a gorge at their confluence as they flow through the Lebombo Mountains at the western boundary of the park. The Olifants is a larger river than the others and is largely bedrock anastomosing sections upstream of the park. The Letaba has a narrow, active channel winding through a wide, flat, sedimented macrochannel.

Anthropogenic Modifications to Flow Regimes

All five perennial rivers have been modified by water abstraction, increased sediment input, and pollution at different degrees of severity. The main sources of flow modification are as follows:

- The Luvuvhu was perennial until the late 1960s but is now seasonal (no-flow conditions approximately 20 percent of the time [O'Keeffe and Uys 1998]), mainly because of commercial forestry, impoundment, and abstraction for irrigation.
- The Letaba was perennial until the late 1960s but stopped flowing approximately 30 percent of the time over the next 30 years (O'Keeffe and Uys 1998) because of large impoundments and abstraction for irrigated agriculture. Negotiations between Kruger's scientists and catchment managers saw perenniality returned in the mid-1990s.
- The Olifants, a very large catchment draining most of the region east of the Witwatersrand, suffers from all the common sources of degradation: multiple impoundments, agricultural and urban runoff, erosion from overgrazing, and mining effluent (Moore et al. 1991).
- The Sabie is considered the least modified of the five rivers, with about 70 percent of its natural mean annual runoff (MAR) remaining in the river. The missing 30 percent is intercepted mostly by extensive commercial forestry and irrigation of fruit farms in the upper catchment.
- The Crocodile River forms the southern boundary of the park, and impoundments in its upper reaches release abnormally high and stable winter flows to support irrigation on the southern banks opposite Kruger. It is heavily infested with water hyacinth (*Eichhornia crassipes*).

Natural Water Quality and Anthropogenic Modifications

Intensive water quality sampling on the rivers of Kruger started only in 1983, so the record reflects the influences of human activities on water quality. Natural conditions to some extent can be inferred from these results. Moore et al. (1991), Kilian and du Plessis (1993), and Muller and Scherman (in prep.) have summarized the water quality problems and trends (Table 21.2).

Under natural conditions, it is probable that the Letaba, Sabie, and Crocodile rivers would have similar water quality because they originate on the escarp-

TABLE 21.2

Variations in total dissolved solids (TDS), major ions, and other components, elevated by human activity in the main rivers of Kruger.

RIVER	SAMPLE SITE	MEAN TDS (MG/L)	90TH PERCENTILE TDS (MG/L)	DOMINANT IONS	ELEVATED
Luvuvhu	Shidzivane	92	—	$CaCO_3$, Na^+, Cl^-	Na^+, Cl^-
Letaba	Mahlangene	227	565	Na^+, Cl^-	NH_4^+
Olifants	Mamba	605	1,300	SO_4^{2-}	SO_4^{2-}
Sabie	Phabene	72	80	$CaCO_3$, Na^+, Cl^-	—
Crocodile	Malelane	221	396	$CaCO_3$, Na^+, Cl^-	NH_4^+, PO_4^{2-}

ment and flow over similar geology. The Olifants would be naturally more saline, with natural sulfate dominance, because it originates much further from Kruger, in the highveld. The Luvuvhu would naturally have low total dissolved solids (TDS) but could have different ion dominance because it originates further away on the escarpment, with predominantly sandstone geology.

Despite the low TDS concentrations in the Luvuvhu, it is slightly mineralized, with an increase in Na and Cl but with no trajectory of change discernible in the short record. Irrigation return flows probably are the main reason that the Letaba is described as moderately mineralized, with occasional high NH_4^+ concentrations and increased turbidity caused by erosion in the upper catchment. Agricultural return flows are leading to a trend of increasing salinity levels during low-flow periods, and flash floods from the Klein Letaba generate high silt loads in summer (Moore et al. 1991). Muller and Scherman (in prep.) suggest that high density of hippos in the Engelhardt dam on the Letaba River is the cause of a trend of increasing phosphates, another illustration of the side effects of river manipulation.

The Olifants is heavily affected by mining and irrigated agriculture, and is currently highly mineralized, although there has been no discernible trajectory of change over the short monitoring record. Perhaps the most significant water quality effects for Kruger are from the mining complex at Phalaborwa, on its western boundary, which together with the Phalaborwa Barrage causes sporadic very high sediment loads, which result in fish kills, and high salinity, particularly of sulfates.

The Sabie River is slightly mineralized, with no recorded trajectory of change. Water quality and the biota in the Sabie have recovered remarkably since the 1940s, when gold mining dumps that had sterilized the river were cleared up (Pienaar 1985). The Crocodile River is moderately mineralized, with an increasing trend in NH_4^+, PO_4^{2-}, and salinity.

The anthropogenic alterations of the physical characteristics of the lowveld rivers have been wide ranging in type and impact. Alterations in flow regime have been most widespread and have the most potential to affect river biodiversity and heterogeneity. However, this should not overshadow the more isolated impacts of water quality in the Olifants and Crocodile rivers that are likely to become more serious as development in the catchments proceeds.

Biological Heterogeneity

In this section, the present knowledge of the diversity of three main groups of biota—fish, benthic invertebrates, and riparian vegetation—is briefly reviewed.

Fish Communities

The present state of the biota in the five rivers is a consequence of the natural differences in the climate and hydrology, geology and resulting soils, the gradients, and geomorphology, overlain by the different levels of anthropogenic modification in sediment regime, nutrients and salinity, and flows. The fish communities of the Sabie River are the most diverse, with 45 naturally occurring species recorded from the whole system, of which 37 have been found in the park. These include the endemic Lowveld largemouth (*Serranochromis meridianus*), indicating a degree of biogeographical isolation. Probable reasons for this diversity are the heterogeneity of habitats and hydraulic conditions and the steep gradient, providing a wide range of temperatures over a short distance. The communities of the other rivers in the park between the 1960s and the 1980s were as follows, according to Russell (1997): Luvuvhu, 35 species; Letaba, 32 species; Olifants, 34 species; and Crocodile, 38 species.

For all five rivers, 46 fish species were found in the park during the surveys of the 1960s and 1980s. Thirty of the species were found in all rivers until the 1960s, but in the 1980s survey, only 15 were common to all rivers. This reduction in commonality (i.e., different species were absent from different rivers) appears to be a result of different responses to anthropogenic disturbances in different rivers (Russell 1997). Russell attributes the absences to decreases in flow and consequently in vegetation cover in the Luvuvhu, to increasing no-flow events and impoundments built within Kruger in the Letaba, to high silt events and pollution levels in the Olifants, and to a mixture of low-flow conditions and the encroachment of floating plants in the Crocodile.

The Crocodile River had four species not found in the other rivers, the Sabie and Luvuvhu had two each, and the Olifants had one. Fish in the Letaba were all common to other rivers. Nine of the species are endemic to a single river, and four species were found in only two rivers.

Benthic Invertebrates

There have been very few studies of invertebrates in Kruger rivers, at least in part because of the dangers involved in sampling crocodile-infested waters. O'Keeffe and Uys (1998) compared the similarities and differences between a number of Kruger's rivers, using the invertebrate data collected by Moore and Chutter (1988). They showed that the Luvuvhu and Sabie rivers contained the most taxa: 150 and 134, respectively. The Mutale had 104 taxa, and the Letaba was the least diverse, with only 60.

The communities of different rivers were similar, with 51 taxa out of a possible 60 (the total for the least diverse river, the Letaba) common to the Luvuvhu, Sabie, Mutale, and Letaba rivers. This indicates that the community of the Letaba River is made up largely of common hardy species, able to survive in most conditions. The Luvuvhu exhibited the greatest number of exclusive taxa (28), with 16 found only in the Sabie and 5 each found only in the Mutale and Letaba.

The two rivers with the most modified flow regimes, the Luvuvhu and the Letaba, had the highest and lowest invertebrate taxonomic diversity, respectively. The low diversity in the Letaba probably is a consequence of the severe flow modifications and generally lower habitat diversity. This was also reflected in the fish community, which was the least diverse and the least endemic. In the Luvuvhu, the added habitat diversity of the gorge and floodplain may provide additional habitat opportunities, and the less frequent no-flow events may allow the persistence of the perennial invertebrate community and the intermittent colonization by species favoring temporary river conditions.

Riparian Vegetation

In cross-section, Kruger's river valleys are generally wide, with well-developed channel shelves. A wide range of rock and sediment structures provides for a variety of habitats for vegetation, ranging from sparse patches of herbaceous plants, to dense reed stands and bush patches, to riparian forest (Carter and Rogers 1995).

The proximate determinants of the structure of riparian vegetation are strongly correlated with a gradient of elevation above the main channel and diversity of fluvial landforms (van Coller et al. 2000). Species at low elevations need traits enabling them to survive high stress and disturbance, whereas species at higher elevations generally need traits that confer competitive advantages. These generalizations are modified at the smaller scale by many variables, such as the geology of bedrock and different geomorphic features. They are also modified by time-dependent variables such as the time since the previous flood event and by larger-scale variables such as the distance downriver.

The macrochannel floor can be described as a complex mosaic of six patch types (water, sand, rock, reeds, shrubs, and trees), the proportions of which vary across the spatial hierarchy. A central hypothesis to emerge from the Rivers Research Programme is that at any one time, the floor of each river is somewhere on a biophysical trajectory from a bare bedrock template to a complex mosaic of sedimentary features colonized by heterogeneous plant communities. The trajectory of change clearly would be in response to many previous events, but the 1925 and 2000 floods probably have played a major role. Intermediate stages of the succession have two easily identified trajectories, depending on the channel type engraved in the underlying bedrock template. Both trajectories are facilitated by plant growth that traps sediment.

In the first trajectory, sand bars, which form in reaches of low slope, are colonized by the reed *Phragmites mauritianus*. Subsequent colonization by bushes and then trees leads to a patchy forest of river bushwillow (*Combretum erythrophyllum*), the common cluster fig (*Ficus sycomorus*), and water elder (*Nuxia oppositifolia*). In the second trajectory, seedlings of the tree *Breonadia salicina* establish in cracks of steeper rocky outcrops and subsequently trap sediment. Matumi (*B. salicina*) remains the dominant species, but the trapped sediment facilitates colonization of other trees such as water berry (*Syzygium cordatum*) and water pear (*S. guineense*) (van Coller et al. 2000).

Analysis of aerial photographs taken at about 10-year intervals showed that these generalized successional sequences were far from linear, and multiple pathways of state change were very common (Carter and Rogers 1995). Furthermore, given the distinctive catchment conditions of the rivers, the combined effects of all events and processes before 2000 had led to a set of very different river ecosystems. The Sabie River had deep pools and was well forested; the Olifants River had a wide, shallow bed and a high proportion of reeded areas; the Letaba had a very sandy, unvegetated bed interrupted by a few rocky outcrops; and the Luvuvhu was the only river with a well-defined, forested floodplain.

The events and processes over the last 50 years, including the 1992 drought and 1996 flood, had different effects on heterogeneity between and within the different rivers, resolvable at different scales. The 1992 drought was widespread across all rivers but caused the death of riparian trees only in the Sabie and Luvuvhu rivers. These rivers are the most and least bedrock-influenced rivers, respectively, but both have large proportions of their small catchments in the semiarid lowlands. At smaller scales, tree deaths were patchy and confined to species whose roots could not track the falling water table and to geomorphic units in which bedrock limited root movement or alluvium was very deep and water tables fell rapidly toward the end of the season.

The 1996 floods, on the other hand, were confined to the Sabie and Olifants rivers in the south. In the Sabie River, effects were small scale and patchy because they were confined to poorly vegetated geomorphic units and the upstream ends of steep bedrock sections. This event highlighted the importance of tree life history characteristics, which operate at the small scale of individual trees, in creating differential responses of species to flooding (Mackenzie et al. 1999). For example, species lowest on the elevation gradient such as water berry (*S. cordatum*) and water pear (*S. guineense*) have flexible trunks and branches and leaf clusters that fold up and reduce resistance to flow. The terrestrial species that colonized the riparian zone between large floods had rigid stems and many thorns that trapped debris and increased resistance to flow. These species were much more susceptible to uprooting by floodwaters than the low-elevation species with flexible stems and no thorns.

In the large-catchment Olifants River, where long stretches of the macrochannel floor were colonized by reeds rather than trees, the flood effects were very different. The riverbed was stripped of sediment and vegetation, exposing the underlying bedrock template.

Implications of Large Infrequent Disturbances

The 1996 floods in the Olifants River and 2000 floods in general were much bigger than any other event studied in the history of these rivers, and their impact and legacy are still largely unknown. They will provide many opportunities to increase our understanding of large, infrequent disturbances (LIDs), to predict their role in ecosystem structure and functioning, and to incorporate the understanding and predictions into effective management strategies.

In their synthesis of LIDs, Turner and Dale (1998) highlight four key concepts that pose research challenges. First, there is evidence that even the largest disturbances do not wipe the slate clean but rather leave heterogeneous imprints, or mosaics, on the landscape that are difficult to describe. Second, predictions of system response to disturbance are not easy to scale up from patch to mosaic. This is, at least in part, a result of the effects of spatial pattern in the mosaic on patch dynamics (Pickett and Rogers 1997). In rivers we can expect that trajectories of patch evolution will be confounded by the spatial configuration of the patch mosaic because it influences hydraulics and sediment erosion and deposition. Third, residual organisms and propagules dictate the initial competitive environment for subsequent colonizers. Fourth, subsequent events determine the future trajectory or state of the system.

The rivers of the Kruger provide opportunities to test all these concepts to further our understanding of event-driven ecosystems and their management.

- First, early observations show that the imprint left by the 2000 floods is extremely heterogeneous between and within rivers and across scales.
- Second, a scaled knowledge of the historical template can be combined with a scaled description of the imprint mosaic to test hypotheses of how predictions of disturbance effects can be scaled up.
- Third, the numbers of residual organisms are visibly different on different channel types, and large infestations of exotic weeds will provide a very different pool of early colonizers from that experienced in the past.
- Fourth, there have been major shifts in the main agents of change in these systems since the major flood in 1925 probably provided the imprint of mosaic we have been studying. Animal densities, especially of large herbivores, have increased many-fold in response to sustained management; similarly, fire regimes, flow regimes, and sediment transport dynamics have all been altered as land-use and management practices have changed. The future ecosystem will also be faced with the many implications of global climate change.

The next generation of river research in Kruger stands to make major contributions to the understanding and management of fluvial heterogeneity in savanna systems in particular and rivers in general.

The understanding generated over the last decade has already been put to good use in innovative moves toward determining environmental flow requirements (EFRs) for biodiversity management and broader cooperative efforts of catchment management.

Managing for River Heterogeneity

Full achievement of Kruger's vision of "maintaining biodiversity in all its natural facets and fluxes" probably is not possible for the rivers. The rivers upstream of the park represent very valuable resources for domestic water supply, for irrigation farming, and for industrial and mining development. Distribution of water access during the apartheid years was far from equitable, and major initiatives are under way to redress this inequity. The fact that Kruger does not control the upper catchments of the main rivers that flow through it makes it inevitable that other interests will have important voices in the management and exploitation of those resources.

The highest-level reference to rivers in the Kruger objectives hierarchy recognizes all this and requires scientists and managers to "maintain, and whenever necessary, restore river ecosystem health and biodiversity particularly through promoting integrated catchment management."

There are two important issues in this objective. First, it accepts that the rivers are not entirely unmodified but does not accept the status quo. The objectives hierarchy records unacceptably high impacts on biodiversity in most perennial rivers as a consequence of dams, abstraction of flow, pollution, sedimentation, and invasion by alien plants. Downstream of the park there are large dams on the Sabie and Olifants rivers.

Second, this objective recognizes that Kruger cannot manage the rivers for its own ends, or in isolation, but must interact with other stakeholders in the catchments to achieve best possible use of the rivers in a developing nation.

Kruger has set its goals for river biodiversity and heterogeneity in the objectives hierarchy and can audit achievement against an array of thresholds of potential concern that determine its comprehensive monitoring program (Chapter 9, this volume). Kruger's management has a clear overall objective—to maintain natural biodiversity—and a good understanding that that biodiversity is governed largely by heterogeneity at multiple spatial and temporal scales. The national water management agencies (Department of Water Affairs and Forestry and future catchment management agencies [CMAs]), on the other hand, have a more confusing and often conflicting set of objectives in that they must both use and protect the water resources of the country (Rogers et al. 2000). Reliable use of the water resources implies constant supply. Management of water supplies often is governed by the volume of water available during droughts. Ensuring water supplies in high-demand catchments entails regulation of flow: for water suppliers, floods are damaging and must be intercepted to increase supply, and droughts cause water shortages and must be augmented. These aims conflict directly with the need to provide variable flow regimes to protect the heterogeneity of rivers and ensure their integrity and health.

The 1998 Water Act and the National Water Resource Strategy recognize that nationally, there must be a balance between use and protection. Demand management and water conservation are prominent among the strategies for meeting future water needs (Department of Water Affairs and Forestry 2002), and the same document emphasizes that "we will need to match the demands with the available water"(p. 10). This is a revolutionary change from the implicit historical policy of making water available (by whatever means) to meet the demands. Nevertheless, because the absolute priorities are to supply clean water and sanitation to the millions in South Africa who still lack them and to meet the need for economic development to alleviate poverty, there is no doubt that demand will increase in the future. This applies to all the rivers that feed into Kruger. Careful management can improve their condition, but they will not be restored to their natural state. A policy of concurrency (the achievement of multiple aims) is embodied in the 1998 Water Act and is the best hope for sustainable development of these water resources.

So there are two very important steps toward advancing these goals:

- Scientists and managers must design and implement a flow regime that can best achieve biodiversity goals but remain cognizant of development needs in the catchments.
- Scientists and managers must engage other stakeholders to ensure the most efficient and equitable use of river resources.

Fortunately, new South African legislation provides excellent context and guidance for achieving these two aims.

Environmental Flow Allocations and the Ecological Reserve

Under the 1998 South African National Water Act, the quantity and quality of water needed to protect the ecological functions on which humans depend (the ecological reserve) must be assessed and maintained in natural water resources. The EFRs for the ecological reserve have been assessed for the perennial rivers of Kruger but have yet to be implemented.

The primary method used for these assessments has been the building block methodology (BBM; King and Louw 1998). The BBM uses a group of specialists to assess different aspects of the river and to recommend a series of flows designed to retain the essential flow variability of the system in order to achieve or maintain a predetermined set of environmental objectives.

Briefly, the BBM comprises the following steps:

- A geomorphological assessment of river reaches is conducted to identify homogeneous sections within which to identify critical sites for flow evaluation.
- Habitat integrity and conservation importance are assessed to derive ecological objectives.
- Biological, social, and water quality surveys are used to assess the present state of the system.
- A hydrological and hydraulic analysis is completed.
- At a multidisciplinary workshop, a wide range of specialists identify the habitat characteristics necessary to maintain the instream and riparian biota, and the flow characteristics necessary to maintain essential processes such as sediment transport and water quality. Biologists use hydraulic cross-sections and calibration graphs to identify the depths, current velocities, and marginal habitats needed to maintain ecosystem function at each site (Figure 21.2). The hydraulic information is then converted into discharges for the ecological reserve. This workshop is the central mechanism for achieving consensus between scientists on the flows needed to ensure that Kruger's rivers are managed to achieve the park's management vision.

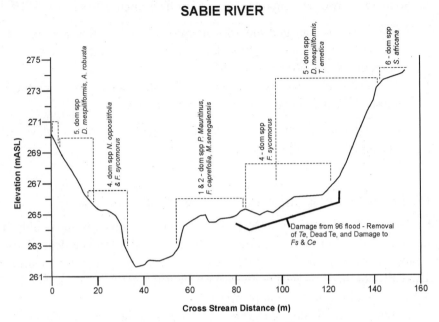

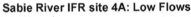

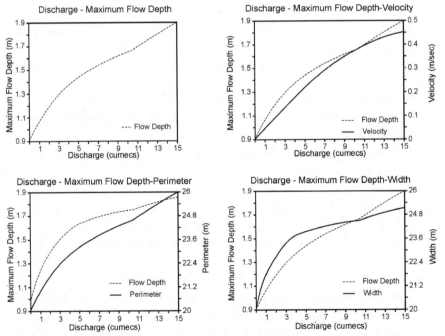

FIGURE 21.2. An example of a hydraulic cross-section and calibration, as used to assess environmental flows for rivers. Once habitat needs have been described in terms of water depth, current velocity, and margins to be inundated, the calibrations are used to convert this information into recommended discharges.

- The specialists consider the needs for the various components and processes for low and high flows, for the wet and dry seasons, during maintenance years (when reasonable rainfall is expected to provide flows adequate for the full range of river functions), and during drought years.
- The necessary characteristics (in terms of depths, velocities, and wetted perimeter) are converted to flows via the hydraulic analysis and expressed as time series and exceedance curves.

This process has been completed for all the major rivers at differing levels of detail. Flows for the Letaba River were assessed in 1994, when the BBM process was at an early stage of development and could undoubtedly be improved. For the Sabie-Sand system, the BBM was undertaken in 1996 and should be revisited in light of the major flood in 2000, which modified the channel morphology. Detailed water quality guidelines have been assessed for the Olifants but not for the other rivers.

It is important to understand that the EFRs, or in legislative terms the ecological reserve, represents a best judgment on the part of the country's best experts. Neither our understanding of river ecosystems nor our methods of determining flow needs are good enough to make judgments with a high degree of precision. However, South African river scientists are prepared to accept the responsibility of providing best estimates so that South Africa can implement its innovative water law. They recognize the risks associated with this bold step and expect water resource managers to work with them into the future. The recommended flow regimes can in no way be considered immutable, something to be implemented in perpetuity. They are hypotheses that must be tested in an adaptive management experiment that monitors the response of the rivers to the changed flow regime and adjusts the flow regime to better achieve the ecosystem properties.

None of the EFR recommendations have been implemented to date, a source of concern to scientists and managers. However, there have been two important catchment initiatives that hold promise:

- A major exercise to achieve public participation in and recognition of the ecological reserve for the Olifants River.
- A set of operating rules and decision support models for water resource management, including the ecological reserve, for the Sabie River (Department of Water Affairs and Forestry, in press). If these rules are implemented, they will represent one of the first implementations of EFRs in South Africa and provide the opportunity to implement strategic adaptive management at the catchment scale.

Both these promising initiatives have been spearheaded by the Department of Water Affairs and Forestry, but the final reports were not available at the time of publication. River scientists and managers will need considerable energy to move from talk and reports to implementation; above all, they will have to ensure acceptance of their initiatives by other stakeholders.

Cooperative Partnerships for Catchment Management

The South African Department of Water Affairs and Forestry is energetically pursuing the multiple aims of the 1998 Water Act and has recently produced a draft National Water Resource Strategy (Government Gazette No. 23711, August 8, 2002). The strategy emphasizes the changes that must be made and the need for an adaptive process: "Implementation [of the 1998 Water Act] has already started, and will continue in a step-by-step fashion over the next 20 years or so. Given the size and complexity of many of the activities, the proposed programme is not rigid" (Department of Water Affairs and Forestry 2002, p. 14).

Under South Africa's 1998 Water Act, the Department of Water Affairs and Forestry must devolve most of the catchment management issues to CMAs that include representatives of local interest groups and relevant national and provincial government agencies. One of the first responsibilities of the CMA would be to develop and implement a catchment management plan. Such a plan must recognize all the legitimate users of the water resources and take account of their needs. In terms of the 1998 act, the only water rights are those for the reserve: water for basic human needs and water for environmental protection. All uses of water other than basic human needs must be licensed. (The water needed for the ecological reserve is considered not a water use but a component of the resource.)

Implementation of environmental flows in the Sabie will present the first challenge of this kind for Kruger. The Inyaka Dam on the Marite tributary is about to come into operation, and operating rules will include the release of water to augment the ecological reserve (Department of Water Affairs and Forestry, in press). The natural MAR of the Sabie at the Mozambique border is 710 million m^3, and the Inyaka impoundment will have a storage capacity of 123 million m^3, which will increase the system yield of the Sabie to 141.3 million m^3/year for domestic, industrial, and agricultural use and for proposed transfers to the Sand tributary. The ecological reserve requirement is for a mean long-term annual volume of 170 million m^3, only 23 percent of the natural MAR but 20 percent greater than the yield of the system. This illustrates that the reserve, although usually a small percentage of the average flow of the river, often will impose restrictions on use during droughts, when environmental needs may be the same as the total flow in the river.

The National Water Resource Strategy emphasizes an adaptive implementation process, but the Department of Water Affairs and Forestry has no experience in adaptive management and no history of managing river ecosystems.

Its experience has been in managing dams to ensure stability of off-channel water supplies. CMAs will have no direct experience in either, and they will be tempted to implement the reserve through bureaucratic issuing of licenses to use the remaining water resource rather than by actually managing flows for environmental purposes (Rogers et al. 2000).

Many challenges lie ahead for management of the lowveld rivers in general and Kruger rivers in particular. New structures and processes for cooperative management involving all stakeholders will be needed to implement innovative new catchment-scale adaptive management systems (van Wilgen et al. 2002). The new constitution and legislation in South Africa provide such opportunities. Collectively, they require the following:

- Accountable and cooperative governance of the resource to provide equity, efficiency, and sustainability of its use
- A healthy environment protected for present and future generations
- Empowerment of citizens to become masters of their own futures and participate in managing natural resources

All this means that partnerships must be formed in which the regulators (government at all levels) and the people who use the resource cooperate in its management. In this case, rivers will be managed to ensure that the ecological reserve delivers the ecological functions on which humans depend. The imperative is that stakeholders, including Kruger, recognize the ecosystem (not only the water) as the resource and form a partnership that is collectively responsible for defining the type and level of ecosystem services to be delivered by the aquatic ecosystems they use. Cooperative stewardship (balanced cooperation and self-interest, and balanced protection and exploitation) of this ecosystem will ensure that it delivers the desired services and provides an organizational system that can be trusted to allocate the ecosystem services across the user community.

Scientists and managers involved with Kruger's rivers have not been inactive in this cooperative catchment management arena and have been centrally involved in the Sabie River Forum and other initiatives in the Olifants, Letaba, and Crocodile river catchments. These bodies are seen as the forerunners of CMAs. During the 1991–1992 drought it became clear that the Sabie River would stop flowing for the first time in history if drastic action was not taken soon. Kruger staff joined in the formation of the Sabie River Forum to negotiate a collective response to the drought. As many riparian trees began dying, banana and mango growers voluntarily stopped irrigating one day a week to reduce stress on the river within Kruger. They lost many fruit trees and income that year but saved their community, Kruger, and the river from much greater and possibly more permanent losses.

The Sabie River Forum was an excellent example of a cooperative management system, but it was informal and lacked the broad constituency needed for the new CMA. The Kruger Rivers Research Programme and subsequent research initiatives have begun to develop the understanding and means to formalize cooperative management systems. Success is a long way off, but the collective vision developed by a wide range of stakeholders in one exercise suggests that they are on the right track (van Wilgen et al. 2003, p. 2):

> We are proud custodians of our rivers. They sustain our economy and heritage. We protect and manage them so that they can continuously bring benefits equitably to our people, the nation and our neighbours.

As encouraging as this seems, there is a highly complex and inequitable legacy of resource use in the catchments to be understood and rationalized (Chapter 20, this volume). Until a CMA is established for the Sabie and a formal cooperative partnership is functioning effectively, interim decisions are likely to be piecemeal and unsatisfactory. Much remains to be done, and much of the initiative and responsibility will have to come from Kruger and its scientific partners.

But looking upstream to resolve river management issues will not be enough. Downstream of Kruger is Mozambique, a country that is just beginning to develop its resources after the turmoil of civil war. Large dams that have been built just downstream of Kruger on the Olifants and Sabie rivers constitute major barriers for biota and discontinuities for river processes. Demand in Mozambique for the water stored in these dams will cause a reexamination of the way in which water is allocated in South Africa. Mozambique's needs may have positive effects on Kruger's rivers because additional water for Mozambique implies increased flows through the park. But even this scenario will require much effort to ensure that the water is delivered in a manner conducive to biodiversity management rather than to ensuring the stability of water levels in Mozambique's dams.

References

Carter, A. J., and K. H. Rogers. 1995. *A Markovian approach to investigating landscape change in the Kruger Rivers*. Report No. 2/95. Centre for Water in the Environment, University of the Witwatersrand, Johannesburg, South Africa.

Davies, B. R., M. C. Thoms, K. F. Walker, J. H. O'Keeffe, and J. A. Gore. 1994. Dryland rivers: their ecology, conservation and management. Pages 484–512 in P. Calow and G. Petts (eds.), *The rivers handbook*, Vol. 2. Oxford, UK: Blackwell Scientific Publications.

Department of Water Affairs and Forestry. 2002. *Using water wisely: a national water resource strategy for South Africa*. Information document. Department of Water Affairs and Forestry, Pretoria, South Africa.

Department of Water Affairs and Forestry. In press. *Sabie River Catchment. Operating rules and decision support models for management of the surface water resources.* Draft report of the Department of Water Affairs and Forestry, Pretoria, South Africa.

Heritage, G. L., A. W. van Niekerk, B. P. Moon, L. J. Broadhurst, K. H. Rogers, and C. S. James. 1997. *The geomorphological response to changing flow regimes of the Sabie and Letaba river systems.* Water Research Commission report No. 376/1/97. Water Research Commission, Pretoria, South Africa. Online: http://www.wrc.org.za.

Hildrew, A. G., and P. S. Giller. 1994. Patchiness, species interactions and disturbance in the stream benthos. Pages 21–62 in *Aquatic ecology: scale pattern and process.* The 34th Symposium of the British Ecological Society, with the American Society of Limnology and Oceanography, University College, Cork. Oxford, UK: Blackwell Scientific Publications.

Hildrew, A. G., P. S. Giller, and D. G. Raffaelli. 1994. Introduction. Pages ix–xiii in *Aquatic ecology: scale pattern and process.* The 34th Symposium of the British Ecological Society with the American Society of Limnology and Oceanography, University College, Cork. Oxford, UK: Blackwell Scientific Publications.

Hynes, H. B. N. 1975. The stream and its valley. *Verhandlung der Internationalen Vereinigung fur Theoretische und Angewandte Limnologie* 19:1–15.

Kilian, V., and B. J. du Plessis. 1993. *Kruger: water quality data inventory of the 6 main river systems for the hydrological years 1983–1992.* Report No. N/0/0/REQ/1393. Department of Water Affairs and Forestry, Pretoria, South Africa.

King, J., and D. Louw. 1998. Instream flow assessments for regulated rivers in South Africa using the building block methodology. *Aquatic Ecosystem Health and Management* 1:109–124.

Leopold, L. B., M. G. Wolman, and J. P. Miller. 1964. *Fluvial processes in geomorphology.* San Francisco: W. H. Freeman.

Mackenzie, J. A., A. L. van Coller, and K. H. Rogers. 1999. *Rule-based modelling for management of riparian systems.* Report No. 813/1/99. Water Research Commission, Pretoria, South Africa.

Moon, B. P., A. W. van Niekerk, G. L. Heritage, K. H. Rogers, and C. S. James. 1997. A geomorphological approach to the ecological management of rivers in the Kruger: the case of the Sabie River. *Transactions of the Institute of British Geographers New Series* 22:31–48.

Moore, C. A., and F. M. Chutter. 1988. *A survey of the conservation status and benthic biota of the major rivers of the Kruger.* Contract Report. National Institute for Water Research, Pretoria, South Africa.

Moore, C. A., M. van Veelen, P. J. Ashton, and R. D. Walmsley. 1991. *Preliminary water quality guidelines for the Kruger rivers.* Kruger Rivers Research Programme Report No. 1. University of the Witwatersrand, Johannesburg, South Africa.

Muller, W. J., and P.-A. Scherman. In prep. *Similarities and differences in the rivers of the Kruger National Park.* Pretoria: Water Research Commission.

Naiman, R. J., D. G. Lonzarich, T. J. Beechie, and S. C. Ralph. 1992. General principles of classification and the assessment of conservation potential in rivers. Pages 93–124 in P. J. Boon, P. Calow, and G. E. Petts, eds. *River conservation and management.* Chichester, UK: John Wiley & Sons.

O'Keeffe, J. H., D. A. Hughes, and R. E. Tharme. 2002. Linking ecological responses to altered flows, for use in environmental flow assessments: the flow stressor–response method. *Proceedings of the International Association of Theoretical and Applied Limnology* 28:84–92.

O'Keeffe, J. H., and M. Uys. 1998. Invertebrate diversity in natural and modified perennial and temporary flow regimes. Pages 173–184 in A. J. McComb and J. A. Davis (eds.), *Wetlands for the future*. Adelaide, Australia: Gleneagles Publishing.

Pickett, S. T. A., and K. H. Rogers. 1997. Patch dynamics: the transformation of landscape structure and function. Pages 101–127 in J. A. Bissonette (ed.), *Wildlife and landscape ecology*. New York: Springer-Verlag.

Pienaar, U. de V. 1985. Indications of the progressive desiccation of the Transvaal lowveld over the past 100 years, and implications for the water stabilisation programme in the Kruger. *Koedoe* 28:93–165.

Reice, S. R., R. C. Wissmar, and R. J. Naiman. 1990. Disturbance regimes, resilience and recovery of animal communities and habitats in lotic ecosystems. *Environmental Management* 14:647–660.

Resh, V. H., A. V. Brown, A. P. Kovich, M. E. Gurtz, H. W. Li, G. W. Minshall, S. R. Reice, A. L. Sheldon, J. B. Wallace, and R. C. Wissmar. 1988. The role of disturbance in stream ecology. *Journal of the North American Benthic Society* 7:433–455.

Richter, B. D., J. V. Baumgartner, R. Wigington, and D. P. Braun. 1997. How much water does a river need? *Freshwater Biology* 37:231–249.

Rogers, K., D. Roux, and H. Biggs. 2000. Challenges for catchment management agencies: lessons from bureaucracies, business and resource management. *Water SA* 26:505–511.

Rountree, M. W., G. L. Heritage, and K. H. Rogers. 2002. *In-channel metamorphosis in a semi-arid, mixed bedrock/alluvial river system: implications for instream flow requirements. Hydroecology: riverine ecological response to changes in hydrological regime, sediment transport and nutrient loading*. International Association of Hydrological Sciences (IHAS) Publication 266.

Russell, I. A. 1997. *Monitoring the conservation status and diversity of fish assemblages in the major rivers of the Kruger*. Unpublished Ph.D. thesis, University of the Witwatersrand, Johannesburg, South Africa.

Townsend, C. R. 1989. The patch dynamics concept of stream community ecology. *Journal of the North American Benthic Society* 8:36–50.

Turner, M. G., and V. A. Dale. 1998. Comparing large, infrequent disturbances: what have we learned? *Ecosystems* 1:493–496.

van Coller, A. L., K. H. Rogers, and G. L. Heritage. 2000. Riparian vegetation-environment relationships: complementarity of gradients versus patch hierarchy approaches. *Journal of Vegetation Science* 11:337–350.

Vannote, R. L., G. W. Minshall, K. W. Cummins, J. R. Sedell, and C. E. Cushing. 1980. The river continuum concept. *Canadian Journal of Fisheries and Aquatic Sciences* 37:130–137.

van Wilgen, B. W., C. M. Breen, J. J. Jaganyi, K. H. Rogers, D. J. Roux, T. Sherwill, E. van Wyk, and F. Venter. 2003. *Principles and processes for supporting stakeholder participation in integrated river management: lessons from the Sabie-Sand catchment*. Water Research Commission Report 1062/1/03. Water Research Commission, Pretoria, South Africa.

Chapter 22

Integration of Science: Successes, Challenges, and the Future

HARRY C. BIGGS

This chapter is about the functional linking of different areas of scientific knowledge generated in and around Kruger. If the ultimate aim is likened to a completed jigsaw puzzle, this chapter shows the few connected pieces and isolated patches present at this time. It discusses the Kruger experience in offsetting what Bell (1995), in a landmark review of the Scientific Services Section, pointed out as compartmentalization of the knowledge produced by them. He indicated that Kruger's Rivers Research Programme (RRP) stood out as an example of how to achieve wider interdisciplinarity. In this chapter I use examples from the RRP and other emergent successes in integration. Frameworks are stressed, as is the use of models. I discuss the meaning of this science integration for management and for the perception of heterogeneity. The need is to now integrate social and economic spheres into our biophysical thinking.

Why and at What Level to Integrate?

The core aim of integration is to improve understanding and thereby develop the ability to respond in a systemic, coordinated manner. A world of fast technological and social change is placing pressure on science initiatives to deliver more integrated services and products. Here a wide range of drivers of integration are discussed; the specific ones often are recent phenomena. The demanding and often paradoxical nature of successful integration is outlined as it is understood today. In addition, our own science paradigm, partly under the same pressures, has evolved toward the viewing of ecosystems as complex and adaptive (Levin 1999), a construct demanding a more integrative understanding.

What do we integrate? The paradoxical drive toward specialization on one hand and integration on the other (Box 22.1) necessitates a healthy balance between component studies and system views, respectively. At this stage, our overall research initiative in Kruger is still skewed in favor of compartmental-

BOX 22.1
Integration as a Process
Mark Dent

Integration is a human phenomenon involving individual and organizational behavior. Integration is driven by exceptionally powerful forces and is better described in terms of its attributes and requirements than by a formal one-line definition.

Universal Drivers of Integration

The need for greater integration is influenced by all the major categories of drivers in the values, social, technological, economic, ecological, and political (VSTEEP) framework, particularly by

- Globalization, which spawned international environmental conventions
- Increased consciousness of environmental externalities
- Decreased societal tolerance for pollution externalities
- Increased consciousness of interdependence
- Declining natural resources
- Mobility of capital, including intellectual capital
- Speed of dissemination and resultant impact of ideas
- Paradoxical forces in society, pushing both specialization and integration
- Inability of compartmentalized systems to deliver solutions in complex, interdependent systems
- Need to balance developmental and environmental pressures
- Improved ability to manage complexity
- Electronic communication networks
- High cost of dysfunctional conflict

Attributes of Integration

- It preserves individual identity and simultaneously transcends it ("fruit salad"); it is not a melting process producing a bland average.
- It thoughtfully and sensibly relates components into a functional whole.
- Integrative processes continuously reconcile tensions of conflict.
- It transcends barriers to communication.
- It enables people to live with paradox and ambiguity and therefore accept heterogeneity.
- It enables people, parties, disciplines, and organizations to collectively visit the consequences of past, present, and future actions and influence each other.
- It entails multiscale, systemic thinking.

- It takes extensive practice.
- It entails individual and organizational change; it cannot happen painlessly or by avoiding conflict or power shifts.
- It does not seek a single answer but rather a range of options.
- It recognizes that solutions themselves often cause new problems.
- It is an ongoing process driven by changing needs.

ized knowledge, with a backlog of integration, as in most environmental research settings. Holistic and reductionist approaches both need to contribute to a larger overall knowledge production system. The attributes of integration (Box 22.1) imply that integration will not happen without challenging and sometimes painful change. Integration is expensive, is not in itself a panacea, and does not need to happen at all levels (Figure 22.1). Only the connectors need to have the larger overall picture and connecting tools, with translators unifying more closely related fields and the operational level often left unaffected. When a backlog in integration is eliminated, ongoing change continues to drive a need for further integration.

The Development of Integrated Understanding: From Single Studies to Frameworks

Understandably, most early and even many recent studies in Kruger were unidisciplinary in nature, often single-species studies. Responsibilities were allocated to park researchers by taxonomic group. Several of these projects reached out into other disciplines; for example, most antelope studies incorporated habitat studies in some degree of detail (Chapter 14, this volume). An ecozone-style classification system and later a land classification system (Solomon et al. 1999) each produced a template that brought out a wider range of ecosystem interconnections. The former described relationships between animal and plant associations and the landscape; the latter integrated macrogeology, geomorphology, and broad vegetation types down to the level of terrain units (Chapter 5, this volume). Studies on anthrax (Chapter 17, this volume) used a comprehensive array of fields impinging on the biology and epidemiology of the organism, making the approach multidisciplinary. Several experienced scientists (e.g., Joubert 1986) gave interpretations of the Kruger ecosystem that showed intuitive integrated understanding of overall system function in a scientific milieu where scale, heterogeneity, and other concepts supporting fuller integration were still poorly developed.

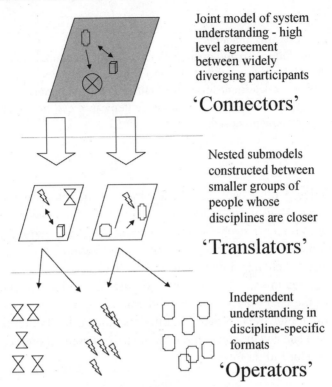

FIGURE 22.1. Conceptual diagram of how integration may be handled at different levels. These levels represent organizational strata in a system or project as much as they do the format of data being summarized in an upwardly compatible way.

Organizational structures of the time may have limited diffusion of knowledge (see Box 4.2, this volume). Therefore, few of these integrative attempts measure up to the more challenging demands of the attributes in Box 22.1. It was only with the sociopolitical changes of the early 1990s in South Africa, global ecological paradigm shifts, and the accumulation of rich datasets from earlier work in the park that integration as outlined here became an imperative in Kruger, with the RRP, as Bell (1995) pointed out, taking a lead.

No generally accepted overall system schemas, along the lines of, say, the flow diagrams depicting major ecosystem interconnections and flows in Serengeti (Sinclair and Norton-Griffiths 1979), were used in Kruger before the 1990s. This compromised the otherwise good joint understanding between researchers and between researchers and managers in Kruger (see Box 4.1, this volume). The RRP began its second phase with a central integrating framework (Breen et al. 1994) of the riverine environment. This spawned several detailed network diagrams of subsystems, and the full suite of framework dia-

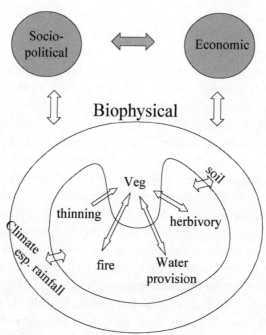

FIGURE 22.2. Example of a high-level framework for terrestrial ecosystem function. This particular framework, simplified here to show only the core elements, was worked out for the Agricultural Research Council in a workshop and written up by M. J. Peel of the Range and Forage Institute for its particular purposes. This institute collaborates with Kruger scientists regarding the many private nature reserves along the western boundary.

grams (a detailed example is shown in Figure 22.4) played an important role in building integration. On the terrestrial side, such frameworks in use in Kruger are still very general (Figure 22.2). Even in this simplistic form, they represent a basis for seeing where any particular project or idea fits into the bigger picture, identifying gaps and reducing conflict. They allow unpacking into finer-level diagrams of subsystems, as was carried out by the RRP (Figure 22.4) using expert knowledge in workshop-type settings. These new frameworks allow stakeholders to identify missing links, research them, and eventually put a more complete picture together. Frameworks help identify levels at which common understanding is necessary and act as a nidus for formulating and examining hypotheses of overall system function. They are indispensable tools in the explicit integrative understanding of ecosystems.

The predominance of mostly unidisciplinary veterinary and large mammal studies in Kruger at the time of Bell's (1995) review was counterbalanced only by aquatic and (some) vegetation studies, the former thanks to the emergence of the RRP at that time. The 12 other subdivisions he featured made very small contributions when viewed as a percentage of projects in the research initiative. This distribution of effort immediately shows that certain areas of major importance in the ecosystem (Figure 22.2) were receiving little or no attention. For instance, soil nutrients and soil moisture studies might be seen as fundamental but were underrepresented.

Another type of framework, less process oriented and more goal oriented, is the objectives hierarchy (Chapter 4, this volume). In the last 5 years, the existence of the Kruger objectives hierarchy has made a major difference in the directions followed in research, with several previously unattended areas receiving major attention because the park had made their intentions explicit. Research institutes often responded by offering work in fields where gaps were apparent (Chapter 4, this volume). The fact that the park vision that underlies this entire objectives hierarchy emphasizes heterogeneity has led to heterogeneity aspects taking a more central position in Kruger's research profile.

Aquatic and terrestrial frameworks generally exist separately, apart from partial integration in some hydrological and erosion models. When savannas are defined in textbooks, rivers and drainage systems usually are not included as essential features. The River-Savanna Boundaries Program under way in Kruger strives to find common currency between these two major interacting components of a landscape, heterogeneous at multiple scales. Several initial frameworks are in place, linking aquatic, riparian, and terrestrial subsystems through various described boundaries. Understanding the joint functioning of the intermeshed terrestrial and aquatic systems in savannas will be a major advance.

Scale as an Issue in Integration

Many ecological debates can be settled almost as soon as scale is made explicit. There are different senses in which the word *scale* is used (extent, grain, or resolution) and different modes of application (observation, analysis, or ecosystem process). In this chapter it always refers to time or space domains (Chapter 3, this volume), as opposed to *levels*, which I use for other aspects, such as life forms and their components, or for steps in a hierarchy. In integration, the key issue is whether extrapolation across scales is valid in the particular cases where it is applied because data themes or layers are never all available at the same scales, and it cannot be assumed that they can always be appropriately scaled up or down to match. There are methods to link different scales (Petersen and Parker 1998); those carrying out integration should be aware of scale and always make it explicit. Scale-neutral conclusions are either extremely general and powerful or, more often, dangerous as input to further steps and reasoning in integration. Results of integration initiatives usually need to be expressed at multiple scales, and a hierarchical framework is appropriate (Chapter 9, this volume).

Current Clustering of Research Initiatives and What They Mean for Integration

Research initiatives in Kruger in the last decade have tended to group themselves into programs, cohesive assemblages of projects addressing related issues. This clustering, with the relationship that results from informal yet meaningful overlap between programs, has enhanced integration (Figure 22.3). The Kruger objectives hierarchy has had a definite influence on positioning of research interests at Kruger, particularly through a volume (Freitag and Biggs 1998) showing gaps, responsibilities, and opportunities for participation. Today's lineup of research projects is increasingly the result of the objectives hierarchy but also includes vested drives from earlier, more species-centered approaches. The organizational form that was followed in response to the resetting of objectives in Kruger in 1997 was a division into three themes: genes, species, and communities; system ecology; and human impacts. The species and community projects tend to act as building blocks of knowledge that are assembled in various ways to be fed into further studies and into an understanding of system ecology, whose outputs are in turn fed through a management orientation filter in the human impact program (Figure 22.3). Occasionally, and especially in the case of individual species conservation issues, a shortcut is taken, with knowledge flowing directly from the species to the management (human impact) domain. There is sufficient distinctiveness yet sufficient overlap in this three-way division to make it useful in promoting integration in that all themes have a distinct yet related role to play. For a discussion of the Kruger monitoring programs, see Chapter 4 (this volume).

Modes of Integration Influencing Kruger

With the acceptance that frameworks or conceptual schemas (such as Figure 22.2) are central to integration and that the Kruger objectives hierarchy and the drive toward program management (Figure 22.3) assist this unification, there are several other important ways in which integration in science is influenced:

- Progress in ecosystem science and in dealing with interdisciplinarity. The last decade has seen the advancement of integrated research

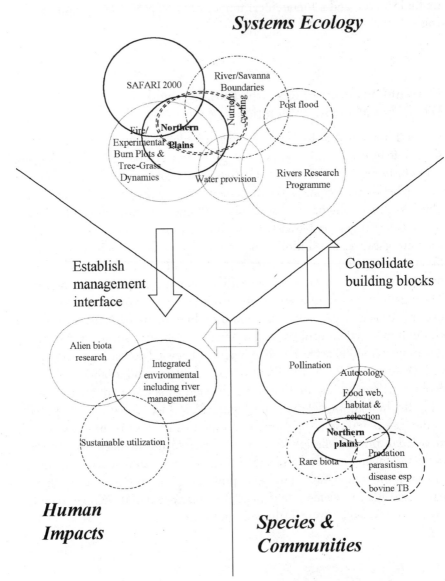

FIGURE 22.3. Schema of clustering and overlap in recent research programs in Kruger. The programs represent formal, informal, closing off, and developing programs. The relationships are necessarily simplified and should be considered dynamic as perceptions change and opportunities for other forms of overlap present themselves.

approaches in ecosystem science and hence the explicit development of concepts and methods to understand and promote this synthesis (Pickett 1999). Several fields now appearing exist only because of prior integration, such as sustainability science (National Research Council 1999).
- Institutional and interinstitutional arrangements and protocols. Funding organizations such as the South African Water Research Commission increasingly include theme areas promoting interdisciplinarity, such as the research underlying integrated catchment management, an activity that is inherently multi-institutional. Movements such as the Long Term Ecological Research (LTER) initiative (http://www.ilternet.edu) have arisen not only because of the longer-term needs but also because of growth, also between institutions, in interdisciplinarity. Kruger's interest in and involvement with nascent South African LTER development is promoting science integration. Kruger National Park scientists and collaborators have also been involved in many South African environmental flow requirement assessments for rivers, multi-institutional interdisciplinary initiatives using a joint protocol (Chapter 9, this volume).
- The use of models to assist integration. Model users often strive to enhance understanding of cause-and-effect links and to gauge what effects changes in one or more components bring about in the system as a whole rather than to produce accurate predictions, although this is sometimes possible. They partially satisfy the need for a unified (integrated) understanding, often in a data-sparse domain. Most case studies in this chapter focus on the use of models and their role in attempted integration. The particular way in which models are chosen and used is crucial. If approached from a cookbook point of view with only one's own discipline or a predetermined outcome in mind, models can even stifle integration. This underlines the nature of integration as a behavioral phenomenon (Box 22.1).

Case Studies and Their Influence on Integration

Case studies 1 and 2 deal with attempted use of two complex modeling systems: SAVANNA (Coughenour 1993) and Hydrological Simulation Program Fortran (HSPF; Bicknell et al. 1996). In both cases there were multifarious expectations, and lessons from their use are discussed jointly. The remaining case studies are dealt with individually: case study 3 describes the use of simple models, and case study 4 describes the integrating value of a suite of models in the context of the RRP. Case studies 5 and 6 deal with the integrating effects of a large outside program and a small in-house program, respectively. All six cases contributed to some extent to integration of science in the Kruger experience, and none are intended as assessments of the respective models or programs.

Case Study 1: The SAVANNA Model, Kruger's First Introduction to an Integrated Representation of the Terrestrial System

The SAVANNA model, after a brief introduction in Kruger by its author, was developed for use here by Kiker (1998), who contrasted its use with simpler systems and with modified hydrological inputs. He needed a large number of parameter estimates and data layers, and much of this information was not in a readily available format. The main modules of SAVANNA are weather, fire, soils, a series of four vegetation and soil submodels, a series of four ungulate submodels, and predator and culling components. Its basic framework has much in common with the introductory terrestrial ecosystem framework (Figure 22.2), reflecting a common perception of ecosystem flows at this high level. The model was used to evaluate the usefulness of the concept of carrying capacity in the large Kruger system. A byproduct was delineation of the geographic extent of terrain units (about which detailed nonspatial information was available; Chapter 5, this volume), which have since acted as a fundamental input to further heterogeneity work in Kruger. Kiker modified input and output of the model to use height classes of trees, for instance, which suited Kruger monitoring endpoints. He spent much time working with individuals and giving group presentations to scientists in Kruger, which promoted the idea that all relevant information regarding major activities or features in the ecosystem could be sewn together to understand specific likely outcomes. These included evaluating the influence of denser distributions of water-points over the landscape and of higher herbivore populations on vegetation. The exposure of Kruger scientists to these model runs enhanced notions of integration and complexity and allowed them to see the carrying capacity concept more critically, in wider perspective. Apart from one isolated case, however, no further use was made of the model after Kiker's thesis, despite ongoing contact between the champion and Kruger scientists. The model is being used in other savanna settings in southern Africa for consultation work, and another study is planned in Kruger, using SAVANNA to examine the emergent topic of riparian-upland links.

Case Study 2: HSPF, the Promise of Understanding the Whole River in Unison

The HSPF model was brought to Kruger via the RRP. On the test river, the Sabie, it was able to generate most major water quality and quantity outputs after exhaustive parameterization. It was attached to a center of excellence in

South Africa, whose staff gave follow-up training to centers such as Kruger that intended to use it. In practice, it was not used further in the Kruger catchment context (and very little elsewhere in South Africa); ostensible reasons were that the model was too difficult to run, could do everything with fair accuracy and nothing well, performed more poorly than a range of other specialized models despite being modular and linkable to other models, and used the wrong functional basis for certain phenomena.

Clearly many of these reasons are subject to different interpretation by proponents and nonproponents of the model, and they probably reflect a lack of comprehension of what is needed to make such a model succeed more than any qualities of model itself. It has an impressive track record in other countries in promoting integrated understanding and reducing conflict (e.g., in Chesapeake Bay; Donigian et al. 1991), both of which are important offerings in the Kruger context. No all-embracing alternative was put in its place in the lowveld catchment zone, although much demand exists to deliver an overall understanding of the effects of catchment management processes outside the park on the water flows and quality coming into the park. However, a simpler, focused range of models (case study 4) met certain park needs elegantly.

Lessons from Case Studies 1 and 2

Large, complex models are more difficult to understand, initiate, code, parameterize, and maintain than the simpler alternatives proposed by Starfield and Bleloch (1991). Therefore, they are often perceived as being hungry for detail or confusingly complex. Yet these models offer some measure of integrated understanding and allow representations of heterogeneity at several key scales. If these models are not used, there is a potential cost of reinventing the wheel, which can occur after a well-intentioned attempt to start out with a simple model leads to addition of more and more complexity. These benefits (and costs) may not be model-specific, often accruing from the use (or nonuse) of any complex model in that general class. This signifies that integration is a behavioral phenomenon of humans, having more to do with how they interact organizationally, socially, and scientifically with the model than with the specific model qualities. The implication is that it may be premature to speak of the success or failure of a complex model because people using it may have only taken the first steps in a long process. A second exposure may be automatically more successful than a first, whether it involves the same or a different complex model. Practice from the first usage may prime later relative success.

The salient lessons might be as follows:

- Do not expect such a human-model system to mean everything to everybody. Rather, identify which actual needs of the client organization necessitate this level of complexity and human investment

and establish whether the system structure is appropriate to feasibly deliver it. It may be reasonable to expect SAVANNA to show what the limiting factors are in supporting roan antelope populations in the northern plains of the Kruger but unreasonable to expect it to produce realistic estimates of each factor.
- Use simpler models when appropriate, but do not trivialize complex processes for which complex models may be the only workable tools.
- Show institutional persistence with selected models and study practical methods of adoption. Such models necessitate access to an expert or champion, regular demonstration, and practice. Therefore, institutions choosing to maintain even a few such central large models make many kinds of investments, but if approached and used correctly these models will deliver integrative benefits, sometimes only in the longer run. The investment must be justified in terms of benefits, just as payroll or personnel systems must.

Case Study 3: Quick Integration with Spreadsheet-Type Models

Professor Tony Starfield promoted spreadsheet models in Kruger over many years by presenting modeling courses for general biologists and wildlife managers. He stressed simplicity of design and intuitive understanding. Local knowledge of managers and scientists was widely used to structure and populate such models, especially if hard data were not available. This allowed them to explore the consequences of their beliefs and check them for outcomes, consistency, and sensitivity. A buffalo demographic model (Starfield et al. 1992) was used to help understand predation phenomena but not for its intended purpose of advising buffalo culling, which in any event ceased in Kruger soon afterward, for other reasons. I prepared a series of unpublished runs of a spreadsheet model demonstrating simplistic scenarios for the northern plains system (see case study 6) in terms of different levels of fire and of various forms of herbivory, to help management advisors conceptualize different integrated scenarios. Various other forms of the Starfield and Bleloch (1991) modeling approach have been adopted with occasional spectacular success in Kruger, especially in the population and demographics area. For instance, it could be shown that an extremely high percentage of cows had to receive contraception if elephant populations were to be stabilized in this way. Nevertheless, wide general adoption of this style has not developed in Kruger, and again human interaction with the modeling systems warrants consideration. It is still unclear how to create a culture of regular modeling among managers and scientists. Obvious benefits for ongoing integration would result from such regular usage, although criticism was also leveled at oversimplification (including a lack of ability to deal with heterogeneity) and at lack of verification of the intuitive concepts. The rule-based models in case study 4, though more complex than spreadsheet models, are extensions of this Starfield and Bleloch (1991) philosophy.

Case Study 4: The BLINKS (Biotic-Abiotic Links) Suite of Models, Fluent Integration at Last

These models arose toward the end of the RRP as integrating products that strove to meet several new Kruger heterogeneity goals, represented as thresholds of potential concern (TPCs; Chapters 4 and 9, this volume). These goals addressed key ecological processes such as sedimentation and resultant changes in riverine and riparian vegetation and in fish communities. The models linked several of the main areas and scales that had been researched: hydrology at sub-catchment scales; fluvial geomorphology in a nested hierarchy, which through hydraulics was seen as the key to habitat determination; and various biota (e.g., riparian, vegetation, and fish), typically at local site scales. The challenge for model structure was to place these differing areas into one general setting, which gave a fluent understanding of the major relationships between elements. At the same time, individual models were needed to inform management directly about expectations regarding potentially deleterious changes to the heterogeneity of Kruger rivers. Several interrelated frameworks were used (such as that in Figure 22.4). The level of detail helps identify the major modules being considered for inclusion in the model dealing with riparian plant community structure and change. Trimming down or excluding certain of these provisional modules then occurs, so that the focused aim of delivering the key answer and TPC can be met parsimoniously.

Although there are limitations in implementing BLINKS models (Jewitt et al. 1998), there have been several notable achievements:

- Groups of disciplines have been drawn to work together outside their comfort zones, finding through practice over time a useful level of common understanding, as depicted in Figure 22.1.
- Running programs has been simplified, with clients having been engaged all along as to demands and interfaces, including the embedding of several important thresholds (for details see Chapter 3, this volume).
- The models enjoy credibility in Kruger and are in use.

One of the original RRP goals, predicting the response of systems to natural and anthropogenic factors influencing water supply, is thus being partly met. Several factors in combination appear to have made these BLINKS models more successful than other model-based instruments of integration in Kruger:

- An excellent base of fundamental knowledge of the components, developed by the RRP

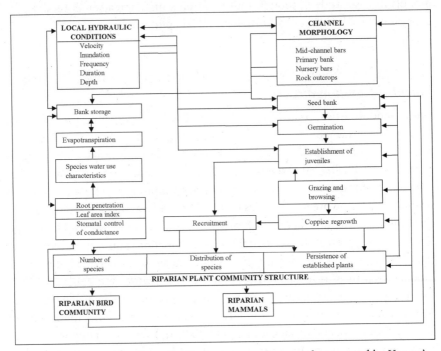

FIGURE 22.4. Example of one detailed subsystem framework generated by Kruger's RRP. This framework was used to inform riparian model development in the BLINKS initiative described in case study 4 of this chapter.

- A well-structured program, committed to interdisciplinarity, with clear working frameworks used from the outset and an explicit scaled understanding as a backdrop, as well as operational recognition of heterogeneity
- A genuine need to integrate components to meet real needs in Kruger and the catchments and willingness and funding to do so
- Models of realistic and appropriate complexity
- Models of adequate usability (again tied in to actual or proposed TPCs in Kruger) and ongoing developer-client interaction

Case Study 5: SAFARI 2000, Integration Spin-Offs for Kruger

In the early 1990s the Southern African Fire and Atmospheric Research Initiative (SAFARI 92), which arose under the auspices of the International Geosphere-Biosphere Program, used a site in Kruger for a managed fire from which emissions were studied. This was a major part of its activities in the wider region, and results filled a special edition of the *Journal of Geophysical Research* (1996, Volume 101). The next discrete initiative in the series was SAFARI 2000 (now named the Southern African Regional Science Initiative), which aimed at a better understanding of the southern African land-atmosphere-human system. SAFARI 92 had successfully characterized, quantified, and validated regional emissions from fires and shown that critical gaps remained in our understanding. These gaps were examined in SAFARI 2000, which examined relationships between physical, chemical, biological, and anthropogenic processes underlying land and atmospheric systems of southern Africa. Particular emphasis was laid on polluting and other (often nonpyrogenic) emissions and their transport, transformation, and eventual deposition, including effects of this deposition on ecosystems. Results are being analyzed and written up and will be widely distributed, also beyond scientific circles. More information is available at http://safari.gecp.virginia.edu and http://eos.nasa.gov.

Kruger played a small role in SAFARI 2000, supporting a flux tower for ground validation of the associated Earth Observing Satellite System and offering an airstrip and ground base for local sensing platforms. Several science studies, some of which were more tightly linked to Kruger objectives (particularly in the nutrient cycling area), were conducted at or near theses sites. Through exposure, Kruger scientists came to share some understanding of biogeochemical cycling. SAFARI 2000, although a subcontinental project with macroscale aims and one from which Kruger expected a low level of direct benefits, brought home to the park the realities of potential deposition of industrial and automotive emissions and possible nitrogen leaks from upstream areas. These are important areas in which conservation authorities can easily remain conceptually disconnected. Without the links being pointed out, they lack integrative insights. It also greatly widened Kruger's access to remote sensing expertise and products. In all these ways integration of science was promoted in Kruger, and even though the field phase is over, SAFARI 2000 has spawned further integrative research in Kruger. Kruger considers itself fortunate that this interaction brought home the concept of an integrated continuum between soil, plants, animals, and the atmosphere, especially given the challenges that face conservation organizations as regional and global environmental changes proceed.

Case Study 6: The Northern Plains Program and Homegrown Integration

This program began as an innovatively home-funded (by the Kruger Park Marathon Club) attempt to understand the decline of roan antelope on the northern basaltic plains of the park, the first project looking at the influence of waterhole placement and closure in this area. As anticipated, closure led to redistribution of zebra that had arrived and built up over the previous decade in response to water provision. However, roan numbers stabilized but did not increase, and other uncommon antelope (sable and tsessebe) declined. Lions appeared not to move off the plains in response to zebra movements.

Because this anchor funding supported local coordination, several more conventionally funded projects were attracted to examine a wider range of issues, such as vegetation recovery after closure. Because the causes of the decline were unclear and conservation of rare antelope was suspected to be at odds with ecosystem management goals, the program recently broadened its focus to examine the general issue of conservation of rare biota, with rare antelope as the particular case, in system management context. For all these reasons, this program is inherently integrative. Because of popular appeal of the program, these integrative views spread. Rare antelope issues have characteristically been contentious in park management (Grant and van der Walt 2000), and much pressure has been placed on researchers to come up with quick-fix solutions, even though some have seemed counterproductive in the longer run. There is growing appreciation that the system is complex and has emergent properties, which is why intervention effects vary over time and with the particular context. For instance, providing artificial water on the plains was successful in the short term, strengthening roan antelope populations; the other system effects unfolded decades later. Explicit frameworks interrelating system factors and hypothesizing some of these bifurcations in behavior have recently become part of northern plains initiatives. The program overlaps with most others in the park (Figure 22.3). It serves a real need, one that demands a wide holistic knowledge base. Conflicting schools of thought—one represented by emotive demands for immediate interventions to "save" rare antelopes, the other by an objective ecosystem perspective—illustrate the ambiguity referred to in Box 22.1.

What Has This Integration Meant for Kruger Management?

The case studies in this chapter have helped to fill in parts of the jigsaw puzzle, yet others, such as the River/Savanna Boundaries Program, may make major future additions. What does this integration of science mean for the manager? Three examples are presented in Table 22.1 to illustrate the view that a move from compartmentalization to partial integration has made a difference in the understanding feeding into management. This move is projected further into a future where we anticipate both wider and tighter integration and increasing inclusion of heterogeneity concepts. Kruger can thus be seen to benefit from this integrative initiative.

Conclusion

Science integration is a complex phenomenon, with a variety of complementary approaches furthering it in Kruger, from the level of visioning and articu-

TABLE 22.1

Three examples of how the extent of science integration influences the thinking behind management recommendations in Kruger.

ISSUE	EARLY COMPARTMENTALIZED VIEW	PARTLY INTEGRATED CURRENT VIEW	POSSIBLE VIEW WITH FURTHER INTEGRATION
Elephant management	Dealt mainly with elephant numbers (carrying capacity), believing that managing them would lead to sustainable vegetation impacts	Tries to more widely assess elephant-vegetation-biodiversity links, with the emphasis now on directly measured (wide) upper and lower impact levels, with planned high- and low-density zones; embraces heterogeneity	Will broaden further by integrating as yet poorly understood interconnections (e.g., between elephants, invertebrates, reptiles, and birds and between surface water, groundwater, and elephant populations)
Fire	Block-burning quasiagricultural approach	Brings in patch mosaic and heterogeneity links, as applied in new fire policy of 2002 (Chapter 7, this volume)	Will add an explicit heterogeneity TPC; will bring underground ecology under consideration
Rare antelope on northern plains	Focused on building numbers up and, in latter stage, on balancing roan with other herbivore (especially zebra) numbers	Roan needs contrast with system integrity needs and system heterogeneity (e.g., in terms of waterhole distribution and its sequelae)	Will broaden understanding in terms of genetic profile, global climate change, and eastern fence removal

lation of objectives through program management, relationships with outside agencies and institutions, and the use of frameworks and models as facilitatory tools. Cardinal issues are as follows:

- Integration serves the learning process; the latter may be more important than any specific outcome. Use of models must be matched to circumstances.
- Human attitudes toward change, knowledge sharing,(see Box 4.2, this volume), and integration must be addressed continuously. Models often make explicit differences between otherwise hidden assumptions, enabling discomforts about differing approaches by different disciplines to be discussed.
- There must be enough time and resources to practice the integration process. Sufficient context must built around an issue, which may

succeed on repeat attempts. A central tool such as a model is unlikely to succeed in isolation.

As a result of an increasingly integrative view, some important gaps in the jigsaw puzzle are receiving attention (e.g., linking understanding of terrestrial and riparian systems; some aspects of underground ecology) or have been identified (decomposition; tree autecology and demography). The Freitag and Biggs (1998) approach, measuring gaps and opportunities against objectives, could be updated and extended to timeline-based flows in which components ripe for synthesis are identified, developed, and related to central questions. Examining the Kruger objectives hierarchy, it is clear that even biophysical understanding is still far from complete.

The choice of examples in Table 22.1 was limited to biophysical issues inside the park and did not incorporate outside components, such as animal rights pressure groups in the elephant management debate. Social and economic factors have rarely been factored formally into Kruger science (e.g., into models), although there are already important ways in which stakeholders are drawn into the management process. Figure 22.2 shows two domains in the top corners (sociopolitical and economic), each interacting with the biophysical. Our interest lies particularly in how the interactions between each of them and the biophysical world operate. This understanding, even if obtained only in broad outline, may aid conservation goals more than increasingly detailed understanding purely in the biophysical arena. Actions satisfying these needs may constitute the biggest growth area in our integrated understanding in the next decade or two, and general frameworks (such as that proposed by Gunderson and Holling 2001) uniting the major domains in these complex, coupled adaptive systems will become central tools in this quest. These coupled systems are even more heterogeneous than the biophysical system viewed in isolation. We need to face this socioeconomic imperative squarely, with ecosystem services as one major intermediary currency between the biophysical world and the real clients in our social contract.

References

Bell, R. H. V. 1995. *A review of the programme of the Scientific Services Section, Kruger National Park.* Warwick, UK: ULG Consultants.

Bicknell, B. R., J. C. Imhoff, J. L. Kittle, A. S. Donigian, and R. C. Johanson. 1996. *Hydrological simulation program: FORTRAN user's manual for Release 11.* Athens, GA: U.S. EPA Environmental Research Laboratory, in cooperation with U.S. Geological Survey, Office of Surface Water, Reston, VA.

Breen, C., N. Quinn, and A. Deacon. 1994. *A description of the Kruger National Park Rivers Research Programme (second phase).* Pretoria: National Research Foundation.

Coughenour, M. B. 1993. *SAVANNA: landscape and regional ecosystem model.* Fort Collins: Natural Resource Ecology Laboratory, Colorado State University.

Donigian, A. S., B. R. Bicknell, A. S. Patwardhan, L. C. Linker, D. Y. Alegre, C. Chang, R. Reynolds, and R. Carsel. 1991. *Chesapeake Bay Program: watershed model application to calculate bay nutrient loadings. Final findings and recommendations.* Annapolis, MD: Chesapeake Bay Program Office, U.S. Environmental Protection Agency, Region 3.

Freitag, S., and H. C. Biggs. 1998. *KNP management plan objectives hierarchy: published work, projects in progress, and opportunities for participation.* Internal Report, Scientific Services, South African National Parks, Skukuza, South Africa.

Grant, C. C., and J. L. van der Walt. 2000. Towards an adaptive management approach for the conservation of rare antelope in the Kruger National Park: outcome of a workshop held in May 2000. *Koedoe* 43:103–112.

Gunderson, L., and C. S. Holling (eds.). 2001. *Panarchy: understanding transformations in human and natural systems.* Washington, DC: Island Press.

Jewitt, G. P. W., G. L. Heritage, D. C. Weeks, J. A. Mackenzie, A. van Niekerk, A. H. M. Görgens, J. O'Keeffe, K. Rogers, and M. Horn. 1998. *Modelling abiotic-biotic links in the Sabie River.* Report to the Water Research Commission No. 777/1/98. Water Research Commission, Pretoria, South Africa.

Joubert, S. C. J. 1986. *Masterplan for the management of the Kruger National Park,* Vol. 6. Unpublished memorandum, South African National Parks, Skukuza, South Africa.

Kiker, G. A. 1998. *Development and comparison of savanna ecosystem models to explore the concept of carrying capacity.* Unpublished Ph.D. thesis, Cornell University, Ithaca, NY.

Levin, S. 1999. *Fragile dominion: complexity and the commons.* Helix Books, Boston, MA.

National Research Council. 1999. *Our common journey: a transition toward sustainability.* National Academy Press, Washington, DC.

Petersen, D. L., and V. T. Parker (eds.). 1998. *Ecological scale: theory and applications.* New York: Columbia University Press.

Pickett, S. T. A. 1999. The culture of synthesis: habits of mind in novel ecological integration. *Oikos* 87:479–487.

Sinclair, A. R. E., and M. Norton-Griffiths (eds.). 1979. *Serengeti: dynamics of an ecosystem.* Chicago: University of Chicago Press.

Solomon, M., N. Zambatis, H. C. Biggs, and N. Maré. 1999. Comparison of classifications commonly used as templates for management, scientific and GIS work in the Kruger National Park. *Koedoe* 42:131–142.

Starfield, A. M., and A. L. Bleloch. 1991. *Building models for conservation and wildlife management.* 2nd edition. Burgess International, Edina, MN.

Starfield, A., M. Quadling, and J. Venter. 1992. *A management-orientated buffalo population model for the Kruger National Park.* Report to the National Parks Board. South African National Parks, Skukuza, South Africa.

Chapter 23

Reflections on the Kruger Experience and Reaching Forward

Michael G. L. Mills, Jane Lubchenco, William
Robertson IV, Harry C. Biggs, and David Mabunda

In this chapter we reflect on the Kruger experience described in the previous chapters and look toward the future. First we describe Kruger as a global treasure, not only ecologically but also economically, aesthetically, and scientifically. Next we look back to some of the areas where we feel mistakes have been made. Then we turn to the present and describe Kruger as a complex adaptive system that South African National Parks (SANParks) is attempting to manage within a strategic adaptive management program. We also address other challenges that are being faced such as data analysis, program integration, and socioeconomic issues. Finally, we look at what we regard as some of the pressing management actions and decisions that must be made.

Kruger's Ecosystem Attributes

Perhaps Kruger's most important attribute is its size: there are few protected areas in the world that are larger than 22,000 km². On top of this is a rich template of abiotic factors (Chapter 5, this volume). Size is fundamentally important because the factors that best account for patterns of biodiversity are scale dependent (Terborgh 1999; Terborgh et al. 2001; Ceballos and Ehrlich 2002; Willis and Whittaker 2002). Although it does not encompass the entire lowveld ecosystem, its 22,000 km² are sufficient to harbor an essentially complete and self-sustaining portion of that ecosystem with almost all plants and animals thought to have occurred there in historical times (Chapters 1 and 18, this volume). Protected areas that can claim to contain an intact large predator guild with viable populations of all member species are rare. The addition of an area in excess of 10,000 km² in neighboring Mozambique and Zimbabwe as part of the Greater Limpopo Transfrontier Park (Chapter 20, this volume) will increase the opportunities and value afforded by all that Kruger has become.

Age is also important, and in human terms Kruger is old, proclaimed more than 100 years ago. The sure thing about ecosystems is that they change. Data series over many decades are needed to understand ecosystems because shorter-term studies often miss or misinterpret important controlling events and processes that are slow, rare, subtle, or complex (Likens 1987; Magnuson 1990; Risser 1991)—just the sort of processes that drive ecosystems such as Kruger (Pickett 1991; Scheffer et al. 2001; Carpenter 2002). Ecosystem research sites and their associated datasets, especially unbroken series, increase dramatically in value with age (Franklin et al. 1987). Over the years, Kruger has largely settled many of the key questions asked and debated about long-term ecosystems research and protected areas: the location is settled, the fact that it is long-term is settled, the purpose is set in the mission statement (which considers ecosystem, human, and wilderness aspects), the ecosystem is monitored, research and science are supported, and the relationship and value of research to management is recognized (Chapter 4, this volume).

Science and Management

A distinctive feature of the Kruger management tradition is the high value placed on science. Shortly after the retirement of Stevenson-Hamilton in 1946 the new warden, Colonel Sandenberg, with the encouragement of a new parks board member, Dr. R. Bigalke, director of the National Zoological Gardens, established a science section in Kruger and began an enduring partnership with academic scientists (Carruthers 1995, 2001). This partnership, although not always as settled as it is today, and the benefits it has provided to Kruger and to science are an important model for protected areas, comparable to the Serengeti National Park, Tanzania (Sinclair and Norton-Griffiths 1979; Sinclair and Arcese 1995). The generally dubious position of science in the U.S. national park system provides a stark contrast (The Conservation Foundation 1972; National Parks and Conservation Association 1988; Sellars 1997).

The science-management-academic partnership in Kruger has been and is evolving into the form described elsewhere in this book (Chapter 4, this volume). The Rivers Research Programme brought together an array of scientific disciplines to study the role of rivers in maintaining the ecosystem. It not only helped develop Kruger's strategic adaptive management program but also contributed to the development of national water policy and law. The Rivers Research Programme and the River/Savanna Boundaries Program include a diverse range of scientific disciplines and academic institutions from South Africa and abroad.

An important attribute of these partnerships is that academic and park scientists are now becoming involved in designing and carrying out the research.

Kruger scientists are encouraged to describe their questions and research and academic scientists are to use their own experience to help design the applied and basic research questions to be addressed. A key benefit of this team approach, which includes graduate students, is that the questions developed, approaches considered, and research planned tend not to be captive to the training of a single or small numbers of disciplines but stretch to include many viewpoints and an overview of the leading developments in many fields. While Kruger scientists pursue their own research projects, they also collaborate on projects with scientists from outside institutions (e.g., universities). Kruger provides access, infrastructure, and funding for some collaborative projects.

Kruger scientists and their academic partners are preparing proposals for research funding from competitive government and private sources. In this way, Kruger is beginning to benefit from cutting-edge ecological research, and the academic scientists are gaining access to research sites, facilities, long–time series datasets, and a Kruger knowledge system that enables them to address increasingly sophisticated research questions. Kruger management benefits by building relationships with key research groups and researchers on whom they can call when advice is needed and decisions must be made. It also gains a broader, more sophisticated research program than could be supported by SANParks funds alone. This new development holds great promise.

Another major feature of Kruger has been its ability to control the devastation caused by elephant and rhino poaching that has been such a feature in much of Africa in the last 50 years (Owen-Smith 1988). Apart from its rich biodiversity, this was arguably the most important reason for Kruger's prominence, especially during the isolation of the apartheid years, probably more than for its management policies and research record during that era.

Ecosystem Services

In addition to the obvious sustainable revenue that Kruger produces from its operation as one of the great national parks, it also provides a panoply of ecosystem services that are valuable in their own right but unquantified. The concept of ecosystem services is new (Daily 1997; Balvanera et al. 2001; Daily and Ellison 2002) and largely unexplored with regard to Kruger. Ecosystem services are benefits provided by the functioning of intact ecosystems and include water purification, water storage and flood damping, reservoirs of biodiversity, moderation of weather and climate, cleaning and maintenance of the atmosphere, aesthetic beauty, and sources of intellectual renewal and spirit. The services provided by Kruger are undoubtedly large, and the new movement to define and quantify them is sure to grow (Daily 1997; Heal 2000; Daily and Ellison 2002). Although it is important to begin work toward elucidating these services, the base argument for Kruger was written by Sawhill (1995): "A society is not only characterized by what it creates but also by what it refuses to destroy." In one of the early pieces of economics and sustainability literature, Robert Solow (1992, p. 14) wrote, "It makes perfectly good sense to insist that certain unique and irreplaceable assets should be preserved for their own sake; nearly everyone would feel that way about Yosemite." If he were South African, Kruger might have been the example.

A well-managed Kruger is sure to provide to the region and the nation more total benefit that is sustainable over time than other possible land uses. Agriculture in marginal farming areas such as the lowveld cannot be sustained economically for long (Daily 1997). One comparative estimate of the value of land use in South Africa, although derived for small areas and therefore perhaps not directly applicable to a vast area such as Kruger, gives the annual yield as US$25/ha from ranching, US$70/ha from farming, and US$200–300/ha for tourism or hunting (Heal 2000). Unlike nonrenewable resources such as gold or diamonds, Kruger will provide jobs, recreation, and all of its other services long after other possible uses are exhausted, and its values and revenues will increase over time (Vitousek et al. 1997). Regardless of these economic considerations, there is a cautionary note to sound about projections of revenue from tourists. The revenue earned from a certain through-flow of tourists cannot be scaled up in linear fashion by simply increasing tourist numbers. Overloading parks with tourists and facilities degrades park resources and ecosystems, diminishing the visitor experience and ultimately driving visitors elsewhere (Bosselman 1978). SANParks is well aware of this problem, and it has been the subject of study (Ferreira and Harmse 1999), but it must be kept in mind by all who would consider the future of Kruger.

Management in the Past

It is generally accepted today that Kruger was overmanaged in the latter half of the twentieth century. In contrast to the other major savanna protected areas in East Africa, where there has traditionally been a tendency toward laissez-faire management, management of South African protected areas has been more hands-on. These differences in approach were brought about by both philosophical differences and practical (economic) constraints. In Kruger, fencing, water provision and manipulation, veld burning, and population and disease control have been the major management tools (Pienaar 1983). Although on their own not all of these activities were always detrimental to ecosystem functioning, the combined effect was one that is currently regarded as overcontrol and a movement toward habitat homogeneity. Also, with a few notable exceptions, there was little contact with outside scientists, except from the University of Pretoria. With the sociological and political changes that first emerged in the mid-1980s and changes in ecological thinking, attitudes toward management also started to change, with less control exerted on the ecosystem.

The distribution of the South African government's wealth in the mid-twentieth century meant that adequate financial resources were available for park management. Kruger management had the money and looked for ways to spend it. In contrast, even if they had wanted to, East African conservation authorities usually did not have the resources to fence reserves, build dams, and cull large animals such as elephants. The strength of political ideology prevalent in South Africa in the 1960s and 1970s was also influential. Two examples are the network of wide asphalt roads running the entire length of Kruger, built in the 1960s, and the acceptance by the National Parks Board to plant a sisal fence along the eastern international border of Kruger (fortunately, this was unsuccessful). The justification for both these activities was that they were in the national interest in case of an invasion of insurgents from Mozambique.

Present and Future Approaches: Viewing Kruger as a Complex Adaptive System

The previous chapters describe relationships that control the Kruger landscape—rivers, uplands, herbivores, predators, pollinators, decomposers, and human activities—all obviously interrelated so that they depend on and influence one another. Although the fact that they are interrelated is clear, the nuances and detail of how they interact and function are not. Kruger is a classic example of a complex adaptive system (CAS) as described and explored by Levin (1999). The essential features of a CAS are diversity and individuality of components, local interactions between the components, and processes that use the totality of the local interactions to control the continuation and evolution of the entire system. A CAS is self-organizing in that it follows rules that determine how the system develops and changes in response to past and present conditions; it does not remain the same, and different development paths and outcomes are possible depending on conditions and chance events (Scheffer et al. 2001; Carpenter 2002). These systems are characterized by heterogeneity; nonlinear interactions between components; hierarchical structure determined and reinforced by interactions and flows of energy, materials, and genes; and surprises. They tend to have modular structure, tight feedback loops, and redundancy. The redundancy buffers the system, provides resilience, and typically results in high diversity. The goal and challenge for Kruger science and management is to understand enough about this ecosystem and the rules that govern its development to allow it to be managed effectively and protected, adapting to changes in social and environmental conditions and persisting in a robust state (Clark et al. 2001).

Adaptive Management

The Kruger strategic adaptive management program is aimed at using the best available information to set thresholds of potential concern (TPCs). It monitors trends and then requires consideration and action when agreed-upon levels are reached. The program is designed to evaluate and modify itself as knowledge is gained and conditions change. Strategic adaptive management, along with the established and developing academic-science-management partnership, although challenging, provides a model way forward for parks, conservation, and science (Folke et al. 2002). Although these links between science and management are a strong feature of Kruger, the numerous TPCs that have been developed must be carefully audited and controlled. Kruger's management aims to keep the ecosystem within desired states and to avoid irreversible threshold effects, but understanding the resilience of the ecosystem to external shocks (e.g., those caused by people, climate change, and exotic diseases and species) and identifying the important controlling variables requires further study. The desired system, and therefore the TPCs, must be continuously redefined as the process of adaptive management proceeds.

Data Analysis

The long history of Kruger, in comparison with most national parks and wildlife research areas, has produced a wealth of information that began with the first systematic game counts and notes on climate and natural history by James Stevenson-Hamilton in 1902 (Carruthers 1995) and steadily intensified (Joubert 1983; Chapter 1, this volume). Kruger has been described as supporting one of the most sophisticated long-term ecological research programs, for which the extensive records are lamentably inaccessible (McNaughton and Campbell 1991). One of the challenges facing Kruger, which is being addressed (Freitag 1998), is capturing the rich historic, ongoing, and future monitoring data and research results to make them more accessible to scientists and managers.

SANParks was prescient in appointing a quantitative ecologist, Peter Retief, who joined the Kruger staff in 1978. He made an invaluable contribution by helping scientists capture and store appropriate data, guiding them in their project analyses, and encouraging the first attempts at integrative analysis. Today Kruger is developing a knowledge environment based on a geographic information system but including nonspatial databases. There is a goal of capturing all new monitoring and research results while systematically mining the data records, field reports, notebooks, and paper files and transferring that information into indexed electronic files. The metadata included in the knowledge environment will make finding information easier, and online tools will facilitate and promote use of the data. There is a commitment and program to extend databases back in time while moving forward and using new tools such as remote sensing. This overall thrust of responsible database management, putting data to productive use, and keeping the intellectual property issues equitable yet sensible will remain critical success factors, as will keeping the equipment, software, and expertise up to date. This program has a Web site and a newsletter. It is a prototype that should be expanded. Ensuring a sensible balance between data collection and use will be a continuing challenge.

Integration of Programs

One of the greatest challenges remaining is to fill in the jigsaw puzzle of savanna function (Chapter 22, this volume). Only a small portion of what can and should be brought together has been accomplished. There is reason to believe that a much larger portion of the jigsaw puzzle will fall into place in the next two decades if research and integration initiatives can be sustained at the currently promising levels. In this regard techniques now becoming available allow a far deeper look into the past. One such example is the use of stable isotopes, carbon, and fossil pollen to explore vegetation trends in Kruger over nearly a thousand years. Better knowledge of the variation of plant communities over a longer time will lead to a better understanding of current patterns and processes and thus the realistic setting of TPCs.

The sciences of ecology and conservation biology involve interactions between many disciplines. Staff complements in Kruger will increasingly need to maintain a balance between specialists and those who understand how to link the efforts of different groups. The River/Savanna Boundaries Program is a good start, but more comparative studies that use interdisciplinary teams, including social sciences, are needed to integrate multiple factors across systems and incorporate greater use of experimentation (Clark et al. 2001). Correlations do not indicate causality but identify the need for appropriately designed experiments (Tilman 1987). This reinforces the importance and need for the science-academic-management partnership that has developed in Kruger. New concepts such as ecosystem services and social ecology have appeared, and others will follow, needing to be further integrated with traditional approaches.

Socioeconomic Challenges in the Future

Despite previous management policies that isolated Kruger from its neighbors, the park is not (and has never really been) an island. People and wildlife move freely across the borders. As with most other national parks, the areas that became Kruger were neither ecologically well defined nor devoid of people. Indigenous inhabitants were dispossessed and excluded (Msimang 2000), and Kruger became seen as the playground of wealthy whites (Carruthers 1995, 2001). A very similar route to land acquisition for national park development was followed in the United States, particularly in the West (Burnham 2000). As Nelson Mandela has said,

> We must also remember the great sacrifices made by rural black people who had to surrender their land to make way for the establishment of the Park. For the best part of the life of this conservation area, successive generations of black people were denied access to their natural heritage—only being suffered to come in to provide poorly rewarded labour. We are grateful that as a result of the struggle to restore the dignity of the oppressed, the ills that afflicted the dispossessed are gradually disappearing. We must also bear in mind that all national parks belong to the nation and are held in trust by SANP on behalf of present-day South Africans and future generations. (Mandela 2000, p. 9)

This statement of responsibility emphasizes what is becoming a new and widely accepted social contract for science and for parks. The new and unmet needs of society include more comprehensive information, understanding, and technologies for society to move toward a more sustainable biosphere, one that is ecologically sound, economically feasible, and socially just. New fundamental research, faster and more effective transmission of new and existing knowledge to policymakers and decision makers, and better communication of this knowledge to the public will all be needed to meet this challenge (Lubchenco 1998). The management-science-academic partnership that has developed in Kruger and the new knowledge capture program position Kruger well to do its part for a portion of this social contract, but what about the rest?

Its social resilience will determine Kruger's future. The Social Ecology Department is a reasoned and ambitious program of SANParks to begin including surrounding communities in the life of Kruger and encourage related economic development (South African National Parks 2000). The initiative is full of promise and hope, but early reaction is mixed, and it is too soon to know how well it will work (Chapter 20, this volume). Symposia and aggressive encouragement to Kruger's scientists to publish their work will promote better communication. Although it can never produce the number of jobs or support the number of families that all would wish, Kruger is moving to provide better opportunities for its neighbors and to help them understand what sustainable development can mean for their future (South African National Parks 2000).

The question of social justice is harder to address. Kruger has moved aggressively to promote social justice and equality within the park and, as a major employer in the region, strives to set a good example for others. With regard to scientists, however, Kruger is in a more difficult position because it must rely on the output of qualified people from the nation's academic institutions. As Lyndon Johnson (1965, p. 2) said of affirmative action,

> But freedom is not enough. You do not wipe away the scars of centuries by saying: Now you are free to go where you want, do as you desire, and choose the leaders you please. You do not take a person who, for years, has been hobbled by chains and liberate him, bring him to the starting line of a race and then say, "you are free to compete with all the others," and still justly believe that you have been completely fair. Thus it is not enough just to open the gates of opportunity. All our citizens must have the ability to walk through those gates.

Even here, however, things can still be done, and Kruger has designed and implemented a national competition for young researchers to draw them into using the park for their research sites. It is working to design a program that will identify young people from previously disadvantaged backgrounds as they emerge from universities to give them work opportunities and a career path to advanced degrees and research positions.

Some Pressing Management Decisions

Sustainable use and the commitment of SANParks to this concept has been a subject of debate for several years. The controversy surrounding elephant culling is a prime example of this debate. Implementing the approved adaptive management experiment with areas of high and low elephant density within Kruger (Whyte et al. 1999) requires that elephants be removed from some areas. The present moratorium on killing elephants, largely as a result of pressure from Eurocentric animal rights groups (Whyte et al. 1999), constrains this program. A solution is to translocate elephants to other areas, such as the Mozambican portion of the Greater Limpopo Transfrontier Park, although it is questionable whether this will be possible at the necessary scale. The alternative strategy of culling could constitute a sustainable use of resources in line with SANParks ecosystem management principles and could also provide tangible benefits to neighboring communities. However, such potential benefits must be weighed against the direct and indirect costs of withstanding international pressure from the anticulling lobby.

Bovine tuberculosis (TB) is an alien disease that probably spread into southern Kruger from cattle 40 years ago (de Vos et al. 2001); since then the disease has steadily spread north, and prevalence rates now exceed 60 percent in some buffalo herds. Although there is no evidence that the size or structure of the buffalo population is being affected, there is evidence that animals in high-prevalence herds lose condition faster in the dry season and carry higher endoparasite loads than animals in low-prevalence herds (Caron et al., in press). Lions in Kruger also contract the disease by preying on infected buffalo and ongoing research is still measuring the impact on the lion population (Chapter 17, this volume). Attempts are being made to contain the disease in the buffalo population by testing and slaughtering all TB-positive animals at the northern limit of the epidemic in Kruger, but this is an expensive operation in terms of both animal lives and money. A clear, sustainable, and pragmatic strategy is urgently needed.

As the greater Kruger system (Chapter 20, this volume) expands to the west and east and through a narrow corridor to the north, more conservation estate is being added to make an even larger system. Some areas need restocking with wildlife and others need more extensive rehabilitation. Additionally, baseline conditions—species, communities, and landforms—must be documented. Although the responsibility for these actions and their success lies with the direct partners, assistance from the nucleus of expertise in Kruger must be forthcoming.

We need to view Kruger through wider spatial and temporal windows. One way to do this is through models or scenarios developed by natural and social scientists to envision possible or likely futures. The lowveld responds to changes in larger weather and climate systems and, increasingly, Kruger management will need to revise its response options to a changing world. From an ecosystem viewpoint the question of climate change is paramount. How is Kruger going to cope with changes in species and populations? Should intervention be implemented to protect species if their habitat changes significantly and their populations are in danger of being extirpated from Kruger? To some extent Kruger already faces this problem with regard to the roan antelope population, although the role of climate change in this case is unknown. Attempts to redress the decline reverted to ecological manipulation (i.e., closing windmills and revising burning policies), but they have been unsuccessful. Roan numbers are increasing while they are held in a protected enclosure, but soon managers must decide how and when to release them. The premise is that ecological conditions will improve through both management strategies and a return to higher rainfall, but if this does not take place (e.g., because of climate change), this species might become extinct in Kruger. The lessons learned might be useful in addressing similar problems, and the broad issue of rare species will surely become more important in the future. They might become indicators of the elements of the system that cannot be managed as a CAS and therefore present cases in which very hard value-based decisions will be needed.

Conclusion

Kruger is a huge national park of immense scientific, aesthetic, and economic value (Balmford et al. 2002), not only to South Africans but also to all people of the world (Pimm et al. 2001; Terborgh and Boza 2002). In this book we have taken a rewarding look at the accomplishments of science in Kruger. The lessons learned, the solutions found, and even the mistakes made have been discussed so that others can benefit and learn from what has been and is being done. As we look forward, Kruger is positioned to deliver a model for stewardship of parks and protected areas through its strategic adaptive management approach and science-academic-management partnerships. However, although Kruger is one of the oldest parks in the world, it is set in one of the newest nations in the world, recreated in 1994 by a society that is involved in one of the most remarkable social experiments of all time. The future challenges will entail continued commitment, new expertise, and increased cooperation between many disciplines and people from all walks of life within and beyond its boundaries. Kruger's custodians will have to work hard at marketing its benefits, defending them against those of alternative land-use options, getting all beneficiaries of Kruger's services to pay for them in an equitable manner (including beneficiaries in other countries), and identifying how the resilience of its various ecosystems can be conserved with less expenditure on management. The way forward involves science and management working together to include the wider society while protecting this spectacular world resource. This is a worthy experiment for all involved: science, management, and society.

References

Balmford, A., A. Bruner, P. Cooper, R. Costanza, S. Farber, R. E. Green, M. Jenkins, P. Jefferiss, V. Jessamy, J. Madden, K. Munro, N. Myers, S. Naeem, J. Paavola, M. Rayment, S. Rosendo, J. Roughgarden, K. Trumper, and R. K. Turner. 2002. Economic reasons for conserving wild nature. *Science* 297:950–953.

Balvanera, P., G. C. Daily, P. R. Ehrlich, T. H. Ricketts, S. A. Bailey, S. Kark, C. Dremen, and H. Pereira. 2001. Conserving biodiversity and ecosystem services. *Science* 291:2047.

Bosselman, F. P. 1978. *In the wake of the tourist.* Washington, DC: The Conservation Foundation.

Burnham, P. 2000. *Indian country, God's country: Native Americans and the national parks.* Washington, DC: Island Press.

Caron, A., P. C. Cross, and J. T. du Toit. In press. Ecological implications of bovine tuberculosis in African buffalo herds. *Ecological Applications*.

Carpenter, S. R. 2002. Alternate states of ecosystems: evidence and some implications. Pages 357–383 in M. C. Press, N. Huntly, and S. Levin (eds.), *Ecology: achievement and challenge.* Oxford, UK: Blackwells.

Carruthers, J. 1995. *The Kruger National Park: a social and political history*. Pietermaritzburg, South Africa: University of Natal Press.

Carruthers, J. 2001. *Wildlife and warfare: the life of James Stevenson-Hamilton*. Pietermaritzburg, South Africa: University of Natal Press.

Ceballos, G., and P. R. Ehrlich. 2002. Mammal population losses and the extinction crisis. *Science* 296:904–905.

Clark, J. S., S. R. Carpenter, M. Barber, S. Collins, A. Dobson, J. A. Foley, D. M. Lodge, M. Pascual, R. Pielke Jr., W. Pizer, C. Pringle, W. V. Reid, K. A. Rose, O. Sala, W. H. Schlesinger, D. H. Wall, and D. Wear. 2001. Ecological forecasts: an emerging imperative. *Science* 293:657–660.

The Conservation Foundation. 1972. *National parks for the future*. Washington, DC: The Conservation Foundation.

Daily, G. C. 1997. *Nature's services; societal dependence on natural ecosystems*. Washington, DC: Island Press.

Daily, G. C., and K. Ellison. 2002. *The new economy of nature: the quest to make conservation profitable*. Washington, DC: Island Press.

de Vos, V., R. G. Bengis, N. P. J. Kriek, A. Michel, D. F. Keet, J. P. Raath, and H. F. K. A. Huchzermeyer. 2001. The epidemiology of tuberculosis in free-ranging African buffalo (*Syncerus caffer*) in the Kruger National Park, South Africa. *Onderstepoort Journal of Veterinary Research* 68:119–130.

Ferreira, S. L. A., and A. C. Harmse. 1999. The social carrying capacity of Kruger National Park, South Africa: policy and practice. *Tourism Geographics* 1:325–342.

Folke, C., S. Carpenter, T. Elmqvist, L. Gunderson, C. S. Holling, and B. Walker. 2002. Resilience and sustainable development: building adaptive capacity in a world of transformations. *Ambio* 31:437–440.

Franklin, J. F. 1987. Importance and justification of long-term studies in ecology. Pages 3–19 in G. E. Likens (ed.), *Long term studies in ecology: approaches and alternatives*. Berlin: Springer-Verlag.

Freitag, S. 1998. The Kruger National Park and the analysis of historic datasets: where are we going? *South African Journal of Science* 94:146.

Heal, G. 2000. *Nature and the marketplace: capturing the value of ecosystem services*. Washington, DC: Island Press.

Johnson, L. B. 1965. *To fulfill these rights*. Commencement Address at Howard University, June 4, 1965. Austin, TX: National Archives and Records Administration, The Lyndon B. Johnson Library and Museum.

Joubert, S. C. J. 1983. A monitoring programme for an extensive national park. Pages 201–212 in R. N. Owen-Smith (ed.), *Management of large mammals in conservation areas*. Pretoria: Haum Education Publishers.

Levin, S. 1999. *Fragile dominion: complexity and the commons*. Reading, UK: Helix Books.

Likens, G. E. 1987. *Long term studies in ecology: approaches and alternatives*. Berlin: Springer-Verlag.

Lubchenco, J. 1998. Entering the century of the environment: a new social contract for science. *Science* 279:491–497.

Magnuson, J. J. 1990. Long-term ecological research and the invisible present: uncovering the processes hidden because they occur slowly or because effects lag years behind causes. *BioScience* 40:495–501.

Mandela, N. 2000. Opening address in South African National Parks, 2000. *Visions of change: social ecology and South African National Parks.* Pretoria: South African National Parks.

McNaughton, S. J., and K. L. I. Campbell. 1991. Long-term ecological research in African ecosystems. Pages 173–187 in P. Risser (ed.), *Long term ecological research: Scope 47.* Chichester, UK: John Wiley and Sons.

Msimang, M. 2000. Address. Pages 11–13 in South African National Parks, 2000. *Visions of change: social ecology and South African National Parks.* Pretoria: South African National Parks.

National Parks and Conservation Association. 1988. *Research in the parks: an assessment of needs.* Washington, DC: National Parks and Conservation Association.

Owen-Smith, R. N. 1988. *Megaherbivores: the influence of very large body size on ecology.* Cambridge, UK: Cambridge University Press.

Pickett, S. T. A. 1991. Long-term studies: past experience and recommendations for the future. Pages 71–88 in P. Risser (ed.), *Long term ecological research: Scope 47.* Chichester, UK: John Wiley and Sons.

Pienaar, U. de V. 1983. Management by intervention: the pragmatic/economic option. Pages 23–36 in R. N. Owen-Smith (ed.), *Management of large mammals in African conservation areas.* Pretoria: Haum Education Publishers.

Pimm, S. L., M. Ayres, A. Balmford, G. Branch, K. Brandon, T. Brooks, R. Bustamante, R. Costanza, R. Cowling, L. M. Curran, A. Dobson, S. Barber, G. A. G. da Fonseca, C. Gascone, R. Kitching, J. McNeely, T. Lovejoy, R. A. Mittermeier, N. Myers, J. A. Patz, B. Raffle, D. Rapport, P. Raven, C. Roberts, J. P. Rodriguez, A. B. Rylands, C. Tucker, C. Safina, C. Samper, M. L. J. Stiassny, J. Supriatna, D. H. Wall, and D. Wilcove. 2001. Can we defy nature's end? *Science* 293:2207–2208.

Risser, P. (ed.). 1991. *Long term ecological research: Scope 47.* Chichester, UK: John Wiley and Sons.

Sawhill, J. 1995. *Conservation by design.* Arlington, VA: The Nature Conservancy.

Scheffer, M., S. Carpenter, J. A. Foley, C. Folke, and B. Walker. 2001. Catastrophic shifts in ecosystems. *Nature* 413:591–596.

Sellars, R. W. 1997. *Preserving nature in the national parks.* New Haven, CT: Yale University Press.

Sinclair, A. R. E., and P. Arcese (eds.). 1995. *Serengeti 2: dynamics, management, and conservation of an ecosystem.* Chicago: University of Chicago Press.

Sinclair, A. R. E., and M. Norton-Griffiths (eds.). 1979. *Serengeti: dynamics of an ecosystem.* Chicago: University of Chicago Press.

Solow, R. 1992. *An almost practical step toward sustainability.* Invited Lecture, Fortieth Anniversary of Resources for the Future 1992. Resources for the Future, Washington, DC.

South African National Parks. 2000. *Visions of change: social ecology and South African National Parks.* Pretoria: South African National Parks.

Terborgh, J. 1999. *Requiem for nature.* Washington, DC: Island Press.

Terborgh, J., and M. A. Boza. 2002. Internationalization of nature conservation. Pages 383–394 in J. Terborgh, C. van Schaik, L. Davenport, and M. Rao (eds.), *Making parks work: strategies for preserving tropical nature.* Washington, DC: Island Press.

Terborgh, J., L. Lopez, P. Nunez, V. M. Rao, G. Shahabuddin, G. Orihuela, M. Riveros, R. Ascanio, G. H. Adler, T. D. Lambert, and L. Balbas. 2001. Ecological meltdown in predator-free forest fragments. *Science* 294:923–1926.

Tilman, D. 1987. Ecological experimentation: strengths and conceptual problems. Pages 136–157 in G. E. Likens (ed.), *Long term studies in ecology: approaches and alternatives*. Berlin: Springer-Verlag.

Vitousek, P. M., H. A. Mooney, J. Lubchenco, and J. M. Melillo. 1997. Human domination of earth's ecosystems. *Science* 277:494–499.

Whyte, I. J., H. C. Biggs, A. Gaylard, and Leo Braack. 1999. A new policy for the management of the Kruger National Park's elephant population. *Koedoe* 42:111–132.

Willis, K. J., and R. J. Whittaker. 2002. Species diversity: scale matters. *Science* 295:1245–1246.

CONTRIBUTORS

Roy G. Bengis P.O. Box 12, Skukuza 1350, South Africa (royb@nda.agric.za)

Tracy L. Benning Department of Environmental Science, 2130 Fulton Street, University of San Francisco, San Francisco, CA 94117-1080, USA (tlbenning@usfca.edu)

Patrick C. Benson P.O. Box 179, Wakkerstroom 2480, South Africa (pbenson_rsa@yahoo.com)

Harry C. Biggs South African National Parks, P.Bag X402, Skukuza 1350, South Africa (biggs@parks-sa.co.za)

William J. Bond Department of Botany, University of Cape Town, University Private Bag, Rondebosch 7700, South Africa (bond@botzoo.uct.ac.za)

Leo Braack P.O. Box 2550, Brooklyn Square 0075, South Africa (braack@mweb.co.za)

Bruce H. Brockett Pilanesberg National Park, P.O. Box 1201, Mogwase 0314, South Africa (bbrockett@nwpg.org.za)

Mary L. Cadenasso Institute of Ecosystem Studies, P.O. Box AB, Millbrook, NY 12545-0129, USA (cadenassom@ecostudies.org)

Jane Carruthers Department of History, University of South Africa, P.O. Box 392 Unisa 0003, South Africa (carruej@unisa.ac.za)

Valerius de Vos South African National Parks, P.Bag X402, Skukuza 1350, South Africa

W. Richard J. Dean P.O. Box 47, Prince Albert 6930, South Africa (lycium@mweb.co.za)

Mark Dent Netshare (Pty) Ltd., Suite 125, Postnet P.Bag X6, Cascades 3202, South Africa (mark@netshare.co.za)

JOHAN T. DU TOIT Mammal Research Institute, University of Pretoria, Pretoria 0002, South Africa (jtdutoit@zoology.up.ac.za)

HOLGER C. ECKHARDT South African National Parks, P.Bag X402, Skukuza 1350, South Africa (holgere@parks-sa.co.za)

LLEWELLYN C. FOXCROFT South African National Parks, P.Bag X402, Skukuza 1350, South Africa (llewellynf@parks-sa.co.za)

STEFANIE FREITAG-RONALDSON South African National Parks, P.Bag X1013, Phalaborwa 1390, South Africa (stefanief@parks-sa.co.za)

PAUL J. FUNSTON Department of Nature Conservation, Technikon Pretoria, P.Bag X680, Pretoria 0001, South Africa (funstonp@techpta.ac.za)

ANGELA GAYLARD River/Savanna Boundaries Program, Shingwedzi Camp, P.Bag X402, Skukuza 1350, South Africa (angelag@parks-sa.co.za)

RINA GRANT Northern Plains Program, Box 106, Skukuza 1350, South Africa (NorthernP@parks-sa.co.za)

JOHN H. HOFFMANN Department of Zoology University of Cape Town, P.Bag Rondebosch 7701, South Africa (hoff@botzoo.uct.ac.za)

IVAN G. HORAK Faculty of Veterinary Sciences, University of Pretoria, P.Bag X04, Onderstepoort 0110, South Africa (ighorak@op1.ip.ac.za)

ALAN C. KEMP Postnet Suite #38, P.Bag X19, Menlo Park 0102, South Africa (leadbeateri@hotmail.com)

PER KRYGER Department of Zoology and Entomology, University of Pretoria, Pretoria 0002, South Africa (pkryger@zoology.up.ac.za)

JANE LUBCHENCO Department of Zoology, Oregon State University, Corvallis, OR 97331-2914, USA (lubchenj@bcc.orst.edu)

DAVID MABUNDA South African National Parks, P.Bag X402, Skukuza 1350, South Africa (davidm@parks-sa.co.za)

GUS MILLS South African National Parks, P.Bag X402, Skukuza 1350, South Africa (gusm@parks-sa.co.za)

SUZANNE J. MILTON P.O. Box 47, Prince Albert 6930, South Africa (lycium@mweb.co.za)

ROBERT J. NAIMAN University of Washington, Fishery Sciences Building, Campus Box 355020, Seattle, WA 98195-5020, USA (naiman@u.washington.edu)

JOSEPH OGUTU Center for African Ecology, School of Animal, Plant and Environmental Sciences, University of the Witwatersrand, P.Bag 3, WITS 2050, Johannesburg, South Africa (josepho@gecko.biol.wits.ac.za)

JAY O'KEEFFE Institute for Water Research, Rhodes University, Grahamstown 6140, South Africa (jay@iwr.ru.ac.za)

LUANNE B. OTTER Climate Research Group, School of Geography, Archaeology and Environmental Sciences, University of the Witwatersrand, P.Bag 3, WITS 2050, Johannesburg, South Africa (luanne@crg.bpb.wits.ac.za)

NORMAN OWEN-SMITH Center for African Ecology, School of Animal, Plant and Environmental Sciences, University of the Witwatersrand, P.Bag 3, WITS 2050, Johannesburg, South Africa (norman@gecko.biol.wits.ac.za)

STEWARD T. A. PICKETT Institute of Ecosystem Studies, Millbrook, NY 12545, USA (picketts@ecostudies.org)

DANIE PIENAAR South African National Parks, P.Bag X402, Skukuza 1350, South Africa (dpienaar@parks-sa.co.za)

STUART L. PIMM Nicholas School of the Environment and Earth Sciences, Box 90328, Duke University, Durham, NC 27708, USA (stuartpimm @aol.com)

SHARON POLLARD Save the Sand Project, AWARD, P.Bag X483, Acornhoek 1360, South Africa (sharon@award.org.za)

ANDRÉ L. F. POTGIETER South African National Parks, P.Bag X402, Skukuza 1350, South Africa (andrep@parks-sa.co.za)

JESSICA REDFERN Department of Environmental Science, Policy and Management, 201 Wellman Hall, University of California, Berkeley, CA 94720-3112, USA (jredfern@nature.berkeley.edu)

WILLIAM ROBERTSON IV The Andrew W. Mellon Foundation, 140 East 62nd Street, New York, NY 10021, USA (wr@mellon.org)

KEVIN H. ROGERS Center for Water in the Environment, University of the Witwatersrand, P.Bag 3, WITS 2050, Johannesburg, South Africa (kevinr@gecko.biol.wits.ac.za)

DIRK ROUX CSIR Environmentek, P.O. Box 395, Pretoria 0001, South Africa (droux@csir.co.za)

MARY C. SCHOLES School of Animal, Plant and Environmental Sciences, University of the Witwatersrand, P.Bag 3, WITS 2050, Johannesburg, South Africa (mary@gecko.biol.wits.ac.za)

ROBERT J. SCHOLES Environmentek, Water Environment and Forestry Technology, P.O. Box 395, Pretoria 0001, South Africa (bscholes@csir.co.za)

CHARLIE SHACKLETON Department of Environmental Science, Rhodes University, Grahamstown 6140, South Africa (c.shackleton@ru.ac.za)

WINSTON S. W. TROLLOPE Department of Livestock and Pasture Science, University of Fort Hare, P.Bag X1314, Alice 5700, South Africa (winfire@eastcape.net)

RUDI VAN AARDE Conservation Ecology Research Unit, Department of Zoology and Entomology, University of Pretoria, Pretoria 0002, South Africa (rjvanaarde@zoology.up.ac.za)

BRIAN W. VAN WILGEN CSIR Division of Water, Environment and Forestry Technology, P.O. Box 320, Stellenbosch 7599, South Africa (bvwilgen@csir.co.za)

FREEK J. VENTER South African National Parks, P.Bag X402, Skukuza 1350, South Africa (freekv@parks-sa.co.za)

JOHAN VERHOEF South African National Parks, P.Bag X402, Skukuza 1350, South Africa (johanv@parks-sa.co.za)

IAN J. WHYTE South African National Parks, P.Bag X402, Skukuza 1350, South Africa (ianw@parks-sa.co.za)

ANDREW J. WOGHIREN Environmentek, Water Environment and Forestry Technology, P.O. Box 395, Pretoria 0001, South Africa

INDEX

Aardvarks, 235
Abiotic template on savanna heterogeneity, influence of the, 119–25
  see also Physical heterogeneity
Acidification of ecosystems, nitrogen saturation and, 144
Adaptive management:
  alien species, invasive, 413
  conclusions, 55–56
  feedback, 66–67
  goals, setting achievable, 48–50
  implementation crisis facing, 45
  institutional design, appropriate, 52–55
  knowledge management, 47–48
  learning by doing, 17
  learning process in conservation, 45–48
  monitoring and appropriate modeling, 50–52
  reflections on Kruger experience, 493
  river heterogeneity, 464–65
  uncertainty, managing, 44
  see also Strategic adaptive management
Adoption and knowledge sharing, 72
Aerial game census, 13
African horse sickness (AHS), 271, 352, 355, 363
African National Congress (ANC), 14
African swine fever (ASF), 352, 364, 365
Afrikaners, 6, 13–14, 392
Agents as key components of heterogeneity, 25–26, 30–31, 81
Agriculture, 429–33, 435–36, 452
Albasini, João, 335
Alcelaphines, 360
Alien plants/animals, 18, 406–14
Alluvial soils, 107, 457
Ammonium, 133
Anglo-Boer War (1899-1902), 6, 7

Animals:
  biogeochemistry: cycling of elements, 141
  fire, grazers and, 150, 151
  piospheres, 182–83
  populations and poaching, managing, 395–401
  river heterogeneity, 205–7, 453–56
  surface water availability, 172, 177–81
  see also Diseases, animal; Species effects on ecosystem characteristics; *specific animal/species type*
Antelope, 12, 76, 93, 179–81, 314, 359, 360, 377–78, 485, 497
Anthrax, 93, 269, 316, 325, 328, 352, 353, 355, 358, 360–63, 471
Anthropods, 229
Anthropogenic influences at ecosystem level:
  alien species, invasive, 406–14
  animal populations and poaching, managing, 395–401
  boundaries and fencing, 394–95
  conclusions, 415, 418
  early humans, impact of, 392, 393
  global environmental change, 415
  overview, 391
  ranking agents of change, 414–17
  river heterogeneity, 453–55
  roads, 403–7
  tourism, 400–3
Apartheid, 13–14, 429–31
Arabian traders, 392
Arid savannas, 256
Arthropods, 286
Artificial water-points, 172, 175–77
Assessments of Impacts and Adaptations to Climate Change in Multiple Regions and Sectors (AIACC), 415
Azolla filiculoides, 410

INDEX

Baboons, 363
Badgers, 228, 272, 363
Bantu peoples, 334
Bantustans, 430–31
Baobab trees, 334–35
Barbets, 227
Basaltic areas, 130, 150, 173, 451
  see also catena in granitic/balsaltic areas *under* Physical heterogeneity
Bats, 226, 270
Bees, 264–66
Beetles, 231, 232, 268–69
Benthic invertebrates, 456
Bigalke, R., 489
Biodiversity and heterogeneity management, reconciling, 41–42
Biogeochemistry: cycling of elements:
  acidification of ecosystems, nitrogen saturation and, 144
  broad-/fine-leaved savannas, 136, 143
  carbon uptake and loss, 143–44
  conclusions, 145
  exchanges between terrestrial pools, 132–35
  land and atmosphere, exchange between
    animals, emissions from, 141
    energy/momentum and water, 134, 137
    fires, emissions from, 138–39
    soils, emissions from, 137–38
    vegetation, emissions from, 139–41
  nitrogen fixation, biological, 142
  overview, 130–31
  pools in savannas, elemental, 131–32
  tropospheric ozone, 144–45
  wet and dry deposition of nitrogen/sulfur, 141–42
Birds:
  electrocutions, raptor, 406
  heterogeneity, contributions to
    carrion, dispersal of nutrients/diseases from, 283–85
      ectoparasites removed from herbivores, 285
      predation pressure on small animals, 285–86
      seed dispersal, 281–83
  insects, feeding on, 270, 286
  overview, 276–77
  road development, 406
  species effects on ecosystem characteristics, 226–30
  territorial and breeding decisions, 277–80
Black people and South African politics, 13–14, 16–18, 429–31

BLINKS models, 481–82
Blowflies, 269, 361
Bluetongue, 271
Bottom-feeders and river heterogeneity, 205
Bottom-up control of ecosystems, 83, 257–58
Boundaries/fences, 10–12, 33–35, 364, 394–95, 431, 433
Bovine tuberculosis (BTB), 359, 361, 363, 365, 366, 496–97
Breonadia model, 213–14
Breonadia salicina, 200–201
Broad-leaved savannas, 98, 104, 125, 136, 143
Browsers, *see* Herbivores; Ungulate population dynamics, rainfall influences on
Brucellosis, 364, 366
Brynard, Dolf, 10
Buffalo, 11, 93, 150, 177, 178, 180, 181, 254, 264, 316, 325, 328, 357, 359–61, 366, 371, 372, 375, 376, 398, 497
Building block methodology (BBM), 459, 463
Bulweni land system, 99, 100, 102, 105
Buntings, 277
Burrowing animals and species effects on ecosystem characteristics, 235
Bush, sickle, 106
Bushmen, 3, 334
Bushwillows, 104, 106, 226, 457

Cactoblastis cactorum, 410–11
Calcium and cation exchange, 117
Canine distemper, 355
Canyons Biosphere Reserve, 434, 439
Carbon and biogeochemistry: cycling of elements, 139, 143–44
Carnivores:
  coexistence between, 372–75
  conclusions, 384–85
  interguild relationships, 378–80
  lion spatial demography, 371–72
  overview, 370–71
  predator-prey relationships and ecological conditions, 375–78
  Serengeti Plains, comparisons with, 380–84
  top-down control of ecosystems, 254
  see also Predation
Carrion, 267–69, 283–85
Catchment management agencies (CMAs), 460, 464–66
Catena, *see* catena in granitic/balsaltic areas *under* Physical heterogeneity
Caterpillars, 252

Cation exchange capacities, 116, 117–19
Change, modes of, 27–29
Channel-in-channel morphology, 452, 453
Cheetahs, 179, 363, 372, 373–75, 380, 381, 383–84
Chemical immobilization of wild animals, 12, 13
Christianity, 6
Chromolaena odorata, 413
Clayey/sandy soils, 112, 114
Climate:
   diseases, animal, 93, 354–56
   global environmental change, 415
   herbivores, 300–304
   lowveld landscape, people and the, 424–25
   physical heterogeneity, 86, 91–93
   vegetation dynamics in relation to, 244
   see also Rainfall
Colonial Period (1836-1902), 6–7
Combretaceae, 97–98
Combretum erythrophyllum, 200
Communal Areas Management Programme for Indigenous Resources (CAMPFIRE), 436
Communal lands in central lowveld, 429–31
Community-based natural resource management (CBNRM), 438, 440
Community Relations Programme, 438
Complex adaptive system (CAS), 490, 495
Conservation, 45–48, 392–94, 401–2, 430, 431–32
   see also Management; Research; Science listings
Constitution, South African, 53
Controllers as key components of heterogeneity, 26, 42, 81
Crocodile River, 5, 10, 433, 449, 452–55, 465
Cuiper, François de, 6
Culling, 11, 337–39, 342–44, 376, 378, 398–99, 496
Cycling, see Biogeochemistry: cycling of elements

Dactylopius opuntiae, 411
Dams, 11, 179, 464
Data analysis, 493–94
Decomposition and nutrient cycling, 267–69, 283–85
de Cuiper, François, 6, 335
Defoliation by insects, 252
Delonix regia, 412
De Villiers Pienaar, Uys, 8, 13, 264

Diffusion and knowledge sharing, 72
Dikes, dolerite, 104–5
Diospyros mespiliformis, 200
Diseases, animal:
   carrion, 283
   climatic conditions, 93
   conclusions, 366
   drivers or responders
      alien diseases, 355
      climatic extremes, 354–56
      environmental degradation, 357
      epidemic potential, 359
      host factors, 355–58
      overview, 352–54
      spread, dynamics of disease, 357–59
   examples/illustrations
      intermediate spread rates, infectious diseases with, 360–62
      rapidly spreading contagious diseases, 360
      slow-spreading contagious diseases, 361, 363
   genetic selection through parasitism and, 270–71
   habitat change, 359–60
   history of, 349, 352
   insects, 269
   lowveld landscape, people and the, 428
   mechanisms of spread and spatial/temporal patterns, 350–51
   overview, 349
   parasitic diversity, 353
   reflections on Kruger experience, 496–97
   resistance, heterogeneity in host, 353
   science, integration of, 469
   species effects on ecosystem characteristics, 230
   surface water availability, 180
   ungulate population dynamics, rainfall influences on, 316, 325, 328
   veterinary controls, 364–66
Disturbances and river heterogeneity, 458–59
   see also disturbance and vegetation dynamics *under* Vegetation
Dogs, wild, 372, 378–80, 383
Drainage lines, 84, 374
Drakensberg Great Escarpment, 118, 424–25
Dropseed, 125
Droughts, 16, 93, 252–53, 372, 376, 433
Dung beetles, 268–69
Dynamic global vegetation models (DGVMs), 255

Ecosystem function *vs.* species-based understanding, 18, 488–91
  see also Species effects on ecosystem characteristics
Ectoparasites, 285, 357
Eichhornia crassipes, 410
Eland, 314
Electrocutions, raptor, 406
Elephants:
  anthropogenic influences at ecosystem level, 398, 399
  census, aerial, 11
  conclusions, 344
  culling, 337–39, 342–44
  disturbance caused by, 250–51
  extinct across much of its former range, 332
  fire, 151, 158
  heterogeneity, impact on savanna, 340–41
  history, population
    colonization, after, 335–36
    game reserve in 1898, after becoming, 336–37
    management era, elephant, 337–39
    pre-European era, 334–35
  low population levels before being shot to extinction, 333
  mistletoes, 282
  policy, what should be the new, 341–44
  population densities causing problems, 332–33
  population studies, 12
  reflections on Kruger experience, 490, 496
  Rivers Research Program, 62
  science, integration of, 485
  surface water availability, 178, 180
  top-down control of ecosystems, 254–55
  ungulate population dynamics, rainfall influences on, 311–12
El Niño-Southern Oscillation (ENSO), 87–88, 244
Encephalomyocarditis, 355, 357
Endoparasites, 357
Engineering, ecological, 31–33
Environmental flow requirements (EFRs), 459, 461–64
Environmental Impact Research, 18
Erosion and deposition, 252
Erosion cycles, 84
Evaporation and surface water availability, 174–75

Feedback in adaptive decision-making loop, 66–67
Fences/boundaries, 10–12, 33–35, 364, 394–95, 431, 433

Fig wasps, 266–67
Finches, 277
Fine-leaved savannas, 98, 104, 105, 136, 143
Fire:
  abiotic template on savanna heterogeneity, influence of, 124–25
  Africa/Northern hemisphere, differing theories from, 244
  biogeochemistry: cycling of elements, 138–39
  burning experiments, veld-, 12, 256
  climatic conditions, 93
  comparison with other areas
    Kakadu National Park, 163, 165
    Masai-Mara Game Reserve, 163, 164
    overview, 161, 163
    role of fire, understanding, 166
    salient features of four large-fire prone areas, 164
    Serengeti National Park, 163, 164
    Yellowstone National Park, 164, 165–66
  conclusions, 167
  diseases, animal, 359
  driver, fire as an ecosystem, 151
  granitic areas, 124
  humidity, 92
  landscape scale
    large fires, is there a difference in, 160–61
    modeling of impacts, 159–60
    observation, interpretation from, 155–59
    research, fire, 154–55
  lowveld landscape, people and the, 431
  management, history of fire, 151–54
  managers, options for, 161, 162
  phenology, 248
  road development, 406
  savanna biome, 97
  science, integration of, 483
  top-down control of ecosystems, 255–57
  variability, sources of fire, 149–51
  vegetation dynamics, 151, 249–50
Fish, 179, 205–7, 453
Flies, 264, 269, 270–71, 349, 352, 359–61, 365, 428, 433
Floods, 16–17, 26, 202–5, 253, 412, 456
Flycatchers, 277
Food webs, 228–29, 270, 271–72, 286, 372
Foot-and-mouth disease (FMD), 352, 355, 358, 363–66
Footslopes, 84, 104, 116
Foraging strategies and species effects on ecosystem characteristics, 222, 228–30
Foxes, 357
Fraser, A. A., 8
Frogs, 270

Gabbro, 106, 451
Game Preservation Era (1902-1925), 7–9
Gazankulu homeland, 430
Genetic selection through parasitism/insect-borne diseases, 270–71
Geographic information system (GIS), 18
Geology:
  abiotic template on savanna heterogeneity, influence of, 119–25
  distribution and characteristics of different rocks, 93–95
  Malelane land system, 104–5
  river heterogeneity, 193–95, 451–53
  Sabiepoort and Klipkoppies land systems, 106
  soils and, strong correlation between, 95, 99
  Vutome and Bulweni land systems, 105
Geomorphology, 194, 196–98, 452
  see also River heterogeneity
Gertenbach classifications, 98, 125, 373
Giraffes, 177, 226, 254, 255, 282, 295–96, 298, 311, 314, 327, 359, 360, 398
Global environmental change, 415
Glynn, H. T., 335
Goals, setting achievable, 48–50, 59
Gonarhezou National Park, 17, 424
Gondwanaland, 84, 242
Granitic areas, 119, 130, 150, 173, 451
  see also catena in granitic/balsaltic areas under Physical heterogeneity
Grass(es):
  Africa/Northern hemisphere, differing theories from, 243
  bottom-up control of ecosystems, 258
  climatic conditions, 93
  diseases, animal, 359
  fire, 150, 249
  nutrient-poor/rich substrates, 97
  phenology, 245–48
  sandy uplands, 104
  Satara and Letaba land systems, 106
  species effects on ecosystem characteristics, 228
  trees, coexistence with, 151, 159
Grazing/browsing animals and species effects on ecosystem characteristics, 234–35
  see also Herbivores
Greater Limpopo Transfrontier Park, 17, 432, 496

Habitats, animal:
  alien species, invasive, 406
  carnivores, 373–75, 379
  diseases, animal, 359–60
  elephants, 333
  herbivores, 293, 297–300, 303–4
  sick habitat syndrome, 357
Herbivores:
  Africa/Northern hemisphere, differing theories from, 242–43
  anthropogenic influences at ecosystem level, 398
  bottom-up control of ecosystems, 257–58
  conclusions, 304–6
  diseases, animal, 354
  ectoparasites from, birds removing, 285
  heterogeneity in savannas and species richness among, 292–93
  literature on feeding ecology, 293–94
  lowveld landscape, people and the, 433
  overview, 292
  rainfall, 150
  Serengeti Plains, 380–81
  spatial heterogeneity, 294–300
  surface water availability, 180–81
  temporal heterogeneity, 300–304
  top-down control of ecosystems, 254–55
  see also Ungulate population dynamics, rainfall influences on
Heterogeneity as the basis for biodiversity:
  components, key
    agents, 25–26
    controllers, 26
    overview, 24–25
    responders, 26–27
    substrates, 26
  conclusions, 36–37
  defining terms, 22–23
  features, three key, 22
  layers of heterogeneity added to physical template
    boundaries, 33–35
    engineering, ecological, 31–33
    patch dynamics, 35–36
  origin/scale/pattern providing context for analysis, 24
  paradigm shift needed to accommodate, 1–2
  process, modeling heterogeneity as a, 27–31
  structure and function, linking, 22
  see also Management; Physical heterogeneity; Reflections on Kruger experience and looking forward; River heterogeneity; Science listings
Hillslopes, 111, 113, 114–16
Hippos, 254, 312, 398
Hippotragids, 360

Historical overview of people/events:
  Stone Ages (2.5 million to 2,000 years before present), 3, 5
  Iron Age (AD 200–1836), 5–6
  Colonial Period (1836–1902), 6–7
  Game Preservation Era (1902–1925), 7–9
  creating a national park (1926-1946), 9–10
  Management by Intervention Era (1946–1990), 10–14
  Black Empowerment Era (1990–2002), 14, 16–18
  conclusions, 19
  timeline, 4
Hluhluwe-Umfolozi Park, 151
Homelands, 430–31
Honeybees, 264–66
Hornbills, 227
HSPF model, 478–80
Humans and savannas, 389–90
  see also Anthropogenic influences at ecosystem level; Lowveld landscape, people and the
Humidity and fire, 92
Hunter-Gatherer period (2.5 million–2000 years before present), 3, 5
Hunting, Colonial period and uncontrolled, 6–7
Hydrology/sediment supply, river heterogeneity and, 190–93
Hyenas, 370, 372, 373, 376, 379–80, 382–84, 398

Impala, 12, 178, 180, 311, 357, 360, 372, 373
Indian Ocean, 84, 86
Indigenous people moved to create parks, 8–9
Inkomati River, 449
Insects:
  birds feeding on, 270, 286
  conclusions, 272
  decomposition and nutrient cycling, 267–69
  defoliation, 252
  diseases, animal, 230
  food webs, 228–29, 270, 271–72, 286
  genetic selection through parasitism/diseases, 270–71
  overview, 263–64
  pollination, 225–26, 265–67
  river heterogeneity, 205
  soil nutrients/moisture, 231–32
Integrative/analytical approach, 208–9, 212–14, 423, 494
  see also Science, integration of

Intertropical Convergence Zone (ITCZ), 91
Invertebrates, benthic, 456
Inyaka Dam, 464
Iron Age (AD 200–1836), 5–6, 392, 393
Irrigation, 435–36
Isoprene, 139, 140

Jackals, 357
Jarman-Bell principle, 298
Journal of Geophysical Research, 482

Kakadu National Park, 163, 165
Kanniedood Dam, 179
Klaserie Private Nature Reserve, 180
Kleptoparasites, 374, 375, 378
Klipkoppies land system, 99, 101, 103, 106
Knowledge management, 47–48, 79
Knowledge sharing, 69–72
Koedoe, 13
Köppen class "BSh," 86
Kruger, Paul, 7
Kruger National Park:
  creation of, 9–10
  geographical overview, 15–16
  map of, 15
  public support for, 11
  Social Ecology Section, 17
  see also Lowveld landscape, people and the; Physical heterogeneity; Reflections on Kruger experience and looking forward; *individual subject headings*
Kudus, 12, 177, 294, 298, 303, 311, 328, 357, 358, 360, 361, 363

Land, see Physical heterogeneity
Lantana camara, 407–8, 410, 412–13
Largemouth, 455
Leadership, 54
Leaf chemistry data, 297–99
Leaf litter, 231
Learning by doing, 17
  see also Adaptive management
Learning institutions, 54–55
Learning process in conservation, 45–48
Lebombo Mountains, 84, 106, 452–53
Lebowa homeland, 430
Legislation:
  Communal Property Associations Act of 1994, 432
  National Environmental Management Act, 53
  National Forests Act of 1998, 435
  National Parks Act (1976), 401
  National Parks Act of 1926, 9, 12

National Water Act of 1998, 53, 436, 460, 461, 464
Promotion of Bantu Self-Government Act of 1959, 430
Restitution of Land Rights Act of 1994, 432
Leopards, 363, 372, 380
Leopold, Aldo, 332
Letaba land system, 99, 101, 103, 105–6
Letaba River, 5, 10, 379, 413, 435, 452–57, 463
Levubu River, 360
Lightning-ignited fires, 151–53, 161
Limpopo National Park, 17, 18
Limpopo River, 5, 6, 107, 360, 449, 452
Lions, 11, 12, 178, 179–80, 254, 327–28, 356, 363, 371–73, 375, 376–77, 380, 382, 384, 398, 483, 497
Lizards, 270
Long Term Ecological Research (LTER), 477
Lowveld landscape, people and the:
  central lowveld: 1900 to present, 428
  communal lands, 429–31
  conclusions, 441
  individual tenure, land under, 431–34
  integrating Kruger into lowveld landscape, 436–41
  integrative/analytical approach, 423
  overview, 422–23
  today, the landscape, 423–26
  topography, 84
  vulnerable landscape?, 436–37
  water, 435–36
  woodland resources, 432–33
  see also River heterogeneity; *individual subject headings*
Luvuvhu River, 5, 107, 435, 449, 452–57

Mabunda, David, 8, 14, 17
Mabuza, Nganani E., 14
Macrochannel bank/floor, 200–201
Madikwe National Park, 358
Maggots, 228, 272
Makuleke land claim, 431, 432, 438, 441
Malaria, 352, 428
Malelane land system, 99–105
Malignant catarrhal fever (MCF), 352, 364
Management:
  animal populations and poaching, 395–401
  biodiversity and heterogeneity management, reconciling, 41–42
  challenges for managers and scientists, 43–44
  diseases, animal, 364–66
  fire, history of, 151–54

integrative/analytical approach better linking science to, 208–9, 212–14, 423, 494
  see also Science, integration of
Management by Intervention Era (1946–1990), 10–14, 337–39
partnerships between scientists/managers, 53, 61–62, 74–76
reflections on Kruger experience, 489–92, 496–97
reintroductions, wildlife, 11, 395–97
research and, Kruger and close link between, 17, 51–52
science, integration of, 484, 485
translocation, wildlife, 364–65, 396, 398, 496
  see also Adaptive management; managing *under* River heterogeneity; Reflections on Kruger experience and looking forward; Strategic adaptive management
Mandela, Nelson, 14, 495
Mange, 356–57
Manyeleti Game Reserve, 13
Mariyeta Project, 436
Marula trees, 106, 340, 341
Masai-Mara Game Reserve, 163, 164
Mdluli land settlement, 436, 439
Medicinal plants, 399–400, 431
Megaherbivores, top-down control of ecosystems by, 254–55
Melia azedarach, 408, 412
Mental models, 54
Mesic savannas, 256–57
Midges, 264, 271
Mimosaceae, 97–98
Mimosa pigra, 412
Mineralization, nitrogen, 131, 133
Mining industry, 435
Mistletoes, 282
Mites, 265–66, 363
Modeling, appropriate, 52, 477, 497
Mole-rats, 235
Mongooses, 228, 229, 357
Monitoring, 50–52, 59
  see also Thresholds of potential concern
Monoterpenes, 139, 140
Moose, 234–35
Mopane, 92, 97–98, 106, 125, 252
Mosquitoes, 271, 352
Mozambique, 87, 394, 398, 466
Msimang, Mavuso, 14
Mutale River, 456

Nagana, 352
Namibia, 358

National Parks Board, 209
National Party, 9, 13–14, 428
National Water Resource Strategy, 460, 464
National Working for Water Program, 409
Nel, T. G., 12
Net primary production (NPP), 143–44
Nicotiana glauca, 413
Nitrogen, 131–33, 142, 231, 234, 282
  see also Biogeochemistry: cycling of elements
Northern Plains Program, 483–84
Nsikazi River, 394
Numbi Hill, 394
Nutrient-poor/rich substrates, 97–98, 151, 183–84
  see also Biogeochemistry: cycling of elements
Nwambiya land system, 99, 101, 103, 106
Nyala, 365
Nylsvley Nature Reserve, 131, 134

Objectives hierarchy, 49–50, 60–63, 72–74, 457, 472
Olifants River, 5, 7, 76, 105–6, 193, 199, 204, 253, 413, 427, 435, 449, 452–55, 457, 458, 460, 463, 466
Opuntia stricta, 408, 410, 412
Organismal/physical agents, contrasting behavior of, 30–31
Oxpeckers, 230, 285

Pafuri Game Reserve, 430
Pafuri land system, 99, 101, 103, 107, 264
Pans, 173
Parasites, 180, 270–71, 353
Patch dynamics, 34–36, 41–43, 114–17, 117–19, 235
Pattern and analysis of heterogeneity, 24
Pedi peoples, 428
People and Parks Project, 438
Periphytivores, 205
Phagmites mauritianus, 200
Phalaborwa land system, 99–105
Phenology, 245–48
Phosphorus, 116, 131, 231, 234
Phylanthus reticulatus, 200
Physical heterogeneity:
  catena in granitic/balsaltic areas
    basalt, example of a catena on, 112
    basaltic hillslope, physical/chemical properties of soils along, 113
      cation exchange capacities, 116, 117
      comparison of ecological aspects of granite/basalt-derived soils, 108–9
    granite, example of a catena on, 110
    granitic hillslope, physical/chemical properties of soils along, 111
    hillslope types, 114–16
    land elements, 114–17
    overview, 107, 110
    sandy/clayey soils, 112, 114
  climate, 86, 91–93
  conclusions, 125–26
  geology, 93–95
  overview, 83–84
  patch dynamics, 117–19
  river heterogeneity, 450–55
  soils and vegetation
    abiotic template on savanna heterogeneity, influence of, 119–25
    alluvial soils, 107
    broad/fine-leaved savannas, 98
    classifications, vegetation/land, 96, 98
    climate, impact of the, 92–93
    geology and soils, strong correlation between, 95, 99
    land systems, 97, 98, 100–107
    morphometric features of land systems, 98
    nutrient-poor/rich substrates, 97–98
    savanna biome, 96–97
    summary of land system characteristics, 99
    woody plants, cover of, 97, 99
  topography, 84–86, 99
  Zambezian Domain of Sudano-Zambezian Region, 95–96
Physical/organismal agents, contrasting behavior of, 30–31
Piospheres, 182–83
Pistia stratiotes, 409, 410
Plant pools, 131
  see also soils and vegetation *under* Physical heterogeneity; Vegetation
Plant Protection Research Institute, 409
Poaching and illegal exploitation, 399–400, 429, 490
Politics, South African, 13–14, 16–18, 427–29
Pollination, 224–27, 265–67, 282
Population studies, 12–13
Porcupines, 235
Portugal, 392
Practice built into reward systems, 55
Predation:
  anthropogenic influences at ecosystem level, 398
  avian predators, 285–86

rabies, 358
ungulate population dynamics, rainfall influences on, 311, 327–28
see also Carnivores
Pretoriuskop area, 401
Pumbe land system, 106
Punda Maria area, 340
Pyrethroid cattle dips/sprays, 365

Quelea, red-billed, 228, 229, 277, 279, 283

Rabies, 353, 355, 357, 358
Radiation, shortwave, 91, 134, 137
Rainfall:
biogeochemistry: cycling of elements, 134, 137
carnivores, 370, 376, 378
distribution of annual, 87–90
El Niño-Southern Oscillation, 87–88
fire and top-down control of ecosystems, 256
herbivores, 150
Köppen class "BSh," 86
lowveld landscape, people and the, 425, 426
nitrogen cycle, 133
phenology, 245–48
road development, 406
soils/vegetation/animals, general influence on, 92–93
summary of annual, land systems and, 99
surface water availability, 172, 174, 175
thunderstorms and cyclones, 86–87, 89, 91, 244
see also Ungulate population dynamics, rainfall influences on
Recreational Opportunities Zoning (ROZ) policy, 402
Reedbuck, 93
Reeds, 457, 458
Reflections on Kruger experience and looking forward:
adaptive management, 493
complex adaptive system, 492, 497
conclusions, 498
data analysis, 493–94
ecosystem attributes, 488–89
ecosystem services, 490–91
integration of programs, 494
management decisions, pressing, 496–97
past, management in the, 491–92
science and management, 489–90
socioeconomic challenges, 494–96
Reintroductions, wildlife, 11, 395–97

Research:
Black Empowerment Era (1990–2002), 18
challenges for managers and scientists, 43–44
fire, 154–55
insects, 263–64
management and, Kruger and close link between, 17, 51–52
objectives hierarchy, 73–74
research section established in Kruger, 12–13
river heterogeneity, 208–9, 214–16, 449
three branches, organized into, 18
TPCs as mediators of science-monitoring-management relationship, 74–76
see also Science *listings*; Ungulate population dynamics, rainfall influences on
Resettlement schemes, 11
Resistance, animal diseases and host, 353
Responders as key components of heterogeneity, 26–27, 81
Rhinos, 11, 254, 255, 312, 365, 396, 399, 400
Rift Valley fever (RVF), 271, 355, 359
Rinderpest epizootic, 7, 359, 360, 433
River heterogeneity:
anthropogenic influences at ecosystem level, 453–55
biological heterogeneity, 455–58
disturbances, large infrequent, 458–59
geographic setting, 447–49
integrative/analytical approach better linking science/management, 212–14
lowveld landscape, people and the, 428, 435–36
managing
catchment management, 459–61, 464–66
environmental flow requirements, 461–64
interdisciplinary understanding, 208–9
overview, 205–8
research, integrated river, 214–16
Sabie River, auditing gemorphological response on, 210–11
thresholds of potential concern, 193, 209–16
vegetation's response to geomorphic change, 211–12
overview, 189, 447–48
physical heterogeneity, 451–55
reflections on Kruger experience, 464–65
soils, 457
surface water availability, 173–75

River heterogeneity *(continued)*
  understanding
    faunal heterogeneity, 205–7
    geology, 193–95
    geomorphic heterogeneity, 194, 196
    hydrology and sediment supply, 190–93
    spatial pattern, 196–97
    temporal pattern, 197–99
    vegetation heterogeneity, 199–205
River-Savanna Boundaries Program, 474, 489, 494
Rivers Research Program (RRP), 18, 62, 189, 209, 457, 466, 489
  see also Science, integration of
Road development, 403–7
Rodents, 355, 357

Sabiepoort land system, 99, 101, 103, 106
Sabie River, 5, 7, 16–17, 119, 193, 194, 202, 210–11, 379, 412, 425, 449, 452–58, 460, 463–66
Sabi Game Reserve, 7, 9, 392, 428, 430, 433
SAFARI 2000, 482–83
Salvinia molesta, 410
Sandenberg, J. A. B., 10, 489
Sand River, 433, 463
Sandy/clayey soils, 112, 114
San people, 3, 334
Satara land system, 99, 100, 102, 105–6, 117, 401
Savanna biome, 96–97
  see also Kruger National Park; Lowveld landscape, people and the; Physical heterogeneity; *individual subject headings*
SAVANNA model, 478, 479–80
Scale and analysis of heterogeneity, 24, 30, 474–75
Science:
  integrative/analytical approach better linking management to, 208–9, 212–14, 423, 494
  new philosophy of, 46–47
  partnerships between scientists/managers, 53, 61–62, 74–76
  reflections on Kruger experience, 489–90
  see also Research
Science, integration of:
  case studies influencing integration
    BLINKS models, 481–82
    HSPF model, 478–80
    Northern Plains Program, 483–84
    overview, 477–78
    SAFARI 2000, 481–82
    SAVANNA model, 478, 479–80
    spreadsheet models, 480–81
  clustering of research initiatives, 475, 476
  conclusions, 484–86
  development of integrated understanding, 471–73
  management, 484, 485
  modes of integration influencing Kruger, 475, 477
  process, integration as a, 470–72
  river heterogeneity, 208–9
  scale as an issue, 474–75
  why and at what level to integrate, 469, 471
Sediment supply/hydrology, river heterogeneity and, 190–93
Seed dispersal, 224–27, 281–83
Serengeti National Park, 163, 164, 487
Serengeti Plains, 370–71, 380–84
Serengeti Research Institute, 12
Shaka, 6
Shangaan peoples, 430
Shingwedzi River, 179
Shingwitsi Game Reserve, 7, 9, 428, 430
Sick habitat syndrome, 357
Skukuza land system, 99–105, 133–34, 264, 401
Snakes, 270
Sociocultural/environmental/economic/political factors (SEEP), 17, 423–24, 430, 436–41, 491, 494–96
Sodic patches in riparian-upland boundaries, 34–36
Soils:
  biogeochemistry: cycling of elements, 137–38
  disturbance, physical, 251–52
  fire, 149–50
  lowveld landscape, people and the, 428
  pools, soil, 132
  river heterogeneity, 457
  species effects on ecosystem characteristics, 230–36
  see also soils and vegetation *under* Physical heterogeneity
South African Department of Water Affairs and Forestry, 464
South African Eden (Stevenson-Hamilton), 9
South African National Parks (SANParks), 17, 71, 391, 432, 434, 436, 438, 439, 493, 495
South African Water Research Commission, 477
Southern African Fire-Atmosphere Research Initiative (SAFARI), 138
Species effects on ecosystem characteristics:

burrowing animals, 235
conclusions, 236–38
feedbacks between soil nutrients/plants/animals, 236
foraging strategies, 228–30
grazing and browsing animals, 234–35
overview, 219, 221–23
pollination and seed dispersal, 224–27
soil nutrients/moisture, 230–36
theoretical/conceptual/empirical developments, 223–24
Species Research, 18
Spirostachys africana, 200
Spreadsheet models, 480–81
Springhares, 235
Starlings, 278
Steenbok, 177, 298, 302, 303, 372
Sterile insect techniques for controlling animal diseases, 365
Stevenson-Hamilton, James, 7–10, 395, 399, 407, 423–24, 430, 433
Steyn, Warden L., 10
Stone Ages (2.5 million to 2,000 years before present), 3, 5, 392, 393
Storks, 277
Strategic adaptive management (SAM):
background, 59–62
challenges for, 77–78
conclusions, 79
core elements
decision-making process, adaptive, 66–68
objectives hierarchy, 60–63
thresholds of potential concern, 63–68
vision statement, 60
forward looking component, 59
objectives hierarchy, derivation and influence of, 70–72
TPCs, 72–76
Substrates as key components of heterogeneity, 26
Suids, 360
Sulfur, 141–42
Sunbirds, 282
Surface water availability:
faunal patterns, 177–81
lessons learned from Kruger experience, 184–86
other ecosystem processes influenced, 183–84
overview, 171–73
patterns and distribution, 173–77
species effects on ecosystem characteristics, 233–34
vegetation patterns, 181–83

Swaziland, 424, 430
Swazi peoples, 428
Systems Research, 18

Tannins in plant tissues, 297–99
Temperature, long-term periodic variations in, 87, 91, 415
Termites, 231–32, 267–68, 272
Theileriosis, 352, 355, 363, 366
Thorns, 104, 106, 125, 200, 226, 350
Thresholds of potential concern (TPCs):
alien species, invasive, 408–11, 413
challenges for, 77–78
decision-making, adaptive, 66–68
defining terms, 50
elephants, 343–44
illustrative examples, 65–66
inductive approach to adaptive management, 63–64
knowledge sharing, 68
mediators of science-monitoring-management relationship, 72–74
outcomes of using, 74–76
reflections on Kruger experience, 493
river heterogeneity, 193, 209–16
strategic adaptive management, 72–76
topics covered, outline of, 64
widely set, 64, 66
Thunderstorms and cyclones, 86–87, 89, 91, 244
Ticks, 230, 355–57
Tinkerbirds, 282
Top-down control of ecosystems, 253–57
Total dissolved solids (TDS), 454
Tourism, 10, 17–18, 400–3, 430, 431
TPCs, *see* Thresholds of potential concern
Trade in the Iron Age, 5–6
Tragelaphids, 360
Transfer and knowledge sharing, 69
Transfrontier Conservation Area, 434, 441
Translocation, wildlife, 364–65, 396, 398, 496
Trees:
alluvial soils, 107
birds and pollination/seed dispersal, 282
bottom-up control of ecosystems, 258
carnivores, 373, 374
droughts, 252–53
elephants, 250–51, 282, 334–35, 340, 341
fig wasps, 266–67
fire, 158–60
grasses, coexistence with, 151, 159
nutrient-poor/rich substrates, 97
phenology, 245–46

Trees (continued)
river heterogeneity, 457
vultures, 284
see also soils and vegetation under Physical heterogeneity
Trichardt, Louis, 6, 335
Tropospheric ozone, 144–45
Tsavo National Park, 335
Tsessebe, 93, 314
Tsetse flies, 264, 349, 352, 359, 360, 365, 428, 433
Tsonga peoples, 428, 430

Uncertainty, managing, 44
see also Adaptive management
Ungulate population dynamics, rainfall influences on:
conclusions, 327–29
methods
census totals, population, 313–14
data transformations, 314
demographic responses, 315
records, rainfall, 312–13
relating annual population changes to rainfall, 314–15
overview, 310–12
results
annual rainfall/seasonal components, responses to, 325–26
demographic responses, 326–27
past rainfall, responses to cumulative, 316, 321–25
variability, rain, 315–20
see also Herbivores
University of Pretoria, 491–92

Vaal River, 433
Vaccinations and controlling animal diseases, 365
Value-focused decision making, 49
van der Schijf, Manie, 12
Vegetation:
Africa/Northern hemisphere, differing theories from, 242–44
alien species, invasive, 406–14
anthropogenic influences at ecosystem level, 399–400
biogeochemistry: cycling of elements, 139–41
birds and seed dispersal, 281–83
bottom-up control of ecosystems, 257–58
climate and vegetation dynamics, 244
conclusions, 258
disturbance and vegetation dynamics
defoliation by insects, 252
droughts, 252–53
elephants, 250–51
erosion and deposition, 252
fire, 249–50
floods, 253
soil, 251–52
fire, 151
geology/soil/fire, interaction of, 119
leaf chemistry data, 297–99
lowveld landscape, people and the, 425
medicinal plants, 399, 431
phenology, 245–48
plant pools, 131
radiation, shortwave, 91
river heterogeneity, 199–205, 211–12, 456–58
species effects on ecosystem characteristics, 227, 228–30, 236
surface water availability, 181–83
temperature, 91
top-down control of ecosystems, 253–57
see also Grass(es); Herbivores; soils and vegetation under Physical heterogeneity; Trees; Ungulate population dynamics, rainfall influences on
Venda homeland, 430
Venter classifications, 98
Verminosis, 357
Vhavhenda peoples, 428
Vision, a collective, 49–50, 55, 60
Vleis, 173–74
Volatile organic compounds (VOCs), 139, 140, 144–45
VORTEX modeling, 381
Vultures, 230, 283–85
Vutome land system, 99, 100, 102, 105

Warthogs, 180, 264, 311, 314, 363, 372
Wasps, 226, 266–67
Water:
carnivores, 377
lowveld landscape, people and the, 435–36
stabilization policies, 10–11
see also River heterogeneity; Surface water availability
Waterbuck, 177, 178, 180, 374
Wildebeest, 93, 177, 178, 180, 181, 247, 295, 311, 357, 360, 372, 374, 376
Wild Life in South Africa (Stevenson-Hamilton), 9
Wind, 92
Woodland resources in lowveld landscape, 432–33

Yellowstone National Park, 164, 165–66

Zambezian Domain of Sudano-Zambezian Region, 95–96
Zebra, 11, 12, 93, 178, 180, 181, 295, 311, 372, 374, 376, 377, 398, 483

Zimbabwe, 394, 438
Zoning used to balance tourism/conservation, 402
Zulu nation, 6

Island Press Board of Directors

VICTOR M. SHER, ESQ., *Chair*
Environmental Lawyer, Sher & Leff

DANE A. NICHOLS, *Vice-Chair*
Environmentalist

CAROLYN PEACHEY, *Secretary*
President, Campbell, Peachey & Associates

DRUMMOND PIKE, *Treasurer*
President, The Tides Foundation

ROBERT E. BAENSCH
Director, Center for Publishing
New York University

DAVID C. COLE
President, Aquaterra, Inc.

CATHERINE M. CONOVER
Chair, Board of Directors, Quercus LLC

HENRY REATH
President, Collectors Reprints Inc.

WILL ROGERS
President, Trust for Public Land

CHARLES C. SAVITT
President, Center for Resource Economics/Island Press

SUSAN E. SECHLER
Senior Advisor on Biotechnology Policy, The Rockefeller Foundation

PETER R. STEIN
General Partner, The Lyme Timber Company

DIANA WALL, PH.D.
Director and Professor, Natural Resource Ecology Laboratory
Colorado State University

WREN WIRTH
President, The Winslow Foundation